A EUROPEAN GEOGRAPHY

A European Geography

---

edited by

TIM UNWIN

Addison Wesley Longman Limited
Edinburgh Gate
Harlow, Essex CM20 2JE, United Kingdom
and Associated Companies throughout the world

Published in the United States of America
by Addison Wesley Longman Publishing, New York

First published 1998

ISBN 0 582 29485 1

British Library Cataloguing in Publication Data

A catalogue record for this book is available from the British Library.

Library of Congress Cataloguing-in-Publication Data

A catalog record for this book is available from the Library of Congress.

Set by 30 in 9/11 Times
Produced by Longman Singapore Publishers (Pte) Ltd.
Printed in Singapore

Contents

Figures

Tables

Contributors

Professor Asbjørn Aase, Department of Geography, Norwegian University of Science and Technology – NTNU, N-7055 Dragvoll, Norway

Professor John Agnew, Department of Geography, University of California, Los Angeles, 1255 Bunche Hall, Box 951524, Los Angeles, California 90095-1524, USA

Associate Professor Roger Andersson, Department of Social and Economic Geography, Uppsala University, PO Box 1003, S-751 40 Uppsala, Sweden

Professor Antoine S. Bailly, Département de Géographie, Faculté des Sciences Économiques et Sociales, Université de Genève, 102 bd Carl-Vogt, CH-1211 Genève 4, Switzerland

Dr Mireia Baylina, Department of Geography, Autonomous University of Barcelona, 08193 Bellaterra, Spain

Eric F. Berthoud, Collège de Géographie, Gymnase français, Débarcadère 8, 2501 Bienne, Switzerland

Professor Roger Bivand, Institute of Geography, Norwegian School of Economic and Business Administration, Breiviken 2, N-5035 Bergen-Sandviken, Norway

Professor Mark Blacksell, Department of Geographical Sciences, University of Plymouth, Drake Circus, Plymouth, Devon PL4 8AA, UK

Dr Isolde Brade, Institut für Länderkunde, Schongauerstr. 9, 04329 Leipzig, Germany

Professor Michael Bradford, Department of Geography, University of Manchester, Manchester M13 9PL, UK

Dr Erlet Cater, Department of Geography, University of Reading, Whiteknights, PO Box 227, Reading RG6 2AB, UK

Professor Tony Champion, Department of Geography, University of Newcastle upon Tyne, Newcastle upon Tyne NE1 7RU, UK

Professor Jean Paul Charrié, Department of Geography, Université Michel de Montaigne Bordeaux-3, 33405 Talence Cedex, France

Dr Roger Collins, Institute for Advanced Studies in the Humanities, University of Edinburgh, Hope Park Square, Edinburgh EH8 9NW, UK

Professor Denis Cosgrove, Department of Geography, Royal Holloway, University of London, Egham, Surrey TW20 0EX, UK

Dr Sarah Curtis, Department of Geography, Queen Mary and Westfield College, University of London, Mile End Road, London E1 4NS, UK

Professor Robert A. Dodgshon, Geography: Institute of Earth Sciences, University of Wales, Aberystwyth SY23 3DB, UK

Professor Panagiotis Doukellis, Department of History, Ionian University, Corfu, Greece

Dr Claire Dwyer, Department of Geography, University College London, 26 Bedford Way, London WC1H 0AP, UK

Dr Paolo Giaccaria, Dipartimento Interateneo Territorio, Politecnico e Università di Torino, Sede di P.za Arbarello 8, 10122 Torino, Italy

Dr James Gow, Centre for Defence Studies, King's College, University of London, Strand, London WC2R 2LS, UK

Dr Derek R. Hall, Department of Leisure and Tourism Management, Scottish Agricultural College, Auchineruive, Ayr KA6 5HW, UK

Professor Ray Hudson, Department of Geography, University of Durham, South Road, Durham DH1 3LE, UK

Professor Brian W. Ilbery, Department of Geography, Coventry University, Priory Street, Coventry CV1 5FB, UK

Dr Nuala C. Johnson, School of Geosciences, The Queen's University of Belfast, Belfast BT7 1NN, UK

Eva H. Karasek, Kamann Karasek Management Consultants, Ubbo Emmiussingel 47-D, 9711 BD Groningen, The Netherlands

Dr Britta Klagge, Research Institute 'Labour and Region', University of Bremen, Parkallee 39, D-28209 Bremen, Germany

Bart Kuipers, Transport Research Centre, Ministry of Transport, Public Works and Water Management, Rotterdam, The Netherlands

Associate Professor Markku Löytönen, Department of Geography, University of Turku, FIN-20014, Turku, Finland

Professor W.R. Mead, Department of Geography, University College London, 26 Bedford Way, London WC1H 0AP, UK

Felicita Medved, Kulturgeografiska Institutionen and Centrum för invandringsforskning (CEIFO), Stockholm University, S-10691 Stockholm, Sweden

Dr Jozef Mládek, Comenius University, Bratislava, Slovakia

Professor Graham Moon, School of Social and Historical Studies, University of Portsmouth, Milldam, Burnaby Road, Portsmouth PO1 3AS, UK

Professor Alexander B. Murphy, Department of Geography, 1251 University of Oregon, Eugene, Oregon 97403-1251, USA

Dr Catherine Nash, Department of Geography, University of Wales, Lampeter, Ceredigion SA48 7ED, UK

Dr Andrew Paddison, School of Management, Middlesex University, Hendon, London NW4 4BT, UK

Professor Ronan Paddison, Department of Geography and Topographic Science, University of Glasgow, Glasgow G12 8QQ, UK

Professor Pamela Pilbeam, Department of History, Royal Holloway, University of London, Egham, Surrey, TW20 0EX, UK

Dr Dmitri Piterski, Institut für Länderkunde, Schongauerstr. 9, 04329 Leipzig, Germany

Dr Jane S. Pollard, School of Geography and Service Sector Research Unit, University of Birmingham, Edgbaston, Birmingham B15 2TT, UK

Dr Pablo Pumares, Departamento de Historia, Geografía e Historia del Arte, Universidad de Almería, Cañada de San Urbano, 04120 Almería, Spain

Dr Jesús del Río Luelmo, Department of Geography, University of Exeter, Amory Building, Rennes Drive, Exeter EX4 4RJ, UK

Dr David Sadler, Department of Geography, University of Durham, South Road, Durham DH1 3LE, UK

Professor André-Louis Sanguin, Department of Geography, University of Angers, 35 rue de La Barre, 49000 Angers, France

Professor Paola Sereno, Dipartimento di Scienze Antropologiche-Archeologiche e Storico-Territoriali, Università di Torino, Via Giolitti 21/E, 10123 Torino, Italy

Dr Matthew Shepherd, Department of Geography, University of Sheffield, Sheffield S10 2TN, UK

Dr Peter Sherwood, School of Slavonic and East European Studies, University of London, Malet Street, London WC1E 7HU, UK

Dr Dušan Šimko, Geographisches Institut, University of Basel, Bernoullianum, Klingelbergstr. 16, CH-4056 Basel, Switzerland

Dr Fiona M. Smith, Department of Geography, University of Dundee, Dundee DD1 4HN, UK

Dr Adam Swain, Department of Geography, University of Nottingham, Nottingham NG7 2RD, UK

Professor Nigel Thrift, Department of Geography, University of Bristol, University Road, Bristol BS8 1SS, UK

Dr Andrew Tickle, Centre for Extra-Mural Studies, Birkbeck College, University of London, 26 Russell Square, London WC1B 5DQ, UK

Dr Tim Unwin, Department of Geography, Royal Holloway, University of London, Egham, Surrey TW20 OEX, UK

Mário Vale, Centro de Estudos Geográficos, Universidade de Lisboa, 1699 Lisboa Codex, Portugal

Professor Andrzej Werwicki, Institute of Economic Geography and Space Organisation, University of Łódź, Al Kosciuszki 21, 90-418 Łódź, Poland

Professor Allan M. Williams, Department of Geography, University of Exeter, Amory Building, Rennes Drive, Exeter EX4 4RJ, UK

Dr Geoff A. Wilson, Department of Geography, King's College, University of London, Strand, London WC2R 2LS, UK

Preface

Europe is an elusive and complex idea. There is no firm agreement on its boundaries, and it includes places as diverse as the mountain zones of northern Scandinavia and the hot dry plains of central Iberia. Its peoples have likewise encountered enormously different cultural and social influences in the past. Today, while some people may see themselves primarily as Europeans, others remain fiercely proud of their national or even regional identities. This book seeks to grapple with the complex issues that are shaping these characteristics of contemporary Europe.

It is avowedly a book on Europe's geography, written primarily by European geographers. Although the majority of the authors are British, more than twenty contributors are drawn from other European countries, with the intention of creating a distinctly European character and flavour to its content. As well as including scholars with long-established research careers on the geographies of Europe, it was deliberately decided to invite many younger scholars to contibute, so that the book reflects different generational perspectives on the issues addressed. Above all, the book is based on the wealth of research undertaken by its authors, and it has been our intention to present this in an accessible and readable style.

As editor, I am extremely grateful to the enthusiasm and energy of the contributing authors, who have prepared their manuscripts with great efficiency and style. I am also grateful to Justin Jacyno for designing and drawing the maps illustrating my own chapters. I would particularly like to thank Sally Wilkinson, with whom the idea of the book was first developed, and Matthew Smith and the staff at Addison Wesley Longman who have seen it to fruition. It is our hope that those who read it will not only find the book to be of interest, but will also enjoy it as much as we have enjoyed writing and producing it.

Tim Unwin
May 1997

Acknowledgements

We are grateful to the following for permission to reproduce copyright material:

Professor Kalevi Wiik for figure 2.3a; Professor W.R. Mead for figure 2.3b; The British Library for figure 6.1; the Louvre, Paris for figure 6.2; Elsevier Science Ltd. for figure 10.1; INRA Editions for figure 10.2; Springer-Verlag Gmbh & Co. KG for figure 11.3; the US Government for figure 12.1; Oxford University Press for figure 15.1; Dr Diane Perrons for figure 17.2; Pion Ltd., London and Sigrid Quack for figure 17.3; Format Partners Photo Library for figure 17.4; and The Norwegian Mapping Authority for figures 18.3 and 18.4.

Figures 1.1, 1.4, 1.7, 2.2, 4.2, 4.3, 4.4, 4.5, 5.3, 5.5, 7.1, 7.2, 10.3, 10.4, 10.5, 10.6, 12.4, 15.2, 19.1, 19.2, 19.3, 20.1, 20.2, 20.3, 20.4 and 20.5, are Tim Unwin's own photographs (c) Tim Unwin 1998.

Figures 5.1 and 5.2 are Denis Cosgrove's own photographs (c) Denis Cosgrove 1998.

Whilst every effort has been made to trace copyright holders, in a few cases this has proved impossible and we would like to take this opportunity to apologise to any copyright holders whose rights we may have unwittingly infringed.

CHAPTER 1 Ideas of Europe

TIM UNWIN

The purpose of this book is to provide a geographical interpretation of Europe. However, both of the terms 'a geographical interpretation' and 'Europe' are highly contested. This introductory chapter therefore seeks to explore their meanings, and in so doing to provide a summary of the conceptual framework within which the book has been created. The majority of contributors are geographers living and working in different parts of Europe. Each brings with them their own cultural interpretations of their research, and it is in this sense that the book is a geographical interpretation; it reflects the theoretical and practical research concerns of European geographers. The styles of the chapters, and the intellectual contexts upon which they draw, are thus an integral part of this particular expression of a European idea. Central to this is a celebration of diversity and difference.

Europe as an idea

Europe has meant many different things to people throughout history. Not only have its physical boundaries been subject to dispute, but so too have the cultural and political concepts associated with the idea of Europe (see, for example, Butlin and Dodgshon, 1998). Indeed, for most of the last two and a half thousand years, Europe has been a fragmented and divided area, with rulers warring against each other for political and economic domination. Nevertheless, at various periods and in particular places, certain uniting themes have emerged which have laid particular claims to being European. With the expansion of the Roman Empire, the idea of a European civilization was born. In the medieval period, Christianity came to prominence as a formative element of a European consciousness, and in the 16th century the Renaissance provided the optimism out of which a modern European world system was forged (Wallerstein, 1974, 1980; den Boer, 1993).

Where and when the idea of Europe first emerged is unknown. In the earliest prose work written by a European, Herodotus in the fifth century BC drew attention to the threefold division of his known world into Asia, Libya and Europe. However, he emphasized that he had no idea of the origin of these names. As he said in Book Four of his *Histories*, 'Another thing that puzzles me is why three distinct names should have been given to what is really a single landmass – and women's names at that; and why, too, the Nile and the Phasis – or according to some, the Maeotic Tanais and the Cimmerian Strait – should have been fixed upon as boundaries. Nor have I been able to learn who it was that first marked the boundaries, or where they got their names from' (Herodotus, 1954: 256). These issues still remain the subject of debate today, particularly with regard to Europe's eastern boundary (Mead, 1982). It is important, though, to emphasize that this division of the world was itself also specifically a European construct. Thus, far to the east, Chinese conceptualizations of the Earth were very different, and paid little cognizance to the western fringes of the Eurasian land mass, concentrating instead on the characteristics of the various provinces of their Empire (Needham and Wang Ling, 1959). Likewise, while later medieval European religious cosmographies were based on T-O designs, dividing a circular world into the three sections of Asia, Europe and Africa, East Asian cosmographies were centred on the legendary mountain of Khun Lun in Central Asia 'west of which are unknown regions, while to the east Korea, China, Indo-China and India form a series of promontory-continents extending into the eastern oceans' (Needham and Wang Ling, 1959: 565).

The ancient contrasts and tensions between Europe and Asia are exemplified by Herodotus in his description of the wars between the Greeks and the

Persians. These are symbolized in his account of the abduction of the king of Tyre's daughter, Europa, by a group of Greeks in revenge for the earlier seizure of Io by Phoenician traders (Herodotus, 1954: 13–14). This story parallels the legend in Greek mythology of how Zeus fell in love with Europa, the daughter of a Phoenician king, and in the form of a bull carried her off to Crete, where she later gave birth to three sons (Ovid, 1986; Guirand, 1968). There is, though, much debate over the precise relationship between the name of the raped princess and the continent. Den Boer (1993) thus comments that while the word Europa might be derived from the Phoenician for 'evening land' or the Greek meaning 'the dark-looking one', neither explanation has been widely accepted and the derivation of the word Europe remains obscure.

The boundaries of Europe according to Herodotus were vague and unknown. As he says, 'About the far west of Europe I have no definite information' (Herodotus, 1954: 221). He mentions that tin, amber and gold are meant to come from the north of Europe, but he was unable to confirm these claims from first-hand experience. He also noted that 'no one has ever determined whether or not there is sea, either to the east or to the north' of Europe (Herodotus, 1954: 256). It was only to the south, where the Mediterranean separated Europe from Libya, or modern Africa, that there was any certainty.

With the expansion of Greek settlement in the Mediterranean, knowledge about the land mass of Europe became much more comprehensive, and by the time Strabo (*c.*63 BC to *c.*AD 21) wrote his *Geography*, the topography of the continent was more clearly established. Moreover, the dominant political power base had also shifted from the Greek world to the city of Rome. Strabo (1949–54) provides a comprehensive account of Europe, in which he emphasizes that it was a highly varied continent, well provided for by nature. It was only in the north, about which information was still lacking, that he considered the continent to be uninhabitable because of the cold.

Until the first century AD, there was no sense of a dominant European culture or political entity. However, with the military expansion of Roman power under the emperors, a new era dawned (Figure 1.1). Although Roman authority was essentially focused on the Mediterranean, which the Romans called 'our sea' (*mare nostrum*), the land mass of Europe to the south and west of the Rhine and the Danube for the first time became incorporated into a single cultural complex (Cornell and Matthews, 1982). While this included peoples from many different backgrounds, and there was no real consciousness of a specific European identity, the expansion of the Roman Empire gave rise to one of the key leitmotifs of Europe's subsequent identity. This was the idea of an urban-based civilization and citizenship. In part derived from earlier Greek conceptualizations of the πολις (city), there emerged a fundamental distinction and separation between the civilized and the barbarian worlds (Fontana, 1995). The word 'barbarian', derived from the Greek βαρβαρος, literally meaning the stammering sound of an unfamiliar foreign voice, was applied initially to all non-Hellenes but was later appropriated by the Romans to refer to all non-Romans. In contrast, the idea of 'civilization' is derived directly from the Latin *civilis*, meaning civil or pertaining to citizens, and it was from the same root as the word *civitas*, meaning citizenship, community or city state. The civilized world was thus essentially an urban world. The idea of citizenship (*civitas*) itself, though, was highly complex. Founded on the idea of shared rights (*ius*), not extended to outsiders, it brought together the concepts of virtue (*virtus*), liberty of conscience and action (*libertas*), piety (*pietas*), glory (*gloria*), loyalty (*fides*) and public position (*dignitas*) (Cornell and Matthews, 1982). These concepts not only found their way into the languages of peoples living within those parts of Europe conquered by the Romans, but they also later came to play a highly significant role in influencing the style and culture of European identities.

With the gradual decline of the Roman Empire from the third century onwards, a new leitmotif was to become a uniting element of European identity in the medieval period. The disintegration of the Western Empire and its conquest by various Germanic tribes led to a fragmentation of identity (Grant, 1976; Starr, 1982; Ferrill, 1986), and it was only with the spread of Christianity that any kind of unifying European idea was to re-emerge. Constantine had made Constantinople, at the eastern edge of Europe, the Christian capital of the Roman Empire in 330. Later in the 4th century, on the death of Theodosius in 395, the Empire was irrevocably split into two halves, each with its own capital and ruler. Although Christianity had become the official religion of the Empire in 391, the destruction of the Western Empire and the fall of Rome in 476 led to a period of profound uncertainty and political fragmentation. Gradually, over the next century, the

Figure 1.1 Remains of the Roman town at Vaison-la-Romaine in Provence, France, once one of Narbonensis's most prosperous cities (*source:* Tim Unwin, 5 September 1990).

Franks, originating from the area of what is now Belgium and the Netherlands, were able to establish a power base in northern France from which they spread southwards and eastwards. Rapidly converting to Christianity, they became the dominant power within the former Western Roman Empire. In their turn, though, the Franks were then challenged in 711, when Islamic forces crossed the Strait of Gibraltar and conquered most of Iberia. Although they gained Toulouse in 721, the Moors were defeated near Poitiers in 732, and from the middle of the eighth century a relatively stable boundary between the Christian and Islamic powers was maintained along the Pyrenees. Significantly, Fischer (1957) notes that a contemporary chronicler from Córdoba described the coalition army led by the Franks that defeated the Moors in 732 as *europeenses* (see also Bloch, 1962).

With the rise to power of the Carolingians as rulers of the Frankish kingdom in the eighth century, a new Christian European alliance was created. The Pope, based in Rome, had sought the protection of Charlemagne in his conflicts with the Lombards, and at Christmas in the year 800 Leo III crowned Charlemagne as Holy Roman Emperor. Although poets referred to Charlemagne as king and father of Europe (den Boer, 1993), his empire was short-lived. The idea of a Christian Europe had nevertheless been forged, and it was to become a binding concept throughout the medieval period. This European identity found much of its resonance through external conflict. Bloch (1962) thus notes that when the Saxon monarch Otto the Great defeated the Magyars in the middle of the tenth century, he was once again referred to as the liberator of Europe. However, it was in the battles against the Islamic states in the eastern Mediterranean and in Iberia that the idea of a Christian Europe was honed. Beginning with the first crusade in 1096, Christian armies from different parts of Europe came together with a common purpose to liberate Jerusalem from the Muslims (Riley-Smith, 1990, 1995). Likewise, in the west in the 11th and 12th centuries, the Christian kings of Aragón, León, Castile and Portugal gradually pushed the Moors southwards in their reconquest of Iberia.

If, as Bloch (1962) argues, the common civilization and human significance of 'Europe was a creation of the early Middle Ages', it was not widely recognized as such until much later. Den Boer (1993: 34) thus notes that 'It was only in the course of the fifteenth century that the word Europe came to be used frequently by a large number of authors', and 'From then on, the identification of Europe with Christendom also became usual'. However, with the Reformation of the 16th century, and the Enlightenment of the 17th and 18th centuries, this idea of Christian unity became much less significant to the meaning of Europe. Instead, it was replaced by the new leitmotifs of economic progress and nationalism. As den Boer (1993: 61) again observes, by the 18th century, 'The growing awareness of a European civilization was based on a tangible increase in the wealth of nations, which could afford to finance costly standing armies and expensive artillery'. Such an increase in wealth had been enabled by the emergence of capitalist relations of production in the core states of what Wallerstein (1974, 1980) has termed a European world-economy.

In Wallerstein's (1974, 1980) formulation, the European world-economy was created from the reorganization of the earlier distinct medieval economies of northern Italy and Flanders, and was enabled by the expansion of the geographical size of the world, the development of variegated systems of labour control, and the creation of strong state machineries in the core states. At its heart was thus a combination of economic and political interests. Although there was continuing conflict between the different states within Europe, the crucial point to emphasize was that each state was increasingly being drawn into an integrated economic system. Furthermore, this European system had at its core a fundamental dynamic of expansion, as it sought ever cheaper raw materials and ever larger markets. As Braudel (1985: 600–1) has so eloquently commented, for the successful development of capitalism it was essential for there to be 'a vigorous and expanding market economy', integrated by the 'liberating action of world trade'. But as he also argues, 'capitalism could only emerge from a certain kind of society, one which had created a favourable environment from far back in time, without being aware in the slightest of the process thus being set in train, or of the processes for which it was preparing the way in future centuries' (Braudel, 1985: 600). It was in Europe that this particular kind of society was nurtured.

By the end of the 19th century, the economic, political and social conflicts inherent within capitalism had given rise to an intense nationalism (Hooson, 1994b), which culminated in the two European wars of 1914–18 and 1939–45, with their massive loss of life and widespread destruction. Moreover, the inter-war period itself was far from being one of peace and stability. Nolte (1991) has thus characterized it as being a veritable European Civil War in which the three ideologies of communism, fascism and liberal democracy were competing for hegemony. As Bugge (1993) has emphasized, there were several contrasting visions of European unity at this time, varying from Naumann's popularization of the Mitteleuropa project to the pan-European ideas of Count Coudenhove-Kalergi and Aristide Briand.

The horror and devastation of the 1939–45 war marked the end of a period when individual European nation-states were in competition for global domination, and at the same time it provided a new beginning for the idea of Europe. The economic strength of the USA, the establishment of international organizations such as the United Nations, and the need to rebuild Europe, all created fundamentally different conditions from those that had prevailed during the first forty years of the 20th century. Moreover, the establishment of Soviet hegenomy in the east gave a new meaning to the idea of western Europe, which for the first time became a distinctly bounded unit, with the Atlantic to the west and the Iron Curtain to the east. Since then, as Hooson (1994a: 1–2) has commented, 'those, who by the luck of the draw, emerged to the west of the so-called Iron Curtain, have moved haltingly but ultimately with inexorable logic, towards integration and voluntary surrender of sovereignty'. This integration reflects a fundamental tension within the idea of Europe, namely that between economic and political interests. Thus, while the economic imperatives of capitalism, reflected in an increasingly global economic system, have led to ever tighter economic integration within Europe, the politics of nationalism have proved more resilient (Smith, 1992). Indeed, the collapse of the Soviet Union during the late 1980s has heightened this tension, leading to the formation of newly independent states in eastern Europe, each seeking to establish its own national identity (Figure 1.2).

Such tensions also reflect very different visions of Europe, and although the signing of numerous treaties by European governments over the last fifty years might give the impression of a linear move towards integration, Wæver (1993: 163) has stressed that 'it skates over different views of European integration and conceals the cut and thrust of European politics as governments wrestled with questions of national self-interest in the context of European economic and political integration'. There is little doubt, though, that the signing of the Treaty of Rome in 1957, the formalization of the European Community in 1967, the Single European Act of 1986, and the Maastricht Treaties in 1991, have all been highly significant in the development both of a particular kind of European project, and also of an increasingly

Figure 1.2 European states in 1997.

European consciousness. The project has been to create a powerful economic and political European entity in the so-called New World Order; the consciousness has been a growing awareness of their common heritage and identity among the peoples of Europe. This, though, represents a fundamental shift of emphasis from past conceptualizations, when the ties and allegiances of most people living in Europe were above all to local places and identities, rather than to any single idea of Europe.

A European idea

The above brief overview of key themes in the meaning of Europe has illustrated not only that Europe has meant contrasting things to different people, but also that the conceptual frameworks in which ideas of Europe have been formulated have varied substantially through time. Thus, in medieval times, one specific religion, that of Christianity, was prominent in the shaping of a European identity, whereas in the modern period it has been economic and political interests that have largely dominated the debate over European integration. However, the ways in which these themes have been formulated are themselves an expression of a specific European idea, by which communities are seen as consisting of four kinds of dimension or activity: the social, the economic, the political, and the religious or cultural.

Early Chinese conceptualizations of human society, for example, were very different from those which emerged in Europe. While there were substantial contrasts within different Chinese philosophies, the concept of *Tao* (The Way) was central to the beliefs of the *Ju Chia* (Confucians) and the *Tao Chia* (Taoists) alike. Needham and Wang Ling (1956) thus emphasize that for the Confucians, *Tao* primarily meant the ideal order of human society, in which social 'man' was inseparable from the whole of Nature. For Taoists, the *Tao* was in contrast the very Order of Nature itself. Both sets of beliefs concerning the relationships between people and the physical environment contrasted markedly with those which had emerged within Europe, and gave rise to an entirely different conceptualization of human life. As Needham and Wang Ling (1956: 9) thus emphasize, 'in early Confucianism there was no distinction between ethics and politics', and Taoists were eager to advocate a kind of 'primitive undifferentiated form of society' (Needham and Wang Ling, 1956: 47).

The Aborigines of Australia likewise have a totally different way of interpreting their lives, in which the concept of Dreaming is central. While it is difficult to capture the meaning of Dreaming in a few words or images, Sutton (1988: 16) notes that 'In traditional Aboriginal thought, there is no central dichotomy of the spiritual and material, the sacred and secular, or the natural and super-natural. While each of the Dreaming Beings and their physical counterparts and manifestations (as animals, plants, water holes, rock formations, or people) are distinguishable, Dreamings and their visible transformations are also, at a certain level, one'. Rose (1992: 43) in her powerful study of the Yarralin people living in the European-named Northern Territory of Australia, similarly captures this sense of the unity of past and present, of spiritual and material, in her observation that for the Yarralin, 'The earth as she is now – covered with vegetation, marked by different land forms, home to a great variety of living things – is the visible and consultable record of origins. Those who know how to look can see in the earth the story of our beginnings. Those who have the knowledge to understand can find in this visible story the meaning and purpose of life'.

The European notion that groups of people can be considered as having distinct economies, societies, political systems and religions is nevertheless very ancient, and can be traced back at least as far as the fourth century BC in the works of Plato (*c*.427–347 BC). In *The Republic*, for example, he clearly distinguishes between these four aspects of the way in which communities are differentiated. Although his main emphasis is on the application of philosophy to political life, Plato (1974) contrasts the varied political systems of timarchy, oligarchy, democracy and tyranny, he discusses the ways in which divisions within society are usually grouped according to occupation or economic activity, and he emphasizes the significance of religious experience in human existence.

While this fourfold division has been of lasting significance in European consciousness, it has also been the source of much debate and controversy. In particular, the precise relationships between each dimension or activity have been hotly disputed, as have the ways in which they interact to produce and reflect specific types of community. Plato (1974) was quite clear about the starting point for his analysis of the ideal state, arguing that an imaginary sketch of the origins of the state should begin with human needs. In more detail, he suggests that 'our first and greatest need is clearly the provision of food to keep us alive' and that 'our second need is shelter, and our third clothing of various kinds' (Plato, 1974: 118). Class divisions within his community are then seen as being derived on an occupational basis from the need for people to contribute the products of their labour for the common good. The starting point for Plato's analysis was therefore very much situated within the economic sphere of production.

This closely parallels Marx's (1971) arguments in the middle of the 19th century that it was the material conditions of human life that provided the origin of such things as legal relations and political forms. In his *Critique of Political Economy*, first published in 1859, Marx (1971: 21) thus provided the clearest expression of his materialist conception of history, when he described the economic base of society as 'the real foundation, on which arises a legal and political superstructure and to which correspond definite forms of social consciousness. The mode of production of material life conditions the general process of social, political and intellectual life'. In the words of Howard and King (1985: 5–6), 'Marx's theory thus asserts the primacy of the "economic structure" in explaining all other aspects of a society, including the prevailing forms of "social consciousness". It is this quality which accounts for its description as *materialist*'. It has also been highly controversial, and has been open to varying interpretations.

Marx never comprehensively systematized his views concerning the relationships between these various structures, and the fluidity of his ideas on the subject has given rise to substantial subsequent debate. In seeking to clarify the issues, Althusser (1969; Althusser and Balibar, 1970) has suggested that in any social formation or grouping of classes, there are three types of social activity: economic practice, political practice and ideological practice (Macintyre and Tribe, 1975). For Althusser, each of these instances or forms of practice is independent, and neither the political nor the ideological can simply be reduced to the economic. However, he suggests that in the last resort, the way in which they are expressed at any particular place and time is conditioned by the prevailing economic practices. He formalizes this by drawing a distinction between what he terms dominant and determinant instances. As Macintyre and Tribe (1975: 8) have succinctly summarized, 'According to Althusser, Marx showed that the economic *or* the political *or* the ideological instance could be dominant in a particular social formation. The role of the economic base, says Althusser, is to determine *which instance is dominant*, by which is meant the instance in which social conflicts are formulated'. The economy therefore provides the arena within which particular religious, political or economic instances are able to become dominant, and it is in this sense that Althusser sees the economy as being determinant.

Until recently, there has therefore been a tendency to place considerable emphasis upon economic explanations of social, political and ideological change. This has been true not only among Marxists, but also of a range of other social scientists, and at the extreme it has given rise to crude economic reductionism (Unwin, 1992). During the 1980s and 1990s such a position has been challenged by a reawakening of interest in the role of culture in social change. This has sought to break away from the crude base–superstructure model outlined above (Williams, 1982; Chaney, 1994), and has found its expression in a new cultural geography, which has sought to explore the role of culture in shaping places, exploring in particular its ideological and symbolic significance (Cosgrove, 1993, 1996; Jackson, 1989, 1996). While some, such as Mitchell (1995), suggest that the idea of culture can best be interpreted as ideology, and thus revert to 'a modified base–superstructure model' (Cosgrove, 1996: 575), there is an increasing awareness that culture cannot simply be reduced to the legitimation of economic contradictions.

At a theoretical level, therefore, the precise relationships between the realms of the social, political, economic and ideological dimensions of European life remain hotly disputed. This is well illustrated by the way in which the different contributors to this book each approach the idea of Europe from varied theoretical and conceptual perspectives. However, the European idea of a fourfold division of human experience does provide a useful framework within which to organize this particular expression of a European geography. The first part of this book therefore explores further the cultural and ideological context of contemporary European life. While culture does in part serve to legitimate contradictions within societies, the chapters included here illustrate that it also plays other significant roles in human experience. The section begins with a historical overview of the peopling of Europe by Bill Mead, which emphasizes the diversity of cultural inheritances within its boundaries. This is followed by Alec Murphy's analysis of European languages, once again highlighting the contrasting linguistic heritages that exist. Language is of central importance to our understanding not only of each other, but also of the world in which we live. There is thus a complex interplay between the structures of our languages and the ways in which we view reality (see, for example, Wittgenstein, 1961, 1967). Given the importance of religion in helping to shape Europe's identity, particularly during the medieval period, the following chapter then seeks to explore the continuing significance of religion in contemporary Europe. The final chapter in this section, by Denis Cosgrove, provides a concluding overview of the way in which the intersection between contrasting ideas of culture and the physical environment have shaped Europe's cultural landscapes.

The second section of the book considers the political dimensions of European life, incorporating both historical and contemporary accounts in order to explore how power has been allocated, implemented and represented. It begins with Nuala Johnson's examination of the ways in which ideas of nations and peoples have found their expression in Europe's geopolitical identity. This is followed by an interpretation of radical traditions of European thought, and the emergence of different kinds of revolutionary idea in the period between 1600 and 1900. The geography of contemporary European politics is then examined in chapters by Mark Blacksell on Europe's political parties, and by Allan Williams on the European Union. The first of these explores the development of Europe's political system as an expression of a particular kind of democratic idea. As well as examining the role of national political parties it also focuses on the part played by regional interests in Europe's political processes. Allan Williams's chapter provides both a historical overview of the emergence of the institutions of the European Union, and an assessment of the future issues facing European governments as they seek to move towards ever closer integration.

The third part of the book examines Europe's economic framework, interpreted through the dimensions of production, exchange and consumption (for a summary of key economic indicators see Table 1.1). One of the most controversial aspects of the European Union has been the implementation of its Common Agricultural Policy, and the section therefore begins with an overview by Brian Ilbery of the complex issues surrounding questions of agrarian production in diverse parts of Europe. This is balanced by Ray Hudson's detailed analysis of the processes of European industrial restructuring in the context of the demise of a Fordist regime of accumulation and an increasingly globalized economic system. The next chapter, by Jane Pollard, continues to pursue this theme of a global economy, but in the specific context of Europe's role in the international financial system. The ways in which increasing economic integration in Europe have occurred through commercial exchange and the development of the communication network provide the focus for Jean Paul Charrié's chapter, which concentrates specifically on the role that the European Union has played in enhancing territorial cohesion and infrastructural linkages. The last chapter in this section, by Ronan and Andrew Paddison, directly addresses issues of consumption through an analysis of the changes that have occurred in Europe's retailing systems over the last twenty years.

Europe's social agendas are explored in the final part of the book, which examines varying dimensions of interpersonal relations. It begins with an overview of population change by Tony Champion, which provides an analysis of the reasons why key demographic variables have varied across Europe in recent years. This is followed by Michael Bradford's examination of the contrasting ways in which different countries have sought to provide social welfare services, concentrating in particular on issues of education. The next two chapters focus on gender and health. Claire Dwyer and Fiona Smith draw upon research in a variety of contexts, to explore how gender is crucial to the structuring and negotiation of social relations in Europe. In so doing, they focus particularly on employment opportunities and on various types of discrimination against women. Graham Moon and Sarah Curtis then address the reforms in the provision of health services across Europe, contrasting the fiscal crises affecting western Europe with the effects of economic restructuring within the east. Particular attention is paid to the access of different groups to health services. The last chapter, by Derek Hall, examines the way in which increases in wealth and leisure time have found their expression in an expansion of tourism across Europe. This has had important social repercussions, and it is on these that the chapter concentrates.

All of these social, political, economic and cultural processes express themselves in different ways in particular locations. It is in understanding this uniqueness of place, interpreted against a broader pattern of national, regional and international change, that one of the key contributions of geographers to an understanding of Europe is to be found. However, the specific character of places is also derived from the interaction of these processes in the context of particular physical environments. The final section of this chapter therefore summarizes some of Europe's most important environmental characteristics in order to provide an overview of the setting within which Europe's geography has been shaped and continues to be contested.

Europe's physical environments

Interpretations of the relationships between human populations and the physical environments in which they live have been at the core of geographical enquiry throughout history (Livingstone, 1992; Unwin, 1992). However, the precise character of these relationships has been hotly debated (Olwig, 1996b). Few would now accept the extreme views of authors such as Semple (1911) and Huntington (1925)

Table 1.1 Europe: selected statistical indicators.

	Real GDP per capita (PPP$)[a] (1993)	Life expectancy at birth (years) (1993)	Adult literacy rate (%) (1993)	Human Development Index[b] (1993)	Urban population as % of total (1993)	Exports as % of imports (1993)	Unemployment rate (%) (1993)	Military expenditure as % of combined education and health expenditure (1990–91)	Televisions per 1000 people (1992)
Albania	2 200	72.0	85.0	0.633	37		19.5	51	88
Austria	19 115	76.3	99.0	0.928	55	83	4.2	9	480
Belarus	4 244	69.7	97.9	0.787	70	95	1.4		
Belgium	19 540	76.5	99.0	0.929	97	90	12.0	20	453
Bulgaria	4 320	71.2	93.0	0.773	70	96	16.4	29	257
Cyprus	14 060	77.1	94.0	0.909					
Czech Republic	8 430	71.3	99.0	0.872	65	96	3.5	17	
Denmark	20 200	75.3	99.0	0.924	85	122	12.4	18	537
Estonia	3 610	69.2	99.0	0.749	73	75	1.9		351
Finland	16 320	75.8	99.0	0.935	63	130	17.9	15	505
France	19 140	77.0	99.0	0.935	73	102	11.7	29	408
Germany	18 840	76.1	99.0	0.920	86	109	8.9	29	558
Greece	8 950	77.7	93.8	0.909	64	39	9.7	71	201
Hungary	6 059	69.0	99.0	0.855	64	71	12.1	18	414
Iceland	18 640	78.2	99.0	0.934	91		5.3		319
Ireland	15 120	75.4	99.0	0.919	57	134	15.6	12	304
Italy	18 160	77.6	97.4	0.914	67	115	10.3	21	421
Latvia	5 010	69.0	99.0	0.820	72	136	5.8		448
Lithuania	3 110	70.3	98.4	0.719	71	143	1.6		375
Luxembourg	25 390	75.8	99.0	0.895	88		2.1	10	267
Malta	11 570	76.2	87.0	0.886	89		4.5	10	744
Netherlands	17 340	77.5	99.0	0.938	89	110	6.5	22	488
Norway	20 370	77.0	99.0	0.937	73	133	6.0	22	424
Poland	4 702	71.1	99.0	0.819	64	74	16.4	30	295
Portugal	10 720	74.7	86.2	0.878	35	63	5.6	32	188
Romania	3 727	69.9	96.9	0.738	55	76	10.6	25	196
Russian Federation	4 760	67.4	98.7	0.804	75	133	1.1	132	370
Slovakia	5 620	70.9	99.0	0.864	58	86	12.7		
Spain	13 660	77.7	98.0	0.933	76	80	22.7	18	402
Sweden	17 900	78.3	99.0	0.933	83	117	8.2	16	469
Switzerland	22 720	78.1	99.0	0.926	60	108	4.5	14	407
Ukraine	3 250	69.3	95.0	0.719	69	134			
United Kingdom	17 230	76.3	99.0	0.924	89	88	10.2	40	435

Notes:
[a]PPP = purchasing power parity.
[b]The UNDP's Human Development Index is an average of its life expectancy index, educational attainment index, and adjusted real GDP per capita (PPP$) index.
Source: Data derived from UNDP (1996). This table shows data only for countries included in that report.

concerning what they saw as the controlling influence of the physical environment on the character of human populations, and yet the physical environment has historically played a significant part in influencing human activities in different parts of Europe. Moreover, while the damaging environmental effects of human activity have become all too apparent in recent years, events such as the widespread flooding in Iceland following the eruption of the volcano under the Vatnajökull ice cap in 1996 serve as constant reminders of the overwhelming physical forces that have given rise to Europe's topography.

The basic structure of Europe's physical environment has been shaped by the mountain-building processes that have occurred over the last 3500 million years (Lumsden, 1994; Stanners and Bourdeau, 1995). The Precambrian shields (before 540 million years ago) and Palaeozoic rocks (540 to 245 million years old) comprise the oldest formations and underlie most of northern Europe (Figure 1.3). Across

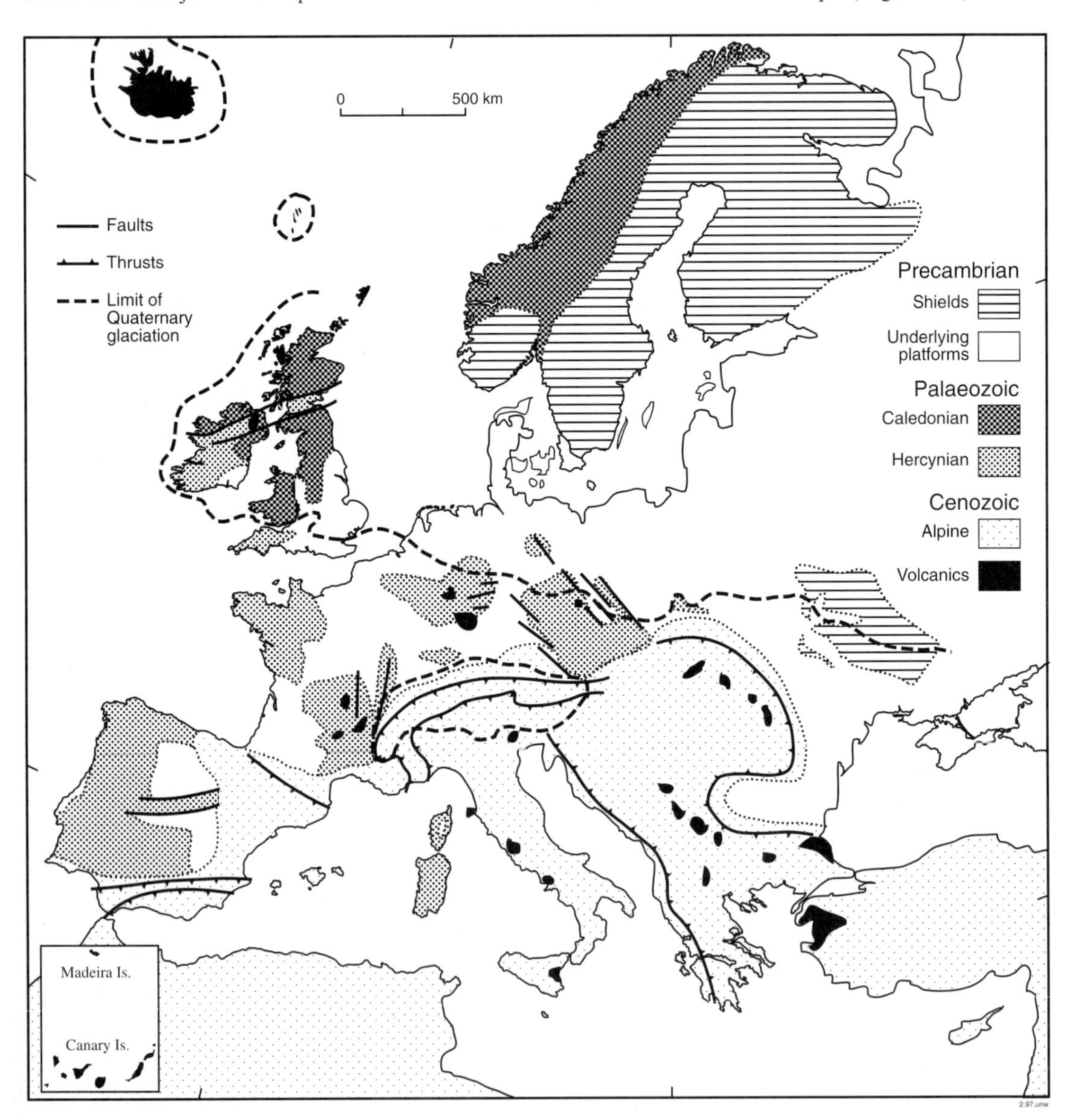

Figure 1.3 Europe's geological structure.

most of France, Germany and central Europe, younger Mesozoic (245 to 65 million years ago) sedimentary rocks have formed basins with a topography of lowlands and scarps. To the south are the mountain ranges and basins dating from the Cenozoic (65 to 2 million years ago), which comprise most of the Pyrenees, Alps and Carpathian Mountains. These mountains are widely seen as having provided 'a permanent divide in the continent for climate, history and trade' (Stanners and Bourdeau, 1995: 19). The sequence of glaciations and interglacials during the Quaternary period (approximately the last 2.5 million years) has subsequently shaped much of Europe's more detailed topography, with most of the soils having been formed since the last main glaciation, which ended some 10,000 years ago.

In very general terms, therefore, Europe's geology has given rise to four main morphological zones: a northern upland and mountainous area, a great central plain stretching from western France across Germany to Russia, a southern mountainous region, and a littoral zone adjoining the Mediterranean Sea (Figure 1.4). Southern parts of Europe are still tectonically quite active, with volcanoes and earthquakes occurring relatively frequently, particularly in the eastern Mediterranean area. At the western edge of the Eurasian plate on the northern end of the Mid-Atlantic ridge, Iceland is also volcanically active. Europe's complex geology has resulted in a wide range of different mineral deposits. In the sedimentary basins of France, Germany and Britain, substantial deposits of coal, oil and gas are to be found, whereas in the northern mountains of Scandinavia and Britain there are widespread deposits of minerals such as iron, copper, lead, tin and zinc. The Alpine mountain-building processes in southern Europe have likewise given rise to considerable mineral deposits, including aluminium, chromium, copper, lead, silver and zinc. Historically, the presence of these minerals has been highly significant in influencing the location of mining activities and the industrial structures of different parts of Europe. Moreover, in recent years the exploitation of oil and gas reserves in the North Sea has contributed significantly to the economies of the countries around its coasts.

In turning to Europe's climates, three broad factors largely determine the different conditions that are encountered: latitude, degree of continentality, and altitude. The south is generally defined as being Mediterranean in climate, the central west as temperate maritime, the central east as temperate continental, and the north as boreal. Latitude has a direct influence on the amount of solar radiation reaching

Figure 1.4 Mediterranean Europe: vineyards surrounding the medieval town of San Gimignano, Tuscany, Italy (*source:* Tim Unwin, 6 August 1987).

different places, with annual sunshine totals ranging from only some 1000 hours in the cloudiest parts of Iceland and the Faeroes to over 3400 hours in southern Iberia (Stanners and Bourdeau, 1995: 20). Furthermore, it is important to note that sunlight regimes vary significantly across Europe, with northern parts of Scandinavia having a very marked seasonal distribution of long dark winter nights and very short summer nights, in contrast to the much more equitable distribution of sunlight in Mediterranean areas. These differences have found important cultural expressions in the past, being reflected for example in contrasting styles of vernacular architecture. They can also, though, be seen as playing a significant role in influencing such things as energy use in different parts of Europe today.

In more detail, the climatic conditions of Europe are strongly influenced by the atmospheric circulation in the Atlantic, with a low-pressure area permanently situated in the Iceland area, and a high-pressure system over the Azores. These systems mean that the dominant wind directions are from the west. When combined with the effects of the North Atlantic Drift, which brings warm water northwards to the western shores of Europe, these give rise to a marked maritime/continental division in Europe's climates. Western coastal areas are thus generally warmer and wetter than eastern and interior zones, with annual temperature amplitudes varying from only some 8–10°C in the west to around 28°C in central Russia. In addition to these aspects of continentality, altitude also plays an important role in determining the patterns of temperature and rainfall across Europe. In particular, Figure 1.5 illustrates the way in which moist westerly air masses give rise to much higher rainfall levels over mountainous areas on the western edges of land masses than are to be found further east. On western Atlantic-facing coasts, for example, rainfall can be well over 3000 mm a year, whereas in parts of central Poland, northern Scandinavia and the Danube basin it is as low as 500 mm.

The combination of Europe's climatic conditions and the variations in underlying geology have given rise to the detailed hydrological regimes, soils and vegetation types encountered in different regions. Most of Europe's land surface drains into one of five main water bodies: the North Atlantic Ocean, the North Sea, the Baltic Sea, the Mediterrranean or the Black Sea (Figure 1.6). These seas have been highly significant in the past as communication routes and have tended to link rather than divide peoples living around their coasts. Moreover, Europe's major rivers, such as the Danube, the Rhine, the Elbe, the Wisla, the Loire and the Rhône have acted as crucial arteries joining together communities along their courses. Europe's seas and rivers today, though, pose considerable problems for environmental management, and it is only relatively recently that international bodies have sought to grapple with the complex issues surrounding their regulation.

In broad terms it is possible to consider Europe's soils as being of four main types: poorly drained, peaty and wetland soils; semi-arid and salt-affected soils; acid, shallow and stony soils; and well-drained black earths, sandy, loamy or clay soils (Fraters, 1994; for a detailed classification, see FAO-UNESCO, 1974). Using such a classification, which largely reflects the degradation potential of particular soils, much of Iberia, southern France, Italy, Greece, Norway and Sweden can be seen as having semi-arid, acid, shallow or stony soils. It is only across the broad band of central Europe, from France through to Poland, that well-drained soils occur, and it is on these that much of Europe's arable production is to be found. Originally, though, this central swathe of Europe was covered with deciduous woodland, varying from the mixed oak woods of western France, through the lime, oak and beech woodlands of eastern France, Germany and Denmark, to the mixed oak–hornbeam woods further east. To the north, the cold climate and poor soils of Scandinavia created an environment best suited to boreal coniferous forests of spruce and pine with birch, whereas in the south the Mediterranean area was dominated by evergreen oakwood forests, xerophytic scrub and coniferous woodlands. In the central European mountains, beech, fir and larch were the main tree species, and in the east, to the north of the Black Sea, steppes and steppe woodland predominated. Little of the vegetation that existed at the end of the last glaciation survives in Europe today, although Van de Velde *et al.* (1994) estimate that some 33% of Europe is still forest, with arable land accounting for 24% of land use and permanent crops about 16%.

Over the last century, Europe's physical environment has been dramatically transformed by human activity. Stanners and Bourdeau (1995: 190) thus note that 'until the last century, biological diversity, in terms of habitat types as well as number of species had in general been on the increase in Europe' but that now 'the trend is reversed'. In particular they draw attention to the increasing isolation of small

Figure 1.5 Average annual precipitation in Europe.

populations which are no longer able to sustain links with the larger genetic diversity of the original ecosystems, thus giving rise to a growing number of endangered species of fauna and flora across Europe. Four main processes have given rise to such environmental degradation: the exploitation of resources, particularly non-renewable resources such as hydrocarbons; the transformation of the physical environment through land reclamation, the construction of new urban areas and the introduction of new systems of agricultural exploitation; the discharge of pollutants into the air, water bodies and directly onto the land; and high levels of waste disposal, including both industrial waste and domestic waste. All of these reflect a complex interaction between Europe's environments and peoples, expressed through their economic, social, political and cultural interests (see Table 1.1). On a more positive note, though, there are now signs that Europe's biodiversity is once again increasing as remedial measures are taken to create more nature reserves and to restore certain valued ecosystems. Somewhat paradoxically, the vast tracts

Figure 1.6 Catchment areas and main surface water currents.

of land set aside for military purposes, particularly in eastern Europe and the former Soviet Union, are also now being recognized as having preserved a rich assemblage of fauna and flora (Figure 1.7).

The ways in which these relationships between people and the physical environment have been expressed and interpreted have themselves, though, varied appreciably over the last 100 years. Thus, at the beginning of this century, many geographers still saw the physical environment as having an overwhelming role in determining human behaviour. As recently as the 1920s, Huntington (1925: 233), for example, was able to write that 'on an average the men of genius in the North Sea countries would be more energetic than those of other regions because they would enjoy better health, even though the medical services were everywhere equally good. They would be continually stimulated by their cool bracing climate, and would feel like working hard all the year, whereas their southern and eastern colleagues in

Figure 1.7 Former Soviet missile base at Keila Joa, Estonia. While much environmental damage was inflicted by such military bases, others helped to preserve a rich assemblage of flora and fauna (*source:* Tim Unwin, 10 September 1993).

either hot weather or cold would be subject to periods of depression which are a regular feature of the less-favored parts of Europe'. That such arguments (see also Semple, 1911) no longer find support today is not only a result of more sensitive intellectual conceptualizations of Europe, but also of the extent to which European societies have come to dominate and control the physical environments in which they live. However, as the above examples have illustrated, the physical environment does indeed continue to play a significant role in influencing human activity across Europe, and there remains a need to examine this interaction in considerable detail. The rise of the environmental movement itself, for example, can be considered as one way in which interpretations of the physical world continue to exert an important influence on European economic, social, political and cultural behaviour today.

A European geography

It is the interactions between people, their inherited cultural, social, political and economic ideas, and these varied physical environments that have given rise to the particular character of different parts of Europe today. As geographers ever since antiquity have stressed, Europe is an immensely varied part of the world. In order to illustrate this diversity, and to interpret the way in which processes operating at the international and regional scales have varying local influences, each thematic chapter of this book incorporates two contrasting case studies drawn from different parts of Europe, ranging from Portugal and Ireland in the west to Hungary and Russia in the east, and from Norway and Sweden in the north to Italy and Greece in the south. The final chapter of the book then seeks to return to some of the broader issues touched on in this introduction, examining the contested meaning of contemporary Europe, its future role in the global economy, and the re-emergence of local cultural identities.

Acknowledgements

I am particularly grateful to Susanna Morton Braund, Denis Cosgrove, Klaus Dodds, Jane Jacobs, Ed Maltby, Lewis Owen and Xing-Min Meng for their comments on parts of this chapter and for pointing me in directions that I might not otherwise have considered.

Further reading

Braudel, F. (1985) *Civilization and Capitalism 15th–18th Century, Volume II: the Wheels of Commerce*, Fontana, London

Davies, N. (1996) *Europe: a History*, Oxford University Press, Oxford

Fontana, J. (1995) *The Distorted Past: a Reinterpretation of Europe*, Blackwell, Oxford

Stanners, D. and Bourdeau, P. (eds) (1995) *Europe's Environment: the Dobříš Assessment*, European Environment Agency, Copenhagen

Wilson, K. and van der Dussen, J. (eds) (1993) *The History of the Idea of Europe*, Routledge, London

PART I European cultural identities

CHAPTER 2 The peopling of Europe

W.R. MEAD

Stones and bones

In his pioneering book on *The Primitive Inhabitants of Scandinavia* (1838: 6) the Swedish archaeologist Sven Nilsson (1787–1883) recalled that as a youth in his native Skåne, he was already collecting 'stones that had evidently been fashioned by the hand of man for some special purpose'. John Frere, with his Hoxne finds in East Anglia, may have been a step ahead of Nilsson in time, but he never achieved quite the same international recognition. Nilsson (1838: 7) went on to describe how he 'made a practice of chipping flint stones and giving them any shape he desired'. Later, he flaked flints with a hammer stone on granite rocks to meet the needs of his flint-lock fowling piece. By the 1830s, Nilsson was collaborating in his prehistoric investigations with C.J. Thomsen, Director of the Ethnological and Archaeological Museum of Copenhagen. Their collaboration led to the classification of Stone Age, Bronze Age and Iron Age, while before his death at the age of 96, the division of the Stone Age itself into Palaeolithic, Mesolithic (*c*.10,000–*c*.3000 BC) and Neolithic had been generally accepted. Nilsson and Thomsen, who was one of the earliest honorary members of the London-based Royal Geographical Society, were recognized by Sir John Lubbock in an address to the Society of Antiquaries in the 1860s as pioneers of archaeological enquiry in northern Europe (Lubbock, 1868). In effect their investigations were basic to an understanding of the early peopling of their homelands.

The work of Nilsson and Thomsen proceeded quite independently of the discoveries that were being made elsewhere in Europe, where bones in some respects took precedence over stones as critical indicators in the dating of settlement sites. In 1856, for example, discovery of *Homo sapiens* in the Neander valley near Dusseldorf was eventually to yield a typological name of primary significance. In France, Boucher de Perthes (1778–1868) and Edouard Lartet (1801–71) had taken a step forward by suggesting a significant association between artifacts and stratigraphy and were laying the foundations for what was being described as 'the science of prehistory'. Lubbock's concept of the Palaeolithic was adopted and subdivided for France by Gabriel de Mortillet in his *Musée Préhistorique* (1881). His systemization, based upon the names of the sites where the most important artifacts were found, endowed them with a romance of their own – Aurignacian (after Aurignac in the Haute Garonne), Solutrian (after Solutré in Seine-et-Marne), Mousterian (after Moustier in the Dordogne) and Magdelanian (after Madeleine in the Dordogne).

The accommodating limestone country of south-western France, with its sheltering rocks and manifold caverns, the availability of stone for implements and the evident abundance of animals for hunting had encouraged the Michelin guide to define the area as 'the capital of prehistory'. Here the quantity of human and animal remains – prehistoric ossuaries – deposited at various levels was accompanied by a range and variety of cave paintings which, when discovered, seized European imagination. At the same time, it encouraged other countries to enquire more deeply into the ancestry of their own prior inhabitants.

The pioneering publication of A.L. Mongait (1955) recalls the new stimulus given to Russian enquiry. It was not generally appreciated in western Europe that Russia had its own antiquarian interests at least as early as those of the Society of Antiquaries in Britain. The Russian Academy of Sciences reported in 1763 that it had received over 4000 replies in response to a questionnaire circulated about archaeological features and that, by the middle of the 19th century, thousands of barrows had been excavated in the Vladimir–Suzdal area alone. The experience of France speeded a Russian concern with 'palaeotechnology' and attracted the attention of its student world. The river terraces of the central Russian plains proved to be scriptural 'valleys of dry bones' waiting to be investigated. Well over 1000 palaeolithic sites were soon recorded. The counterpart of 'Mousterain Man' was discovered to have occupied the Black Sea caves of the Caucasus as well as those of France. Uzbekistan claimed its Neanderthal counterpart. As for the mammoths, the finds of whose bones and tusks had excited West Europeans, no counterparts existed to the mammoth bone dwellings of Russia – 'bone-homes' indeed.

Such dwellings, be they cave or cabin, are primary indicators in the process of settlement. Indeed, the peopling of a place may be deemed to have occurred when mobile hunter–gatherers occupied some primitive form of accommodation which acquired a degree of permanence. Ian Hodder (1990) employs the word *domus* for such a place – a hearth about which humans gathered and the occupation of which was sufficiently prolonged to leave evidence for posterity to analyse.

Enquiries into the peopling of the continent balance the micro and the macro, with the increasingly numerous and often complex discoveries at the micro-level being combined to build up generalizations at the macro-level (Turnock, 1988). And the broad outlines of the colonization of the continent at the macro-level have to be seen in the light of Europe as a peninsula of Asia and as a northern neighbour of Africa – sources of population inflow responding to the changing human appreciation of *la bonne terre* with climatic change. They have also to be observed in the context of (1) the expanding temporal horizon – the long-accepted Biblical thousands of years being extended tenfold, perhaps nearer a hundredfold – and (2) arguments about the continuity or discontinuity, of a hiatus or no hiatus, in the historical settlement of much of the continent.

The study of the peopling of Europe began as a subject of antiquarian concern – a kind of marriage between archaeology (the word first appeared in print in 1607) and anthropology (which predated it by a decade). It has come to command the attention of an increasing range of scientists. To the fieldwork of archaeologists, anthropologists, geologists, glaciologists, pedologists, botanists and others has been added a range of new techniques as well as new methods of analysing stones and bones in relation to their sites of discovery. And to each field of activity belongs its own refined and extending vocabulary.

The laboratory has become increasingly the adjunct of field investigation. Since the 1940s, radiocarbon methods have offered a critical new means of dating artifacts. Simultaneously, pollen grain analysis, especially from peatland, has provided an increasingly accurate means of reconstructing plant assemblages (not least as climatic indicators). Lacustrine sediments through electronic microscopy have their own tale to tell, as have deep sea Ice Age recording techniques. All are a means of confirming and challenging existing theories or calling for revision of received wisdom.

The study of the peopling of Europe is an interdisciplinary study par excellence. Three aspects will be briefly developed in the following pages: (1) the changing geography of Europe as it has affected the distribution of settlement since the Pleistocene; (2) with the emergence of written records, the slow crystallization of names attributed to the peoples of Europe and the places with which they have been associated; and (3) the passing phase of concern with the links between past and present inhabitants of different parts of the continent – the continuities and discontinuities in the history of settlement.

The changing environmental background

A map of the physical geography of the continent as it might have appeared at the time of the maximum of the Weichselian/Würmian glaciation shows, first, that the shape of the continent was totally different from the familiar outlines of today because of the extent of the ice sheet and the lowering of the sea level by some 150 m (Anderson and Borns, 1994). Secondly, the relief of the land also differed from that of today and was subject to more powerful erosional forces. Thirdly, the overall temperatures were probably some 10 to 12°C lower, with the circumstances exaggerated in winter by the continental anticyclone and with significant cyclical and regional variations in precipitation. Fourthly, there were periods of greater aridity, when 'cold desert' conditions prevailed over extensive areas leading to the spread of windborne periglacial loessic deposits. Finally, and perhaps most important from the human point of view, there were not only fundamentally different patterns of vegetation but variations within them. These had critical consequences for faunal distributions.

The major European ice sheet, with its principal centre in Scandinavia, was complemented in the south by lesser ice sheets which covered the higher ranges of the Alpine system (Figure 2.1). Glaciers filled the upper reaches of the valleys with the snow line descending seasonally. Between the northern and the southern ice sheets, there was a zone of permafrost, broad and continuous in the east, narrower and discontinuous towards the Atlantic. The permafrost zone was overlain with loessic deposits, which were scattered in the west and more extensive in the east. Extensive areas supported herbaceous tundra vegetation. Because of the very different daylight–darkness rhythm, it is not easy to equate the mid-latitude tundra conditions with those prevailing in contemporary high-latitude Europe. The flora was different. Given the broad southward shift of vegetational zones, the northern Mediterranean displayed forest tundra features, with temperate woodlands reflecting the precipitation regime covering extensive areas of North Africa.

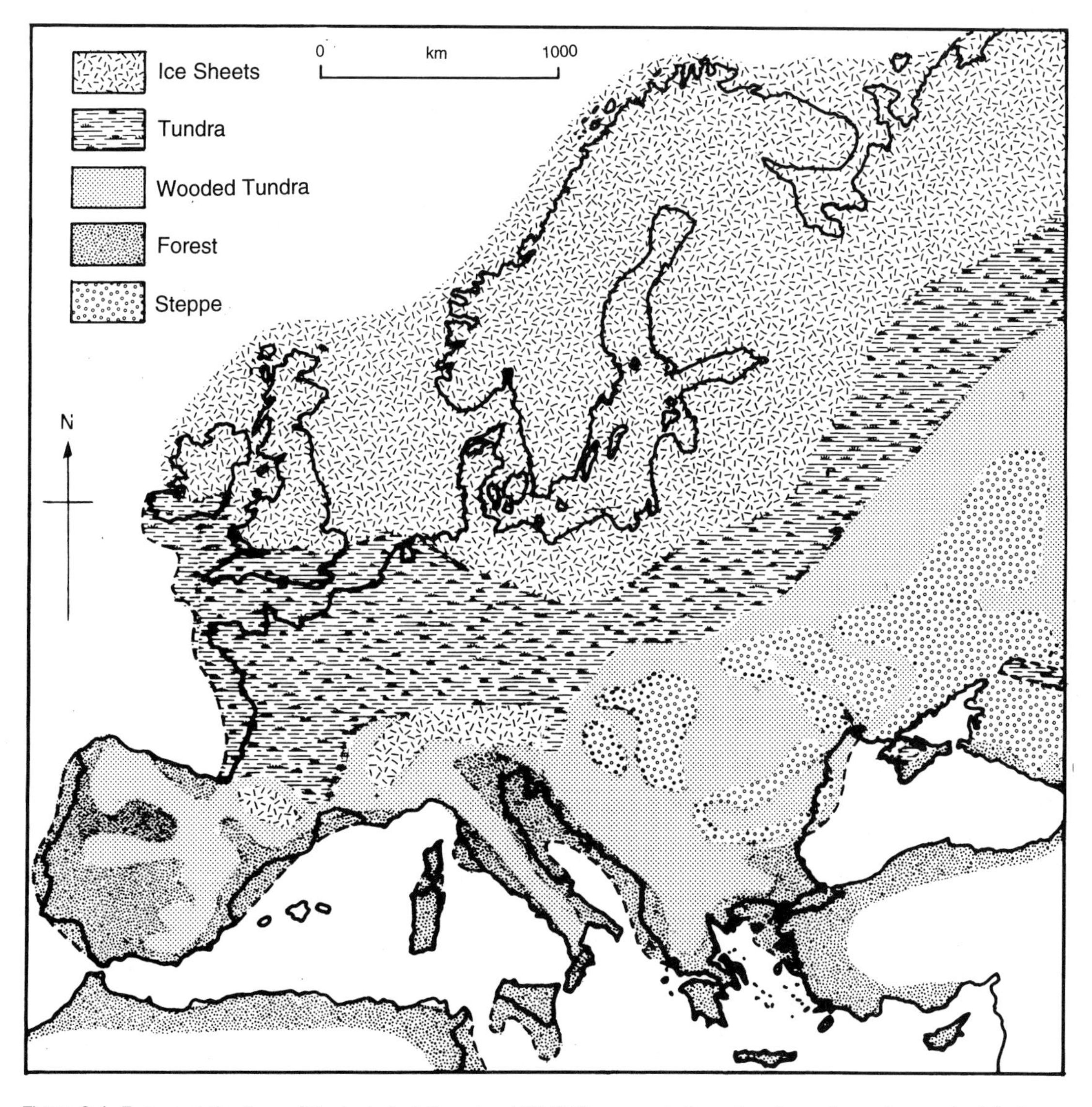

Figure 2.1 Europe at the time of the last glaciation about 20 000 years ago (based partly on the endpapers in C.S. Coon, 1948, *The Races of Europe*, New York).

Within this environmental situation, the limestone and sandstone areas of south-western France were favoured above those of Spain and Italy because of the greater variety of fauna supported by their particular tundra vegetation. Furthermore, the valleys of such rivers as the Dordogne and their tributaries afforded routes along which herbivores travelled from summer upland pasture to winter grazings on the westerly lowlands. The diversity of the fauna has been preserved for posterity in the cave paintings.

At the conclusion of the Würm glaciation, the character of the entire continent as a settlement area changed. The shift of the cyclonic system and its precipitation regimes led to the retreat of the ice sheet and the northward migration of the vegetation zones. In the process, the grass and parkland of North Africa experienced a gradual desiccation and the greater part of the Mediterranean area acquired a character more closely resembling that of the present. From the human point of view, it was either a matter of adjustment to the gradually changing circumstances or of migration in an attempt to find a more familiar or more amenable environment. In the west, migrants from North Africa appear to have moved into Iberia as well as into south-eastern Europe by way of the Near East. They brought with them the techniques of their particular cultures.

Simultaneously, there was probably a northward movement of those whose hunting and gathering activities were based on the faunal habitat of the tundra. However, the broad territory of the ice-edge meltwater streams and rivers with its emerging tundra forest was essential reindeer country. The meltwater plains, with their primordial river valleys, were an alien environment for such creatures as the woolly mammoth, the woolly rhinoceros, the bison and species of wild horse, all of which slowly disappeared from the west European scene. Thus, although the climate might have improved, the familiar sources of food supply diminished or disappeared.

It is probably purely coincidental that one of the areas of Europe for which the postglacial evolution of human settlement has been investigated most thoroughly is the home territory of the pioneering archaeologists Nilsson and Thomsen. The settlement of the Baltic area and its North Sea approaches has been infinitely sensitive to the succession of changes in the physiography of the area and the accompanying succession of climatic changes with their vegetational responses.

The history of the Baltic Sea has been worked out in considerable detail. From an ice-dammed lake (*c.*10,000–8000 BC) it was transformed into the Yoldia Sea (*c.*8000–7000) – so-called after a saltwater mollusc. Isostatic adjustment resulted in a land link across the North Sea to Britain and the conversion of the Yoldia Sea into the Ancylus Lake (*c.*7000–6000). In turn, eustatic adjustments led to its transformation into the Litorina Sea (*c.*5000–2300) – the proto-Baltic.

The succession of climatic changes – and the associated modifications in vegetation – run from cool continental (pre-6500 BC) to warm, dry continental (*c.*6000–5500) and thence to a period of warmer, moister oceanic climate (*c.*5500–2500) which resulted in the intrusion of mixed oakwoods into the pine and birch woods across the entire North European Plain. A return to warmer, drier continental conditions followed.

Into this sequence may be fitted the human settlement of this particular corner of Europe. Until the period of the Ancylus Lake, it would seem that the hunter–gatherers, associated with the tundra or boreal forest margin, persisted. The Ahrensburg site near Hamburg provided the principal initial evidence. Somewhat earlier, but totally different in character, the retreat of the Scandinavian ice sheet in north Norway in the face of warming Atlantic waters encouraged the appearance of a distinctive settlement along the coastal fringes. The remains of the settlement are in Troms county and are defined as belonging to the Komsa culture (*c.*7000 BC). A closely parallel settlement occurred around Fosna at the approaches to the Trondheimfjord. Simultaneously with the transformation of the Ancylus Lake into the Litorina Sea, settlements were established around the littoral of what was to become Denmark. Archaeologists have named them Maglemose, after the site where they were first identified. The subsequent increasing concentrations of settlement, characterized by the rich legacies of shell or midden heaps, were first investigated more intensively in the Ertebolle area (by which name their culture is best known).

In contrast to these minor adjustments in the north-west periphery of the continent, major responses to climatic change were taking place in the territories of what were to become Syria, Palestine, Jordan and Anatolia (*c.*7000–6000 BC). A new range of techniques was being developed – the diversification and improvement of tools, the beginnings of the domestication of plants and animals and, eventually, the management of water systems. Such developments facilitated, indeed demanded, permanent settlement. In general, the new food-producing methods could not be introduced to Europe at large because of the widely differing climatic and pedological conditions. Nevertheless, a variety of adjustments followed – all leading to the longer-term occupation of settlement sites. By *c*.3500 BC established farming systems were scattered over much of so-called temperate Europe. In the process, two broad environmental adjustments began to emerge – a Mediterranean system with crop production in association with sheep and goats, and an Atlantic system in which cattle and swine took precedence, with a greater reliance on fodder.

Migrants were considered to have carried the essentials of these techniques via two routes. One followed the coastlands of North Africa and thence to Iberia and the western Mediterranean. The other, thrusting into western Europe by way of the Danube basin, had its origins in Greece and the Balkans, in Asia Minor (via the Bosphorus and Black Sea coastlands) and in the steppes of south-central Russia.

The introduction of sedentary ways of life based upon animal husbandry and cropping into territories which were hunting and gathering grounds called for varying degrees of adaptation. In the Mediterranean it was not only an elaborate seasonal adjustment, but also acceptance of the natural hazards in the fertile valleys consequent upon their erosional and depositional features. North of the Alps, many settlements reflected the quality and characteristics of the soils – the well-drained and friable loess being the most favoured. Woodland clearance, principally through firing, produced widespread landscape changes, though in some areas there was probably cyclical occupation of settlement sites in response to rotational slash-and-burn cultivation. The denser peopling of the land against the background of these activities is reflected in the greater quality and diversity of archaeological remains.

A new impetus was given to concentrations of permanent settlement by the introduction into southern Europe in the third millennium BC of primitive metal-working. The Aegean hearth of the European Bronze Age and the diffusion of the techniques through the eastern Mediterranean led to the beginnings of urban settlement (*c*.2500–1300 BC). As the techniques spread to Danubia and Iberia, metal-working acquired a greater importance, calling for defensive structures around the centres of operation. But, while strengthening population concentration, it also resulted in local pressures on the land.

The development of iron-working techniques had a much greater impact because of the more widespread occurrence and accessibility of the resource. They also provided implements which facilitated not only the clearance of land but also the cultivation of heavier soils. The development of iron-working is associated in particular with the culture of the Hallstatt people, who were concentrated in Austria and southern Germany (*c*.1000 BC), and subsequently with the Celts (*c*.500 BC), whose centre of diffusion is assumed to have been on the western margins of the Hallstatt culture area in the upper Rhine valley. As the case study indicates, the Celts were significant colonists of north-western France, the British Isles and northern Italy.

It was Europe west of the Don where the sedentary ways of life were being established. Eastwards, the wooded landscapes gave way increasingly to the open steppes, the Russian prairies, with their migrant herdsmen. The maritime climates of western Europe yielded here to continental regimes, with exaggerated effects on the natural vegetation resulting from any cyclical variations. The earlier thesis of Ellsworth Huntington (1924) involving population outpourings from the Asiatic heartlands consequent upon climatic changes may be too simple, but in the context of the changing geography of Eurasia, especially its desiccation during shorter-term arid cycles, it probably contains some element of truth.

Among the geographers who have attempted to map the occupation of the territories north of the Alps, three may be mentioned. Two were concerned with prehistoric times. Robert Gradmann (1931) focused attention on the territories between the Danube and the Moselle, where he suggested a threefold division into those areas that were fully settled, those sporadically settled and those settled in parts and at certain periods. Josef Engel (1970) covered a more extensive territory. His map of the prehistoric settlement area was bounded by the Rhine, the Danube and the Vistula. It showed essentially river valley settlement, with the heathlands, marshes and more heavily wooded areas unoccupied. Otto Schlüter's (1952) map of the woodlands of 'Central Europe' at the end of the first millennium AD reflects the tardiness with which people cleared and permanently settled much of the North European Plain.

As names became associated with the peoples of the continent and the places that they occupied, toponymical enquiry has developed as an adjunct to the study of settlement.

The Celts and the Celtic fringe

Robert A. Dodgshon

The Celts are amongst the most deep-rooted of European ethnic groups. Yet, arguably, few pose more problems of identity. Though stereotyped as *the* culture of the Iron Age, stock interpretations dated their first appearance as a distinct culture to the late Bronze Age, *c.*1200 BC+. The supposed hearth for these proto-Celts lay in the North Alpine Province, especially the tributary valleys that network Bohemia, upper Austria and Bavaria. This was an area of bronze metal-working and salt-mining, items that helped to sustain gift exchange and social hierarchies. By 800–700 BC, bronze-working had begun to shift slowly into iron-working, iron being a metal with more everyday uses. Yet the adoption of iron apart, the transition to the Iron Age was more about continuities than discontinuities. The earliest Iron Age communities, or those of the Hallstatt culture, capture this continuity with the initial phases (*c.*1200+ BC) of their type-site at Hallstatt (Upper Austria) being rooted in the late Bronze Age.

Like their late Bronze Age predecessors, communities associated with the Hallstatt culture appear structured around dominant local chiefs. This aristocratic structure developed further when Hallstatt communities in the North Alpine Province began to interact with Greek city-states on the Mediterranean coast, around 700 BC+, exchanging iron and salt for items like wine. The social impact of this exchange is manifest in the lavish burials of Hallstatt chiefs, including some literally buried with waggon-loads of material goods. In time, the wealthiest sites became those in the middle Rhine area, a shift encouraged by the area's greater accessibility to trade with the Mediterranean that developed along the Rhône. As an intensively competitive society, it is not surprising that it was warlike, with defended sites, including hill-forts, becoming its hallmark. These defining characteristics appear far beyond the Hallstatt's supposed hearth in the Alps with an overall geography that extended southwards into Italy, westwards into France, northwards into southern Britain, and south-westwards into northern Spain.

In a radical review, Clark (1966) argued that there is no basis for assuming that Iron Age cultures like the Hallstatt arrived in Britain via a migration of colonists from a supposed core area. At most, we may be dealing with the movement of only a small warrior aristocracy, horse-based and mobile, extending Celtic culture through a political hegemony over existing communities. Others would now go further and argue that no social movements of any sort were involved in the development of what has been called 'Celticity'.

Traditional interpretations saw Celtic culture as continuing to change in its supposed core area, with the Hallstatt giving way to the La Tène culture by the fifth century BC, a development first apparent in the middle Rhine district and the Marne area of France. Again, the identifying traits of the La Tène culture appear geographically expansive. They extend eastwards and south-eastwards along the Danube, westwards over southern Germany and most of France as well as across to England, and spreading to the most northerly and westerly fringes of the British Isles. The La Tène culture saw an invigoration of exchange with Greek city-states, but now it was a trade that moved via Alpine passes. Further, in chiefly burials, the chariot replaced the waggon, a sign not just of changing taste but of critical changes in warfare. Hill-forts that had first appeared during the late Bronze Age were elaborated, acquiring complex timber-framed ramparts. The function of hill-forts, though, varied. Some were defended encampments for cattle or grain. Others were temporary refuges for people. Many encircled permanent settlements, with trade and craft functions, as at Danebury (Hampshire). The importance of trade and crafts is highlighted by the appearance of *oppida* during the late Iron Age. *Oppida*, like that at Manching (Bavaria), were large enclosures that functioned as major towns. Within their ramparts were concentrations of houses and craft workshops. Though timed later than in the classical world, the barbarian world was clearly progressing towards a more urbanized form when the Romans overwhelmed it.

The relationship between Celtic society and the classical world has a bearing on the debate over Celtic identity. The Greeks were the first to put a name to the Celts, whom they saw as one of the

continued

four barbarian tribes that inhabited the known world beyond the Greek world itself. The name 'Celtic' is derived from 'Keltoi' or 'Celtae', which is how the Greeks thought that the Celts described themselves (Powell, 1958: 21). However, there are doubts over whether any Celtic-speaking culture 'ever called themselves by those names' (Greene, 1964: 14). Indeed, the Greeks may have taken the name for a tribe in the hinterland of Massilia (Marseille) and used it for all the barbarian tribes that lay to the north and west (Renfrew, 1987: 221–4). However, later sources suggest more awareness of internal differences. Ceasar, for instance, distinguished between the Galli, or Celtae, who lived between the Garonne and the Seine, the Aquitani in south-west France, and the Belgae, who – by the late Iron Age – inhabited the north of France and south-east England.

Renfrew (1987) has taken these doubts further. Building on Clark's (1966) ideas, he has argued that we cannot 'restrict Celtic origins in any artificial narrow way to a specific region of the Alps', this applying as much to their language as to other characteristics (Renfrew, 1987: 249). Hitherto, the different forms of Celtic dialect were explained – partly on the basis of an archaeological argument that is now no longer tenable – by a combination of migration from a core area and language divergence. However, linguists were unable to agree over whether the major west European dialects – Gaulish, Goidelic (or Q-Celtic) and Brythonic (or P-Celtic) – developed within the core area of the North Alpine Province before spreading via successive migrations or whether they developed through divergence after such migrations had taken place. Certainly, we can document divergence in historical contexts. When the Scotti spread into the south-west Highlands from Ireland around *c.*500 AD and Cornish-speakers into Brittany at about the same time, it led to separate forms of Celtic, or Scottish Gaelic and Breton, out of Irish and Cornish respectively. However, Renfrew (1987) has proposed that the *basic* forms of Celtic dialect have older, more independent roots, each devolving slowly out of a broad 'undifferentiated' early Indo-European language that spread across Europe with the first farming communities, a conclusion that echoes Evans's (1958) suggestion that early Irish cultural landscapes owe more to the first Neolithic farmers, and their broader Indo-European context, than to anything we can specifically label as Celtic, a term which he used cautiously.

With the growth of the Roman Empire, Celtic languages and cultures were overwhelmed across large areas of western Europe. Only where Roman authority thinned or stopped short in the Atlantic ends of Europe did they survive, either beyond as in Scotland, or within its outer frontier as in Wales. When Roman authority decayed during the fourth and fifth centuries, the Irish especially responded with a vigorous expansion into parts of western Britain. The longer-term survival of Celtic societies, though, was not ensured simply by the collapse of the Roman empire. The spread of Anglo-Saxon hegemony from the fifth century onwards pushed Celtic political control – if not the Celts themselves – back into the familiar Celtic spaces of Wales, Cumbria and Cornwall. Later, when the Anglo-Normans took a more expansive grip in the late 11th–12th centuries, Celtic kingships were pushed still deeper into Europe's Atlantic ends.

Starting in the 16th century, the English Crown worked to secure its control over the British Isles as a whole, with Welsh Acts of Union (1536–42), fresh conquests of Ireland coupled with settlement programmes that dispossessed the native Irish, and a union first of crowns (1603) then of parliaments (1707) with Scotland. In the wake of these moves towards a United Kingdom, legislation was enacted against what remained of Celtic legal custom in these areas. The language of government became English. Legislation, like the Statutes of Iona (1609) with its attempt to compel west Highland chiefs to educate their sons through the English tongue, advanced this process. By the late 19th century, state education served further to extend spoken English in these remote fringes, confining the various spoken forms of Celtic to the hearth and the pulpit. For cultures that were largely sustained through an oral tradition, to silence them in these different ways was to weaken them. Compounding these problems further, all parts of the Celtic fringe suffered huge levels of emigration over the second

continued

half of the 19th century, a movement that projected a Celtic diaspora across most continents of the world but greatly diluted it at home. Accelerated by the Great Famine of 1846–50, this was undoubtedly the prime reason why the number of Irish speakers alone declined from *c*.4 million to only 680,000 between 1835 and 1891.

The 20th century has seen a renewal of Celtic political consciousness. The creation of the Irish Free State in 1922, and the adoption of Irish as its first language by Article 8 of its Constitution, effectively created the first Celtic state. Furthermore, the nature of its creation has meant that Celtic symbols have been used to create a Celtic iconography of statehood, one that values the Celtic identity written into its landscape. Yet whilst a more 'Celticized' vision of statehood has emerged, and whilst educational policies and the Irish Language Board (*Conradh na Gaelige*) have promoted the greater use of Irish, it has not prevented a decline in Irish speakers. Within the *Gaeltacht* – the area where Irish speaking is dominant – only 71% or 56,469 were fluent in it in 1991, although numerically many elsewhere in other parts of Ireland have varying degrees of fluency, including a reported one-third of Dublin's population. In Scotland too, the long-term picture has been a progressive decline in Gaelic speaking as population in the western Highlands and Islands declined. Yet though new initiatives, such as *Sabhal Mor Ostaig* (the Gaelic College on Skye) will conserve an awareness of Gaelic culture, it is doubtful whether it will reverse the decline of Gaelic speaking with only 86,000 still capable of using it in 1991. The current status of Welsh-speaking culture is different. Long-term decline in the area of Welsh speaking, *Cymru Gymraeg* or Welsh Wales as opposed to *Cymru ddi-Gymraeg* or anglicized Wales, a decline hastened in the south by a scale of industrialization unique to a Celtic-speaking area, has now halted. A new defined status for the Welsh language, including a Welsh Language Board (1993), a policy of bilingualism in all government services and the compulsory teaching of Welsh in schools, has meant that Welsh has held its position as a spoken language over the past decade, with 500,000 still able to speak it. Indeed, urban communities like Cardiff have actually shown an increase in Welsh speakers between 1981 and 1991. Of all the surviving Celtic languages, Breton can claim the most speakers. In addition to the 500,000 who still used it on a daily basis in 1989, mainly in western Brittany, many more (1.2 million) have a working knowledge of it.

Further reading

Champion, T., Gamble, C., Shennan, S. and Whittle, A. (eds) (1984) *Prehistoric Europe*, Academic Press, London

Clark, G.J.D. (1966) The invasion hypothesis in British prehistory, *Antiquity*, **40**, 172–89

Cunliffe, B. (ed.) (1994) *The Oxford Illustrated Prehistory of Europe*, Oxford University Press, Oxford

Evans, E. (1958) The Atlantic ends of Europe, *Advancement of Science*, **15**, 54–64

Greene, D. (1964) The Celtic languages, in Raftery, J. (ed.) *The Celts*, Mercier Press, Cork, 9–21

Powell, T.G.E. (1958) *The Celts*, Thames and Hudson, London

Renfrew, C. (1987) *Archaeology and Language: the Puzzle of Indo-European Origins*, Cape, London

'The naming of parts'

The history of place names leads to a fuller understanding of the peopling of Europe. The recording of particular places that were settled and of the areas that were occupied by named peoples began in the Mediterranean. First came the alphabets – Phoenician, Punic, Etruscan, Greek – from *c*.800 BC. Next came the names to be found in the literary inheritance of the Greek civilization. The sprinkling of coastal villages that became πολεις may have sometimes been the homes of peoples of diverse ethnic origin, but they bore Greek names. From a variety of sources, Norman Pounds (1973) has plotted the distribution of named 'cities' with their metal-working and polyculture on the outline of the present-day map and has defined the major regions of the Hellenic world from Thessaly in the north by way of Euboea and Boetia to Attica in the south and

Ithaca in the west. Beyond the core area of the homeland lay the colonial settlements of Magna Graecia (Figure 2.2). Easternmost on the Black Sea coast was Phasis; westernmost on the Mediterranean coast of Spain were Alonia and Emporiae. Massilia was the ancestor of Marseilles. Among others, southern Italy had the poetically named Parthenope, Syborus and Poseidonis, while Sicily had half a dozen colonial settlements. Simultaneously, from their bases in Tyre and Sidon, the Phoenicians established stepping stones along the coast of North Africa – Leptis, Carthage, Icosium – to four named colonies around the 'Pillars of Hercules'. While the named and settled Mediterranean in Greek times was essentially thalassic, there were loosely organized urban settlements inland of non-Greek origin. Those of the Etruscans in the Italian peninsula were among the best known. In general, the mountainous hinterlands of the Mediterranean spilled the surplus population from their agricultural settlements seawards.

As the centre of gravity of the Mediterranean shifted from Greece to Rome, increasing numbers of named people from beyond the encircling mountains were brought into the orbit of control. Within the ordered frame of the Roman Empire, the settlement of Europe acquired a measure of permanence. The peoples of the E31mpire beyond the homeland sometimes had names imposed upon them, but their own names or emerging names were more usually employed, though in Latinized forms. They included the imperial provinces of Thrace and Dacia in the south-east, through Illyria and Pannonia, Noricum and Raetia, to Belgica and Britannica in the west. A twilight zone on the north-west fringes of the continent was occupied by the Frisians, Saxons and Jutes. East of the Rhine and north of the Danube, loose tribal organizations prevailed under the collective name of Germania. By the time that the Emperor Diocletian (245–313 AD) had established his major system of four prefectures, Sarmatia was written across the isthmian lands between the Black Sea and the Baltic to complement Germania in the west.

A primitive synopsis of the peopling of the continent was first provided by Claudius Ptolemaeus, more commonly known as Ptolemy (*c*.90–168 AD). Ptolemy, who was an inheritor of the tradition of Greek mathematicians and astronomers, was resident in Alexandria. He constructed the first significant map of Europe, correcting and improving the work of a predecessor, Marinus of Tyre. Ptolemy's principal contribution was to calculate the latitude and longitude of some 8000 places in the known world of his day. His record of the place names of Europe marked a new stage in the presentation of knowledge about the settlement of the continent. The *Geography of*

Figure 2.2 Temples at the ancient Greek colony of Paestum founded in southern Italy in the sixth century BC (*source:* Tim Unwin, 16 May 1987).

Claudius Ptolemaeus (Stevenson, 1991), though, was lost during the Dark Ages and only came to light a millennium after it was written.

Within two centuries of his time and with the decline of the Roman Empire, the map of European settlement was being substantially modified by the so-called barbarian invasions – named peoples not always of clear provenance. From the fifth to the eighth centuries, the situation was particularly fluid. It was no simple outpouring of peoples, although they did come principally from the north and the east. Germanic peoples – the Goths and Visigoths of the history books – moved into much of northern Gaul, slowly settling beside the prior inhabitants, merging with them and eventually becoming a part of the early Frankish realm. In lesser numbers, but with greater immediate effect, the Lombards penetrated into the Italian peninsula. In Iberia, a Visigoth kingdom was established. The eastern frontiers of what had been the Roman Empire were invaded by steppeland peoples – the Huns in the fifth century, the Avars after them. Immigrants of Slavic, Bulgar and Gothic origin entered the Balkans. Between the seventh and ninth centuries, Slavic groups pushed westwards as far as the Elbe.

Simultaneously, new forces were rising in the Near East which were calculated to affect the peopling of the Mediterranean world. Under the banner of Islam, an Arab Empire emerged, unifying the settlement of the North African coast before conquering Spain in the course of a decade (711–21). Arabs also settled in other Mediterranean islands, including Sicily and the eastern marchlands of the old Roman Empire. The Muslims brought with them their learning – mathematics and astronomy – and their numerals (see also Chapter 4).

The other force which was to have a profound impact was Christianity (see Chapter 4). It confirmed existing settlements through the construction of churches and created new sites through monastic foundations, often in unpeopled wilderness areas. Through proselytes and priests – not least of the great Frankish realm – it brought peripheral parts of Europe into a broad, common cultural system. Furthermore, as Collins (1975) has put it, Christianity was 'the religion of the book'. And the book implied more than the Bible. The church was concerned increasingly with ecclesiastical documentation, records for one purpose or another of where people lived. Place names multiplied apace within the loosely defined German tribal territories.

Paradoxically, Christianity, prospectively a unifying force, became a divisive agent in the continent at large. As the Roman Empire itself was divided with the establishment of the Byzantine Empire, so from the sixth century onwards the Roman Catholic church of the west was increasingly distinguished from the Orthodox church in the east. And the discourse of the two churches was in different languages – the Latin of the west set against the Greek of the east. In the process, the bilingual Mediterranean world was divided into two monolingual halves.

Independently of the classical languages, population migration implied also that language was on the move. It was not unusual for the language of indigenous peoples to yield to that of the immigrant tongues, though place names – or at least the lesser ones – tended to survive the elimination of the original local language. Such was the case in eastern Europe, with Finno-Ugrian place names frequently discernible beneath the Slavic (see the case study of Finland).

By the latter half of the first millennium AD, a broad linguistic division of the continent into Romance, Germanic and Slavic-speaking peoples was recognized. The subsequent subdivision of this threefold grouping marked the further differentiation of the peoples of Europe. And, naturally, each language developed its own vocabulary to describe the process of settlement – the language appertaining to the home and hearth (the *domus* of Ian Hodder (1990)), to village and hamlet, pasture and ploughland, woodland clearance and fenland drainage. To the evidence of the peopling of the continent told in stones, bones and artifacts was added the living word.

'Alas poor Yorick'

As soon as the names of people and the places with which they were associated began to be written down, distinctive attributes began to be ascribed to them. At first, they were descriptive and accompanied the texts of early atlases. Accounts by encyclopaedists such as Diderot in the 18th century followed, to which were eventually added mythologies concerning the origins of Europeans. The brothers Grimm established an appealing model for ethnographic enquiry with their collections of folk tales and folklore. Their example was rapidly followed throughout much of the continent. Historians such as J.S. von Herder (1744–1803) and F.G. Fichte (1762–1814) advanced ideas about the

The peopling of Finland

W.R. Mead

For a long time, the origin of the Finns was a subject of debate among European scholars. *Who are the Finns?* asked R.E. Burnham, the first British university lecturer in the Finnish language, in a book bearing that title (Burnham, 1938). Since then, speculation has waxed and waned within Finland and without. It has waned to the extent that some scholars would like to consider it a non-subject. Nevertheless, the avenues of enquiry into the peopling of Finland that have been pursued are of interest in their own right – archaeologically, anthropologically, linguistically, even medically. So, too, are the sources that enable the reconstruction over the last 400 years of the settlement of the Finns of the territory that now constitutes *Suomen Tasavalta*, the Republic of Finland.

The earliest archaeological evidence of human habitation in Finland belongs to the final stages of retreat of the Quaternary ice sheet (*c*.7500 BC). Archaeologists recognize an increase in hunting and fishing communities with the climatic improvement from *c*.5000–4000 BC. The crude comb-ornamented pottery by which the first 'culture' is identified is associated with settlement sites around Suomusjärvi in south Finland. Remains from the succeeding Boat Axe culture (*c*.2200–1700 BC) and more particularly Bronze Age finds (1500–500 BC) associated with the name Kiukais suggest that there was immigration into Finland from western as well as eastern sources. As in Estonia and Latvia, stock-keeping was established (1000–200 BC). The village burial grounds of the succeeding Roman Iron Age indicate an increasingly intense occupation of parts of south-west Finland. Notable concentrations are preserved in the Åland Islands. They also indicate the steady colonization of the peripheral coastal areas as the land emerged from the sea (Figure 2.3). Palynological evidence of changing land uses (principally as a result of swidden) and the presence of carbonized cereal grains also belong to the first millennium BC (Zvelabil, 1978).

From where did these early colonists derive? Side by side with archaeologists, linguistic scientists have their theories about the sources from which the Finns arrived in their present homeland. Alexander von Humboldt (1769–1859) wrote of languages as 'visions of the world'. The visions are of the past as well as the present. Linguistic taxonomy has been a happy hunting ground for Finnish philologists, who have also sought to reconstruct the earlier movements of their Finno-Ugrian ancestors by analysing words in contemporary vocabularies that appear to have had their origins in areas through which earlier migrants passed.

M.A. Castrén (1852–70), a Finnish anthropologist and ethnographer, was the first to set out on an expedition to Siberia to search for the antecedents of the Finns. It is now generally accepted that the earliest identifiable home of the Finno-Ugrians was between the Urals and the central Volga area, perhaps more narrowly around the confluence of the Kama and Volga rivers. Certain groups of so-called Volga Finns, the Mordvins and Cheremiss in particular, still occupy the same area, with the Votiaks to the north and the Zyrians in the Komi district. Smaller groups of Finno-Ugrians – the Voguls and the Ostyaks – are scattered along the Ob valley to the east of the Urals. Many place names in the middle Volga territory support the view that beneath the Slavic toponymy there is a sub-stratum of earlier Finnish forms (Hajdú, 1976).

Kauko Pirinen (in Jutikkala and Pirinen, 1979) has written of 'the parting of the ways' of the migrating forebears of the Finns and Hungarians several thousands of years ago. Those who moved westwards – the Baltic Finns – encountered other cultures on their journey, absorbing new vocabularies as well as adopting new techniques. In particular, their encounter with East Germanic tribes has left its own residue of identifiable words in the Finnish language. The Baltic Finns also included the antecedents of the Karelians, Vespians, Ingrians and Estonians. The languages of these neighbouring peoples are closely related to Finnish.

While the Baltic Finns moved westwards, groups which now bear the name Lapp moved northwards by way of the Ladoga–Onega corridor towards the Arctic watershed. Here, as reindeer hunters and herders, they met the Norwegian

continued

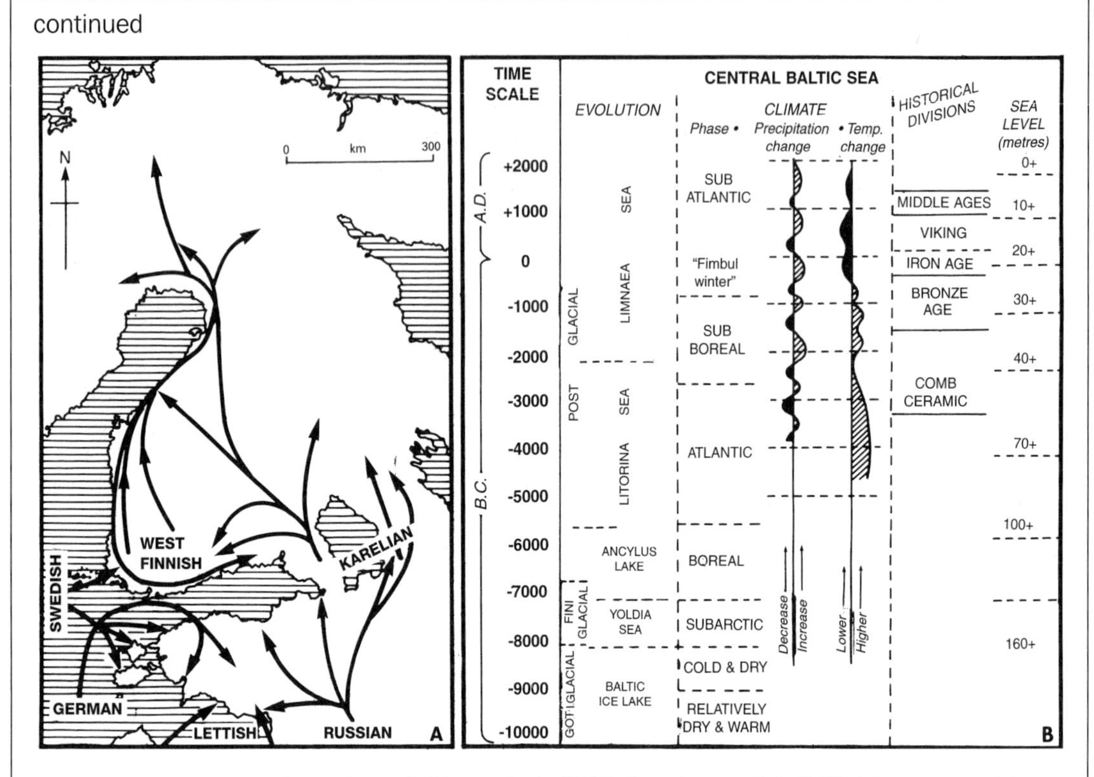

Figure 2.3 (a) Migration in the north-east Baltic area in the Middle Ages (*source:* Kalevi Wiik, *Itämerensuomalaisten kansojen ja kielten syntyksymyksiä*, unpublished manuscript, *Turun yliopisto Fonetikka*, 1995: 7).
(b) The Åland Islands in the central Baltic Sea reflect in fine detail the sensitivity of colonization to isostatic adjustment since the period of the Comb Ceramic culture (*c.*3000 BC) (reproduced by permission of the author from W.R. Mead, 1964, Saltivik, studies from an Åland parish, *Geographical Field Group Regional Studies*, **10**, Nottingham).

coastal inhabitants, who named them Finns – and *Finner* the Lapps remain to the north Norwegians today. Residual Lappish elements remained in south Finland until they were slowly driven northwards by the Finnish forest colonists from about AD 300 until the Viking period. Already before the turn of the millennium it would seem that one stream of Finns had crossed the Gulf of Finland to colonize Uusimaa (Nyland) and then the Häme (Tavasthus) district. The phonetician Kalevi Wiik considers that they brought with them the name *Suomi* from Latvia/Lithuania, where it has its roots. A second stream probably travelled overland at much the same time to establish themselves in south-eastern Finland.

In addition to archaeology and linguistics, ethnographic studies have added their quota to the understanding of the peopling of Finland. The unique folklore of Finland, much collected orally, had its pioneering publication, *Kalevala*, in 1835. Seamus Heaney senses in it 'access to an evolutionary ear'. The content of *Kalevala* certainly enables a picture of society and economy to be reconstructed from about the time that Finland was leaving the pagan for the Christian era. Also during this period, many more of the ancestors of the present Swedish-speaking Finns crossed the Baltic to settle in the archipelagoes and along the coastlands of south-west Finland and Ostrobothnia. One of the world's finest ethnographic atlases, by Matti Salmela (1994), displays the distributional patterns of the distinctive cultural features of the Finnish people – from folk custom to folk art, thereby illustrating the subtle variations of common themes across the length and breadth of the country.

In the early centuries of the first millennium, the permanently populated part of Finland lay south

continued

of a line linking Pori (Björneborg) with Sortavala, where the settlers were eventually to find themselves between the spheres of influence of the Roman church in the west and the Orthodox church in the east. Following the 12th century crusades, proto-Finland became an integral part of the Swedish realm. With the passage of time, the highly efficient systems of record-keeping that Sweden gradually developed applied equally to Finland, so that the peopling of the country can be traced in considerable detail. From the 1540s onwards, the tax rolls enabled the expansion of settlement to be followed closely (Jutikkala, 1973; Soininen, 1961). The pioneering census and survey systems established in the 18th century, which continued in principle when Finland became a Grand Duchy of Russia in 1809, provide additional material for reconstructing the settlement pattern.

In historic times, both the peripheral location of Finland and its general impoverishment discouraged immigrants. For this reason, independently of the distinctive language, the 'racial' homogeneity of the Finns has marked them apart in Europe. This is not to deny that there are variations in physical appearance no less than in dialect between Finns in different parts of the country. Emigration has followed the trend in other parts of Europe, with the 'blood letting' most pronounced in Ostrobothnia and reaching a peak at the beginning of the 20th century. Finland became an independent republic in 1917, constitutionally bilingual, called Finland by its Swedish-speaking minority and Suomi (a word of uncertain provenance) by the Finnish-speaking majority of *Suomalaiset*.

The biggest single short-term change in the peopling of Finland took place in the mid-20th century. At the conclusion of the Russo-Finnish wars (1939–40, 1941–44), Finland ceded a tenth of its territory to the USSR, including the greater part of the province of Karelia. Rather more than 400,000 Finns left the ceded territories to be accommodated in the reception areas in the remainder of the country. This fundamental redistribution of the population was to be followed within a decade by a major rural–urban migration consequent upon fundamental changes in the economy. Simultaneously, the proportion of the Swedish-speaking population tended to decline and the areas in which they have been a majority element in the population have contracted.

So far, the increased mobility of the population has not significantly reduced its biological distinctiveness. Finns, in general, have 'many rare genes and polymorphic traits' in which they differ from other European people (Eriksson, 1973). At the same time links with other inhabitants around the Baltic Sea may be discerned in blood groupings. In addition to these qualities, there is a distinct disease inheritance. The Finnish population has a score of diseases to which it is more susceptible than other Europeans, some of which are confined almost exclusively to people of Finnish stock.

Further reading

Hajdú, P. (ed.) (1976) *Ancient Cultures of the Uralian Peoples*, Corvina, Budapest
Jutikkala, E. (1959) *Atlas of Finnish History*, Werner Söderström, Porvoo, 2nd edn
Jutikkala, E. (1973) *An Atlas of Settlement in Finland in the Late 1560s*, Helsinki
Jutikkala, E. and Pirinen, K. (1979) *A History of Finland*, Heinemann, London
Lehtinen, I. and Kukkonen, J. (1980) *The Great Bear*, Suomalaisen Kirjallisuuden Sevura, Helsinki
Salmela, M. (1994) *Suomen Perinne-Atlas, II*, Suomalaisen Kirjallisuuden Sevura, Helsinki (captions in English)

relationship between peoples, the places where they lived and the nations or emerging nations that they constituted. Carl Ritter (1779–1859) took a step further with his 'classification of nationalities'. By the middle of the 19th century, maps of the peoples of Europe and their languages were being published. The firm of Justes Perthes was incorporating ethnographic maps in the *Hand Atlas* of 1854, probably the best atlas of its kind in Europe. The next step was to equate people with race, Ripley's *Races of Europe* (1891) becoming a standard text. The definition of racial types encouraged enquiry into the pedigrees of people. Ethnography and archaeology were drawn together. A prehistoric dimension was added to the historical.

Among the artifacts linking past inhabitants and present occupants of territories, the skull was presumed to be the most significant. Basic measurements of skulls, expressed as the cephalic index, provided initial clues as to the physical features of the prehistoric peoples who had settled in different parts of the continent and the possible links between them. Craniological titles in the bibliography of C.S. Coon's classical study *The Races of Europe* (1939) illustrate the rise of interest among European scholars – *Crania germaniae* (1863), *Crania helvetica* (1864), *Crania britannica* (1865), *Finska crania* (1878), *Crania boemica* (1891), *Crania romani* (1894), *Crania suecia antiqua* (1900). Each was a study within the frame of an existing or emerging nation-state, though at the same time the references it incorporated revealed the extremely limited extent of the empirical evidence that was available to support the theses of the day.

The distribution of long-headed inhabitants (dolichocephalic) and round-headed inhabitants (brachycephalic) compiled from the location of skeletal remains was to be complemented by head measurements of the inhabitants of many European countries. In the inter-war years, empirical enquiry entered university classrooms. In some, calipers became a part of departmental equipment. The measurement of student heads was for a while a regular procedure, for example at the University of Aberystwyth. The application of craniometry to the student body which, before present-day mobility, was of distinctly Welsh ancestry, revealed a predominantly brachycephalic result. This, coupled with other physical characteristics, suggested a Mediterranean stock. Prehistoric skulls might not be available for comparison, but there seemed little doubt that in this isolated part of Britain, a megalithic ancestry common to that of much of the Atlantic fringe could still be identified.

At mid-century, Coon (1939) produced a generalized map of Europe which confirmed the association of broad-headed or Alpine stock with the Mediterranean and that of long-headed stock with the more northerly lands. In fact, heterogeneity rather than homogeneity tended to characterize the craniology of most European countries. Even Nordic (Teutonic) man, past and present, has displayed a generous mixture of broad heads. Certainly for any particular territory there seems to be little correlation between the form of any prehistoric skulls available and those of the majority of the present inhabitants. The history of migrations suggests that it has been the repeopling of places rather than their peopling that emerges as the significant characteristic fact in the settlement of Europe. Alas for those Europeans who, beset with ethnomania, once sought in the ground beneath them evidence of direct ancestry with its original settlers.

From Genesis to Numbers

Enquiry into the genesis of the peopling of Europe began with stones and bones. There is the book of Exodus to follow – indeed a number of books explaining whither the settlers moved and whence they came. Contemporary historical atlases employ a swirl of arrows to indicate the succession of migrations through the millennia. H.G. Wells in his *Outline of History* (1920) must have been among the first to synthesize the tracks of 'various migrating and raiding peoples during the Dark Ages' with circles representing the places of settlement.

There is also a book of Numbers. A valuable method for assessing settlement is the carrying capacity of the land under different systems of use. In terms of individuals per square kilometre, Ruth Whitehouse (in Collins, 1971) suggested a range of 0.04–0.1 for the Mesolithic population, 1.5–5 for the early Neolithic and 5–10 for the late Neolithic. With the rise of the Mediterranean civilizations, the transformation of European agricultural practices at large and the emergence of new systems of social organization, increasing densities could be supported. The structures and institutions of the Roman Empire offered a new degree of security over the largest area of the continent down to its day and population increased responsively. A figure as high as 50 million has been estimated for Europe at the time of the Roman apogee. A rapid decline followed during the turmoil of the new age of migrations, with a widespread redistribution of population. In turn, a significant rise has been estimated towards the millennium – neither the decline nor the rise to be explained without reference to climatic worsening or improving. For Europe in the Age of Charlemagne, Norman Pounds (1973) quotes a figure (based on a calculation of I.C. Russell) of 25 million. The boundaries of territorial jurisdiction might be drawn around them or shift above them, but with the consolidation of settlement of the Germans and Slavs, under the emerging Holy Roman Empire, the pattern of population distribution in Europe began to anticipate that of the modern age.

Further reading

Anderson, B.G. and Borns, Jr, H.W. (1994) *The Ice Age World*, Scandinavian University Press, Stockholm

Collins, D. (ed.) (1975) *The Origins of Europe*, Allen & Unwin, London

Cornell, T. and Matthews, J. (1982) *Atlas of the Roman World*, Phaidon, Oxford

Cunliffe, B. (ed.) (1994) *The Oxford Illustrated Prehistory of Europe*, Oxford University Press, Oxford

Dennell, R. (1983) *European Pre-History*, Academic Press, London

Gipel, J. (1969) *The Europeans, an Ethnohistorical Survey*, Longman, London

Hodder, I. (1990) *The Domestication of Europe*, Blackwell, Oxford

Mellars, P. (ed.) (1978) *The Early Postglacial Settlement of Northern Europe*, Duckworth, London

CHAPTER 3 European languages

ALEXANDER B. MURPHY

Introduction

Language is one of the most important and most basic features of human society. It is a principal means by which communities share and disseminate ideas and information, and it both reflects and shapes peoples' experiences with the world around them. As such, language is deeply implicated in the history and human geography of Europe. On its surface, learning about the geography of European languages might seem to be a straightforward matter. Everyone knows that a variety of languages are spoken in Europe, and understanding the nature, spatial character and significance of that variety is presumably what a geography of European languages is all about. Yet this assumes that we have a clear sense of what is meant by the term 'language' itself.

Unfortunately, there is no commonly accepted definition of language, or of what distinguishes a language from a dialect. Many dictionaries define language as a set of words and phrases that are understood by a substantial number of people. Yet how substantial does that community have to be? And what do we mean by understood? We commonly refer to German as a language, yet a 'German' speaker living in northern Germany might well be hard pressed to understand the 'German' spoken in a household situated in a small town in northern Switzerland. At the same time, we think of Norwegian and Danish as distinct languages even though a Dane could understand most of what was being said by someone speaking the form of Norwegian that prevails in Oslo.

There is no easy way to resolve such definitional ambiguities, but recognizing the problem draws attention to the fact that any map of languages embodies a subjective perspective on the meaning of the term language. It is sometimes said that a language is a dialect with an army behind it, and there is a certain truth to this since the tongues that we commonly designate as languages are those that have achieved some international status as a consequence of the historical successes of particular peoples in political or economic spheres. This, in turn, is a nice way of illustrating one of the fundamental premises of this chapter: that language cannot be understood in isolation from place. To state the proposition more completely, language is a cultural construct that reflects the changing history and geography of the places where it develops.

A related, but somewhat different, point about language is that it is a dynamic phenomenon. The present configuration of European languages is nothing more than the latest expression of a cultural construct that is in constant flux, in terms of both its essential character and its spatial distribution. Languages evolve over time, changing in type, location, extent and even existence. They spread into new areas when people move, they acquire new characteristics when people are confronted with different circumstances or other people, and they can decline or even die out when speakers of a language find themselves in marginal social and political circumstances. As such, the evolving language pattern of a region through time can provide important clues into the history of population movement and interaction, as well as the power relations that govern interactions between peoples.

This chapter seeks to situate the evolving language pattern of Europe in the context of changing demographic, cultural and political arrangements and understandings. The goal is first to sketch some of the most significant influences on the evolution of the European linguistic map over the past several thousand years. The discussion will then turn to the relationship between language and politics in 19th and 20th century Europe. The chapter will show that language is not simply an objective cultural component of human communities living in particular places; instead it can be an emotionally charged facet of peoples' identities that has significant political and social implications.

The emergence of the major European branches of Indo-European

Most modern European languages belong to a few major branches of one of the world's most important language families: Indo-European. There are great differences among the Indo-European languages of Europe, but they all share some basic structural characteristics, and even some continuities in vocabulary. It is difficult to know exactly where Indo-European tongues originated, but a combination of archeological and linguistic evidence suggests that the hearth area was in the vicinity of the Black Sea – very possibly near the Caucasus Mountains (Gamkrelidze and Ivanov, 1990; Renfrew, 1988). The emergence of the modern European language pattern can be traced to migrations of various peoples from this hearth to the west and north-west that began as early as 6000 BC. In the succeeding several millennia, certain peoples moved west across the Aegean Sea into what is now Greece, others moved north-west into west-central Europe and ultimately to the shores of the Atlantic, and yet others moved north into the Hungarian Plain and beyond.

As groups separated from one another, their languages began to diverge ever more in structure, syntax and vocabulary. Over time linguistic divergence became great enough that distinct branches of Indo-European emerged. These branches can be best understood when illustrated on a language tree (Figure 3.1) and represented on a map (Figure 3.2). The language branch of those peoples who crossed the Aegean Sea and settled in what is now Greece is Greek, also known as Hellenic. Peoples moving to the north-west ultimately split into three branches: a Celtic branch, composed of peoples of uncertain geographical origin who occupied a vast area north and west of the Pyrenees and Alps, but south of modern Denmark (see also Chapter 2); an Italic branch, which encompassed those who occupied the south-central part of the Italian peninsula; and a Germanic branch, which encompassed those who moved into modern Denmark and adjacent parts of Sweden. Those who moved to the Hungarian Plain and beyond split into two branches, a Baltic branch with geographical roots in what is now western Poland, and a Slavic branch with probable geographical roots in the Hungarian Plain itself.

The Indo-European migrants encountered other peoples as they moved into new areas. Although the tongues of the migrants ultimately predominated in most places, this did not happen without some incorporation of words and phrases from pre-existing languages, which furthered the divergence between the different Indo-European branches (Jordan, 1996). Little is known about the languages encountered by the Celtic migrants because they were not written languages. Some argue that modern Basque may be a descendant of one of these languages, since it is a tongue bearing no relation to Indo-European (Hualde, *et al.*, 1996). The linguistic situation in the region at the time of the Celtic encounter, however, is not well understood. More is known about Etruscan, a non-Indo-European language encountered by the Italic peoples as they spread through the Italian peninsula. Although Etruscan did not survive the Roman period, the fact that it was a written language meant that a tangible record of the language was available to future generations. The Slavic and Baltic peoples displaced speakers of the Uralic language family who had moved into eastern Europe several thousand years earlier, but the new immigrants did not overwhelm Uralic speakers everywhere. Uralic speakers have hung on in northern Europe and can still be found in northern Scandinavia, Finland, Estonia and northern Russia to this day (see also Chapter 2).

The current language pattern of Europe is the result of particular historical developments that played out on this largely Indo-European base over the past 2500 years. The easiest way to make sense of the pattern is to recognize that most modern European languages fall within three major branches of Indo-European: an offshoot of the Italic branch called Romance, the Germanic branch and the Slavic branch (Figure 3.3). Understanding how these branches of Indo-European came to dominate in south-west, north-west, and eastern Europe, respectively, goes a long way towards explaining the general character of the modern European language map.

The Romance branch

The Romance branch traces its roots to the extraordinary ancient history of the speakers of one of the

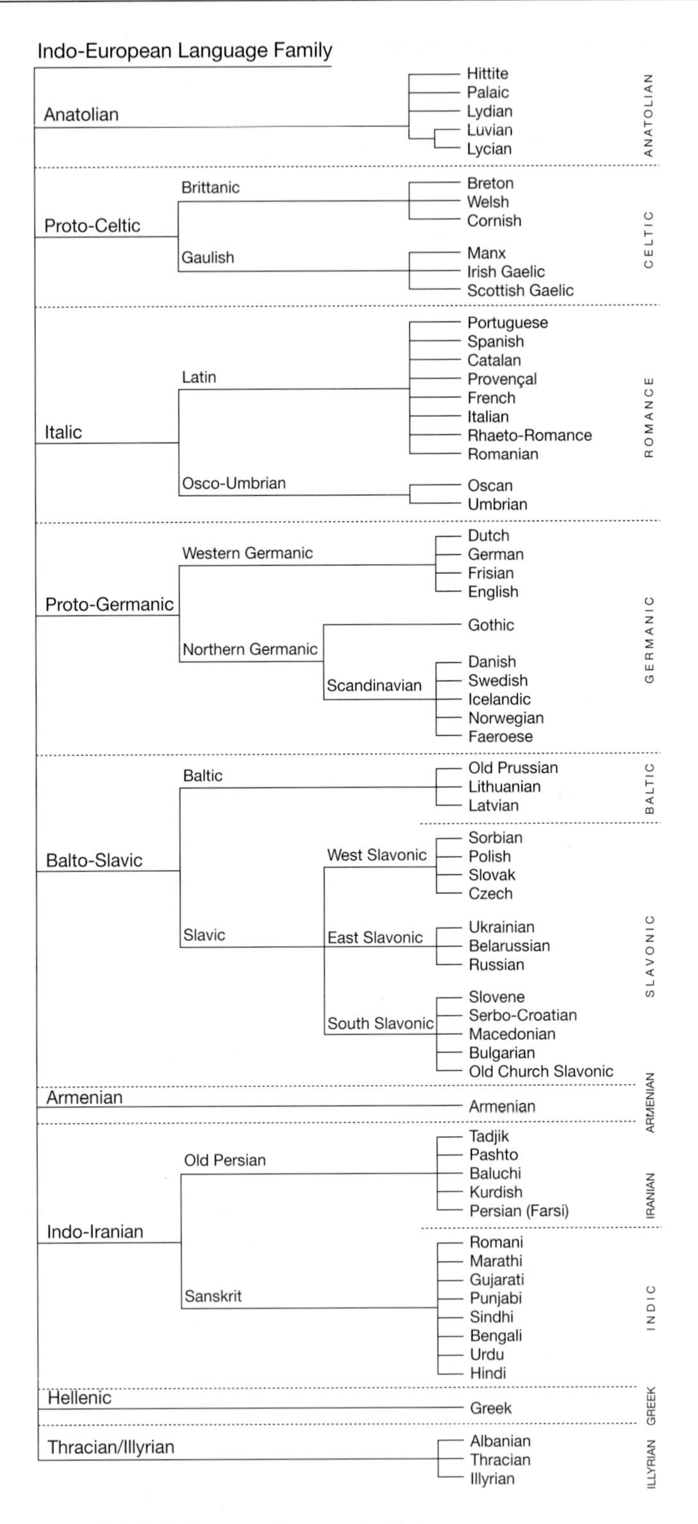

Figure 3.1 Indo-European language family tree.

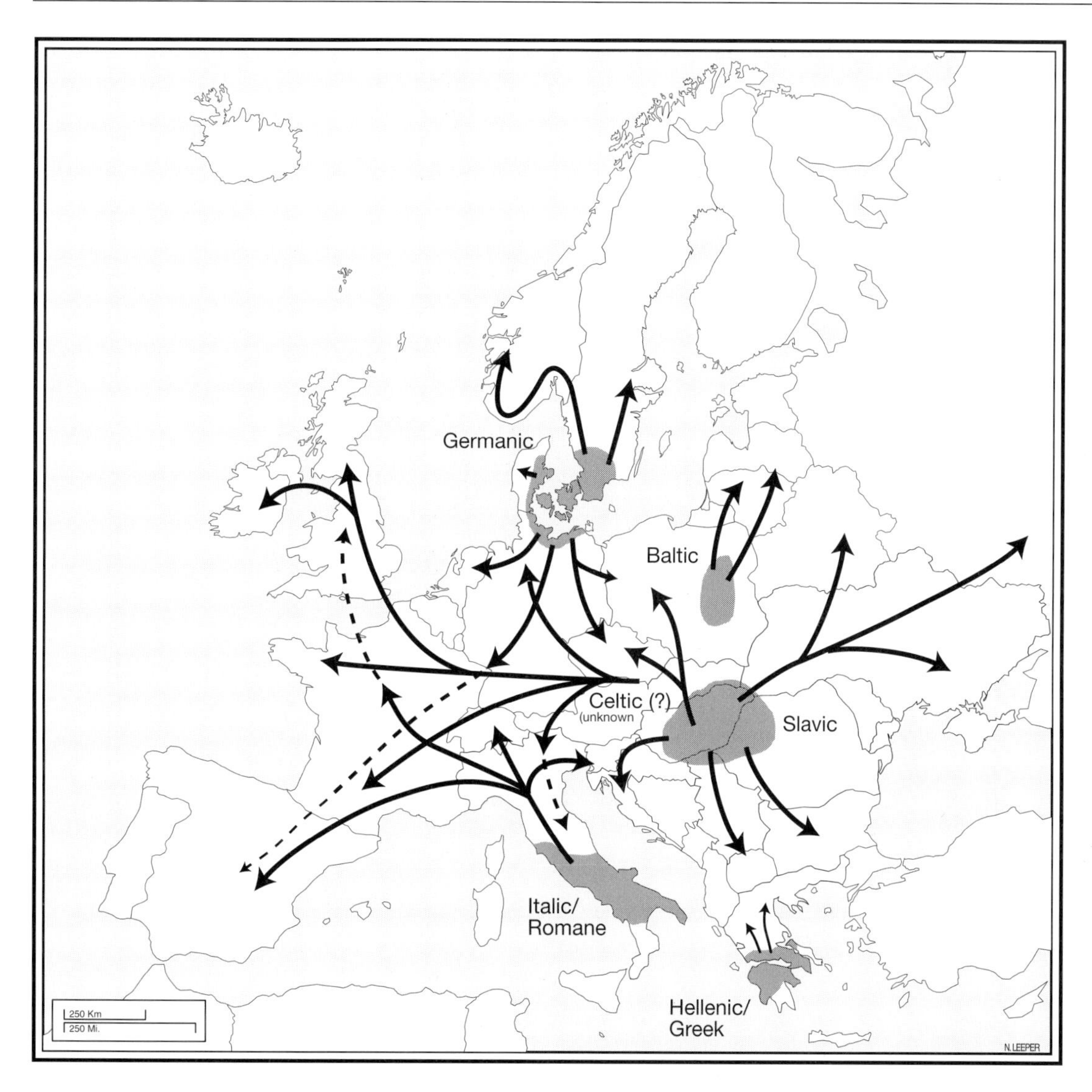

Figure 3.2 Possible hearth areas of major Indo-European language branches and directions of dispersal.

Italic languages, Latin (see generally Elcock, 1975). The speakers of Latin – the Romans – lived in the central part of the Italian peninsula, and ultimately built an empire that stretched from the shores of the Mediterranean to southern Scotland in the northwest, and from the Atlantic Ocean to the Black Sea in the east. The Romans did not force their language and culture on conquered peoples, but they did impose an administrative and trade system based on Latin. As such, those who were tied in with the formal operation of the Empire came to speak Latin. In those areas where the Roman Empire held sway for long periods of time, Latin assumed a sufficiently prominent position that it became the primary tongue of the upper classes and an important medium of communication throughout the Empire. The major exception was in Greece, where the cultural and social institutions that had evolved prior to the rise of the Roman Empire were sufficiently robust to withstand the influence of Latin.

The collapse of the Roman Empire in the fifth century AD was precipitated in part by invasions from the north and east of speakers of Germanic tongues. Throughout much of the eastern and northern parts

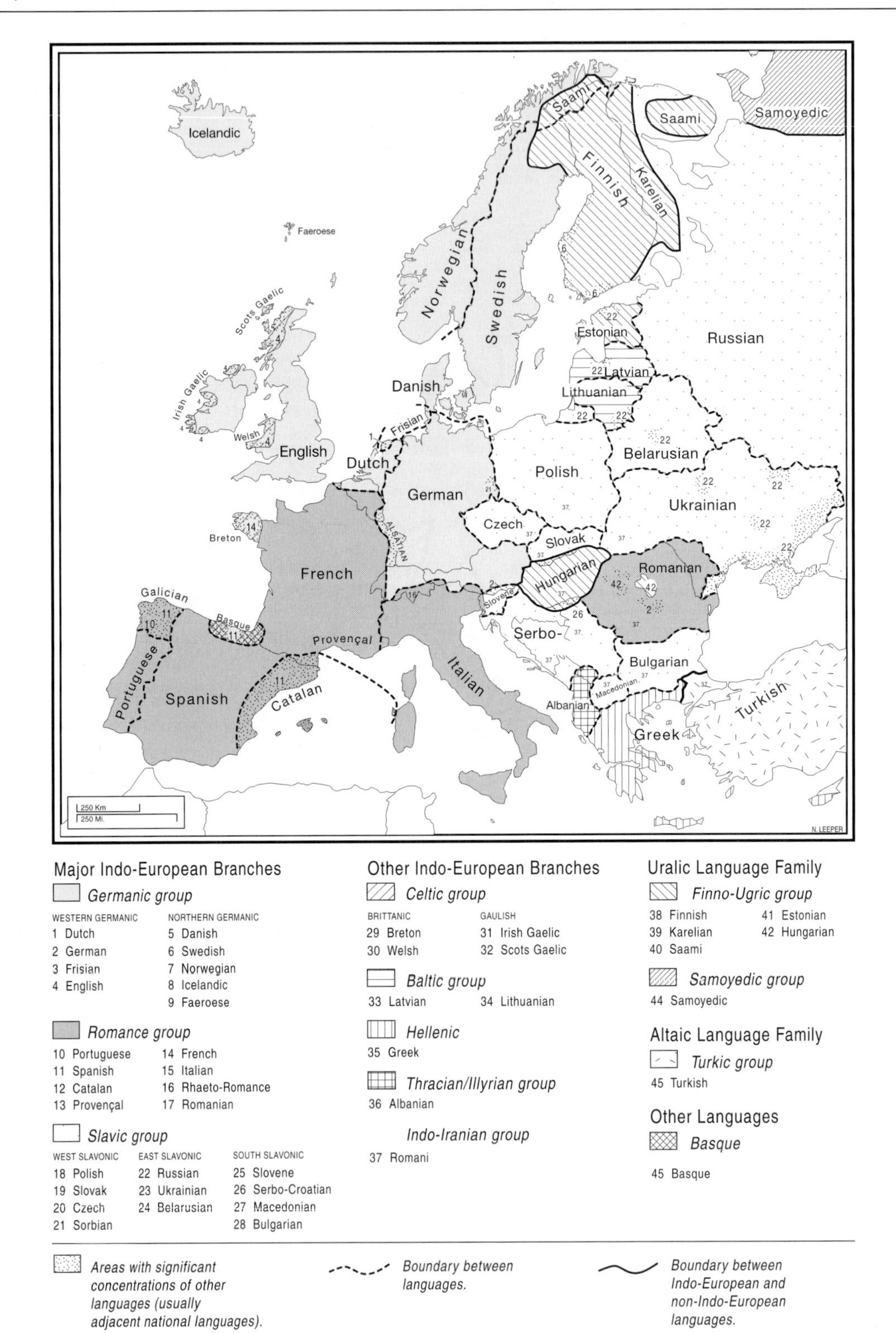

Figure 3.3 Linguistic map of Europe (after Jordan 1996, and other sources).

of the Empire, these invaders overwhelmed the mixture of Celtic and Roman peoples, in the process erasing the linguistic influence of Latin. Only the Latin terms for certain places survived, as with Trier, the modern name for the Roman town Augusta Treverorum, the capital of the Treveri. In the east the collapse of the Empire opened the way for Slavs to move into south-east Europe. This Slavic influence was sufficiently great that it largely displaced pre-existing tongues. A partial exception was in an area on the western shores of the Black Sea, where a Latin-based tongue either survived the Slavic invasion or was reintroduced by migrating Latin-speaking peoples called Vlachs from the north-western shores of the Adriatic. This explains why a Romance tongue – Romanian – is spoken in the area today (Rosetti, 1973).

The western and southern European parts of the Roman Empire were less profoundly affected by external linguistic influences. Instead, as Europe entered the so-called Dark Ages, tongues were emerging in these areas that reflected a mixture of Latin with older Celtic and Italic languages. These tongues shared in common a significant Latin influence, and therefore came to be known by a term that reflected their origin in the Roman Empire: Romance. Since the mixture of Latin with older tongues was different from place to place, Romance Europe was characterized by considerable linguistic fragmentation. Fragmentation was furthered by the lack of any large-scale political–territorial framework that could facilitate interaction between peoples.

Romance tongues did not entirely displace older tongues in all parts of western Europe. Basque hung on at the western edge of the Pyrenees, and Celtic tongues continued to be spoken in places. Moreover, the arrival of the Moors in the Iberian peninsula in the eighth century brought new linguistic influences into south-west Europe. Nonetheless, tongues influenced by Latin ultimately rose to dominance throughout most of the area shown on Figure 3.3 as the Romance language area. The particular languages that emerged – Portuguese, Spanish, Catalan, French, Italian and several less widespread tongues – were those of groups that ultimately were the most successful in asserting political and economic hegemony over particular territories.

Latin itself continued to play an important role in Europe in medieval and early modern times. In the period of upheaval that followed the fall of the Roman Empire, one of the few havens of relative stability and safety were the Christian monasteries that had survived in peripheral areas (Cahill, 1995). The majority of the European population was illiterate, but those in monasteries kept learning alive through the vehicle of the Latin language. Thus, as Christianity gained ground in Europe during the Middle Ages, so did the Latin of the Church (see Chapter 4). Because the Church played a key role in society and Latin was the linguistic medium for that role, it became the language of high culture and learning, and its role endured beyond the Middle Ages. Many of the great works of science and literature in the 16th through 18th centuries were produced in Latin, and the language continued to be used in governmental affairs, particularly in those parts of Europe where Roman Catholicism held sway.

The Germanic branch

The Germanic branch of Indo-European spread from its relatively small base in southern Scandinavia to dominate northern Europe. The first expansions took place before 500 BC through southern Scandinavia and along the shores of northern continental Europe. Several centuries later the Germanic peoples began to move south, in the process overwhelming the Celtic languages of the area (see generally van der Auwera and Konig, 1994). The Germanic expansion continued farther to the south and south-west until it was checked for a time by the northward-expanding Roman Empire. The *limes* or frontier of the Roman Empire in Germany was a dynamic zone for many generations, but in the declining days of the Empire, Germanic invasions penetrated beyond the *limes* with increasing frequency. Some Germanic peoples moved into what is now France, and some went as far as present-day Spain and Italy.

Germanic languages did not replace the mixture of Latin and indigenous tongues that dominated throughout much of the southern and western parts of the Roman Empire. However, their cultural impact was great; the very name for the inhabitants of the modern country of France can be traced to a West Germanic group called the Franks that migrated westward in the fourth century AD. Germanic influence was greatest along the northern frontiers of the Empire; it was in these areas that Germanic tongues took root. The boundary between Romance and

Germanic Europe thus loosely corresponds with a modestly retracted northern boundary of the Roman Empire (Figure 3.3). The major exception is in the British Isles, where invasions – first from what is now Germany and Denmark and later from other parts of Scandinavia – led to the almost complete displacement of Latin-based tongues by Germanic languages.

As the Germanic peoples spread out, Celtic tongues continued to decline in northern Europe. Many of them ultimately died out, and those that hung on did so in peripheral areas (see Chapter 2, and generally MacAulay, 1992). The story throughout the rest of the Germanic world also mirrors that of the Romance world; the particular languages that emerged out of the Germanic linguistic soup – principally German, Dutch, Danish, Norwegian, Swedish, Icelandic and English – were those of peoples who were able to establish a degree of political and economic control over a substantial area.

As a general rule, the Germanic tongues have substantially different sentence structures and vocabularies from their Romance counterparts. A partial exception is the Germanic tongue of the British Isles, because the Norman Invasion of 1066 brought speakers of an early form of French into England. As conquerors, the Norman invaders assumed a dominant role in the social and political structure of the early English state. They were too small in number to change the essential character of the Germanic tongues that they encountered, but they introduced a new language that coexisted, and ultimately blended, with the Germanic tongues of the pre-Norman inhabitants of the region (McCrum *et al.*, 1986). As a result, English has two words for many things that are described by a single word in other languages. In English people both sweat and perspire; they are tired or fatigued; they seek help or aid. English speakers also refer to an animal in the field as a pig, but when the meat of that animal is consumed they call it pork. These features of English reflect the blended or composite character of the language. They even tell us something about the social structure of Norman England. The terms that have a more everyday feel to them – sweat, tired, help, pig – are Germanic in origin, whereas the terms that connote sophistication – perspire, fatigued, aid, pork – are related to the early French of the Normans. This is a clear reflection of the social situation in Norman England, where the speakers of a Romance tongue were the élite.

The Slavic branch

The largest branch of Indo-European in Europe in terms of numbers of speakers is the Slavic branch. There are three major subgroups of Slavic: South Slavonic, West Slavonic and East Slavonic (see generally Comrie and Corbett, 1993). These subgroups have their origins in the splintering of the Slavic branch that occurred as Slavic Indo-Europeans spread throughout the eastern portion of the North European Plain and farther south into the Balkans beginning in the third century AD. The collapse of the Roman Empire created an opportunity for the South Slavonic peoples to move into the Balkans in the fifth and sixth centuries. These peoples ultimately overwhelmed the vestiges of Latin and older tongues in most of south-east Europe. In the process, the South Slavs came to occupy the dominant cultural position in the region. Their tongues eventually crystallized into modern Serbo-Croatian, Bulgarian and Macedonian (Figure 3.3).

Other Slavic peoples stayed in the Hungarian Plain or moved north, west and east. Those Slavs that moved to the east and north-east formed the Eastern Slavonic offshoot of Slavic. Their languages evolved into modern Russian, Ukrainian and Belorussian. The Western Slavonic peoples stayed in the Hungarian Plain or migrated to the north and west. Those moving to the north settled in what is now the Czech Republic and western Poland, pushing the Baltic peoples farther to the north. The westward migrants moved into modern Slovenia. Collectively, these peoples spoke tongues that were the antecedents of modern Polish, Czech, Slovak, Slovenian and Sorbian. The in-migration into the Hungarian Plain of Magyars speaking a tongue in the Uralic language family in the ninth and tenth centuries occurred at the expense of the Western Slavonic peoples (see Hungarian case study). Those in the Hungarian Plain were assimilated or pushed out, and those in Slovenia were cut off from their northern cousins (Figure 3.3).

As the Western Slavonic peoples spread to the west, they encountered Germanic peoples. In the Middle Ages the boundary between Germanic and Slavic speakers lay approximately along the Elbe and Saale Rivers (Jordan, 1996). German expansion to the east eventually pushed that boundary back, and more recent political developments led to its consolidation along the current Polish–German boundary.

Hungary, Hungarians and the Hungarian language

Peter Sherwood

The terms Hungary, Hungarians and the Hungarian language have meant different things at different times, and the varying nature of the overlap between them has been a notable source of tension throughout the history of the region. Here we focus on each in turn, in order to illuminate a problem that is now thought of in mainly linguistic terms, but which will continue to affect European integration well into the next millennium.

The Kingdom of Hungary was established in the year 1000, barely a century after the arrival of the pagan Hungarians in the region, by St Stephen, whence its more precise technical designation as the Lands of the Crown of St Stephen. Its territory included Upper Hungary (now the Slovak Republic) and swathes of land which are now parts of easternmost Austria, north-eastern Slovenia, Vojvodina (now Serbian), as well as westernmost Ukraine (Transcarpathia); and from early times, Hungary had a special relationship with Croatia. Transylvania, though historically much smaller than the present-day province of Romania bearing that name, was also intimately linked with Hungary. Southern and central Hungary was occupied by the Ottoman Turks in the 16th and 17th centuries, and from this time Hungary was in thrall to the Habsburgs, whose rule was not dislodged in the revolution of 1848–49, but with whom an accommodation was reached in the form of the Austro-Hungarian, or Dual, Monarchy between 1867 and what became World War I. That war brought to an end an unprecedented economic boom and also, in the Treaty of Trianon of 1920, the multinational Kingdom of Hungary (only about 50% Hungarian-speaking in 1914), which lost half its population and some two-thirds of its territory. Admiral Horthy's interwar authoritarian 'regency' tried to secure the return of the lost Lands of the Crown of St Stephen, but being on the losing side in World War II only reduced Hungary's territory even further.

Hence Hungarian as applied to a person can be a problematic label. For centuries the *natio Hungarica* was defined largely in terms of social class, consisting only of the nobility, which comprised almost 5% of the population, and county officials and other professionals; serfs, town burghers and even the aristocracy did not count as members of the Hungarian nation. Until the 18th century, this is the most usual sense of 'Hungarian' (*hungaricus*) as applied to a person, though the Latin term was also used in the sense of 'from the Hungarian kingdom'. The roots of the Enlightened sense of 'Hungarian nation', that is of a notion of *magyar nemzet* as the 'community of all Hungarian speakers', must reach into the 18th century, but it is likely that the full modern sense blossomed only as the achievements of the language renewal movement made possible the enunciation of the nobility's programme of political reform in the first decades of the 19th century. This sense of Hungarian ('member of the Hungarian nation') exists to this day and must be carefully distinguished from the post-1945 sense 'citizen of the Hungarian Republic' ('People's Republic' in Communist times), even if 95% or more of the latter are also the former. Indeed, the fall of communism has allowed the difference between these two definitions to come to the fore.

Table 3.1 Numbers of Magyar speakers living outside Hungary in the 1990s.

Slovak Republic	566 000
Ukraine (Transcarpathia)	180 000
Romania (mainly Transylvania)	2 000 000
Former Yugoslavia (mainly Vojvodina, in the 1980s)	400 000
Slovenia	9 500
Austria (Burgenland only)	4 000

Magyar (Hungarian) is today the first, and often only, language of over 10 million people in the Hungarian Republic. It is also spoken by more than 3 million as their first or best language in those former Lands of the Crown of St Stephen which formed, or were secured by, neighbouring countries. Bilingualism is common, but needs to be systematically investigated. Table 3.1 gives carefully estimated figures of Magyar speakers living outside Hungary. These Hungarian speakers all live in communities located alongside or near the respective borders with Hungary, with one very important exception: the Hungarian speakers of

continued

Romania. Here two groups must be spotlighted. First the Szeklers (Hungarian: *székely*; Romanian: *Secui*), who are an ethnographically (and perhaps originally ethnically) distinct part of the Hungarian-speaking community in multicultural Transylvania, settled by the monarch along the easternmost fringe of the Kingdom of Hungary as its guardians, in return for certain privileges of which an awareness still remains. More than three-quarters of the population of the Carpathian counties of Harghita (Hungarian: *Hargita*) and Covasna (Hungarian: *Kovászna*) are still Hungarian-speaking. Especially amongst 'mainland' intellectuals in the Hungarian Republic, the Szeklers have been mythicized as the essence of Magyarhood and their speech is often claimed as the purest and most attractive variety of the language. For many mainlanders, the non-Szekler Hungarians of Transylvania also bask in the Szekler glow, with even important 20th century novels about the past of Transylvania being entitled 'Fairy Garden' (Zsigmond Móricz, *Tündérkert*, 1922). Interestingly, however, when a wave of Transylvanian Hungarian emigration swept over Hungary in the early 1990s, many Szeklers complained that their dialect was a definite disadvantage in securing employment in Budapest, a stark contrast between myth and reality. Romania's other group of Hungarian speakers to have a distinctive designation are the 40,000 to 100,000 originally Roman Catholic speakers of the heavily Romanian-influenced dialect-cluster known as Csángó Hungarian. The Csángós live mainly in the villages and small towns around Bacău (Hungarian: *Bákó*) in Romanian Moldavia.

Thus, at least one in four of all Hungarian speakers in the region lives outside Hungary proper, and form numerically the largest minority in Europe, apart from the total number of Russians living outside Russia. Magyar is a unitary language, in that (apart from Csángó) there are no great dialect differences; certainly, the distances between the regional varieties of British English are much greater. The language itself is unusual in a number of ways. It is genetically Uralic, with its closest kin in north-west Siberia, and although a distant cousin of Finnish and Estonian, it is unrelated to any of the languages that surround it today. The discovery of its remote kinship, the story of the language's survival through centuries of the trek from the vicinity of the Urals, its endurance and adaptation to European conditions through further centuries of Christianity and Islam – all these factors helped strengthen Magyar's role as the outstanding component of the new definition of nation, more potent than other elements – such as the constitution, folklore, religion (and holy relics, like St Stephen's crown), music and even national dress – which the emerging élite used to articulate a cohesive Hungarian past.

Magyar survived Herder's prophecy of 1791, that *nach Jahrhunderten wird man vielleicht ihre Sprache kaum finden* (in a few centuries their [i.e. the Hungarians'] language will perhaps hardly exist). Though it might easily have been fulfilled earlier in the century, by the time it reached Hungary the movement to renew the language (the *Nyelvújítás*) was in full swing. This is not to say that the nightmare vision of the death of the nation (*nemzethalál*) did not haunt the Romantics and others for a long time; but it haunted them, and their readers, in Magyar. For a few crucial decades, every literate person in the land debated the future of the language: grammars were written to prove that Magyar was just as precise and well-ordered a language as Latin or German; spelling was standardized; historians of the language and writers fought bitterly over the creation of new words; Latin (the Hungarians' 'father-tongue' and the language of public life until the 1840s) and German were supplanted in their technical domains; poets wrestled with the beauty or ugliness of new coinages; and all the while the friction, and the translations, and just the sheer increase in the volume of writing helped, by the end of the century, to hone this Uralic language into the flexible, all-purpose European instrument that we know today. A still-current slogan from the mid-19th century sums up the intimacy of the link: *Nyelvében él a nemzet* (A nation lives in its language).

However, not all Hungarian speakers have benefited to the full. The story of those Hungarian speakers left in the successor states remains to be written, but it is certain to be a mainly unhappy tale, if seen in terms of human (and especially linguistic) rights of the Hungarian minorities. In the

continued

largely rural western areas (Burgenland, Slovenia) there seem to be few linguistic problems beyond natural attrition. Elsewhere in the region, however, the Hungarian speakers are embattled. Vojvodina has been devastated by the Balkan war and many of its Hungarian speakers are homeless or are emigrating. In both Transylvania and the Slovak Republic, complex laws regulate the public use of language, restricting the use of Hungarian to certain geographical areas and/or spheres of activity. Regulated in this way are parliamentary discussion, place, street and even personal names, the language of the media, telecommunications and transport, commerce, manufacturing and services, and the language of schooling, where even in the few minority schools that exist some subjects or classes are taught in the state language by law or at the request of non-Hungarian parents. In the Slovak Republic, for example, the new Language Law, enforced by language police from 1 January 1997, provides stiff penalties for such offences as displaying bilingual signs which do not have the Slovak text first, or failing to address a new customer in Slovak.

At the Battle of Merseburg in AD 933 it is recorded that from the still-pagan Hungarians *vero turpis et diabolica* hui, hui *frequenter auditur* (a truly dreadful and diabolical *hui hui* could often be heard) (Jakubovich and Pais, 1929: 12). A millennium later the Hungarian language is still powerfully heard in central Europe, and continues to pose a problem for European integration.

Further reading

Czigány, L. (1984) *The Oxford History of Hungarian Literature*, Clarendon Press, Oxford

Frič, P. *et al.* (1993) *The Hungarian Minority in Slovakia*, EGEM, Prague

Gal, S. (1979) *Language Shift: Social Determinants of Linguistic Change in Bilingual Austria*, Academic Press, New York

Hajdú, P. (1975) *Finno-Ugrian Languages and Peoples*, André Deutsch, London

Ignotus, P. (1972) *Hungary*, Ernest Benn, London

Schöpflin, G. and Poulton, H. (1990) *Romania's Ethnic Hungarians*, Minority Rights Group Report, London

The Eastern Slavonic group, by contrast, pushed back or assimilated the Uralic and Altaic peoples living in what is now the Ukraine and Russia. With the subsequent expansion of Russia, the speakers of the most important Eastern Slavonic language moved east, ultimately crossing Siberia. From a linguistic standpoint, then, the notion of Europe ending at the Urals is nonsensical. There is a continuous distribution of Russian speakers from the shores of the Baltic to the shores of the Pacific Ocean.

The coming of the Magyars in the ninth and tenth centuries had significance for Slavs beyond those living on the Hungarian Plain. Since the area the Magyars occupied lay between the Romance language zone east of the Carpathians and the Germanic zone of the eastern Alps, the Magyar invasion effectively cut off the South Slavs from their linguistic cousins to the north. Under the circumstances, one might have expected a great divergence of tongues in the Slavic world. Considerable divergence took place, but it was slowed in the east by the use of a standard form of a South Slavonic tongue known as Old Church Slavonic in the texts and liturgies of the rapidly expanding Eastern Orthodox Church (Comrie, 1990a).

Other language groups

Those European languages that do not fall within one of the three dominant branches of Indo-European were nonetheless heavily influenced by the spread of Romance, Germanic and Slavic tongues (Figure 3.3). The language of the Basques may represent the last remnants of pre-Indo-European speech in Europe. Celtic speakers came to be increasingly marginalized in Europe, and now occupy relatively limited areas along the western margins of the continent. Speakers of Uralic tongues dominate in Finland and Estonia, as well as in some of the more peripheral parts of northern Scandinavia and Russia. The Baltic peoples were pushed to the north by the Slavs and now occupy a position on the eastern shore of the Baltic Sea in modern-day Latvia and

Lithuania. The Uralic-speaking Hungarians (see case study) are one of the few groups outside the three major Indo-European branches that occupy an area of central political and economic significance. The Hungarians displaced Indo-European speakers, not the reverse, and the good agricultural land they occupied helped them to prosper (Fodor, 1982). Romani, a language of the Indo-Iranian branch of Indo-European (see Figure 3.1), is spoken by the Roma, otherwise known as Gypsies. The 2–3 million European Roma are scattered throughout south-east, central and south-west Europe. Their language is related to the Indic tongues of south Asia.

The other languages in Europe that fall outside the three major Indo-European branches are concentrated in the south-east (Comrie, 1990b). Modern Greek is a much-changed descendant of ancient Greek, and therefore falls in the Hellenic branch of Indo-European. Albanian is a language in the so-called Thracian branch of Indo-European that traces its roots to migrants from Anatolia. It has been much affected by neighbouring languages and has split into two main dialects – Gheg in the north and Tosk in the south. Turkish, an Altaic language, has some presence in modern Greece and Bulgaria. Turkish is also one of several languages from North Africa and south-west Asia that found its way into northern and western Europe when guest-worker immigrants moved into the region beginning in the 1960s.

The emergence of modern European languages

Despite the high degree of linguistic diversity found in modern Europe, there are far fewer languages and dialects today than there were 1000 years ago. This is not just because speakers of tongues belonging to one of the three main Indo-European branches in Europe assimilated other peoples or pushed them to the side. It is because the large number of dialects and languages in the major branches have given way to a relatively modest number of tongues. Thus, as recently as 200 years ago a number of different Romance languages/dialects could be found in what is now southern Belgium, including standardized French, Walloon, Picard, Liègeois and Lorrain (Baetens Beardsmore, 1981). Today, French dominates throughout the region, and speakers of regional dialects are relatively few in number.

The process by which certain tongues emerged at the expense of others is firmly rooted in the intersection of politics, technology and literary culture. In particular, the splintering of languages that characterized much of European history was reversed by the development and dissemination of literary texts (Eisenstein, 1979). The reversal occurred in places with both the means to produce standardized texts and the infrastructure to ensure that those texts could be widely disseminated and used. In the Middle Ages, only certain religious institutions and political communities had the ability to produce and disseminate such texts. Religious institutions played this role in some parts of eastern Europe through the use of Old Church Slavonic. Elsewhere in eastern Europe, certain regional languages gained prominence over others in places characterized by relative political stability and sufficient economic resources to support the development of a literary culture. In the western part of Europe, religious institutions played a minimal role before the Reformation because religious texts were written primarily in Latin. Instead, a few organized political communities provided venues where literary cultures could develop and flourish.

The 16th century signalled the beginnings of a major shift in the linguistic geography of Europe. That century saw an expanding use of the printing press and the growing ability of certain political leaders to consolidate control over wider territorial domains. The printing press, originally developed in the 15th century, provided an economical means of producing written documents on an unprecedented scale (Eisenstein, 1979). Those with access to the technology had a vast advantage in the linguistic arena. Not surprisingly, those working in prosperous cities were usually the ones who had the greatest access to the printing press and its products. When such cities were at the heart of expanding political realms, they were the key nodes for the spread of standardized linguistic norms, as the infrastructure of territorial control facilitated the dissemination of printed texts.

The sorts of texts that became vehicles of linguistic standardization varied from place to place, but in many areas religious texts were of pre-eminent importance. This was particularly true in the parts of northern and western Europe where the Reformation took root (see Chapter 4). The resultant production of vernacular bibles had an extraordinary influence on the development of German and English, respectively. Both were widely disseminated and both helped to make a particular variant of a language the standardized, dominant form. The publication of literary works, and later dictionaries, furthered this process.

With the coming of the age of nationalism (see Chapter 6), language came to be even more closely tied to the political arena. Nationalist ideology was premised on the idea that distinct nations have the right to control their own affairs. Language, in turn, was seen as a key ingredient in defining nationhood. Against this ideological backdrop, language and politics became inextricably intertwined. National movements based their claims on real or perceived linguistic continuities, and political leaders of emerging states saw the promotion of linguistic uniformity as a key to their survival. The map of languages in Europe today is a testimonial to the winners and losers in the struggles over nation creation and nation building during the last two centuries.

Each language on the modern European scene developed in a distinctive way, but a brief examination of the cases of French and Polish shows the importance of the influences outlined above. What we now call French traces its roots to the Francien dialect of the Paris region (Rickard, 1989). That dialect was one of many that could be found in the Middle Ages in what is now French-speaking Europe. With the emergence of the Paris region as an increasingly coherent, politically powerful entity in the 12th century, the Francien dialect assumed greater importance. That importance grew as the rulers of the expanding French state worked to create a highly centralized political entity with Paris at its core. By the 15th century, the language of the Paris region was seen as a key unifying force in a realm that stretched from the North Sea to the Mediterranean and encompassed Flemish speakers in the north, Celtic speakers in the north-west, and speakers of various dialects of Occitan, sometimes called *langue d'oc*, in the south. A clear statement of the importance placed on the French of the Paris region came in 1539, when the authorities in Paris adopted the Edict of Villers-Cottêrets making Parisian French the official language throughout the so-called royal domains (Green, 1994).

The Edict of Villers-Cottêrets coincided with the development of a variety of texts that facilitated the spread of a standardized form of French. These included the first grammar of French and the first French dictionary, both published in 1531 (Kibbee, 1994). The printing press allowed for these and other texts to be disseminated widely, and the state backed this endeavour. Thus, the French of the Paris Basin gained ground rapidly. By the late 18th century, French nationalists treated French as a defining characteristic of national identity, and the French language was affirmed as a symbol of national identity and culture by the post-Revolutionary Convention of 1794. The subsequent development of a national education system solidified the position of French throughout modern-day France. Until recently regional languages were strongly discouraged, and any language other than French was not recognized in official circles.

The story of Polish shows some of the same influences at work as in the French case, albeit against the backdrop of a very different political history. During the eighth and ninth centuries, what is now Poland was home to a variety of West Slavonic languages, including a so-called Lechitic tongue that was the ancestor of modern Polish (Stone, 1990). The latter half of the tenth century saw the rise of a dynasty in the area between the Oder and Vistula Rivers dominated by a tribe called the Polanie, a name meaning 'inhabitants of the plains'. The Polanie, who had converted to Roman Christianity, established control over a variety of other tribes, and in the process, their particular Lechitic tongue as well as their religion came to dominate the core of the early Polish state.

With the split of the Western and Eastern churches in 1054, the Polanie saw themselves as an outpost of western culture, and the leaders of the early Polish state were therefore much concerned with advancing their language and religion in the region. They also sought to create a realm that could withstand the eastward pressure of Germanic peoples. Capitalizing on their wealth and strategic position, as well as their ties with the western Church, the Polish rulers constructed one of the best integrated political entities of late-medieval Europe, in the process solidifying the dominance of their particular tongue – Polish – throughout their realm (Davies, 1982). They were helped in this regard by the introduction of printing technology into Poland in 1513, which helped to fuel a Golden Age of Polish literature in the 16th century.

The first partition of Poland in 1722 signalled the beginning of a difficult period for the Polish language that was to last until the restoration of Polish independence after World War I (Stone, 1990). The partitioning powers (Prussia, Austria and Russia) sought to reduce the social and economic functions of Polish, but despite substantial discrimination against Polish speakers they were unable to obliterate a language that had already fostered a rich literary tradition and that was deeply rooted in local culture. Indeed, with the dawn of the Age of Nationalism in the 19th century, the Polish language became one of the focal points around which the movement for an independent Poland was organized. That movement finally achieved its goal after World War I, when the framers of the post-war European political map saw the Poles as a distinct linguistic and national community entitled to self-determination.

Poland in 1918 was established within boundaries that were different from those of today; both its eastern and western boundaries were farther east than their present equivalents, and most of what is now northern Poland was under German control (Pogonowski, 1988). The re-emergence of Poland within these boundaries provided a secure political–territorial foundation within which the Polish language could once again flourish. Many minorities lived in post-World War I Poland, however, including Germans, Jews, Ukrainians, Belorussians and Lithuanians. These minorities found themselves marginalized in a state whose identity was tied up with the Polish language and with Roman Catholicism.

World War II saw yet another partition of Poland, and a Nazi-led campaign against certain minorities. When Poland was recreated within its present boundaries, it was seen as the state of the Poles – a people defined in part by their language. Consequently, most non-Polish speakers (particularly Germans) left the country. At the same time, many Poles who found themselves outside the new state moved to those areas being vacated. Thus, post-World War II Poland came to be a relatively homogeneous state, with upwards of 98% of its population speaking Polish (Barnett, 1958). To this day the boundaries of the state define the boundaries of the language region, and the state itself functions as a key institution in the continuing reproduction of the Polish language.

Language and politics in contemporary Europe

The contemporary linguistic geography of Europe has been heavily influenced by the rise of the modern state system. However, this does not mean that a single language is spoken in most countries. Minority languages are present in almost all European states, and in some states – most notably Belgium and Switzerland (see case study) – at least a third of the population speaks a language other than the tongue spoken by the majority. Given the close tie between language and nationalism, it is not surprising that speakers of minority languages have not always fared well in 19th and 20th century Europe. From the Hungarian speakers of Romania to the Gaelic speakers of north-west Scotland and Ireland, minority language communities have struggled against linguistic assimilation by dominant national language communities.

There are a few states that have long adopted a generous attitude towards their linguistic minorities, most notably Finland and the Netherlands (Stephens, 1978). These, however, are the exceptions. Before the 1960s, most states treated linguistic assimilation as a stated policy objective. Schools were not allowed to offer instruction in minority languages, and few provisions were made for the incorporation of minority languages in public affairs. Such policies eventually led to the disappearance of some languages, such as Manx on the Isle of Man, and the marginalization of others, as with Breton in France. Organized reaction to such policies started to grow in the 1960s, and governments gradually responded (see Weinstein, 1990). The last few decades have seen the adoption of policies permitting the use of minority languages in schools. In some cases there has even been public support for publications and radio and television broadcasts in minority languages.

In the case of states where more than one language is of national significance, there is a potential for language conflict to challenge the very stability of the state. In such instances the character of internal political geographic arrangements can be a key to the maintenance of harmony. Switzerland and Belgium present an interesting contrast in this regard. In the Swiss case, language has always been a matter that is dealt with at the level of the canton, the basic administrative unit of the country, and linguistic politics do not therefore challenge the integrity of the state (see case study of Switzerland, and McRae, 1983; Steiner, 1983). Belgium, by contrast, is made up of three primary language regions, Dutch-speaking Flanders, French-speaking Wallonia, and the bilingual Brussels region. The state's internal political geography is a function of relatively recent conflicts between the language communities (Murphy, 1988). With powers over a variety of issues vested in units that correspond loosely to language boundaries, the

The geography of languages in Switzerland

André-Louis Sanguin

One of the essential legal principles on which Switzerland is founded is the equality of the rights of national languages. Indeed, in Article 116 revised in 1938, the Federal Constitution stipulates: 'German, French, Italian and Romansch are the national languages of Switzerland. German, French and Italian are declared as the official languages of the Confederation'. Moreover, at the cantonal level, Bern, Fribourg and Valais have two official languages (French and German), while Graubunden possess three (German, Romansch and Italian). The distinction between national languages and official languages is highly important. The national language is the language of the people and recognized as such, with its status being the constitutional establishment of its existence amongst an ethnic community integrated into the state. The official language is the state language, that is to say, the one used by all state organs and organizations. Language is a cultural carrier and is therefore of major interest for geographers concerned with spatial distributions. Switzerland represents, in miniature, the linguistic kaleidoscope of Europe. In this perspective, the distribution of languages in the Confederation, the interactions between different types of speech, and other spatial variables of an environmental or cultural order are of fundamental interest.

The Swiss linguistic mosaic is often linked to the physical environment, with the distribution of land types being seen as reflected in the linguistic map. Thus mountain and hill zones often delimit linguistic frontiers, with parts of the Alps forming divisions between German, French and Italian-speaking zones. For example, the St Gotthard Pass is German-speaking on its northern slope but Italian on the southern slope. However, conversely, linguistic limits also cut across plains. In the Upper Rhine and where side valleys join the Rhône, a spatial sliding of linguistic boundaries has been observable for several centuries. The mountains can equally, though, be considered as places of linguistic refuge, explaining for example the survival of the Rhaetoromansch islet in Graubunden. These illustrations highlight the central problematic of Swiss territorial–linguistic limits, as a consequence of which arises the question of the extent to which the linguistic factor acts as a barrier within the Confederation. The relationship between physical and human elements, in terms of the political geography of multilingualism, cannot simply be reduced to the point where a physiographic map can answer all of the questions. Contrary to other countries, Swiss multilingualism is essentially a *juxtaposed* multilingualism, in the sense that the linguistic domains are placed side by side, but are not copied and do not interpenetrate. The only departure from this rule is the Rhaetoromansch situation. Indeed the Romansch-speaking space is not compact, and in numerous places it imperfectly perforates the German-speaking Graubunden territory. Here it is possible to speak of a *superimposed* multilingualism.

Swiss multilingualism, for the majority, takes on a sense of spatial juxtaposition of linguistic areas. It would thus be incorrect to consider Switzerland as a polyglot state. Nowhere are the four national languages and the three official languages mixed or superimposed. Swiss multilingualism is all the more unique because of its precise link with federalism, and one of the most significant accidents in Swiss history has been the spatial non-coincidence of language and religious frontiers. The combination of religious and linguistic structures has created a political framework of great complexity. Swiss multilingualism finds its roots in a long tradition of a local sense of community identity; the principle of linguistic equality is the consequence of the principle of a cantonal equality. It was the Helvetic Republic (1798–1803) which marked the birth of today's multilinguistic Switzerland. This, for the first time, placed German, French and Italian-speaking citizens on the base of complete equality before the law. By the decree of 20 September 1798, the laws of the Republic were published in the three major languages. The Mediation Act preserved this equality, which disappeared temporarily to the use of German under the 1815 Pact, reappearing permanently in the Federal Constitution of 1848.

The federal consequences of the linguistic context cannot be overstressed. By Article III of the

continued

Federal Constitution, the cantons (local level) are sovereign insofar as the Constitution does not limit their power in favour of central government. Some specialists thus speak of the sovereignty of the cantons (*Sprachenhoheit*), through which they have the right to determine everything concerning linguistic matters in their territories. This has given birth to another concept, the principle of territoriality (*Territorialitätsprinzip*), according to which all cantons or all linguistic domains have the right to preserve and defend their linguistic character against all exterior elements which might distort or endanger them. In this way the *Sprachenhoheit* prevents the linguistic unit of a region from being put in danger by the immigration of people with another language. This can only be stopped by their linguistic assimilation. This principle applies to parts of the cantons linguistically differentiated, but especially to unilingual cantons. The *Territorialitätsprinzip* is accepted everywhere in Switzerland as an acquired fact, one with no return. As a consequence, there is an obligation for immigrants of another language to assimilate into their new milieu and to enrol their children in local schools. This results in remarkable stability of linguistic frontiers. With the exception of a few specific places, the boundaries have changed very little since Switzerland's creation. For more than 3000 municipalities counted at present, only six have changed their linguistic regime since 1848.

Swiss multilingualism raises some fascinating consequences for the structure of its society. The German-speakers (Alemanics), even if they form the majority group with nearly 75% of the population, are nevertheless divided into 26 different dialects. These show a tendency to become more uniform and to create a common idiom, the *Schwyzertüütsch*, which is clearly differentiated from literary German (*Hochdeutsch*). Here is a typical case of *diglossia*. Indeed the cultural and social situation of the Alemanics demands that one sphere of life be conducted in *Hochdeutsch*, while the other sphere is the domain of a different manner of speaking, the *Schwyzertüütsch*. The situation in French-speaking Switzerland is completely different; the dialects (patois) of local country people have disappeared, and the spoken and written language has become very close to universal French. The speech of the Romansch people (50,000 people) is not homogeneous, but is divided into five literary variants (sursilvan, sutsilvan, surmiran, puter, vallader) sufficiently distinct that the Graubunden authorities publish primary school alphabet books in all five idioms.

Multilingualism is particularly interesting when Swiss from different origins converse between themselves. When people master two distinct languages they are said to be bilingual. Unfortunately official statistics have never tried to measure the extent of bilingualism practised by Swiss citizens. Empirical observations, however, show that a significant part of the population is capable of conversing in two, three or even four languages. When two or three languages are territorially confined, it is not rare to meet *pidginization* phenomena. By proximity and reciprocal contamination, two languages may be altered and lose their purity. The *pidgin* often adopts the vocabulary or the basic grammar of the first language, but borrows the construction and thinking order of the second. This is a phenomenon that one sees at Biel/Bienne, the only pidginized town in Switzerland.

To perceive physiographical limits as the only linguistic cleavage lines is a gross over-simplification of the problem. The linguistic map of Switzerland can also be explained in terms of the movement of ethnic groups, their initial settlement, their subsequent colonization from outside, and their mutual reactions when they enter into contact. Accordingly, it is possible to draw up a fourfold typology of *isoglosses* (linguistic boundaries). A first boundary category can be called *physiographic*, and is found when a linguistic boundary coincides with a ridge-line, a mountain chain or a river. It is illustrated by the frontiers between the Italian-speaking valleys of Southern Graubunden and the rest of the canton (Moesano limited by the San Bernardino Pass, Val Bregaglia separated by the Maloja Pass, and Val Poschiavo by the Bernina Pass). The same occurs for Ticino, cut off from Alemanic Switzerland by the Nufenen, St Gotthard, Lukmanier and Greina Passes.

The break in the longitudinal profile of the Valaisian Rhône provides a transition with the second type of boundary, namely the *historico–administrative* isogloss. Here, between the fifth and

continued

sixth centuries, the limit was fixed between French-speaking Lower Valais and German-speaking Upper Valais, between the Romands and the Walser. With the exception of the aforementioned physiographic elements, the German/French-speaking isogloss is essentially of the historico-administrative type. The linguistic limit between Vaud and Berne exactly follows the cantonal boundary, and the same observation can be made for the isogloss between Neuchâtel and Berne fixed on the Zihl Canal connecting Neuchâtel Lake with the Bielersee.

A third category is the *confused* isogloss. This is the case for the linguistic frontier in the cantons of Berne and Fribourg, where extreme confusion reigns, particularly on the northern bank of the Bielersee. Here, the French/German isogloss has overcome the physical obstacle of Lake Biel from Twann (Douanne), Schäfis (Chavannes), Ligerz (Gléresse), Tüschez (Alfermée) up to the northern outskirts of Biel/Bienne (Magglingen/Macolin, Leubringen/ Evillard). The same observation applies to the isogloss between the Délémont (French-speaking) and Laufen (German-speaking) districts. Often municipal limits are not sufficient to fix it.

A last category is the *perforated* isogloss. This is the case of the Rhaetoromansch zones of Graubunden in osmosis with the German-speaking majority of the canton (57.6% of the resident population). This situation is above all true in the Oberland/Surselva (*Rein anteriur*), Domleschg/ Sutselva (*Rein posteriur*) and the Oberhalbstein/ Surmeir (*Julia*) The sursilvian, sutsilvian and surmiran villages float like islets in the middle of an Alemanic sea. This is less true in Lower Engadine, where the *ladin* homogeneity (puter and vallader languages) is territorially stronger. The same phenomenon is seen in the Gondoschlucht and the Zwischbergental (Valais on the Italian side south of the Simplon Pass).

This typology should not, though, be regarded as rigid. Combinations between different types are possible. For example, the Lower Valais/Berne and Neuchâtel/Berne isoglosses are as much physiographic as historico-administrative. One can also notice the contrast existing between isoglosses and cantonal boundaries. The latter are perforated with territorial disconnections, whereas the isoglosses never give place to linguistic exclaves, with the exception of the German-speaking village of Bosco-Gurin (Ticino). Hence, according to the present synchronic use of languages, the break is generally clear-cut between two language communities. But from the historical and toponymical viewpoint, there are broad zones of reciprocal influence and interference from each side of a separation line which reflect continuing cultural exchanges and demographic mobility.

Further reading

Federal Department of Internal Affairs (1989) *Quadrilingualism in Switzerland*, Federal Chancellery, Berne

McRae, K.D. (1984) *Conflict and Compromise in Multilingual Societies: The Case of Switzerland*, Wilfrid Laurier University Press, Waterloo

Rougier, H. and Sanguin, A-L. (1991) *The Rumanschs or the Fourth Switzerland*, Peter Lang Publishers, Berne

Sanguin, A-L. (1983) *Switzerland, An Attempt at Political Geography*, Ophrys Publishers, Paris

administrative structure of Belgium discourages the development of political cleavages that might cut across language lines. Instead, most political issues are structured by a political–geographic framework that pits one language community against the other (Murphy, 1995). Given these circumstances it is not surprising that language has been a much more volatile issue in Belgium than in Switzerland.

Language and politics are also closely linked at scales above the state. The European Union (EU) is a case in point. French and English are widely understood by those involved in EU matters, and one or the other of those two languages tends to be used for many day-to-day operations of the EU (see Chapter 9). At the same time, the EU operates under the principle that all major Member State languages should have official status, and representatives from Member States frequently insist on using their national languages in public settings (Krause, 1991). As a result, over 40% of the administrative budget of the EU supports language services, and almost half of all EU employees work in that arena (Henriksen,

1990). Thus, language continues to be seen as an important element of national identity; it is accorded such political and social importance that the participants in the EU are willing to devote enormous resources to ensure that national languages are not displaced by English or French.

Beyond the intergovernmental arena, the growing use of English as a *lingua franca* is seen by many as a problematic development. Steps are being taken in France in particular to purge the French language of English vocabulary, and in many other quarters of Europe serious questions are being raised about what might be lost if local languages disappear or are radically changed by the rush towards English. In eastern Europe, Russian functioned as the *lingua franca* in the post-World War II era, but the fall of the Iron Curtain has led to a rapid shift towards English and German. German has the benefit of tying the former satellite states of the Soviet Union to the economic powerhouse of central Europe, but there is still resistance to German based on memories of the Nazi era.

As Europe heads into the 21st century, its linguistic character will continue to evolve. Languages are dynamic phenomena that do not exist in some space unto themselves. Indeed, the very nature of the tongues we think of as languages may change as political and social developments unfold. This is happening before our eyes in the Balkans, where some Croat nationalists are arguing that the true language of the Croats is not some generic Serbo-Croatian, but a pure Croatian that is purged of various Arabic and Serbian expressions and that incorporates words coined by Croatians many centuries ago, or new words coined by Croatian nationalists (Hedges, 1996). This struggle serves as an important reminder that the study of the geography of languages provides a window into a wider world of conflict and compromise, oppression and liberation. As such, it is a necessary component of any attempt to understand the nature and complexity of place.

Acknowledgements

Research for this chapter was supported by the US National Science Foundation under grant no. SBR-9157667. The author is grateful to Joanna Kepka, Sarah Shafer and Anthea Fallen-Bailey for research assistance, and to Nancy Leeper for drafting the figures.

Further reading

Asher, R.E. (ed.) (1994) *The Encyclopedia of Language and Linguistics*, volume 3, Pergamon Press, Oxford

Comrie, B. (ed.) (1990) *The Major Languages of Eastern Europe*, Routledge, London and New York

Hualde, J.I., Lakarra, J.A. and Trask, R.L. (eds) (1996) *Towards a History of the Basque Language*, J. Benjamins Publishing Co., Amsterdam

Van der Auwera, J. and Konig, E. (ed.) (1994) *The Germanic Languages*, Routledge, London and New York

CHAPTER 4 Religious dimensions of European culture

TIM UNWIN

Religion has played a profound role in shaping European culture and society. This chapter first addresses the complex issues surrounding the meaning and significance of religion, before providing a historical introduction to the contemporary distribution of religious practices (Sopher, 1967; Gay, 1971; Park, 1994). It concludes with brief examinations of three specific themes: the influence of religion on landscape; the imperial legacy of minority religions; and the role of religious sites as places of cultural tourism.

Meanings of religion

There are enormous difficulties facing anyone seeking a widely acceptable definition of religion (Byrne, 1988; Turner, 1991). Significantly, though, as King (1987: 282) has noted, 'The very attempt to define *religion*, to find some distinctive or possibly unique essence or set of qualities that distinguish the "religious" from the remainder of human life, is primarily a Western concern' derived from a Jewish, Christian and Islamic inheritance.

In the sociological literature, much of the debate concerning the meaning of religion can be traced back to the late 19th century in the contrasting, but not entirely unrelated, views of Emile Durkheim (1915) and Max Weber (1965, 1992). For Durkheim (1915), collective ritual practice and social thought were central to the idea of religion, which he defined as a unified system of beliefs and practices concerning sacred things, set apart and forbidden. In contrast, Weber's (1965, 1992) central concern was with meaning and the way in which religions provide frameworks in which individuals live and make sense of their existence. Taken together, these two positions, a concern with the social functions of the sacred and an interest in questions about the meaning of life and death, underlie many subsequent definitions of religion (although for a comprehensive critique, see Turner, 1991).

While it is difficult to provide a comprehensive definition of religion, there is general agreement that there is a religious element in almost all cultures that refers to 'some sort of ultimacy and transcendence that will provide norms and power for the rest of life' (King, 1987: 286; see also Byrne, 1988). Religions are belief systems and practices through which people seek to understand and act in the world in which they live. These beliefs and practices also usually give special value to particular places. Such religious systems contain some or all of the following characteristics: they involve myths designed to express what cannot be explained otherwise; they differentiate between the sacred and the profane, frequently referring to a deity, or the holy, that is in some way separate from the material and mundane world of human existence; they are expressed and represented through the use of symbols and icons; they often include concepts of another world, to which adherents have access through salvation; and they are usually traditional in that they seek to maintain themselves by reference to their earliest forms as models of purity.

While debates over definitions of religion are important (Geertz, 1966; Berger, 1969; Bell, 1980), they should not detract from the central aim of this chapter, which is to examine the complex role that religions have played in shaping European identities (Turner, 1991). This role has changed considerably in its influence and importance through time. This is particularly evident in the distinction that has emerged between religion and science since the 17th century. Both religion and science seek to provide some kind of certainty, but they do so in very different ways. Scientific knowledge is usually seen as being grounded in reason and empirical experience, whereas religious knowledge derives its basis from a faith which non-believers see as irrational and ungrounded. Religion is thus frequently thought of as dealing with questions which science is unable to solve. This division between science and religion emerged most clearly through the formulation of a new kind of science in 17th century Europe by Descartes, Leibniz and Newton. When combined with the philosophical inspiration of John Locke, this provided the underpinnings of the Enlightenment (Cassirer, 1951), which 'promoted the values of intellectual and material progress, toleration and critical reason as opposed to authority and tradition in matters of politics and religion' (Wood, 1987: 109).

Significantly, though, Locke himself suggested in his *Essay Concerning Human Understanding* (IV.XVII.24) that reason and faith were not incompatible. He thus argued that 'There is another use of the word *Reason*, wherein it is opposed to *Faith*. ... Only ... *Faith* is nothing but a firm Assent of the Mind: which if it be regulated as is our Duty, cannot be afforded to any thing, but upon good Reason' (Locke, 1975: 687). Despite Locke's comments, the fulfilment of the Enlightenment in the 19th and 20th centuries has seen a growing separation between science and religion, derived in part from the increasing success of science as an explanatory system, in part from the strength of the materialist critique of ideology (Foucault, 1967, 1972; Marx, 1976; Marx and Engels, 1975b; Engels, 1977), and in part from the failure of established religions to adapt to social, political and economic change. Paradoxically, one recent outcome of this has been the increased attention paid by many people to alternative, fundamentalist and New Age religions, as they continue to seek to grapple with the failings of modern European society and issues that science has been unable satisfactorily to explain (Heelas, 1988; Lawrence, 1990).

Religions in Europe

Figure 4.1 provides a broad overview of the distribution of the main religions in each country in Europe. In interpreting this map, it is important to recognize that there is great variability in actual practice within any one religion (Fogarty, 1957). It should also be noted that it fails to illustrate the large number of minority religions that are encountered throughout Europe. Despite these shortcomings, the overwhelming dominance of Christianity as the main or official religion of most European states is readily apparent. It is also evident that different branches of Christianity dominate in different places: Roman Catholicism in the west, various Protestant faiths in the north, and Orthodox Christianity alongside Islam in the east.

Christianity is a relatively recent introduction to Europe, having become widespread only on its adoption as the official religion of the Roman Empire during the fourth century AD. Prior to the advent of Christianity, the dominant religions of Greece and Italy were largely derived from the male-dominated Olympian pantheon, headed by Zeus (Ferguson, 1980; Burkert, 1985). However, these religions were never monolithic or static, and in the Roman case it is more appropiate to think of a mosaic of different influences, including contributions from previous Etruscan, Syrian, Persian and Egyptian religious elements (Guirand and Pierre, 1968). Indeed, within the Roman Empire 'all local religions were respected on the basis of assimilating their gods to their supposed Roman equivalents' (Fontana, 1995: 13). Preceding the Olympian pantheon, the chief deity of the Aegean in the second and third millennium BC was a feminine Great Goddess or Universal Mother (Guirand, 1968). Above all, this goddess symbolized the fertility of plants, animals and humans, and can be seen as but one expression of a common and fundamental religious concern throughout western Asia and much of Europe in the prehistoric period. Elsewhere in Europe, this early concern with fertility of the animal and human world can be found, for example, in the cave paintings and statuary from the Palaeolithic period in areas such as the Dordogne in France and Cantabria in Spain (Ucko and Rosenfeld, 1967; Leroi-Gourhan, 1968; Bahn and Vertut, 1988).

In Britain, parts of France and Iberia, the Olympian pantheon and its Roman successor were never dominant, and a separate Celtic mythology evolved (see Chapter 2). Likewise, in northern and eastern Europe the Teutonic and Slavonic peoples had developed their own distinctive religious beliefs (Alexinsky, 1968; Davidson, 1982). In addition, the magico-religious practices of shamanism were widespread among the Finno-Ugric peoples of the east (Eliade, 1964; Lewis, 1988). When Christianity spread into Europe, first through the Roman Empire, and then later through missionary and crusading activity in the north and east, many characteristics of these religions survived. Some aspects were adopted into Christian ritual, whereas other elements remained widely practised, particularly in rural areas, where ancient traditions continued to be followed

Figure 4.1 Distribution of the main religions in Europe in the 1990s (based on data in Barrett, 1982; *World Almanac 1993*; *Philip's World Atlas & Gazetteer*, 1996).

largely in the pursuit of successful harvests. Among many other examples, Frazer (1981) thus notes the Estonian belief in the forest spirit Metsik, to whom prayers and sacrifices continued to be offered in return for protection of the cereal crop and cattle. In general, such traditions tended to survive longest in those parts of Europe most recently converted to Christianity (Smart, 1989).

The emergence of a Christian Europe

Glacken (1967) has emphasized the difficulty of generalizing about these early religions, but with the rapid spread of Christianity, particularly from the fourth century, and its adoption as the official faith of the Roman Empire, there was a conscious movement to suppress alternative beliefs. Fontana (1995: 29) thus comments that 'When a persecuted faith was transformed into a stable Church, what was intended was not only that it should have a presence in society, but that it should control society. For this to happen imperial public order had to be christianized and people's habits and customs had to be changed'. Critical among these changes according to Fontana (1995: 30) were a new sense of time and space, a reordering of the city, and the modification of 'a whole series of rules for life, such as those concerned with sexuality'.

Two other significant effects of the introduction of Christianity as the dominant religion were the way in which it conceived of the relationship between people and nature, and its emphasis on patriarchal relationships. The Judaeo-Christian teleological belief in God's creation and care for the universe thus had important ramifications for subsequent medieval cosmology, in particular through the ideas that people and nature were separate, and that it was through human disobedience of God in the Fall that disorder entered the natural world (Unwin, 1992; Glacken, 1967; Doughty, 1981). Likewise, Christianity, along with Judaism and Islam, was patriarchal in its practice and theology, and this helped to create 'a social system of patriarchal power in which women were assigned to and inscribed in inferior and subordinate positions' (Turner, 1991: xv). In Christianity this view was in part derived from the belief that it was Eve who first transgressed in the Garden of Eden, although this was to an extent balanced in the Roman Catholic Church through the emphasis placed on the figure of the Virgin Mary.

The threefold division of European Christianity into Catholic, Orthodox and Protestant reflects the complex theological and political disputes that have taken place over the last 2000 years, particularly the division between the Western and Eastern Roman Empires at the end of the fourth century, and the Reformation in the 16th century. By the third century, five Christian centres had emerged in the cities of Rome, Alexandria, Antioch, Carthage and Ephesus, each with its own independent interpretation of the faith. With the conversion of the Emperor Constantine to Christianity, the persecution of Christians within the Empire was halted by the Edict of Milan in 313. Constantine then sought to reconcile the many differences within the Church at the Council of Nicaea in 325. This, however, failed to resolve either the differences between the East and West, or the theological disputes within the East itself, particularly concerning the precise relationships between Christ and God. Constantine's dedication of Constantinople as a Christian city and his capital in 330 further exacerbated the differences between East and West. In the East, Church and State subsequently became closely integrated, whereas in the West the Papacy remained separate from secular control and jealously proud of its independence. The next major attempt to find a compromise position and 'a common interpretation of the faith acceptable to both Greek and Latin Christianity' (Frend, 1988: 165) was at the Council of Chalcedon in 451. Although accepted by both Constantinople and Rome, this was eventually to exacerbate the divisions between Orthodox East and Catholic West, and it also led to a schism between Constantinople and the other eastern Churches.

The sack of Rome in 476 and its political separation from Constantinople further divided Eastern and Western Christianity. However, the Papacy managed to retain its power in the West, and the conversion of the Germanic tribes to western Christianity at the beginning of the sixth century ensured a continued role for the Catholic Church. Harper-Bill (1988: 194) emphasizes that 'shrewd calculation cannot have been absent from the minds of both Germanic chieftains and Catholic bishops. The invaders were attempting to establish their control over a sub-Roman population by whom they were heavily outnumbered'. One way in which Clovis, the leader of the Franks, was able to gain acceptance and legitimacy was through his conversion to the Roman faith, and at the same time the bishops realized that a strong and stable government would enable them to expand their missionary activities. This rapprochement between Rome and the Germanic tribes reached its culmination in 800, when Charlemagne, King of the Franks, was crowned Emperor of the Romans by the Pope.

The establishment of this Carolingian Empire caused further conflict with the East, and as Wybrew (1988: 173) has noted, it 'increased the political rivalry between the two very different worlds which were becoming self-contained and self-centred'. Meanwhile, the expansion of Arab Islam in the early seventh century effectively separated the other eastern Churches from Constantinople, and thereafter provided a constant challenge and threat to the Eastern Byzantine Empire. Despite this threat, Orthodox Christianity began to spread gradually through the Slavic tribes which had settled in the Balkans in the sixth century, and in the ninth century Orthodox Christianity became firmly established among the Bulgars and Serbs. Meanwhile the dispute over icons in the eighth and ninth centuries further divided the Eastern and Western Churches. As Cândia (1987: 1) has noted, 'Christians in the Greek-speaking areas of the Byzantine empire carried forward their pre-Christian traditions in which religious images played a significant part in divine worship and, therefore, venerated icons with a zeal that made them suspect of idolatory'. Further missionary activity led to the baptism in 988 of Vladimir, Prince of Kiev, and through the combined interests of Church and State, Orthodoxy eventually became the established faith of the Russian peoples (Smart, 1989). Disgruntled with the rapprochement between Constantinople and Rome in 1439, the Russian Orthodox Church declared itself independent from Constantinople in 1448; five years later in 1453 Constantinople finally fell to the Islamic forces of the Ottoman Turks.

The Reformation

Intellectual and political ferment in northern and central Europe in the 16th century laid the basis for the Reformation (Hillerbrand, 1987; Cameron, 1991) and a split of the western Christian tradition into two parts, Catholic and Protestant. The emergence of Renaissance humanism in northern Europe when combined with the corrupt state of the established Church, and the emerging strength of the Habsburg Empire in Germany, provided the context against which Europe was to be divided upon both religious and political grounds. In 1517 Martin Luther posted on the castle church at Wittenberg his 95 theses critical of the Catholic practice of selling papal indulgences by which, it was claimed, people could be forgiven their sins in exchange for certain payments. His theology soon developed into the doctrine of the 'priesthood of all believers' (Gilley, 1988; Smart, 1989), which effectively challenged the role of priests as intermediaries between people and God. The German princes, eager to suppress the peasant uprising of 1525 and to claim church property, were among Luther's strongest allies, and helped to develop a Lutheran ecclesiastical organization in northern Germany in the 1520s. The term Protestant arose from the protests by these Lutheran princes at the Diet of Speyer in 1529 (Gilley, 1988), and Lutheran Protestantism then spread rapidly into Scandinavia and the Baltic states by the end of the 1530s. Likewise, the English King Henry VIII in his dispute with the Papacy, and eager to appropriate the wealth of the Church, turned to the Lutheran Thomas Cranmer for theological support in the mid-1530s, paving the way for an independent Anglican state–church. Subsequently, important refinements and changes to Luther's views were proposed by the Swiss Ulrich Zwingli and the French-born John Calvin, whose ministry was based in Geneva. This led to the emergence of Calvinism, which became the Scottish National Church in 1560, and which also gained considerable support amongst the Huguenots in France and the inhabitants of the northern part of the Netherlands. Divisions between the Protestants, and an effective Counter-Reformation established by the Catholic Church from the middle of the 16th century, subsequently prevented further spread of Protestant doctrines. By the beginning of the 17th century, therefore, a pattern of religious beliefs was established in Europe which would broadly survive into the 19th century.

Christianity was not the only significant religion in Europe during the medieval and early-modern periods. Indeed, many of its political and social characteristics were developed specifically within a context of opposition to other religions that were competing with it in the minds of the people. This is illustrated in the case study which examines the significance of Judaism and Islam in European culture.

Religion and European culture

The previous section has provided a broad account of the emergence of different religious practices within Europe (Figure 4.1). While adherents of Christianity, Judaism and Islam have usually emphasized their differences, there are nevertheless, as Turner (1991: 7) stresses, 'very strong sociological continuities between them. The orthodox core of these religions exhibits a common set of components: a high god, scriptural tradition, prophetic revelations, salvationism'. They share a common Abrahamic tradition, and their patriarchal nature has reinforced the social framework of male-dominated societies.

The above account has also illustrated the complex interaction between religious, political, social and

Judaism and Islam

Tim Unwin

While Christianity was the dominant religion of medieval and early modern Europe, two other religions, Judaism and Islam, also played a significant part in shaping the people's cultural and intellectual background. The sack of the Temple in Jerusalem by the Romans in AD 70 and the prohibition of Jews from entering the city following the abortive revolt under Bar Kochba in AD 132–135 led to the spread of Jewish people throughout the Roman world. Babylonia replaced Palestine as the main seat of learning and Jewish culture, but Jews also settled widely in other areas of the Roman Empire, and particularly in North Africa, the Balkans and Iberia. While Christians generally persecuted the Jews, particularly from the 11th century, the Arab Islamic world was initially less antagonistic to them. Friedlander (1988: 117) thus notes how the Arab conquest of Iberia 'gave the Jews entrance into commerce, the professions, literature, science and philosophy. Islam and Judaism met, and there was much to share, including the Greek philosophers who had been preserved in Arabic texts'. Gradually, two different European traditions of Judaism emerged, each with its own languages and cultures: a Sephardic tradition in the Mediterranean and Iberia, and an Ashkenazi one in Germany and central Europe.

From the 11th century, and particularly with the commencement of the Crusades in 1096, increasing persecutions of Ashkenazi Jews took place at the hands of Christians. Christian prohibitions on usury meant that one of the few occupations open to Jews was money-lending, and Christian indebtedness, combined with hatred on religious grounds, further exarcerbated tensions between Christians and Jews. In 1290 the Jewish community was thus expelled from England, and then in 1306 they were also expelled from most of France (Friedlander, 1988). Eager to escape the violence and death caused by the Crusaders on their way to the eastern Mediterranean, many of those expelled sought refuge in eastern Poland and Lithuania. In Iberia, the Christian reconquest and Inquisition in the 15th century once again also led to violence and hostility against the Jews, and they were eventually expelled at the end of the century (Kamen, 1985; Monter, 1990). Many of those who survived the persecution fled to northern Africa, Italy, the eastern Mediterranean, and once again to Poland. But even here in eastern Europe, despite the granting of self-government to the Jews by King Sigismund II Augustus of Poland in 1551, they did not remain unpersecuted. It was not until the 19th century, under the influence of the Enlightenment and North American Puritanism, that Jews throughout Europe were to be emancipated. By the 1930s, Jews accounted for less than 1% of the population of most western European countries, but their historical presence in the east meant that they were relatively much more important there, accounting for between 3% and 10% of the populations of Austria, Hungary and Romania, and just over 10% of the population of Poland. The rise to power of the Third Reich in Germany, the Nazi pursuit of racial purity, and the desire to focus on a scapegoat for the defeat of 1918, all led to renewed persecution of the Jews in Germany during the 1930s, culminating in the Holocaust of the 1939–45 period in which some 6 million Jews were killed. While some groups did actively seek to save Jews during this period, the majority Christian voice in Germany remained silent, and it is salient to note also that Pope Pius XII failed openly to condemn the Nazi extermination of the Jews (Gilley, 1988).

The other major religion to have had an important influence on European culture is Islam. Like Judaism and Christianity, it was not indigenous to Europe, but it played a significant role, most notably in Iberia and the south-eastern Balkans (Coles, 1968). This role was twofold: on the one hand, there was a direct influence in areas conquered and ruled by Islamic forces; but on the other, Islam provided an external entity against which European Christian polities could forge their own identities. Islam emerged in the early seventh century in the Arabian peninsula, and particularly under the military strength of the Umayyad Dynasty spread rapidly westwards, eastwards and northwards. Attacks on the Byzantine Empire and the siege of Constantinople in 718 proved fruitless, but in the west the forces of Islam invaded the

continued

Kingdom of the Visigoths in Iberia in 711, and within two years had conquered the peninsula. In subsequent campaigns the Islamic forces pushed the Visigoths back into what later became France, and it was only the Franks under Charles Martel who eventually defeated them in 732 at the battles of Tours and Poitiers when they had reached as far north as the Loire. For three centuries under the Ummayadas a flourishing Islamic culture centred on Córdoba existed in Iberia, but with the collapse of the Caliphate in 1031 and its disintegration into a number of small separate Islamic kingdoms, the scene was set for the Christian reconquest to begin. In 1085 Toledo was captured by the Castilians, and subsequent Christian victories meant that by the mid-13th century all of Iberia with the exception of Granada had been reconquered. Granada nevertheless remained a centre of Islamic cultural efflorescence for a further two centuries, until its eventual conquest in 1492 (Figure 4.2).

Figure 4.2 The Alhambra, Granada, Spain (*source:* Tim Unwin, 13 January 1997).

As Islamic power in western Europe declined, the establishment of the Ottoman Empire in 1301 heralded the advance of Islam in the east. Military expansion in the second half of the 14th century saw most of the Byzantine empire fall to the Ottomans, and in 1389 Christian resistance in the Balkans was broken at the battle of Kosovo. Constantinople eventually fell in 1453, and under Sultan Suleyman, who came to power in 1520, the empire expanded yet further to include parts of Hungary following the Ottoman siege of Vienna in 1529 (Inalcik, 1973; Sugar, 1977). It was only in the early 19th century that Serbian and Greek uprisings began to achieve freedom from Ottoman rule in this part of Europe, with final independence for many people not being gained until the redrawing of the Balkan boundaries following the 1914–18 war. In Iberia following the Spanish reconquest, and particularly under the Inquisition established in 1479, Muslims and Jews were either forcibly converted to Christianity, expelled or killed. However, with the collapse of the Ottoman Empire in the 19th and early 20th centuries, Islam survived in parts of south-east Europe leaving substantial numbers of Muslims in Bulgaria and former Yugoslavia (see Figure 4.1).

The Islamic conquests of Europe introduced not only a new religion, but also considerable economic and cultural benefits. Their agrarian expertise enabled flourishing urban centres to be established in Iberia, and it was through their agency that many attributes of Greek and Roman science and thought were reintroduced to Europe. Nevertheless, Islam provided an economic, political and ideological threat to the new kingdoms of northern Europe, and from 1096 this found its response in the emergence of the crusading movement (Riley-Smith, 1995). As Fontana (1995: 54–5) notes, the relationship between Christians and Muslims 'was from this time presented as a warlike epic that was to give a new sense to the whole history of medieval Europe'. The motivations of the crusades were highly complex, but Riley-Smith (1990: xxviii) has captured their essence in noting that 'A crusade was a holy war fought against those

continued

perceived to be the external or internal foes of Christendom for the recovery of Christian property or in defence of the Church or Christian people'. At their heart was a religious conviction in the concept of a just war fought on behalf of God by Christian people. If the early crusades were directed against the Muslim control of Palestine, the crusading idea also played a highly significant role in the Spanish reconquest, in the crusades of the Teutonic Knights against the Prussians and Livonians in the Baltic, and in the persecution of the Jews that occurred as early as 1096 in Germany (Riley-Smith, 1990). Religion not only legitimated these violent, arduous and expensive conflicts, but more importantly it also initiated them. The religious fervour that inspired the crusades was also directed towards the suppression of heresies within the Christian world. Typical of the crusades against such heresies was that against the Cathars of southern France, beginning in 1209. Although it failed to destroy the Cathars, who survived until the Inquisition of the early 14th century (Le Roy Ladurie, 1978), the 13th century crusades against the Cathars were carried out with the same violence and brutality as that associated with the crusades against the Muslims (Riley-Smith, 1990). Similar violence was meted out to those decried as witches, and particularly women, throughout the medieval and early modern period (Kors and Peters, 1972; Levack, 1987).

Further reading

Coles, P. (1968) *The Ottoman Impact on Europe*, Thames & Hudson, London

Friedlander, A.H. (1988) Judaism, in Sutherland, S., Houlden, L., Clarke, P. and Hardy, F. (eds) *The World's Religions*, G.K. Hall & Co., Boston, Mass, 111–41

Kamen, H. (1985) *Inquisition and Society in Spain in the Sixteenth and Seventeenth Centuries*, Weidenfeld & Nicolson, London

Riley-Smith, J. (ed.) (1995) *The Oxford Illustrated History of the Crusades*, Oxford University Press, Oxford

economic structures in Europe's past, and the difficulty of disentangling precise notions of cause and effect between them. While religions have thus played a very important role in defining Europe's moral codes and ethical systems, these codes and systems have then frequently been used by those in power to legitimate and maintain their positions. Moreover, the very formalization of such codes has involved political intrigue and negotiation. For the most part, it was the Christianization of Europe during the medieval period that provided the foundations for its ethical values in the ensuing centuries. Education in the early medieval period was largely in the hands of the monasteries, and the various subsequent interpretations of the mid-sixth century Rule of St Benedict, the founder of western Monasticism, provided a framework of prayer, study and manual work which was to be a model for Christian life.

Christianity not only provided a legitimation and justification of Europe's political conflicts, but it also helped to define the basic order of society. Throughout the medieval period, society was essentially divided into three broad categories of people: those who laboured, those who fought, and those who prayed (Bloch, 1962). Such a division is clearly illustrated, for example, in William Langland's classic 14th century poem *Piers the Plowman*, where Piers, in response to the knight's agreement to help him, agrees 'for my part, I'll sweat and toil for us both as long as I live, and gladly do any job you want. But you must promise in return to guard over Holy Church, and protect me from the thieves and wasters who ruin the world' (Langland, 1966). The Church not only legitimated a feudal society in which the knights expropriated a profit from the unfree villeins, but it also sanctioned the violence and warfare practised by the knightly class through the code of chivalry (Keen, 1984). Moreover, this accepted ordering of society was to last long after the medieval period, surviving for example in France in the form of the Three Estates (the Church, the nobility, and everyone else) through to the events of the 1789 Revolution (Pilbeam, 1990).

While the Enlightenment and 19th century secularization have greatly reduced the role of religion in society, its influences have persisted and can be found in many aspects of contemporary life. The very different teachings of the Roman Catholic and

Protestant churches over birth control, for example, have played a significant part in influencing the demographic profiles of different European countries (see Chapter 15). Likewise, the contributions of Protestantism to the expansion of capitalism have been widely discussed ever since Weber (1992) argued at the beginning of this century that Puritanism's emphasis on individuals' obligations to fulfil their duty in the affairs of this world was crucial to rational capitalist behaviour. Despite the role of materialist reason and science in leading to a decline in contemporary religious practice in Europe, the lasting significance of past practices in European culture can be seen through three sets of examples: their influences on the landscape; the practices of religious minorities; and the role of religious sites as places of cultural tourism.

Religion and the landscape

Over the centuries, Christianity has helped to shape and stabilize the European landscape both through its formalization of a parish and diocesan organization, and through its creation of important symbolic elements in the landscape (Park, 1994). The continuing importance of traditional Christian festivals and celebrations, particularly in Catholic Europe, is also a constant reminder of the significance of religion in the everyday lived experience of people's landscapes (Figure 4.3). In its earliest European expression Christianity was largely urban-based, but the need soon emerged to provide people everywhere with access to churches and with a system through which ecclesiastical taxation and dues could be appropriated. While many early churches were closely linked to the most important properties of great magnates, a system of parishes gradually emerged to fulfil these functions across Christian Europe. The boundaries of many of these parishes closely followed those of previous secular estates, but once established they provided a fundamental element of landscape continuity that has survived to this day.

Places of religious worship have not only been an important focus for communal life, but they have also been central symbolic elements of the European landscape. So too have cemeteries, which in their landscapes of death, reveal much about the cultural beliefs of different religious groups (Park, 1994). In medieval times, when the majority of buildings were constructed of insubstantial materials, the built fabric of stone-built churches was one of the dominant elements of the visual and spiritual landscape. Even today, in many rural parts of Europe, the church remains the central and most visible element in most villages (Figure 4.4). Often

Figure 4.3 A religious *festa* at the village of Santa Maria Avioso in north-west Portugal (*source:* Tim Unwin, 4 August 1983).

Figure 4.4 The church of Arcos in the Minho, north-west Portugal, dominating surrounding houses and fields (*source:* Tim Unwin, 9 July 1981).

situated on the highest ground, with its tower or spire raising the eye 'heavenwards', the built fabric of the church served as an ever-present reminder of the importance of Christianity. In urban areas too, churches and cathedrals acted as central features of the built environment (McLeod, 1995) (Figure 4.5), although today in most areas they have been eclipsed by the modernist tower blocks of contemporary capitalist society. Islamic mosques and Jewish synagogues have also left their imprint on the architectural heritage of Europe's towns and cities, with in several cases, particularly in Iberia, the same building actually having been used and transformed by different religions.

The living landscape of contemporary Europe also reflects the continued importance of religious experience in the numerous *fiestas* and *festas* that occur annually in Catholic countries, and provide a rich imagery of colour and symbolism (Figure 4.3). Many of these local festivals reflect the surviving importance of folk religion, and the way in which Christianity in its earliest forms subsumed previous religious expressions into its culture. Brettell (1986) thus notes the continued role of the *festa* in north Portuguese life, and the celebration of many vine- and wine-related festivals connected with Christian figures such as St Vincent provides ample testimony to the lasting popularity of such practices in European culture.

Religious minorities and the legacy of empire

Figure 4.1, in illustrating the dominant religious practices of contemporary Europe, fails to emphasize that numerous other religions, as well as various Christian sects, have come to play a significant part in the day-to-day lives of Europe's peoples. The role of the Waldensians in north-west Italy, for example, is highlighted in this chapter's second case study. The increasing secularization of society since the 19th century has, though, provided a context within which there has generally been much greater religious tolerance and freedom of practice. Moreover, the 19th century imperial policies of European states, most notably those of Britain, France and Germany, provided the agency through which other religions were introduced into the heartlands of the imperial powers (Said, 1978). Although European Christian missionaries took their message to the far corners of the world, there was also a small return flow of people with other beliefs. The need for additional labour in the economic expansion of the post-1945

Figure 4.5 The imposing early Gothic cathedral of Laon, northern France (*source:* Tim Unwin, 23 March 1988).

period considerably enhanced this flow of people from other religious backgrounds into Europe. France thus experienced significant levels of Muslim immigration from North Africa, Britain received large numbers of Hindus, Sikhs and Muslims from India and Pakistan, as well as Rastafarians from the Caribbean (Cashmore, 1979), and Germany experienced substantial Turkish Muslim labour immigration. In general most of this migration has been to urban areas, manifesting itself in the construction of a new wave of mosques and temples in many European cities. While welcomed during the economic boom of the 1960s, these foreign minorities, distinguished not only by religion but also by colour, language and race, have nevertheless been subject to considerable discrimination and racial abuse during the recessions of the late 1970s and early 1990s. In particular the rise of the New Right and extreme nationalist tendencies, particularly in France and Germany, have provided serious threats to these immigrant minority communities.

Another aspect of the continued survival of minority religions is reflected in the position of Christianity itself across Europe today (Smart, 1989). Throughout much of Europe, particularly in the Protestant north, regularly practising Christians are but a small minority of the population. Somewhat paradoxically, Christianity has actually retained much of its dynamism in the areas where it has been most persecuted. Thus, in the former Soviet Union and communist countries of eastern Europe, where Christianity was actively discouraged, it has flourished (Bociurkiw and Strong, 1975; Ramet, 1984). Gilley (1988: 239) thus notes that 'Christianity has shown an impressive resilience as a spiritual refuge from secular totalitarian state power' in eastern Europe. Indeed, one of the driving forces behind the Solidarity movement in Poland was the Catholic Church, strengthened in part by the election of a Polish Pope, John Paul II.

A third feature of the post-1945 period has been the increasing significance of new cults and religions as people have sought alternatives to the modernist lifestyles of late-20th century capitalism. The dehumanizing experiences of capitalist society have led increasing numbers of people to return to questions concerning the meaning of life, which lie at the heart of most religions. Clarke (1988: 907) thus notes that 'Over four hundred "new" religions have emerged in Britain alone since 1945 and many of these and others are to be found in the rest of Western Europe'. Some of these have been derived in part from other religions, such as Buddhism or Hinduism, but others have been the creation of specific charismatic leaders (Beckford, 1986). Most tend to be developed in opposition to rational argument, and are based above all on human experiences (Clarke, 1988). Furthermore, most tend to reject the Christian view of the Fall of humanity, and instead believe in the perfectibility of human nature in the present.

Religious sites and cultural tourism

The concern with alternative religions described above has also reawakened an interest in ancient European religious traditions, and in particular with pre-Christian Celtic and Druidic practices. Thus important religious sites such as Stonehenge in

The Waldensians

Paola Sereno

It is the night of 17 February 1997. As I write these pages here in Turin, a little more than 30 km away in the 'Waldensian Valleys' the beacons, or freedom fires, light up the mountains as they have on this night every year for the past 149 years, since that night of 17 February 1848 when they carried the *bonne nouvelle* from valley to valley and from village to village. On that day, in Turin, King Carlo Alberto had finally signed the edict conceding freedom of religion to the followers of the Waldensian faith. These were the descendants of those who, at the Synod of Chanforan in 1532, had decided to remain with the Protestant Reformation but who had been followers long before of the 'heresy' which had grown from the preaching of Waldo, the Lyonese merchant excommunicated by the Council of Verona in 1184. At the end of the 12th century, the 'cult' had spread first into southern France, then Bohemia and northern Italy. The crusade against the Albigensians, ordered by Pope Innocent III, also hit hard against the Waldensians of Provence. The survivors sought refuge in the Alps, some on the Italian side, particularly the Cottian Alps, where large groups settled in the old Valle di Luserna (today Val Pellice), Valle di San Martino (or Val Germanasca) and Val Chisone, which would all go on to become the 'Waldensian Valleys'. In years to come, many families would leave from there to found Waldensian colonies in southern Italy, especially in Apulia and Calabria, and also to bring the Waldensian church back to nearby Provence.

The Waldensians therefore make up a religious minority, Evangelical, Presbyterian and non-Catholic, distributed on both sides of the western Alps, along that shifting border which from the beginnings of modern time has separated the Kingdom of France from those territories of the Duchy of Savoy 'on this side of the mountains', roughly the present day Piedmont (Morland, 1658; Leger, 1669). In this old Italian state, Waldensianism has written itself not only into the history of religion, but also into the historical geography of Europe. Persecutions by other Catholic states, particularly France, against the Waldensians tended towards the objective of dispersing them. In contrast, the policy followed by the Court of Savoy chose as its instrument forced concentration, with the idea of better controlling them and isolating them from the realm of the State. The end of the first persecution, decided by Emanuele Filiberto with the Treaty of Cavour in 1561, was already oriented towards this idea. The same policy was reconfirmed and made even clearer by his successors, starting with the Edict of 28 May 1602. In this proclamation, Carlo Emanuele I ordered the Waldensian population in his kingdom to settle in the two alpine valleys that already had a strong Protestant majority, Valle di Luserna and Val San Martino. In Val Chisone, where there was a mixed Catholic and Waldensian population, they were to confine themselves only to the right bank of the river. These became the definitive borders of a full-blown ghetto, outside which it was actually forbidden for Waldensians to own land, practise commerce, contract marriage or even stay after sunset without special permission.

This ghetto quickly became the 'little homeland' of the Waldensians and the village of Torre in Val Luserna its capital. Its existence, however, did not protect its inhabitants, either from the actions of anti-reformist missionaries or attempts at Catholicization directed by the Court, and above all from episodes of violent repression such as the *Pasque Piemontesi* (The Piedmont Easter) of 1655 and the anti-Protestant policies initiated by Vittorio Amedeo II, in order to please Louis XIV, between 1686 and 1689, following the revocation of the Edict of Nantes. These events had the specific effect of strongly rooting the Waldensian population in the three valleys. They zealously defended themselves, even resorting to guerilla warfare, supported by a topographical knowledge of their territory, and their places of battle or shelter and worship gained powerful symbolic value, which has been transmitted through until today. Not only small, alpine villages such as Rorà or Angrogna, but also field names of uninhabited places such as Pra del Torno and La Balziglia, became central places in the geography of Protestant Europe, frequently mentioned in Chancelleries and Courts from England to Switzerland, Germany to the Netherlands.

The name 'Waldensian Valleys' was earned through a history of hard, bloody defence,

continued

paradoxically, of this ghetto, symbolically reinterpreted as a place for the affirmation of 'otherness' and cultural identity, liberty and tolerance (Muston, 1851). This name had so far been expressed more as a literary than as a geographical category. At first ignored by cartography and geography, it circulated mostly through the books and drawings of English travellers who, in the 19th century, in the Grand Tour of Italy and their discovery of the Alps, included among their destinations this little Protestant land hidden in the mountains of Piedmont (Tourn, 1994). History had already provided a model for their geographic identity. The massacres of the *Pasque Piemontesi*, halted by the intervention of Cromwell and eulogized by his poet Milton, marked a change in the attitude of the Court of Savoy in the confrontation with the above-mentioned Reformists. This change could be read in the gradual transformation of the vocabulary used in legislation regarding the Waldensian question, a clear sign of an ideological shift. The Waldensians went from being 'heretics, professing the reformed religion', to slowly becoming in the following years 'most evil bandits', then 'barbarians' and finally 'worse than the Saracens', associating them with the popular image of ancient legends of Saracen incursions in the Alps.

The definition of 'barbarians' was connected to an ideology regarding the discovery of the New World. The Waldensians were reduced to the level of beasts, wolves and tigers, in the official position of the Society of Jesus. The Jesuits were given the mission of 'this holy war' so that they could demonstrate at home their capabilities, which were already proven in the 'New World'. This was not, coincidentally, an attitude which foreshadowed the policy introduced after the revocation of the Edict of Nantes: execution of rebels, deportation of those Catholicized, confiscation of all their lands, which were then redistributed to the Piedmontese and Savoyards in a great plan of frontier colonization of the Valleys. Once again, the intervention of the Protestant European countries managed to provide exile for some, negotiated their later return to the homeland, *La Glorieuse Rentrée*, and the restitution of confiscated lands to the few survivors (Arnaud, 1710). But, overall, the reconquest of 'their fathers' heritage', both spiritual and material, carried a price high enough to transform the ghetto from a concentration camp to a symbol of freedom. After their return, they defended themselves by instituting a triple strategy of territorial control: demographic reconstruction by means of a great effort to increase their birthrate; spatial representation of land tenures through the renewal, sometimes fictitious, of kin networks as a system of landed transmission; and finally, a specific subsistence and common resources policy. They thus resolved the problem of the need for food in the face of the low numbers of working-age adults and the voluntary high birthrate through an expansion in the numbers of chestnut trees, the 'bread tree', which became a symbol of freedom in the Waldensian Valleys, as it also was in the valleys of Mondovi, in Cevennes and in the Corsica of Pasquale Paoli (Sereno, 1988, 1990).

The borders of the ghetto were never really sealed. By way of mountain trails, pastors reached the religious and cultural educational centres of Protestant Europe and returned from there to the Valleys, bringing back ideas, books, the latest debates and even economic aid. Later, in the second half of the 19th century, along the trail of transoceanic emigration from the Alps weakened by a subsistence crisis, consistent segments of the ancient Waldensian lineages refounded their villages elsewhere, particularly in South America, frequently conserving the same place names and never losing their own historical/geographical memory. It is probably also this cultural and geographic dimension, both European and cosmopolitan, which makes the people of this ghetto a 'nation', aware of their own identity but never nationalistic: an important lesson for the present to learn from the history of this small population of Waldensian people in the mountains of Piedmont.

Further reading

Comba, A. (1990) *Gilly e Becwith fra i Valdesi dell'Ottocento*, Società di Studi Valdesi, Torre Pellice

Pascal, A. (1937–68) *Le Valli Valdesi negli anni del Martirio e della Gloria*, Claudiana, Torino

Sereno, P. (1990) Popolazione, territorio, risorse: sul contesto geografico delle Valli Valdesi dopo la Glorieuse Rentrée, in De Lange, A. (ed.) *Dall'Europa alle Valli Valdesi*, Claudiana, Torino, 293–314

Tourn, G. (1977) *I Valdesi. La Singolare Vicenda di un Popolo-chiesa*, Claudiana, Torino

southern England have become places of renewed veneration and religious symbolism, while at the same time bringing their worshippers into conflict with the secular authorities. Elsewhere in Europe, an increased concern with human origins and identity has led to further interest in the earliest known Palaeolithic sites, where cave drawings and rock carvings are seen as expressing a deep human desire to understand and commune with the natural world.

These are but two examples of a reassertion of the importance of European religious sites in the latter part of the 20th century. At the same time, there have been dramatic increases in leisure time and affluence for many people, which have enabled the tourist industry to become one of the most flourishing sectors of the European economy (Shaw and Williams, 1993) (see Chapter 19). While much of this tourism has been developed around Mediterranean coastal resorts, there has also been a significant increase in cultural tourism in other places, and particularly to the numerous churches and monasteries that scatter Europe's urban and rural landscapes (Fladmark, 1994). Such features figure significantly in the many tourist guidebooks, not only of western Europe but also of the newly independent countries of the former Soviet Union (for an Estonian example, see Huma, 1996).

Travel to religious sites, though, is nothing new and builds on a long tradition of Christian pilgrimage. Nolan and Nolan (1989) have thus identified some 6150 Christian shrines in western Europe, of which at least 830 receive more than 10,000 pilgrims a year (for a case study of Lourdes, see Rinschede, 1986). In the later medieval period full remission of sins was granted to visitors to numerous places that had been designated as being of particular holiness. These included sites of international importance such as Rome and Santiago de Compostela in north-west Spain, for which there were well-defined and specified pilgrimage routes, but also many smaller sites of local importance, a visit to which could bring salvation that much nearer. The 20th century has also witnessed the emergence of a number of new sites of importance for pilgrimage, notably resulting from claimed visions of the Virgin Mary, as at Fátima in Portugal in 1917. As Gilley (1988: 238) has pointed out, this 'not only reinforced Portuguese popular religion, but gave a supernatural sanction to Catholic hostility to Communism'.

Europe's religious heritage

The above examples illustrate that religion not only has played an important part in shaping Europe's cultural identity in the past, but also continues to play a significant role at the present. While some see religion merely as false consciousness, or the result of sexual repression, many Europeans continue to believe in miracles and in the Creed agreed at the Council of Nicaea in AD 325. These positions are not easily reconciled. Christianity has been of central importance in shaping Europe's philosophical debates, its moral and ethical codes, and its cultural landscapes. However, it is essential to acknowledge the significance of other religions, notably Judaism and Islam, in Europe's cultural heritage. Moreover, it is also important to emphasize that religions have been used throughout history to legitimate the social and political actions of the powerful. The horrors of the Spanish Inquisition, the Holocaust of the 1939–45 period, and the 'ethnic cleansing' that followed the break-up of Yugoslavia in the 1990s, have all owed much to emotions fuelled by religious ideals.

Acknowledgements

I am particularly grateful to Chris Park for his comments on an earlier version of this chapter.

Further reading

Fontana, J. (1995) *The Distorted Past: a Reinterpretation of Europe*, Blackwell, Oxford

Nolan, M.L. and Nolan, S. (1989) *Christian Pilgrimage in Modern Western Europe*, The University of North Carolina Press, Chapel Hill

Park, C.C. (1994) *Sacred Worlds: an Introduction to Geography and Religion*, Routledge, London

Sutherland, S., Houdlen, L., Clarke, P. and Hardy, F. (eds) *The World's Religions*, G.K. Hall & Co., Boston, Mass.

Turner, B.S. (1991) *Religion and Social Theory*, 2nd edition, Sage, London

CHAPTER 5 Cultural landscapes

DENIS COSGROVE

Landscape expresses the sense of a harmonious, social and aesthetic unity between land and human life, visible across a defined geographical area. Landscape is a richly evocative word in English, with a complex history and multiple layers of meaning. It can refer to specific, named regions – actual land areas – displaying a certain unity in their physical features, forms of settlement and community life: for example the Vosges mountains of France, the Burren limestone region of western Ireland or the polderlands of Holland. Equally, it can refer to specific scenes, framed and composed as sources of aesthetic delight or awe and represented in a garden design, a painting, photograph or film, in poetry or prose: a view, for example, of the Roman Campagna by Claude Lorrain, a panning film shot across the avenues and lakes of Versailles, or a description of Denmark's Jutland Heath in a novel by Steen Steenson Blicher.

The English word *landscape* has equivalents in all the other European languages, although the emphasis given to different aspects of its meanings varies, especially between those languages which owe their origins to Latin and those with roots in Germanic speech. European societies are by no means the only ones to have expressed a sense of social and aesthetic attachment to land, and to have represented it artistically; China and Japan have highly developed traditions of landscape art, while Europeans themselves derived much of their garden aesthetics from Persian, Arab and Mughal Indian models. But the unique cultural experiences of a half-millennium of social and environmental modernization have yielded characteristically European ideas, experiences and expressions of landscape, with associated design principles by which their landscapes have come to be recognized and evaluated. Landscape thus provides one of the perspectives through which Europe's geography and European environmental meanings and relations may be understood.

This chapter is divided into five sections. The first is devoted to a discussion of *landscape meanings* in Europe, paying particular attention to relations between experience and aesthetics, and to Germanic and Latin cultural inheritances. The second section considers how these meanings converged in the context of the social, economic and geographical changes associated with early European modernization to yield a specific *idea of landscape*. The third part develops this historical study by tracing aesthetic *expressions of landscape* in European art and design from their regional origins in upper Italy, southern Germany and Flanders to their influence on the ways we appreciate and conserve landscape today. The fourth part examines the role that *landscape ideology* has played in constructions of territory and identity in Europe, paying specific attention to visions of Romantic landscape and ideals of urban design respectively in the discourse of 19th and 20th century European nationalism. The final section examines the continued significance in European societies today of *landscape as container and medium of collective memory*, and as an economic and environmental resource for a tourist and heritage society.

Landscape meanings and European culture

The history of geographical change across Europe since Neolithic times has been dominated by the spread of settled human communities, exploiting favourable environmental conditions for the permanent cultivation of cereal grains (see Chapter 2). Changes in climate, technology and political ideology, and catastrophic events such as war and plague, operating at different scales, have produced advances and retreats in the areas devoted to field cultivation, in the balance between livestock and grains, in the area of forest and woodland, and in the spread of uncultivated 'waste' lands on the one hand or reclaimed lands on the other. The ways in which cultivation is organized, the selection and balance of crops, the traditions of cultivation, the implements used, the construction materials, tools and techniques employed in building, all vary over European space and time. And for the past three centuries the overwhelming majority of Europeans have left the land for towns and cities, becoming urbanized at least in so far as their employment lives are concerned. But even today the idea and the image of a community of people producing their own means of life, bound together by tradition and attached to a definite area of land, *villa*, parish or *commune*, remains a powerful model of social form for Europeans. It testifies to the historical depth of settled agriculture as the defining feature of European social experience and social memory. It is this which underlies the imaginative appeal of *landscape*.

'Landscape' is the English rendering of a composite Germanic word: *Landschaft*. *Land* signifies a bounded territory composed of the different uses needed to support a community: cultivated fields, pasture meadow and woods. Thus England, Jutland, Scotland are the territories of the Angles, Goths and Scots respectively. *Land* has thus more than a purely spatial and environmental meaning, it is a social and legal entity: 'such a land is defined by its customs and culture, not by its physical characteristics, though it may conform to an area of dry land, e.g. an island' (Olwig, 1996a: 633). The suffix *schaft* emphasizes the collective social features of the legal entity which is *Land*. *Schaft* implies the shaping or crafting of a social unity out of a collection of disparate elements, and is connected to the English suffix 'ship', as in friendship or fellowship. In the Germanic language community, therefore, landscape denotes a form of spatiality: expressing the experience and intention of a social group tied by bonds of custom and law to a determined territory. The customs and laws that traditionally bound such communities across Europe, determining the allocation of plough and pasture rights, fixing dates of planting, haymaking and harvest, distributing ceremonial tasks by gender and age, were concerned above all with ensuring good order and productivity in the social management and cultivation of land. And as Ambrogio Lorenzetti's mid-14th century image of the Sienese countryside illustrates, good order and productivity is *visible*, it makes for beautiful landscape (*bel paesaggio*).

If the spatiality of Germanic *Landschaft* emphasizes locality and small-scale intimacy, an alternative spatiality which has powerfully influenced the environmental organization and design in Europe comes from Rome's expansion of imperial order across the south and west of the continent (see Chapter 1). This is a spatiality of distant control radiating from a powerful centre along straight roads, mapping its colonial appropriations in surveyors' grids and marking its power in military camps, mileposts, arches and columns. Since the fall of Rome, the rational geometries of Latin landscape have provided Europeans with a vocabulary for inscribing central authority across their territories. The junctures of early Germanic and Latin cultural influences within peninsular Europe are variously apparent in its cultural landscapes. There are no sharp or consistent lines demarcating this all-important cultural meeting, and material boundaries are few: the Antonine Wall in Scotland, and the occasional stones marking the *limes germanicus* in the Rhine or Danube valleys. But the heritage of Roman colonial influence can still be traced in language, in legal tradition, in local custom and even sometimes in traditions of construction and decoration across the continent. The historian Marc Bloch used such indicators to map the cultural legacies of Rome in the French feudal landscapes: plough forms, field and settlement patterns, property and inheritance laws, farmhouse types and roofing materials. These distinct traditions of social and environmental organization also find linguistic expression in the different words for landscape in northern and southern Europe. The closest equivalent to the Germanic *Landschaft* in the Latin-influenced languages is *paysage* (in French), *paesaggio* (in Italian) or *paisaje* (in Spanish). All derive from the Latin *pagus*, which, like its German equivalent, denotes a socially defined area of land legally connected with a settled agrarian community. The sense of close attachment between the human inhabitants and the natural environment conveyed by this term survives in the English words *peasant* and *pagan*: one who worships the natural world rather than the Christian God.

Both 'peasant' and 'pagan' carry negative connotations in English: of uncouth, primitive ignorance. This is the perspective of the urbanite, of central authority. It dates to the union of imperial political and Christian religious power in the later Roman Empire and recalls the much more tenuous autonomy of local communities in the lands colonized by

Rome than in the Germanic lands. The continued authority of urban centres over rural communities after the collapse of imperial control, especially in the Mediterranean regions, sustained traditions of canonical law and a market in land in those parts of the continent, while the influence of Germanic customary law and personal obligation sustained local collective use rights over land much more strongly in other areas. While the shared meanings of the Germanic *Land* and Latin *pagus* derive from a common experience of local collective cultivation of land by a legally recognized and settled community, the precise nature of that experience, the modes of cultivation and the legal forms of recognition and regulation have thus varied widely across the continent. The medieval Carolingian Empire sought to combine both traditions within feudal structures designed to coordinate the local lives and production of agricultural communities with more distant authority in order to support its militarized class of knights and squires. The system met with varying degrees of success, and further hybridized the experience and meaning of landscape across its territories.

These complexities of European agrarian history and land law are significant in two ways. First, they account as much as does variation in the physical geography of Europe – climatic, topographic, geological and geomorphological – for the traditional appearance of European agrarian regions, *Landschaften* or *pays*. One of the defining projects of 20th century geographical scholarship in most European countries has been to identify, map and describe the characteristic features of local landscapes, seeking to explain their emergence from long historical continuities of local settlement and interaction between land and life (Buttimer, 1994). French geographers, for example, lovingly described the distinctive *pays* of the Île de France, the wide horizons, compact villages and severe courtyard farms of the Beauce, the more accidented and tapestried pastures of the Brie, the *limon*-covered chalk slopes and gothic towers of Picardy, or the plateau *bocage* of hedges and paddocks with half-timbered, thatched farms in the Pays de Caux (Claval, 1994). In Germany 'landscape indicators', including field shape, village form, roof pitch and constructional materials, for example, allowed geographers to identify characteristic settlement types, such as the *Rundling* or the *Strassendorf*, and to relate these to histories of land clearance and settlement (Mayhew, 1973; Denecke, 1992; Sandner, 1994). British geographers, too, distinguished regions by their field patterns and settlement nucleation, relating these to the fine grain of geological and geomorphological variation acoss the islands and to the successive waves of early settlement, while in Italy and Spain semi-nomadic traditions of transhumance on the one hand and Islamic cultural influence on the other were called upon to distinguish individual *paese* or *paisaje* from one another, and from urban territories controlled and shaped by sharecropping leases and polycultural cultivation. Much of this work was driven by an ideological imperative to identify and celebrate the unique heritage and beauty of nationalist landscapes.

The second reason why variations in agrarian history and land law are important is that they help account for the different ways that modernization has affected European landscapes: both the experience and visible evidence of land/life relations in different parts of the continent, and the meaning of landscape, which modernization has made enormously more complex. The geographical project of identifying, mapping and describing regional landscapes derived not merely from national pride, but also from a certainty that the continued existence of local variation and the traditional look of the land were threatened, and that such loss was socially, politically and aesthetically dangerous. Geographical science was thus responding to certain *ideological* imperatives that connect to landscape, and it is these which have further complicated its meanings as a European cultural concept.

Modernization, space and the idea of landscape in Europe

In modern English usage it is not the close association of land and life, region and community that is first evoked by the word landscape, but the sense of a view, a vista, scenery. Landscapes are primarily pictorial. We may visit a specific landscape, walk, drive or cycle through it, even choose to make our home in it, but in common usage landscape implies something outside and beyond ourselves, to which we relate aesthetically. This meaning of landscape expresses a quite distinct spatiality from that denoted by the original meanings of *Landschaft* or *pagus*, and the historical relations between these two sets of meanings is vital for understanding contemporary European landscapes. Landscape as scenery, as an aesthetic experience, draws upon older meanings of landscape as connection between community and territory for much of its force, but it also represents a different, and characteristically modern, way of relating to land and to the natural world more generally. The evolution of this second meaning of landscape reveals how this is so.

In the same areas and in the same late-medieval years that feudal structures of economic and social life in Europe were first yielding to modern ones, through long-distance commercial trade, organized production of goods and agricultural products for the market, independent city government, urban capital investment and territorial organization in surrounding rural areas, we may observe the emergence in art and literature of aesthetic landscape representations and the design of actual landscapes as sources of pleasure and delight. In upper Italy, where commercial cities such as Siena, Florence, Bologna and Verona had amassed great wealth for their citizen classes and established control over the surrounding countryside, in the upper Rhine and Danube lands, where cities such as Frankfurt, Nuremburg and Ulm held similar positions north of the Alps, and in the textile trading towns of Flanders, such as Bruges, Ghent, Roubaix and Antwerp, for example, a new and more pictorial meaning of landscape had become established by the turn of the 16th century (Cosgrove, 1997). By Shakespeare's time, Jacobean Englishmen had begun to use the word *landskip* to describe a genre of painting by then well established across the North Sea, and to purchase drawings and paintings of 'inland scenery' by Flemish and Dutch artists to decorate their homes. In the 17th century, wealthy English landowners imported Dutch gardeners to design their parkland estates as they imported Dutch hydraulic engineers to design drainage canals and fields out of regions of fen and marshland level. European aristocrats from France and England to Russia and Scandinavia were drawing also upon Italian garden and architectural traditions, themselves revived from Classical textual and archaeological sources, to transform their estates into elaborate tapestries of planted avenues, reflecting lakes, waterworks and parterres. New-rich Englishmen, combining design principles learned on continental 'Grand Tours' with nationalist principles of 'natural liberty', synthesized a new art of informal 'landscape gardening' in competition with continental ideas of large-scale scenic parkland to be found at Karlsruhe, Vaux le Vicomte or Potsdam. The informal, 'picturesque' English manner of landscape gardening to be found at Stowe, Blenheim or Castle Howard consciously sought to blur the boundaries between designed garden spaces intended for pleasure and relaxation, and the inhabited, productive countryside of cultivated fields, farms and woods beyond its borders: the local 'landscape' (Figure 5.1). The productive territory of the nation as a whole – the 'countryside' – thus came to be seen and evaluated aesthetically. Such 'landscapes' came to characterize not merely local areas but whole territories, sources of national identity and pride, a common heritage to be understood, protected and

Figure 5.1 Rousham Garden, showing landscape park merging into cultivated rural landscape (*source:* Denis Cosgrove).

admired as a badge of citizenship. In this way landscape completed a cultural process that had begun in the 15th century paintings of the Bellini family, Albrecht Altdorfer, Jan van Eyck and the Limburg brothers: aesthetic renderings of humanized land escaped from the confines of the artist's easel and the poet's pen, to determine the appreciation of actual scenes and design of 'landscape' itself (Warnke, 1995). A new meaning of landscape, expressing a changed relationship with the land and a new spatiality, was fully grafted onto the traditional sense of local attachment between life and land, now expanded to the scale of the nation.

The geography of this changing landscape spatiality is significant. Demand for aesthetic views of scenery did not emerge from among local agrarian communities, but among the sophisticated citizens, often important landowners, for whom landed property was a source of income and status. These were people for whom the distanciated and political sense of landscape, with its vocabulary of power and authority inherited from Classical models, held appeal. If such people were tied economically or legally to specific lands it was not as cultivators and dwellers but as property owners. In this sense they were outsiders, exerting a measure of external control over the environmental and social destinies of the lands they owned, able to appreciate a distanced perspective. From such a perspective, landscape indeed becomes a 'way of seeing', one which is closely connected to the development in Europe of modern urban, commercial life wherein the countryside is increasingly commodified as a means of production and exchange rather than a totally encompassing element of local social life and identity (Cosgrove, 1997).

As increasing numbers of Europeans have left agrarian communities and occupations and migrated to cities, other regions and continents, so this secondary meaning of landscape has come to be the most meaningful to them, obscuring the earlier one in all the European languages; obscuring but not replacing. A powerful aspect of the popular and continuous aesthetic appeal of landscape representations – if no longer in progressive art, certainly in movies and advertising images – lies in their nostalgic evocation of pre-modern forms of collective life in productive nature. Imagined worlds of agrarian simplicity and social harmony are captured, not only in Constable's Suffolk scenes or in Vasilii Yakovlev's images of traditional Russian life painted at the height of Stalinist collectivization, and reproduced on household walls and hotel bedrooms, but equally in the refurbished farmhouses, mills, barns and stables set in conserved fields and protected countryside from Lake Vättern in Sweden, through Hesse and Bavaria to Tuscany and Umbria. Continuing processes of modernization, robotics, the telephone, fax, computer and Internet, releasing people from industrial production into services as completely as they were once released from agricultural into manufacturing production, allow increasing numbers of Europeans almost unlimited choice over the locations in which to conduct their lives. More and more of them choose to live out a form of rural life which draws upon the nostalgia for the social and environmental intimacies and attachments associated with the primary meanings of landscape. Thus they demand strict controls over the 'look' of the land, in order to preserve an aesthetic record of imagined attachments, imposing across European space the secondary meaning of landscape, whose characteristically modern feature is its connection to vision and the techniques of seeing.

Techniques of representation and European landscape expression

In early modern England, landscapes were often called 'prospects'. This is a word which captures both the experience and the techniques whereby the modern spatialities of landscape became established historically. *Pro-spect* implies directional and intentional looking, across both space and time. Europeans began to look at land in new, calculating ways as it became a source of capital accumulation. If the potential of land is to be realized as capital it must become 'property', a measurable commodity able to be bought and sold in the market. Survey, cadastral documentation and mapping are the techniques by which this is achieved, and unsurprisingly, modern methods of survey and land measurement were pioneered in the very regions of Europe where landscape art itself emerged. The techniques of the old Roman *agrimensures* were first revived in the Italian city-states, and in those regions of the continent where lands were drained and improved for commercial agriculture, such as Lombardy, the Rhine delta lands and the Venice *terraferma* (Cosgrove, 1993) (Figure 5.2). These environmental modifications involved a combination of sightings from vantage points using instruments for calculating angles accurately, and measurement by pacing, use of

Figure 5.2 Polder landscape from Veneto, Italy (*source:* Denis Cosgrove).

chains and geometrical calculation. The perfection of lenses and their use in sophisticated sighting instruments in 17th century Holland improved and lengthened the visual prospect.

Prospects over the land allowed the Euclidean geometry of the Classical world to be reinscribed across European territories, as during the heathland reclamations of Frederick the Great on the North German Plain; or in the English enclosure of common lands, which produced new patterns of hedgerowed and fenced fields across the Midlands. It is visible in the die-straight roads and barge canals of France constructed by his engineer Vauban for Louis XIV, or in the dykes, ditches and reticulated polderlands of the Valpadana, the Vendée and Holland, or again in the stellar revets and earthworks of garrison towns such as Willemstadt in Holland, Neuf Brissac in Alsace or Palmanova in Friuli. The same geometrical principles underlay the invention and development of linear perspective, which, from the mid-15th century, allowed Italian painters to create the illusion of three-dimensional space within a picture frame. This 'prospective' technique was the foundation for landscape painting in the European tradition of 'pleasing prospects' that runs from Giorgione through Titian, Claude, Poussin and Turner to the French Impressionist landscapes of the late 19th century.

More than perspective geometry was required to produce aesthetically satisfying landscapes. To meet the lusts of the eye, nature itself had to be selected, managed, framed and composed. Cultivation of the land for crops and animal husbandry both depend upon domesticating the natural world. But landscape art is nature domesticated for visual pleasure. Thus the boughs and leafy branches of trees are bent to frame the scene as curtains to a window or the wings of a theatre set, the eye is led through space towards the horizon by alternating or overlapping bands of topography coloured and shadowed to emphasize distance, and along sight-line features that may be orthogonal paths or serpentine streams, past incidental features and figures from the natural or human world ('staffage') which mark out and fill the intermediate space. The whole is composed by human artifice so as to blend harmoniously into a unitary vision of order. Rules and conventions have been developed from painting, theatre and gardening to create a 'discourse' of landscape which has been communicated and debated across Europe through original and reproduced works of artists and designers, and in theoretical writings on landscape. This discourse continues to affect not only environmental design, for example in new airports or industrial or entertainment parks and corporate headquarters commissioned from professional landscape architects, but also the ways that ordinary people enjoy landscape, cultivate and design their domestic garden spaces, and record places in photographs, videos or sketches. It also affects the political pressures they place on the authorities charged with caring for existing landscapes.

Romantic nature and capital cities: landscape ideology and European nationalism

So far in this discussion I have focused largely upon cultivated landscape and the aesthetics associated with rural scenery. This responds to the origins of landscape meanings which I have traced in the association between land and life in agrarian communities. Prior to the 18th century few Europeans took aesthetic pleasure in the wild or uncultivated scenes of mountain, moorland, marsh or rocky coastline. Certainly such scenes would not have been termed 'landscapes', for they did not support human communities. Those who did make a living in such environments were conventionally regarded in literate European society as more wild and uncouth even than 'peasants', scarcely meriting the label human, and certainly not living in a 'community', a term reserved for agrarian forms of economy and society. It is not surprising therefore that European landscape painters and writers from the 15th to the 19th century persistently illustrated scenes from Classical authorities such as Ovid and Virgil, in whose poetry a cycle of natural and social evolution, paralleling that of organic life generally, was reflected in ideal landscapes which evolved from wilderness nature through cultivated garden to lordly city before an inevitable collapse and return to the state of uncultivated savagery. The 'perfect' landscape within this cycle was that of an imagined 'Golden Age', a harmonious balance between collective human life and a productive nature, associated with youthful fertility: the *Arcadia* of Greek shepherds, or the Sicily of Georgic cultivators. Although Flemish and Dutch landscapists such as Peter Bruegel or Jacob van Ruysdael were less influenced than their Italian, French or later English peers by Classicism, preferring to celebrate in their own genre scenes of local land and life as it passed through the seasons of the agricultural year, their scenes celebrated a similar 'middle landscape' of collective human life in nature.

But the social forms and spatial orders lying either side of this imagined middle landscape, the wilderness and the city, have both been drawn within the circuits of landscape discourse, in particular by one of the processes of modernization which I have not so far discussed: the construction of the European nation-state (see also Chapter 6). Growing initially out of the strongly centralized feudal kingdoms of France, England and Aragon, the territorial state has emerged since the 17th century as the primary geographical expression of spatial allegiance and identity for modern European societies. In this sense the state has come to replace local landscape as the political expression of collective socio-spatial attachment. The native 'land' from which contemporary Europeans derive their citizenship is a territorial state whose shape and meaning is the expression of the political, ideological and military struggles which have dominated modern European history for the past three centuries, from the Treaty of Utrecht ending the Thirty Years' War in 1650, through Britain's 'Glorious Revolution' (see Chapter 7), the French Revolution of 1789 with its Napoleonic aftermath of national state formation in Greece, Italy and Germany, to the 20th century European wars fought to determine the boundaries of existing European states and create new ones such as Romania, Croatia and the Czech Republic.

The spatial scale of even the smaller European states, so much greater than the local 'land' that gave original meaning to landscape, makes collective, face-to-face community impossible. Allegiance and identity have therefore to be constructed within an 'imagined community' (Anderson, 1983). Together with language, landscape has been a key pillar in this ideological construction of nationhood. Since a key task of ideology is to render a state of affairs 'natural', the idea of a geographical unity of people and land in which the community is literally born out of the natural word they inhabit has powerful potential. Thus 'Romantic nationalism' in 19th century Europe commonly celebrated at the scale of the nation what the French sociologist Frederick Le Play called the unity of 'place, work and folk', and sought iconic representations of this bond in visible landscapes (Matless, 1992). An example is the art of the German painter Caspar David Frederick, whose lone individuals facing the awesome power of nature across misted mountain peaks, or kneeling before a cross among snowclad pines, consciously linked the German *Geist* to a characteristically German natural world. John Constable's and J.M.W. Turner's landscapes have served the same purposes for England and Britain during times of intensified patriotic self-consciousness beginning in the Napoleonic wars. Landscape paintings provide good examples because they are intense and concentrated individual icons. But iconic national landscapes have been built up through a wide range of representations, including poetry, folk music and song, theatrical setting, photography and film (Johnston, 1994; McCannon, 1995). Norwegian national identity, for example, draws heavily on connected images of mountain snows, icy fjords and polar light as a setting for muscular manhood on skis and mystical nordic spirituality and promoted through the music of Grieg and the paintings of Edvard Munch. In struggling for independence from Britain, Irish nationalists constructed as the pure native landscape a mythical Western world of whitewashed cottages collected into *clachan* hamlets, and set under grey Atlantic skies against windswept limestone moorland on the glaciated shores of Galway or Kerry (see case study of Ireland) (Figure 5.3). It is common for nations to select regions that appear – at least to the urbanized eye – untouched by modernity, closest to natural wilderness, as those in which the original spirit of the nation is to be found. Thus in Fascist Italy, emphasis was placed on the Alpine border landscapes of *monti, boschi, torrenti* (mountains, woods and streams) as the testing ground of national heroism, and Italians were encouraged to visit these border regions: vacationers following the maps and guides published by Touring Club Italiano, schoolchildren as campers and scouts (Zanetto *et al.*, 1996).

More virile landscapes tend to be promoted by aggressive patriotism in preference to the more domestic, settled landscapes of cultivation favoured by less militant forms of national pride: the water meadows of south-east England, for example, or the *bel paesaggio* surrounding Italian towns. Even these milder and more domesticated 'middle' landscapes can still serve powerful ideological functions for an aggressively nationalist state. Thus the highly tended domestic *Landschaften* of decorated wooden farm buildings, hedged fields and carefully managed woodland which had been the subject of so much attention by German settlement geographers from Meitzen to Walter Christaller became a weapon in the struggle for legitimacy by the National Socialist Party in the interwar years and a weapon of race domination in the eastern territories after Germany's invasion of Poland in 1939 (Denecke, 1992). The routes of the early German *autobahnen* were carefully 'landscaped' to follow the long, curving contours of local topography and to open out views across 'typical German' landscape, encouraging pride among drivers of the new *Volkswagenen* (Rollins, 1995). And landscape geographers were employed by Heinrich Himmler to impose a 'national' settlement landscape across those parts of the occupied eastern lands deemed to be integral regions of the *Reich*, while a rational economic landscape designed according to the precepts of Central Place Theory was to be inscribed onto the 'free planning zones' cleared of their 'inferior' population of Slavs.

Rational landscape, patterning human occupancy according to iron laws of spatio-economic efficiency, is modernity's particular contribution to European landscape. Its expressions were a common feature of post-war reconstruction across Europe, especially in regions utterly devastated by modern warfare, and in much of the eastern part of the continent, where Stalinist central planning dominated for half a century. Early examples of the former are to be found in the reconstructed zones of northern France, where by 1918 virtually all evidence of pre-existing human settlement had been replaced by a landscape of trenches, barbed wire, gun emplacements and cemeteries (Clout, 1996). Within this zone consciously designed fragments of the 'national' landscapes of the combatant nations are still to be found in the military cemeteries and monuments to fallen soldiers: the white limestone headstones and lawns of English war graves, the pines, granite and iron crosses of the German. After 1945, destroyed areas seemed to offer *tabulae rasae* over which utopian designs of scientific modernity could be realized. The invasion zones of Normandy, areas of heavy fighting in the Rhine valley and industrial regions of Germany offer examples in villages and towns constructed in post-war concrete according to modernist designs, simple field geometries marked by posts and wire. But even where no military devastation had occurred, modernist landscapes have been created by state planning. East Germany offers a prime case of collectivization changing all aspects of former rural landscape: field patterns and sizes, farmhouses, storage buildings, routeways (Ritter and Haidu, 1989). Russia, Ukraine and other areas of the former Soviet Union reveal similar landscape change, a process currently being reversed as collectives are broken up and lands reallocated to private proprietors. Even in the most democratic regions of Europe, on the

Narratives and names: Irish landscape meanings

Catherine Nash

Figure 5.3 Killarney National Park, Kerry, Ireland (*source:* Tim Unwin, 4 February 1995).

The representation and organization of places within many Western landscape traditions have been primarily concerned with promoting a certain way of seeing. However, in Ireland the symbolic meaning of landscape has been deeply tied to issues of language and narrative – stories from history, mythology, folklore, of local events and personal memory. This has been the case in the way certain regions in Ireland have been symbolically important as *national landscapes*, as the physical and human geography of particular places have come to symbolize the collective culture as a whole, and in the associations of the countryside more generally. In the 19th and early 20th century strong claims were made about the value of Irish culture and the cultural differences between Ireland and England, in order to justify and promote Irish cultural and political independence. Irish culture and identity, it was suggested, was based in the Irish language and in the Irish soil. Nationhood came to be defined through the twin frames of land and language as the Land League sought to improve the conditions of Irish agricultural tenants, and eventually allow tenants to own their own land, and as the Gaelic League promoted the Irish language after years of colonial suppression.

For many cultural nationalists, urban areas were most affected by English culture and were therefore least Irish. Thus, the west of Ireland became of great cultural significance as it was the region in which most Irish was still spoken, and furthest from English influences (Figure 5.4). Many Irish artists who had trained in Paris and painted the French countryside returned to paint the region and brought with them their sense that in regions like Brittany and the west of Ireland, antidotes to the ills of modern, urban and industrial life could be found. In the literary and artistic representation of the region, European cultural primitivism meshed with Irish cultural nationalism as the landscape and its people were acclaimed as uncorrupted by modern life and English influences. But these ideas of the deep links between land, language and nationhood also led to a more general sense that in the soil and stories of rural landscapes true Irishness could be found, and

continued

Figure 5.4 Signpost in Ireland (*source:* Catherine Nash).

moreover, that the meaning of landscape in Ireland is more to do with its stories than with scenic qualities. These counter-discourses of native meaning turned to themes of language and narrative, in order to challenge English landscape aesthetics and the casting of the Irish landscape in 19th century romantic melodrama, in which dramatic topography and changeable weather in Ireland were featured as a reflection of the charming sentimentality but unruly nature of the Irish (Figure 5.3).

Thus, in deliberate opposition to English landscape aesthetics, the cultural significance of the Irish landscape often resides in the stories associated with particular places and in the place-names that connect language and locality. Many of the 'big houses' of landed estates in Ireland were destroyed or allowed to fall into ruin in this century because of their associations with colonial power and economic inequality. Even the houses of the aristocracy which survived do not have the kind of popular appeal that they hold elsewhere. While ruins of cottages are signs of a once heavily populated countryside and recall the kinds of suffering which led to their desertion, the importance of images of cottages as symbols of Irish rural life and traditional culture in paintings, postcards and tourist promotion material does not mean that old cottages and farm buildings are always preserved. The building of new bungalows beside older, abandoned cottages and farmhouses, and the slightly unkempt nature of the Irish countryside, seem to suggest not that the landscape lacks significance but instead, speaks of an approach to landscape that is different from the English visual tradition of celebrating and preserving tidy cottages, villages and vernacular architecture. The contrast between English landscape traditions and the way in which landscape is understood in Ireland through the stories in mythology, folklore, family memory, local history and poetry that name places, and the stories that place-names hold in their references to topographical features, land use and people, reflects a particular history of colonial representation and anti-colonial efforts to define Irishness through a different way of considering landscape.

Naming places was also part of 19th century English colonial policy. The British Ordnance Survey mapping of Ireland in the 19th century imposed English value systems and cartographic conventions, and anglicized Gaelic place-names into forms acceptable to English map users. The English authorities imposed new English names

continued

on towns and counties in Ireland, such as Queenstown for Cobh, or Kings County for Offaly, and derived simplified versions of Irish place names, as with Ballinlavan for Baile an leamháin (town of the Elm) and Killeagh for Coill liath (grey wood). This toponymic editing implied the superiority of English culture, the naturalness of English political authority in Ireland and the inferiority of Irish culture and Irish people. These actions, it is argued, furthered the decline of the Irish language. Not only did the maps mis-record Irish names but they also left many names unrecorded, and 19th century Ordnance Survey maps are described as empty in contrast to the cultural landscape with its dense references to mythology, folklore, history and local habitation. Many have expressed a sense that the loss of Irish as a widely spoken language has estranged people from the local histories and meanings encapsulated in older place-names in Irish. In the lyrics of Irish songs and poetry written in English, place-names can be markers of cultural difference. Their evocative force does not depend on the listeners' knowledge of the specific places, but on a collective sense of the value of detailed local knowledge and deep attachment to place.

After independence in southern Ireland the original Irish versions of place-names have been reaffirmed, through published lists of official place-names, the production of maps in Irish, and bilingual road and street signs (Figure 5.4). Yet finding the most appropriate Irish version is not always an easy task, since in the predominantly oral culture of the mid-19th century there were frequently different spellings of one name. The search for the original name and by extension the search for a true pre-colonial Irish culture raise more complicated problems about how landscape and identity in Ireland are understood. Searching for the original version of a place-name can be a project of cultural retrieval but can also speak of a longing for a lost authentic Irish identity, rooted in land and language. The importance given to names and narratives in the Irish landscape, and debates about Irish aesthetic and representational traditions, are clearly tied to questions of culture and identity. Reading the rural landscape in Ireland is not simply a question of tracing in its fields and vistas a reflection of a history of successive invasions, of systems of land ownership, inheritance and use, internal and European legislation or evidence of 'global' cultural influences. The ways these historical and cultural geographies are interpreted depends on particular notions of culture and cultural change.

Claims to a deep and true culture through a close identification of people with land and landscape were useful in resisting colonialism. Yet these kinds of claims tend both to essentialize and to simplify identity. Rather than seeing culture, like language, as something continuously made and remade as people create, communicate and contest meanings and ideas, they imply that culture and people's identities are stable, unchanging and inherited. The idea that Irish national identity is a product of a marriage of land and language tends to overlook the diversity of Irish identities in Ireland in the past and present, and results in impossible searches for pre-colonial authenticity and wholeness. In recent debates in Ireland, many have pointed out that focusing on cultural loss implies that contemporary Irish culture is contaminated and diluted, rather than vibrant and dynamic because of the diversity of its influences. When criticisms are made of new street names in suburbs for their lack of reference to Gaelic culture, or when people find an anglicized place-name changed to an Irish form despite its personal significance to them, questions of identity and identification with places become fraught. Which is more important – the personal and local associations tied to a contemporary name or adherence to an older, 'authentic' name? The importance of local names has been reflected in the opposition to the Post Office's scheme to introduce postcodes in addresses in Northern Ireland since the early 1970s. By naming rural roads and allotting numbers to the houses along them, addresses would change from the format of house or farm name, townland name and parish, to house number, road name, nearest postal town and post code. This new system, it was felt, would make the traditional townland names redundant and diminish the importance of these local sources of identification. Here the issue was not necessarily to recover the original form of the townland names but to keep the townland names as they are now known in common currency. This

continued

issue of naming mobilized cross-community support in Northern Ireland, but when claims that affiliation to place in Ireland depends upon a deep and inherited connection to land and language resurface, place-names can be enlisted to mark out exclusive rights to belong. The ways and extent to which particular places, such as early medieval ecclesiastical buildings, or kinds of places, such as planned Plantation towns, feature in the tourist and heritage industries are often tied to particular versions of Irish history, culture and identity. Yet new initiatives to increase revenue from tourism are leading to re-evaluations of what counts as a valued landscape and a shift away from a predominant emphasis on the rural.

Yet, though language and land have been tied in dominant discourses of a unified culture which have tended to dismiss cultural differences in Ireland, this does not mean that references to the rural, to particular places like the west of Ireland, or to the Irish language or cultural traditions, automatically reinforce simple and essential ideas of Irish identity. Many contemporary artists and writers tap the charge of Irish cultural traditions in order to suggest ways of viewing Irish identity, and ways of acknowledging the value of these traditions without fixing what they mean. Landscape imagery and the Irish language were often deployed to define Irish identity in restrictive and conservative ways, and especially for women, as cultural purity and continuity were seen to depend on their actual presence in the rural landscape and conservative versions of motherhood and femininity. Yet, many artists and writers suggest that progressive attitudes to gender, sexuality and cultural identity are not incompatible with attachments to traditional Irish culture. Much of the vibrancy and originality of culture in Ireland derives from its mixing of old and new traditions, respecting this inheritance without searching endlessly for an impossible return to an impossibly stable pre-colonial culture or defining Irishness through one kind of aesthetic or one approach to landscape. Thus, in Ireland questions of how landscape looks, how it is represented and what it means, are tied to a specific history of cultural change, but are also inseparable from the ways in which culture itself and cultural change are understood.

Further reading

Gibbons, L. (1987) Romanticism, realism and Irish cinema, in Rockett, K., Gibbons, L. and Hill, J. (eds) *Cinema and Ireland*, Routledge, London, 194–257

Gibbons, L. (1996) Topographies of terror: Killarney and the politics of the sublime, *The South Atlantic Quarterly*, **95**, 23–44

Graham, B. (1994) Heritage conservation and revisionist nationalism in Ireland, in Ashworth, G.J. and Larkham, P.J. (eds) *Tourism, Culture and Identity in the New Europe*, Routledge, London, 135–58

Hammer, M. (1989) Putting Ireland on the map, *Textual Practice*, **3**, 184–201

Hannan, R. (1991) *An Ball Uaigneach Seo:* Attachment to Place in Gaelic Literature, *Eire-Ireland*, **26**, 19–31

Nash, C. (1997) Embodied Irishness: gender, sexuality and Irish identity, in Graham, B. (ed.) *In Search of Ireland*, Routledge, London, 108–27

Sheenan, P. (1988) *Genius Fabulae:* the Irish sense of place, *Irish University Review*, **18**, 191–206

Ijsselmeer polders of the Netherlands, for example, rational, modernist landscape testifies to the widely shared utopian belief in central planning in the middle years of the 20th century, consciously ignoring local tradition and custom and seeking an international language of space for a mechanized society.

Resistance to planned landscape since the 1970s has not arrested the process of European landscape change so much as redirected it. While less aggressively nationalistic than earlier in the century, the desire to maintain local and national distinction in the forms of the visible landscape remains powerful throughout Europe. A struggle is thus engaged between the demands of highly capitalized and subsidized agricultural industries for larger fields, monocultures, buildings in modern materials that can efficiently house machinery, and better transport networks on the one hand, and the increasingly political (and even profitable) demand for traditional landscape forms, for 'heritage' and tourism, for environmentally gentle methods of cultivation and land use. Within this, the 'appearance' of the landscape is a powerful weapon, wielded generally on the side of tradition and identity.

The landscapes where European national identities have been most self-consciously articulated are those of the continent's cities. If the domesticated landscape of small-scale farming has traditionally stood for the quiet heart of the nation, and wild nature for its sturdy loins, the capital city is its crowned head. Elements in the landscape of every European capital stand as metonyms for the nation as a whole: the Eiffel Tower, Big Ben, the Colosseum, the statue of the Little Mermaid or that of the urinating cherub, the Brandenburg Gate (Figure 5.5) and the Parthenon. Every capital city calls upon a similar vocabulary of spatial forms to create a landscape of public performance for the nation's rituals: avenues, squares, fountains, statues, monuments and decorated building façades. Commonly these call upon either the architectural language of ancient Greece (see case study of Greece) and Rome, which invested cities during the period of nation building in the 19th century with the aura of European high imperialism, giving historical depth to the cultural references in urban landscape, or upon a supposedly 'national' architectural tradition, as in the gilded onion domes of Habsburg European capitals, or the northern 'gothic' of London's Parliament and law buildings. The landscapes of Europe's capital cities are genuine palimpsests where previous patterns have been partially erased, either by natural processes, as in Lisbon, whose earthquake-wrecked central zone was reconstructed as a severe Enlightenment grid of streets and squares, or by ideological *fiat* as in the boulevards and *places* of Baron Haussman's Paris or Benito Mussolini's Rome. Since the ruins of ancient Rome were first explored by Renaissance scholars, archaeology has served as a science of European nationhood in capital city as well as countryside, exposing past landscapes to be fenced off for public view, or reconstructed to offer visions of former grandeur more in keeping with contemporary ideology. Berlin's Museum Island housing the treasures of ancient Troy and Pergamon, the British Museum, and the many Egyptian obelisks gracing squares, crossroads and river embankments across Europe are all testimony to the power of the past as a language in the contemporary landscape of European capitals. And monumentalism continues today as an element of competition between European cities: in the form of those *grands projets* through which French presidents deem it necessary to record themselves in the landscape of Paris, or of competing office towers through which the financial districts of Frankfurt and London vie for competitive edge (see also Chapter 12).

Figure 5.5 The Brandenburg Gate, Berlin (*source:* Tim Unwin, 3 July 1983).

Mentalités, ideologies and modern Greek, landscape management

Panagiotis Doukellis

Landscape management has become a central point of focus for much basic and applied interdisciplinary research. The ongoing search for new usable space on the part of developed and developing societies alike has raised the urgent question of managing landscapes that have already been greatly downgraded by previous usage. This issue can only be resolved through systematic cooperation between scientists from various fields, including urban planners, geographers and historians. This case study examines the viewpoints concerning sites and landscapes with archaeological interest that have been held by the official Greek state from the mid-19th century to the present.

The establishment of the modern Greek state and the need to substantiate the birth of the nation, mainly for the benefit of other Europeans, led both the Greek intelligentsia and the Greek state to draw arguments from classical antiquity. It had to be demonstrated that the Greeks of the 19th century were directly descended from the Hellenes of the classical period, the progenitors of modern European civilization. Through this claim, the Greeks were entitled to demand recognition of their national identity and the establishment of an independent nation-state. Incontrovertible witnesses to this national identity were, among others, the archaeological monuments scattered over and under Greek soil, whose protection has been a constant concern of the state. However, the various laws enacted have always given precedence to the decisions of the State Archaeological Service. Thus charting any physical policy, determining land use for industrial, agricultural, developmental or other zones, has presupposed the opinions asserted by the state archaeologists. There are, though, exceptions to this rule, particularly when powerful foreign economic interests have obliged a different solution.

The main point to be made here is that until very recently, both among the intelligentsia and at the political decision-making level, it was only the concept and existence of the *monument* itself that was perceived and acknowledged. Gradually the notion of the *surrounding area of a monument* has been added to this, but only recently has the idea of *landscape* been adopted, although not yet at the legislative level. This point takes on particular significance if it is seen over time, within the particular historic and geographic context of the evolving needs and goals of Greek society, and also in terms of results, with respect to aesthetics and the reproduction of new concepts and messages given out by the monuments themselves to both Greeks and foreign visitors.

In this context, two observations concerning the role attributed to antiquities and their surrounding landscapes by the modern Greek state are particularly pertinent. First, from the establishment of the Greek state up to the recent past, the problem has been focused almost exclusively on urban areas. Systematic physical planning interventions have in general been restricted to cities and large communities. The behaviour of the Greek state regarding the remains of past generations is defined, naturally, by their perception as evidence of racial continuity. However, this is related to the dominant thesis, which, until recently, has cut off the monuments, buildings and even the removable finds from the functions and the needs of the era and the society that produced them, and which perceives them exclusively as self-existent works of art, depriving them of their own historicity. As a result, there has been an indifference towards the surrounding landscapes of each monument, and even towards the functional associations of nearby monuments. This theoretical framework is indicative of the long-established urban planning policies which have resulted in the oppressive urban development around archaeological sites. Besides, any alternative could only be exceptional, because of the small plot sizes characteristic of Greece, particularly in urban areas.

Second, the climate has changed only in recent years, and this change has been both quantitative and qualitative, as technological and economic development have created simultaneous opposing

continued

and offsetting tendencies. Physical planning zones are being broadened, a more systematic restructuring policy is being implemented for urban, farm, forest and other zones, and measures are being taken to protect the environment. At the same time, new views of cultural objects are gaining ground. From the monument we have come to the cultural landscape and the need to protect it. New scientific opinions which have ceased to view monuments merely as objects of art and have begun to regard them as cultural objects and witnesses to a past way of life are gradually influencing the attitude of scholars and the state towards cultural heritage. The bearer of this heritage is no longer seen to be the monument-as-art object, but rather the cultural landscape, a clearly defined space-as-bearer of features from the collective memory.

These shifts in ideas and ideologies are connected with the appearance of two new trends in Greek society. First, the development of initiatives and planning of environment-related activities by the Greek state, the European Community and NGOs has had an impact on policy regarding cultural landscapes. Cultural landscapes and archaeological parks in particular are concepts which are now beginning to appear more frequently in texts on physical planning interventions. One of the best examples of this is the proposed unification of the archaeological sites in Athens, a plan which provides for the creation of an uninterrupted zone of green space and monuments in the centre of the Greek capital. Archaeological and other monuments are being divested of the role of national symbols, the space around them is being upgraded, and they are being associated with policies of improving living conditions and the environment.

There are similar proposals for establishing archaeological parks in various other parts of the country, as with the merging of the archaeological sites of Vergina-Dion-Pella in Macedonia, and with the site of ancient Nicopolis in Epirus. These involve considerable reclassifications of land use, and of such things as the existing road network. Such policies aim not only to improve standards of living, and the landscape heritage is at the same time acquiring additional value in an economic sense. The integration of cultural landscapes into the fabric of modern societies, economies and tourist needs, together with the undertaking of basic infrastructure works to attract visitors, all lend a new dimension to cultural goods. It can now be seen that they are exploitable economically as well as politically and nationally.

A second apparent trend is that the ideas concerning environmental/cultural landscapes that have been outlined here acquire particular significance when viewed at the level of regional policy. The upgrading of the role of local government in the past decade and the transfer of jurisdictions from the central government to the regions have created new relationships between local societies and their local histories, and consequently with the monuments and cultural objects of their area. The identity and special features of a region are made known by publicizing their historic monuments and sites. The status of historic symbols may remain, but it is increasingly taking a back seat to social and economic agendas.

Further reading

Biris, K. (1995) *Athens (from the 19th to the 20th Century)*, Melissa, Athens, 2nd edition (in Greek)

DEPOS (1988, 1990) *Land Use Structure and Characteristics of the Built Environment of Large Urban Centres, Volumes I and II*, DEPOS, Athens (in Greek)

Kotsakis, K. (1991) The powerful past: theoretical trends in Greek archaeology, in Hodder, I. (ed.) *Archaeological Theory in Europe*, Routledge, London, 65–90

Provelegios, A. (1974) *The Spirit of the City*, Athens (in Greek)

UN (1996) *UN Conference on Human Settlements, Habitat II: National Report of Greece*, UN, Athens

Memory and landscape in modern Europe

National monuments are the most obvious signs of the past in landscape. They are public inscriptions intended to celebrate and record specific individuals, groups and events. Memory is one of landscape's most consistent and vital functions. Burial of the corpse in the earth is the most common and enduring European funereal tradition, linked to the Christian belief in the eventual resurrection of the uncorrupted body, but the style of graveyards and plot markers is so varied that they are one of the most sensitive indicators of local cultural variation to be found across the continent. Even when modes of rural life and village populations have been wholly transformed by the processes of modernization, traditional cemetery landscapes and forms of inscription will often be maintained relatively unaltered. Thus the highly polished three- or four metre-marble squares of Flemish family graves, gathered shoulder to shoulder along gravel paths against the squat parish church, contrast strongly with the walled necropoli of chambered tombs lying among palms and cypresses outside a Murcian agrotown, or again with the beautiful calligraphic etchings that record on simple black slates ordinary Leicestershire lives among the yews and oaks of a Charnwood churchyard.

But landscape's memorial role goes far beyond the funereal. The historian Simon Schama (1995) has explored the powerful and enduring role of nature myths in the construction of European social identities since the dawn of modern times. We have touched upon many of these in the discussion of landscape and nationalism. Such intimate ties between the material landscape and social identity have to be carefully negotiated in contemporary Europe. Two examples will suffice to make the point. In the local bookshop of any European town or city it is easy to purchase a guide to local flora and fauna. Dating from the interwar years, there remains a tradition of teaching schoolchildren through field visits to recognize such features of their immediate landscape as trees and wild flowers, birds, animals and insects. Recognition is believed to lead to care, protection and a sense of citizenship, cast today more in terms of sustaining ecological balance and environmental heritage than protecting the national landscape, as might have been the rhetoric fifty years ago. Then the threat was felt to come from the invading soldiers of other European nations. Today it comes from rather vaguer outside forces of change: from 'polluters', corporate state, business or European interests, in some cases from immigrant minorities, but always from forces that threaten a sense of balance and harmony between local land and life, remembered from the past and desired for the future, of which the landscape is the physical record (Ashworth and Larkin, 1994; Buttimer, 1994). Here, the root meaning of landscape endures in local communities across Europe.

A second type of book that may be found in such a shop is a collection of reprinted photographs of the local area, including individuals and groups of local people, scenes of daily life and views of buildings and places now lost or unrecognizably altered. Here too the images serve the purpose of sustaining a sense, however superficial and invented, of rootedness in locality measured through landscape memories. Here is the second sense of landscape, as a visual construction of something always outside and beyond ourselves, distanced in space and time from the place we now occupy.

For Europeans, the landscapes occupied today across the continent are recognizably hybrid. This is blatantly apparent in the ruthlessly contemporary and internationalized landscapes of the 'new Europe': airports, motorways, fast food outlets, mass tourist coastal and skiing zones and commercial retail or financial services centres that service the consumer demands of a continent which is fabulously rich by any historical or global geographical measure. Picture in this context EuroDisney, rising incongruously out of the *paysage* of the Île de France so minutely detailed by French geographers, and drawing upon a pastiche of folk narratives, fairy castles and enchanted forest landscapes on which 19th century German Romantics erected the substantive foundations of their national identity. Disneyscape mixes this with American myth, cinematic memory and futurist fantasy. But this may be no more a hybrid than the Chianti landscapes of Tuscany or the English Lake District, conserved by tight legislation so as to appear in the forms that the late 20th century imagination pictures their past: cleared of such historical inconveniences as broken carts, the smell of slaughtered beasts and rutted dirt roads, and recast as heritage. In a savagely ironic text, the French writer Jean Baudrillard (1988) represented American culture and landscape as a simulacrum: the realized image of a reality that never existed. Implicit in his critique was the idea that European cultural landscape by contrast is more authentic, more rooted, more historically real. As the third millennium beckons, that is a difficult claim to sustain. Europe's geographical landscapes are not – and never have been – static pictures of changeless relations between land and life.

Further reading

Béguin, F. (1995) *Le Paysage*, Dominoes-Flammarion, Paris
Cosgrove, D. (1993) *The Palladian Landscape: Geographical Change and its Representations in Sixteenth-Century Italy*, Leicester University Press, Leicester
Cosgrove, D. (1997) *Social Formation and Symbolic Landscape*, 2nd edition, Wisconsin University Press, Madison
Hooson, D. (ed.) (1994) *Geography and National Identity*, Blackwell, Oxford
Schama, S. (1995) *Landscape and Memory*, Alfred Knopf, New York
Warnke, M. (1995) *Political Landscape*, Reaktion, London

PART II Political dimensions

CHAPTER 6

Nations and peoples

NUALA C. JOHNSON

Introduction

On 15 September 1996, Umberto Bossi, leader of the Northern League in Italy, made a declaration of independence and sovereignty for Padania – the name of the northern state that he seeks to create. The name derives from the Padano plain through which the Po river runs on its 650 km journey from the Alps to the Adriatic. Bossi gave the Italian government one year in which to meet his demands for the division of Italy between north and south. The precise geographical boundaries of Padania, however, remained vague. It would stretch broadly from the border in the north of Italy to within some 100 km of Rome and it would comprise a population of about 31 million.

In his discussion of the political geography of the Northern League, Agnew (1995) points out that its political rhetoric has changed geographical scale from claims for an ethnoregional cultural identity to macro-regional economic units (see also Italian case study in Chapter 8). Although the basis of political support for the Northern League may be localistic, the division between local and national interests breaks down (even among parties which are constituted locally) as they invariably become embroiled in the larger spatial context of the national state. While Bossi was rallying support in Venice for his 'declaration of independence', pro-unity demonstrations were also taking place in cities throughout Italy. In Milan, for instance, an estimated crowd of 150,000 people demonstrated in support of unity. The far-right leader Gianfranco Fini claimed that 'Italy is here. It will not be insulted and it will not be divided' (*Guardian*, 18 September 1996). The national president of Italy also warned that Bossi could face criminal charges if he incited illegal acts. In this context, therefore, Padania is as much a place of the imagination as it is a geographically bounded space. It is this conjunction that is crucial to understanding the relationship between Europe's peoples and the nation-state idea.

The changes to the political map of Europe over the past decade underpin the fragility of the relatively stable political boundaries that characterized the Cold War. Not only have the political and cultural complexities of what was eastern Europe begun to emerge since the fall of communism, but in western Europe the fragility of the territorial nation-state has also been challenged; from below through an emphasis on the 'politics of difference', and from above through processes of globalization and a focus on large transnational groupings, such as the European Union. The tensions between geographical units such as the nation-state and the peoples which comprise them, however, is in no way a recent phenomenon. Divisions of the Earth's territory into spatial units such as the parish, the village, the landed estate or the nation-state have had to compete with pre-existing or newly emerging loyalties and political practices.

The division of Europe into a series of discrete territorial 'nations' can be situated in a specific historical context, where one can sketch the contours of the processes involved in promoting nationally based cultures. In this regard, we must be mindful of establishing a European exceptionalism, that is to say suggesting that Europe alone invented the national state and exported it uniformly around the globe. Indeed the role of interrelationships between European peoples and those living in other parts of the Earth in the evolution of the ideology of nationalism should not be underestimated. The influence of European discourses of colonialism and empire had significant effects on the development of non-European independence movements in South America, South Asia and Africa, and indeed on the development of nationalism at home.

This chapter addresses three main issues. First, it highlights some of the principal explanations of

European nationalism, and by examining interpretations of nation-building in France, Britain, Italy and Czechoslovakia it amplifies the problems of finding a single theory of nationalism. Second, it examines the gendered nature of European discourses about nationhood, emphasizing the historical feminization of the ideal of 'nation' and commenting on the gendering of nationalist discourse in contemporary Europe. Third, it addresses the challenges to the nation-state that are present in the debate about European integration.

Theorizing nationalism

> 'A nation is primarily a community, a definite community of people. This community is not racial, nor is it tribal. The modern Italian nation was formed from Romans, Teutons, Etruscans, Greeks, Arabs, and so forth ... Thus a nation is not racial or tribal, but a historically constituted community of people'
>
> (Joseph Stalin, cited in Hutchinson and Smith, 1994: 18)

Like culture, nation is a notoriously difficult term to define; its associated terms nationality and nationalism are similarly enigmatic. The concept of nation has varied historically and geographically. Although deriving from the past participle of the Latin verb *natio*, to be born, and the noun *nationem* connoting breed or 'race' (Connor, 1978), Stalin's quotation above highlights the difficulty in using etymological origins as the basis for understanding the concept of the nation. Max Weber suggests that '... a nation is a community of sentiment which would adequately manifest itself in a state of its own; hence, a nation is a community which normally tends to produce a state of its own' (cited in Gerth and Wright-Mills, 1948: 179). Weber's definition of nation moves beyond a cultural account to a territorial aspiration – a nation seeks or becomes a state. Giddens (1985: 119) extends the territorial element in his definition, by claiming that 'A nation, as I use the term here, only exists when a state has a unified administrative reach over the territory over which its sovereignty is claimed'. Focusing on the administrative and bureaucratic apparatus of the state, Giddens claims that the nation-state is a 'bordered power container' which structures both internal regulation of territory and people, and external relations with other states. While Giddens rightly emphasizes the territorial and military structures around which the nation-state is maintained and contained, his analysis underemphasizes the cultural bases of support for the nation, and the popular allegiance that it wins. In this respect Anderson's (1983: 15) assertion that a nation is an 'imagined political community – imagined as both inherently limited and sovereign' is forceful. He emphasizes that national communities are imagined because even in very small nations not all members will know each other, nor will they ever meet each other. Nevertheless, in their imagination they share a common cultural base. Anderson acknowledges that all sorts of communities (apart from small villages) do not experience face-to-face contact and thus are also imagined. Anderson's (1983: 15) critical observation, therefore, is not to judge the nation by emphasizing its inventedness and thus implicitly seeking to undermine its legitimacy, but to recognize that 'Communities are to be distinguished, not by their falsity/genuineness, but by the style in which they are imagined'. In this context, then, it is a bounded space, with finite, albeit changing, boundaries and a sovereign government, which encapsulates the imagined community of the nation.

Most analysts agree that nationalism, the ideology linking nation to state, is a distinctly modern phenomenon. It has not existed from time immemorial, and the emergence of nationalism can be located in a particular time and in specific places. While some authors focus on the *ideas* underlying a nationalist ideology, others stress the *material* conditions which acted as preconditions for the evolution of the nation-state. Geographers have tended to draw together the ideological, cultural and territorial dimensions of nationalism in their accounts.

Theorists interested in the ideology underlying nationalism focus on the emergence, in the late 18th century, of a notion that human beings are autonomous, free-thinking individuals. With the gradual rejection of religious sources of belief, individuals moved towards political ideals as the basis of their salvation. Influenced especially by the writing of the German thinker Johann Gottfried Herder, who emphasized the natural division of the Earth into separate language groups, a new educated, secular middle-class began to channel their energies into some new centre of cultural and political power. As Kedourie (1960: 102) puts it, 'Such a need [to belong] is normally satisfied by the family, the neighbourhood, the religious community. In the last century and a half such institutions all over the world have had to bear the brunt of violent social and intellectual change, and it is no accident that nationalism was at its most intense where and when such institutions had little resilience and were ill-prepared to withstand the powerful attack to which they became exposed'.

Other theorists have focused on the material changes that were emerging in European society from the late 18th century onward which contributed to nationalist thinking becoming the driving force of political organization. Gellner (1983), for instance, links the rise of nationalism to modernity and the transformation of European society from an agricultural one to an industrial one. The 'tidal wave of modernization' necessitated the homogenization of cultures. This went part and parcel with the emergence of an intelligentsia who sought to accommodate the modernizing forces of industrial society. Marxist thinkers have similarly focused on the material transformations of European society. They link the rise of nationalism with the development of a capitalist political economy. Nairn (1977: 341), for instance, suggests that nationalism is the response of peripheralized people to the uneven development of capitalism, where nationality is employed as a means of defence against core industrial economies – 'Real, uneven development has invariably generated an imperialism of the centre over the periphery ... these peripheric areas have been forced into a profoundly ambivalent reaction against this dominance, seeking at once to resist it and to somehow take over its vital forces for their own use'.

Historians generally locate the emergence of nationalism in the period corresponding to the rise of parliamentary democracy – usually from the French Revolution onwards (see Chapter 7). They have generally focused on the processes involved in state building. This emphasis on the structural changes associated with democratic forms of government, according to Mossé (1975), underestimates the popular bases of support underlying a nationalist politics. He claims that 'Historians have stressed parliamentarianism as being decisive in the political formation of the age ... the study of the growth of a new political style connected with nationalism, mass movements and mass politics has been neglected, not only so far as the nineteenth century is concerned but also as a necessary backgound for fascism' (Mossé, 1975: 3). It is here that Anderson's (1983) account is most helpful. By combining an analysis of the cultural and ideological underpinnings of nationalism through the concept of an imagined community, with an analysis of the decline of dynastic and religious sources of legitimacy and the rise of new ways of recording time through the calendar and the clock, Anderson links together effectively cultural, economic and political processes. The significance of the clock in creating a simultaneity of time across space rests in its capacity to effect common bonds between people who are spatially disconnected. The advent of the printing press similarly had the effect of connecting populations over wide geographical areas. Anderson (1983: 44) claims that 'The coalition between Protestantism and print-capitalism, exploiting cheap, popular editions, quickly created large, new reading publics'. While publishing's initial market was for Latin readers, the saturation of that market had been achieved within 150 years of the invention of the printing press, thus necessitating the printing of books for the mass of monoglot European peoples. The Reformation accelerated this process when Martin Luther had his theses translated into German in 1517 and became the first known best-selling author. According to Anderson (1983), these new print languages were fundamental for the emergence of national consciousness (Figure 6.1), for three reasons. First, they geographically connected speakers of, for example, huge varieties of English, Spanish and German, and

Figure 6.1 A German printing press of 1522 (reprinted by permission of the British Library).

made known to them the existence of those who shared the same language group. Second, print-capitalism fixed language by having the capacity to produce and reproduce the same text over space but also through time, thus linking contemporary generations with past ones. Third, print languages gave priority to one dialect over another and thus awarded the chosen dialect political power.

Anderson's (1983) account provides a persuasive overall framework for locating European nationalism in time and space. Although the ideology of nationhood had a profound effect in creating the contours of European nation-states, nation building itself, that is the homogenization of culture within the boundaries of the state, usually evolved through processes both of consent and of coercion. Not only was the administrative apparatus of the state mobilized but the population forming the nation was acculturated into a set of symbols and traditions that won popular support. Hobsbawm and Ranger (1983) refer to this process as the 'invention of tradition', that is '... a set of practices, normally governed by overtly or tacitly accepted rules and of a ritual or symbolic nature, which seek to inculcate certain values and norms of behaviour by repetition, which automatically implies a continuity with the past' (Hobsbawm, 1983: 1). Examples of these include the introduction of Bastille Day in France in 1880, the public display of national flags, the initiation of national sports (Giro d'Italia 1909) and the development of international competition between nation-states such as the revival of the Olympic Games in 1896.

Making European identities: old nations

The processes involved in the formation of nation-states were by no means always peaceful or consensual. Even in the case of countries that have had a long territorial existence, the particularities of their histories reveal the ways in which creating a common identity inside the boundaries of these states was contested (for a good example of this, see the case study of the Basques). In France, Girardet (cited in Lebovics, 1992: 3) has exposed the myth of a 'kind of geographic predestination of the French nation' – the notion that a virtual France existed prior to the evolution of a historic France. He claims that this myth was being perpetuated by the authors of 19th century school textbooks. The political boundaries of France were relatively modern. Brittany came under the rule of the French crown in 1532; in 1659 the Catalan regions of south-west France became politically French; Flanders was annexed during the Revolution of 1795; Corsica came under French dominion in 1768; Algeria came under French rule in 1830; Savoy and Nice transferred from Italian to French rule in 1860; and Alsace-Lorraine's political status changed hands four times from the Franco-Prussian War until the end of the Second World War. While the territorial definition of France changed through the centuries, the cultural geography of the state also underwent a series of transformations.

By the middle of the 19th century linguistic diversity still prevailed in France in the form of patois, dialect and entirely separate languages, such as Basque and Breton. While peasant society retained its own forms of expression, standard French gradually infiltrated local dialect especially as the written word became more widespread. Weber (1976) highlights some of the mechanisms involved in linguistic transformation. Pointing specifically to the role of roads, railways, schools, commercial centres, the army, education and the circulation of goods and services, Weber charts how regional preoccupations were gradually replaced by national ones and the masses were acculturated. While one does not wholeheartedly have to accept Weber's thesis of transforming 'peasants into Frenchmen', he does illuminate some of the processes involved in nurturing a sense of a French identity. He emphasizes the role played by schools and schooling in the middle of the 19th century in advancing a concept of a united France. Maps of France, for instance, began to be widely distributed by the state after the Franco-Prussian war (1870–1). By the latter decades of the 19th century few classrooms were without a map, and especially in rural areas they contributed to nurturing a sense of being part of a larger whole – the national state.

While Weber deals with 19th century nationalization, Lebovics (1992) examines the ways in which the search for a French identity was articulated from 1900 until after the Second World War. By focusing on the work of anthropologists, folklorists, museum curators and others involved in cultural reproduction, he charts the search for a 'True France' which would transcend regional and local identities. He points out how the concept of nationhood in France was constantly under revision and subject to negotiation as France's internal and external relations changed. He focuses in particular on how the contested region of

The Basques

Roger Collins

The Basques are unique amongst the peoples of western Europe in having enjoyed continuous occupation of a distinct territory without ever once having established any form of unitary state on it (Gómez-Ibañez, 1975). Indeed, the lands they occupy on both faces of the western Pyrenees have not been ruled by a single political authority since the end of the Roman Empire in the fifth century, other than for a 15-year period in the early 11th century (Collins, 1990a). Even at this point the Basque-speaking population was only one component of a short-lived 'empire' that extended beyond their lands, and which was in no sense consciously Basque in either language, culture or political self-expression. This lack of a previous historical phase of political autonomy has not, however, affected their sense of identity as an ethnic group distinct from those around them to both north and south. Nor has it inhibited modern Basque nationalism from taking the form of a struggle, often violent, to create a Basque nation-state (Heiberg, 1989).

The roots of the Basque identity and the factors that have kept it in being across a span of over 2000 years can reduce themselves fairly succinctly to the continuous occupation of a territory and the possession of a distinctive language. The two need to be combined, in that the chronologically and geographically widespread Basque diaspora is marked by the rapid disappearance of the use of the language in all contexts away from the western Pyrenean and Biscayan homelands and the concomitant blending of émigré Basques into the wider cultural and ethnic identities of their new domiciles (Collins, 1990a). The only partial exception to this rule can be seen in Nevada in the USA, where the modern interest in 'roots' and the fractionalizing of the American identity have led to interest in the Basque origins of a significant percentage of the population of the state and even an attempted revival of the language (Douglass and Bilbao, 1975; *Basque Studies Program Newsletter*, 1968 onwards).

For the late 19th century Nationalist movement, which grew out of opposition to the Liberal government in Madrid, whose 1876 Constitution had abolished the region's distinct legal and political rights and institutions, the antiquity of the people was a vital characteristic that marked them off from the inhabitants of the increasingly intrusive Spanish and French states under whose authority they lived (San Sebastián, 1984). In the present century this came to be linked to a supposed genetic purity, distinguishing the Basques from the 'mongrel' nature of the neighbouring populations (Altuna, 1983). More soundly based research has confirmed a distinctive haematology amongst the inhabitants of the western Pyrenees and adjacent areas, and recent study of their DNA has indicated that the occupation of this region by a population genetically related to the present one goes back about 10,000 years (de Mouzon *et al.*, 1980; Sykes *et al.*, 1996). Thus, the earliest detectable ancestors of the modern Basques would seem to have established themselves in the western Pyrenean and east Biscayan area in the late Neolithic, and they have remained there ever since, under the political domination of successive outside powers from the Roman Empire onwards.

The ethnic distinctiveness of this population is confirmed by study of its language. While arguments that would see Basque as once being more widely spoken, at least in the Iberian peninsula and southern Aquitaine in the pre-Roman period, can be neither confirmed nor fully disproved, it is clear that the language has been totally isolated for at least the last 2000 years. The only linguistic parallels that are even remotely close are with some Kartvelian languages in the Caucasus (Collins, 1990a). However, it is not primarily this linguistically exotic nature of the Basque language that makes it a vital determinant of ethnic separateness. Possession of a language different enough to be incomprehensible to members of the neighbouring populations, whom in all other cultural respects the Basques now resemble, is sufficient in itself. It is this survival of a language that has preserved a Basque ethnic identity long enough for it now to be articulated in additional ways, not least in the modern aspirations towards political independence and nationhood, neither of which have any antecedents in the actual history of the people, or were ever sought prior to the late 19th century.

continued

The Basques are in many ways a classic example of an upland or mountain-dwelling people. A pastoral economy, principally related to sheep, has conditioned their relationships with their neighbours and their own internal social organization for centuries, or even millennia (Wickham, 1985; Pallaruelo, 1988). Just as they have produced surplus livestock that could be traded with adjoining lowland regions, so they have produced continuous surpluses in human population greater than the resources of their own territory could support. The history of the Basques, from their earliest appearances in the archaeological and historical record, is essentially one of the various ways in which they as a society have tried to deal with this problem (Collins, 1990b). The recognition that the land itself and their pastoral lifestyle was only capable of supporting a proportion of the population seems to underlie the long-established custom of the non-partibility of inheritances. Within the rural community today, as in the past, the all-important family house, its land and the winter and summer grazing rights in the upland and lowland pastures of the village, are passed to one heir, chosen by family discussion and not by automatic right of primogeniture (Lafourcade, 1983; Granada Television, 1987; Martínez-Montoya, 1995).

While additional family members might be useful at various points in the annual cycle, this kind of pastoralism is far from labour-intensive, and with different households within a village community cooperating to take turns in herding the flocks in the winter and summer pastures, alternative strategies for the maintenance of excess population have always been necessary (Collins, forthcoming). The most obvious solution, that of territorial aggrandisement, was inhibited by the lack of complex political organization and by the greater strength of adjacent lowland societies, at least from the Iron Age onwards. Only in certain limited periods, notably during the repopulation of the upper Duero valley in the later ninth and tenth centuries and in the less well documented northwards expansion into the area that would become Gascony, was it possible for large-scale movements of Basque population into areas contiguous with their own regions to take place. Even in these cases it is notable that the use of the Basque language and sense of a common identity with the Pyrenean core population disappeared within a century or two at the most.

Other strategies included recruitment into the armies of the major powers within whose political frontiers the Basques were located. Seaborne trade and whaling, which could take the form of an oceanic version of the pastoral cycle of different winter and summer activities and locations, were also useful in diversifying Basque economic activity and in spreading the surplus population from the land into new areas or into alternative means of generating wealth (Tuck *et al.*, 1985). It is also not surprising to find that while Basque seamen took an important part in the overseas expansion of the Spanish Empire, this was as nothing compared with the role of the Basques as emigrants and settlers in the new territories in South, Central and North America. Only in the later 19th century did it become possible for the first time for much of the surplus Basque population to be absorbed within their original European heartlands, thanks to the development of labour-intensive heavy industry in the eastern Biscay area. This industrialization of the east Biscay coastal area, which had its origins in a late medieval iron-working tradition in the region, proved initially so successful that it sucked in non-Basque immigrants from other regions of the peninsula, subsequently intensifying nationalist sentiment (Harrison, 1983; Díez de Salazar, 1983).

Modern Basque nationalism, which was in origin a conservative and Romantic movement that aimed at the recovery of traditional Basque institutions, notably the use of the *Fueros*, the medieval laws which had come to be regarded, albeit anachronistically, as the touchstone of Basque liberties, received new impetus from the urban economic decline that set in from the 1920s onwards (Apalegui, 1975; Lafourcade, 1983). It also gained ever-growing popular support from the not only politically but also culturally repressive measures taken against the Basques and their language under the Franco regime of 1939–75. It had been a paradox of the 1930s that the politically and religiously conservative nationalist party in the Basque region, the Partido Nacionalista Vasco (PNV), had at that time entered into alliance with the ultra-Republican government in Madrid because it was only the latter which was willing to contemplate giving the Basques a statute

continued

of regional autonomy (Fusi, 1979, 1984). In consequence the Basque nationalists found themselves on the losing side in the Civil War of 1936–39, and becoming an object of deep suspicion and the subject of repression thereafter. This in turn led directly to the formation of the modern Basque terrorist movement, ETA, with its 'political wing' in the form of the party called Herri Batasuna, and the embracing of far more radical left-wing ideologies by such extreme nationalists (Güell, 1987; Sullivan, 1988; Clark, 1994). The PNV remains the dominant voice of Basque nationalism, with the more limited but economically more realistic objective of the fullest possible regional autonomy rather than political independence as its aim (Ardanza, 1991a, 1991b). Even so, the sense of a common Basque identity continues to outweigh other, perhaps more pragmatic, considerations in popular consciousness and in the policies of the PNV. Thus, in the autumn of 1996 the PNV threatened to bring down the recently formed national coalition government if its demand that all Basque political prisoners were to be kept exclusively in gaols in the Basque region was not met (Jacob, 1994).

Further reading

Collins, R. (1990) *The Basques*, 2nd edition, Blackwell, Oxford

Heiberg, M. (1989) *The Making of the Basque Nation*, Cambridge University Press, Cambridge

Sullivan, J. (1988) *ETA and Basque Nationalism: the Fight for Euskadi 1890–1986*, Routledge, London

Alsace-Lorraine was culturally inscribed as French despite its German-speaking population.

In a British context also, the territorial boundaries of the state emerged earlier than in much of the remainder of Europe. Indeed the foundational trappings of the state were in place by the 17th century. In England at this time, loyalty to the crown and to the state was beginning to be declared. While there may not have been any substantial collective sense of a people as a nation, the extension of centralized control over the territory of the kingdom was leading to nation-building practices. The museum, the school, the art gallery and the landscape were all playing roles as agents in the transformation of Britain, from a multicultural mosaic of peoples, into a quasi-homogeneous national state. Key episodes in the past, such as the Gunpowder Plot, were assuming national importance through their incorporation into a national imagination. Unlike other 'old' states which commemorate their founding revolutions, in Britain there was no national anniversary upon which to focus national consciousness in the same way as, for example, Bastille Day in France (14 July).

In English memory, the discovery of the Gunpowder Plot on 5 November 1605 has been a most enduring symbol. For nearly 400 years the unsuccessful attempt by Guy Fawkes to blow up the Houses of Parliament has been celebrated. While this event may not have been the most important historical moment of the Tudor and Stuart period, it has been the most widespread and systematically remembered episode in the collective imagination. From the outset it was legislated that 'An act of public thanksgiving to Almighty God every year ... to the end this unfeigned thankfulness may never be forgotten' (cited in Cressy, 1992: 71). The discovery of the plot was initially seen as a confirmation of God's covenant with England and an endorsement of Protestantism on the island. The rituals of bell-ringing, bonfires and the burning of effigies of Guy Fawkes (later to be replaced by effigies of the Pope or the 'Devil') were as Cressy (1992: 73) observes '... an act of loyalty as well as piety, with national, dynastic and religious connotations'. Although the meaning of the annual ritual has changed over time it has had the capacity to respond to the political realities of the day, and its influence rests in its popular appeal and its ability to mobilize a national memory. Cressy (1992: 87) puts this in context: 'Over almost 400 years it has been associated with a creative festive tradition, with shifting sponsorship, varying intensity and periodic reinfusions of meaning'. While in the latter stages of the 20th century the 5th of November may have diminished in significance, its role in forging especially a sense of English identity cannot be underestimated. But what of the other groups that made up British society, for instance the Scots, Welsh and Irish? How were they incorporated into a British national culture?

The issue of British identity as a whole received scant historiographical attention until after 1945. Much research has adopted a 'four-nations' model of the state, where British identity is seen to be the result of the interaction of four separate, albeit related, cultures living in the bounded geographical space of two islands (Kearney, 1989). In some respects like France, the geographical boundaries were seen to be coterminous with the cultural ones. But the problems with this analysis are twofold. First, this model underestimates the weakness of some of the actors. Since the 16th century much of Ireland has sat uncomfortably with the Protestant Reformation and thus could not be easily integrated into a British identity which focused on its common dissent from Catholicism. Second, the four-nations model ignores the complex array of regional and social loyalties that preceded and coexisted within the context of an increasingly centralized state. Colley (1995: 65) claims that '... men and women often had double, triple, or even quadruple loyalties, mentally locating themselves, according to the circumstances, in a village, in a particular landscape, in a region, and in one or even two countries'. While the four-nations model does highlight internal differences and at times points to the hegemonic influence of England in nurturing an identity across the two islands, it omits to place British identity within the broader context of an emerging European nationalism and an expanding overseas empire.

By the 18th century British identity was being forged through a common Protestantism (with the exception of Ireland) against a mainly Catholic Europe, through wars with European neighbours, particularly France, and through the acquisition of an overseas empire. By 1820 over 26% of the world's population was under British dominion (Bayly, 1989). After American independence most of that empire lay in the east: a population which did not speak English, were non-Christian, were not 'white' and who lived in a physical environment regarded as exotic. The treatment of colonial subjects as different from Britons in the world's largest empire enabled the people of the two islands to develop a common identity – 'Britons could join together *vis-à-vis* the empire and act out the flattering parts of heroic conqueror, human judge and civilising agent' (Colley, 1995: 74). The unifying effect of this process did not completely eliminate internal differences (especially with respect to Ireland) but it did provide a strong foundation for the cultivation of common bonds of identity. In the late 20th century, with the loss of the empire, some of these older fissures have re-emerged and the contested nature of contemporary British identity around issues of 'race', ethnicity and gender in particular are resulting in the renegotiation of national space.

Making European identities: new nations

Newer national identities, formed over the last 100 years, reveal even greater strains in the meshing of territorial and cultural aspirations. The path to securing a unified territorial unit along the Italian peninsula was fraught with difficulties. The combined strategies of consent and coercion eventually led to the unification of Italy by the 1870s. After unification the prime minister of Piedmont observed 'We have made Italy: now we have to make Italians' (cited in Seton-Watson, 1977: 107). The 'making of Italians' proved to be a difficult task, one in which linguistic and regional difference persisted. Even Mazzini, a firm supporter of unification, '... wanted Italy to be "made from below" through popular struggle' (cited in Agnew, 1996a: 37). Consequently the regional tensions that have haunted Italy from the outset may account for the persistence of animosities between north and south and the rise of the Northern League alluded to at the beginning of this chapter.

Since the cataclysmic events of the late 1980s, the political map of Europe has undergone immense revision including the reunification of Germany, the break-up of the Soviet Union and the renegotiation of Balkan national spaces (see case study of Slovenia). This turmoil has witnessed the resurrection of ethno-regional conflict, particularly in central Europe. The break-up of Czechoslovakia indicates the precarious basis upon which national territories can sometimes exist. Created after the First World War in 1918, Czechoslovakia emerged from the remnants of a collapsing Austro-Hungarian empire. Comprising Czechs, Slovaks and a variety of numerically smaller ethnic groups, the initial proposal for the new state was a loose confederation of the different cultural groupings. While the Slovaks enjoyed greater cultural autonomy under the new regime than they had under Magyar rule, Kaiser (1995: 213) points out that '... Czechoslovakia was constructed primarily by Czech political élites as a much more unitary state controlled by Czechs. This caused rising reactive national separatism among the Slovak and

Slovenia

Felicita Medved

Slovenia emerged as a separate state on the European political map in June 1991, when it unilaterally declared its independence from the Yugoslav socialist federation. Since then, it has achieved international recognition, has avoided the wars waged by its former partners in the Balkans, and has become one of the most prosperous states to emerge from former communist Europe.

For centuries, Slovenes have lived in an area of considerable geopolitical significance. Their homeland is the meeting point of the Alpine, Pannonian, Dinaric and Mediterranean worlds, and also of four major European cultures: Germanic, Romance, Slavic and Finno-Ugric. Language was crucial to the development of Slovene self-awareness, and the first Slovene written records date back to the tenth century. Most historians argue that the Slovene people emerged during the Great Migrations between AD 400 and 600. At the end of the sixth century, Slavs began to settle the valleys of the Sava, Drava and Mura rivers and under pressure from the Avars reached the shores of the Adriatic and Black Sea, the Danube and Lake Balaton, where they formed a series of medieval states. The first of these was Carantania, with its core in the present-day Austrian Carinthia. Although Carantania soon became a part of the Frankish empire, it later took on considerable mythical value in the period of intense European nation-building in the 19th century, when it provided a territorialization of collective memory for aspirations of Slovene state independence.

Between the 13th and the 16th century, almost all Slovene lands came under the rule of the Habsburgs. Slovenes were divided between six political jurisdictions including the city of Trieste, with some Slovenes also living across the Mura river in the Hungarian part of the empire, and yet others inside Venetian Friulia. Each of these areas gradually formed a distinct cultural, linguistic–dialectic and economic-interest unit. Moreover, the cleavages between the 'Carniolian', 'Styrian' and 'Carinthian' Slovenes were further heightened because Slovenes fell under separate ecclesiastical jurisdictions. With the exception of the short Napoleonic intervention, this situation remained essentially the same until 1918.

In spite of these divisions, foreign rulers were unable to impose their language on the peasantry. Imperial rule was imposed through taxation and conscription, but there was no real sense of national identity among the mass of the population. People identified locally, with their village and parish, and it was primarily the Church that linked local society with the wider world. The process towards an all-Slovene identity was neither fast nor smooth. However, once a language is written it is preserved, and once printed it can create a new kind of power. The Slovene language, which survived amongst the peasantry, gained considerable cultural and political significance during the Reformation. The industriousness of the Protestant reformers produced over fifty books in the Slovene language, and although Protestantism was suppressed, the early seeds of a national movement were laid.

The development of a 'high' Slovene language, which could provide a formal basis for the emergence of a national consciousness, had to await the introduction of compulsory school attendance in the 19th century. The French Revolution and the short Napoleonic regime (1809–13) united parts of the Slovene and Croatian lands into the 'Illyrian provinces', each of which adopted the colloquial language of the peasants as a legitimate vehicle for official communication and documentation. Subsequently, Slovene national identity emerged through resistance towards German, Italian and later pan-Slavic cultural imperialism. The movement started with a tiny group of intellectuals who had access to education and financial support. They recognized in their *ethnie* (Smith, 1986) an embryo of the Slovene nation. Influenced by Romanticism and the enthusiasm of the Enlightenment, they started building what in Herder's terms could be called *Kulturnation* (Medved, 1995). Eventually, by the second half of the 19th century, the concept of a separate Slovene language had gained broad support, and was able to provide a unifying linguistic and social force.

When revolution came in 1848, Slovenes were more a nation than ever before. The first political programme for a 'Unified Slovenia' in 1848 called

continued

for unification of fragmented territory into one crownland within the empire and the introduction of the Slovene language in schools and administration. However, despite their involvement in the economy, Slovenes were deprived of opportunities for higher public office as well as higher education, both of which were monopolized by those with a German cultural background. During the second half of the century, increasing spatial integration occurred as a result of the improvements in land communication resulting from the construction of the railways. Exchanges of ideas, goods and people were greatly facilitated. Likewise, the growth of popular movements, libraries, theatres, reading clubs and sports societies, all of which stimulated a sense of national identity, could not be ignored by the imperial authorities. The political parties could also for the first time be organized on a national basis, and all Slovene parties stood for Trialism, a transformation of the dual Austro-Hungarian monarchy into a trialistic entity with Slavs as equal partners.

Slovenes were incorporated into the state of Yugoslavia only in the aftermath of the First World War. The South Slavs of the Austro-Hungarian empire had proclaimed independence in October 1918, while Italy was already occupying the western Slovene lands assigned to it by the secret London Treaty of 1915. In panic, this self-proclaimed state asked Serbia for military assistance and within two months joined the Kingdom of Serbs, Croats and Slovenes (SHS), subsequently named Yugoslavia. Within Yugoslavia, the Serbian centralist view of the state at once came into conflict with the Croat and Slovene federal views. Although the drive towards total Yugoslav integration became the guiding principle of the dictatorship, Yugoslavia was to a large extent beneficial for Slovenes. Slovenia was by far the most industrialized part and became one administrative unit in 1931.

Following the Second World War, communist power and peace were imposed at a high price. The national problem in the new Yugoslavia was to be solved partly through the federal state system, and partly through a minority policy which gave smaller groups substantial rights. From the 1974 constitution, the six republics and the two autonomous provinces functioned virtually as independent states. According to the post-war peace accords, the Slovenian republic gained areas in the west, including a coastal strip along the Adriatic Sea, and Slovenes were quite successfully integrated into the Yugoslav political system.

During the 1980s, the democratization process gave a specific and important political role to intellectuals: taboos were opened, the history of the Second World War was questioned, demands emerged for transition from the one-party system to multi-party parliamentary democracy, and solutions to questions of nationalism were sought. Liberalization and orientation towards the market economy, which were the immediate consequence of the reforms, however, soon led to very different views concerning the future of the country. These focused primarily on the varying character of national ideologies. Slovenians saw the Serb attempts to change the constitutional arrangements and the electoral system as a threat to the republic's survival. Moreover, as the most prosperous of the republics, Slovenes were interested in a different economic agenda.

Defence of the charismatic national language became a matter of honour and political prestige. One of the key events in the emergence of an independent Slovenia was the military trial of the so-called 'Slovene Four' in Ljubljana in 1988. The 'Slovene Four', accused of handling a secret military document, insisted on but were denied a trial held in Slovene as opposed to Serbo-Croat, which was the language of the Yugoslav military forces. In these circumstances, the Slovene language became a symbol of political and national oppression, as well as a symbol of liberty and human rights. The trial became conceptualized in terms of the 'militarization of the Yugoslav state' and of the 'civil society' of Slovenia. More and more nationalists, from artists and university professors to high-ranking communists, voiced a case for an independent Slovenia, whose 'natural' place was in the heart of central Europe as opposed to its peripheral location in an artificial and alien context of Yugoslavia. In 1989 Slovenia proclaimed a new constitution and the collapse of communist regimes across eastern Europe removed the last rationale for even a socialist regime. At the end of

continued

1990 a referendum was held on 'independence and autonomy', and negotiations concerning the status of Slovenia within the federation failed to produce an acceptable compromise.

The creation of a Slovene national identity has always taken place within the context of larger political units. The Slovenian tradition has been to use the medium of a distinctive language to create sub-state level institutions that would stand between the state and the individual. Although their precise form has varied over time, they have always acted as the focus of common action and identity. It is precisely this political culture that made Slovene nationalism so strong in post-war Yugoslavia. In the most recent period, the 'new' generation of Slovene political and professional élites created a movement which sought to integrate the social and economic struggle with nationalist aims. They had no 'ethnic' territorial claims, but 'nationalized' the territory of the post-war republic. They defined citizenry in territorial terms, thereby asserting jurisdiction and authority evenly throughout territory and population. They extended protection and promotion rights for Italian and Hungarian national communities, including guaranteed seats in the elected local bodies in the border regions and in the republican parliament. People from other parts of Yugoslavia who had immigrated to Slovenia in the post-war period and who constituted around 12% of the population were given an option to apply for Slovenian citizenship. All of these constitutional and legal measures, however, are not enough to guarantee a democratic state. Neither do they mean that nationalism has ended. While Slovenia has emerged as an independent state in the 1990s, aspirations to join the European Union by the end of the millennium are now creating a whole new series of questions over national identity and the meaning of being a Slovene.

Further reading

Gosar, A. (ed.) (1996) *Slovenia, A Gateway to Central Europe*, The Association of the Geographical Societies of Slovenia, Ljubljana

Medved, F. (1995) A path towards the cartography of Slovene national identity, *Razprave in gradivo (Documents and Treatises)*, Institute for Ethnic Studies, Ljubljana, 177–210

Nationalities Papers (1993) Voices From the Slovene Nation (1990–1992), *Nationalities Papers*, Special Issue, **21**(1)

Ruthenian nationalists'. While this élite had hoped to assimilate the Slovaks through a nation-building programme – Czechoslovakism – for many Slovaks this translated into a Czechization of Slovak cultural identity. Although Slovaks would have favoured a dual polity in a federal-style state, the unitary policies of the Czech leaders contributed to the fuelling of a separatist politics and to the movement in Slovakian education and language policy to mobilize and maintain a separate sense of Slovak nationhood. Added to these tensions was the existence of other minority groups including Germans and Magyars, geographically concentrated in border regions adjacent to their 'homelands', where irredentist nationalism emerged in the interwar years.

After the bitter struggles of the Second World War, the national space of Czechoslovakia was reconfigured. Lands in the east were lost to the Soviet Union and the internal ethnic diversity was all but eliminated. If in the first decades of the state the Slovaks and Czechs remained suspicious of each other, after the war nearly all minority groups had been effectively removed. The majority of Jews had been exterminated in the death camps, gypsies were similarly treated, Germans were expelled and the Magyar and Polish populations dramatically declined. The state, *de facto*, evolved into two regions, each occupied almost exclusively by Czechs or Slovaks.

In the years since 1945 Slovak separatism has persisted as a political goal. On 1 January 1993, with the end of the Cold War, independence was declared, creating the Czech Republic and Slovakia. The relatively easy path to partition has resulted in this process being labelled the 'Velvet Divorce', that is to say, a no-fault separation. Nevertheless, opinion polls taken just prior to partition indicated that neither a majority of Slovaks nor of Czechs desired complete separation. The declaration of independence, in turn, has redefined national space and has necessitated a new phase of nation building – the reassertion of new national communities and the need to integrate cultural minorities into each of the new states. Kaiser (1995: 232) warns, however, that 'Each state (particularly Slovakia) is being restructured around the idealised image of the nation state envisioned by nationalists, and this leaves little room for the accommodation of ethnic "others"'. Indeed, the fate of 'Others' throughout Europe at the turn of the millennium remains unclear as the national state is challenged from above by the European Union and from below by regionalist movements.

Gender and the construction of national space

This chapter has so far considered the evolution of European nationalism mainly from the point of view of ethnic identities. However, the role of gendered identities has also been highly significant in forging the national state (see also Chapter 17). Analyses of nationalism are often presented as gender-neutral discourses, where the desire to create a national imagined community disguises other sources of identity such as gender or class or 'race'. The imperatives for creating a unified voice frequently obfuscate other voices, especially those of women. The focus on ethnic identity around issues of language, tradition and historical landscape has tended to underestimate the role of gender in the articulation of nationalist politics. Nevertheless, as Marina Warner (1985) has persuasively documented in her study of the allegory of the female form, women are often seen to embody the ideals and aspirations of the nation. Pictorial representations of the European nation have been encapsulated through female allegorical figures. The image of Marianne in France, found in statues and paintings, uses the female form to represent Liberty, the Republic and France. While the meaning attached to Marianne has changed over time and in space (Agulhon, 1981), the picturing of the nation as female has remained fairly uncontested. In Eugène Delacroix's famous painting *Liberty leading the people*, the path to freedom in revolutionary France is led by a scantily clad female waving the tricolour (Figure 6.2). She is not a specific, named woman – she is an idealized woman representing the nation.

In Britain, too, the allegory of Britannia derives not from a historically constituted woman but from a mythological figure – Athena, Goddess of War. This image of Britain has enjoyed popularity, especially during the height of empire and periodically during Margaret Thatcher's premiership. Warner (1985: xxi) suggests that one of the reasons that the female form has represented concepts such as liberty, justice or the national community is '... the common relation of abstract nouns of virtue to feminine gender in Indo-European languages'. While a linguistic analysis offers some explanation for the feminization of national iconography, in addition this can be set alongside an assessment of the material circumstances in which such gendering occurs: 'Often the recognition of a difference between the symbolic order, inhabited by ideal, allegorical figures ... depends on the unlikelihood of women practising the concepts they represent' (Warner, 1985: 3). At the height of European nation building, towards the end of the 19th century, a discourse of national self-determination can be contrasted with the lack of voting and property inheritance rights for women. Even though suffragists were working to secure those rights, they sometimes competed with the ideals of a nationalist politics.

The following account of the relationship between women and the state in the construction of national citizenship in part illuminates the ways in which nation building practices have constructed men and women differently. Women's participation in the development of ethnic and national consciousness has taken several specific forms. First, women have served as the biological reproducers of members of the nation. Encouragement to reproduce often uses religious and nationalistic arguments about women's duty to motherhood. Second, women can be seen to have reproduced the boundaries of a national group. Inter-ethnic marriage may be discouraged so as not to dilute the 'purity' of the nation and penalties for such behaviour can be imposed by the state. Extreme forms of this racialized discourse can be seen in Germany under the Nazi regime. Third, women have played a role in the defining of the nation by acting as the ideological purveyors of the nation's culture through their role as transmitters of that culture. Women, as the main socializers of small children, may thus acculturate the offspring into the specified ethnic identity, and may also act as conveyors of the rich heritage of ethnic symbols associated with the nation. Fourth, the ideological construction of the nation around ideas of femininity has been embodied through women. The nation as a mother-

Figure 6.2 'Liberty leading the people' by Eugène Delacroix, 1830 (*source:* Musée de Louvre, Paris, reproduced by permission).

land to be protected from rape by a hostile group has figured large, particularly in national liberation struggles, where men are asked to fight in defence of actual women and metaphorically in defence of a gendered land. Ironically, too, of course, the gendering of land as virginal, open for conquest in European colonial discourse, contrasts with the protective imagery used with respect to the national homeland. Finally, recent research has highlighted the very active role played by women in physical combat. They often play this role as makers of munitions, nurses and carers for the injured but also as actual combatants in guerrilla warfare.

In her discussion of gender and nationhood in eastern Europe, Sharp (1996) points out how women's reproductive role has re-entered the public discourse since the demise of socialist and communist rule in the 1980s. Abortion, for instance, was outlawed in Croatia in 1992 by the nationalist party for fear that the stability of the Croatian nation would be threatened by a falling birthrate. In the Bosnia–Herzegovina war the systematic rape of minority women by the Bosnian Serb army underscores the relationships made between the female body and the nation, both literally and metaphorically. In eastern Europe under communism, while women's right to work outside the home was guaranteed by the state, this practice did not necessarily represent the liberation of these women so desired by western liberal feminists. Many held ambivalent views about their role in the workplace under the strict regulatory regime of a command economy.

Nor did their participation in waged labour release them from the responsibilities of home. Since the fall of communism, nationalist movements have held more reticent views about the role of women as wage earners because this form of emancipation carries with it memories of the rhetoric of the older communist regimes. Consequently, ideas about nationhood are being renegotiated and in some instances women have been reassigned traditional roles of wife and home-maker. The precise lineaments of the relationship between gender and the resurgence of ethnic nationalism in the newly formed spaces of eastern Europe have yet to be fully developed, although preliminary evidence suggests that these issues will be hotly contested.

Challenges to the nation-state: the European Union

If eastern Europe is experiencing the effects of the collapse of communism, in western Europe some of the challenges to the state system are being channelled through efforts to achieve greater European integration and through the ascent of regionalism. While the inspiration for the creation of the European Economic Community can be traced back to post-war reconstruction, more recent developments have focused on 'deepening and widening' the Union. Efforts to achieve integration through the development of the Single European Act (1986) and the Maastricht Treaty (1991) reflect the desire to unify the peoples of Europe. Together these innovations in European Union (EU) policy seek to strengthen economic union through the creation of a common currency; to redefine senses of European citizenship; to extend EU responsibility for a suite of social, environmental, health and cultural policies; to deepen the powers of the European Parliament and to develop a more coherent common defence and foreign policy (Jones and Budd, 1994).

For the European system of nation-states these processes have led to the evolution of at least two opposing views of the effect of integration on the efficacy of the current state system. On the one hand it is suggested that the processes of European integration will strengthen individual state power by evolving a system of European-wide cooperation on central issues such as the restructuring of agriculture and industry in a global economy, while at the same time providing external support for weaker states within the Union. This view of integration regards it as a form of intergovernmental cooperation rather than supranational control. In this interpretation the power of the nation-state is not threatened, but rather it is enhanced by the focus on common interests and problems. An alternative view of integration, however, suggests that it is eroding the sovereignty of the national polity by weakening the authority of national governments to legislate and control internal affairs (Keating, 1995). Underlying this view also is the notion that integration will dissolve the cultural and political diversity of European peoples.

While both views represent ideal types, the processes of European integration coupled with the rise of regionalism (both economic and political) will have significant consequences for reconceptualizing the modern nation-state at the end of the 20th century. Balibar (1991: 17) suggests that 'All the conditions are therefore present [in Europe] for a sense of *identity panic* to be produced and maintained. For individuals fear the state ... but they fear still more its disappearance and decomposition'. While the EU may be moving towards greater administrative and economic interdependence, the issue of nurturing a collective cultural identity remains problematical. Early indicators suggest that efforts to mobilize a pan-European identity have important consequences for minorities, most especially non-European migrants. Harris (1995) notes that western European states since the 1970s are increasingly seeking to deter non-European immigration and to encourage repatriation of existing immigrant groups within Europe. Changing definitions of citizenship have accompanied changing European attitudes towards immigration as the new world labour market collides with the conventional parameters of identity around which the national state was built.

Current debates about European integration need to address these issues of identity head-on so that differences as well as common interests are reflected in any new attempts to reconfigure the cultural as well as the political map of Europe. The 'rebirth of history' that is taking place in some parts of central and eastern Europe, as well as the regional conflicts that have persisted in western Europe, all speak of the contested and at times contradictory identities that have marked the making of Europe's peoples and nations.

Acknowledgments

Many thanks to David Livingstone for his helpful comments on an earlier draft of this chapter.

Further reading

Anderson, B. (1983) *Imagined Communities: Reflections on the Origin and Spread of Nationalism*, Verso, London

Hutchinson, J. and Smith, A.D. (eds) (1994) *Nationalism*, Oxford University Press, Oxford

Jones, B. and Keating, M. (eds) (1995) *The European Union and the Regions*, Clarendon, Oxford

O'Dea, M. and Whelan, K. (eds) (1995) *Nations and Nationalisms: France, Britain, Ireland and the Eighteenth-Century Context*, Voltaire Foundation, London

Warner, M. (1985) *Monuments and Maidens: the Allegory of the Female Form*, Picador, London

CHAPTER 7 A revolutionary idea

TIM UNWIN

The political and economic changes in the former Soviet Union and eastern Europe during the early 1990s have widely been interpreted as being of revolutionary character (Callinicos, 1991; Lieven, 1994). They were, though, very different from the more common usage of the term in the 20th century to refer to the overthrow of capitalist governments by Marxist revolutionaries. The word 'revolution' has thus been used to describe a whole range of economic as well as social and political changes, from the so-called 'industrial revolution' of the 19th century to the Russian Revolution of 1917; the most famous revolution of all was the French Revolution beginning in 1789 (Schama, 1989). In all of these examples, the key meaning of revolution is that of rapid and significant change, but there is enormous debate over precisely how fast and how significant such change must be to be called revolutionary (Tilly, 1993). This chapter explores the way in which this idea of 'revolution' has emerged within Europe. In particular, it builds on the suggestion that the *idea* of revolution itself has been used to serve specific social, political and economic ends.

Understanding the revolutionary idea is central to our understanding of Europe. This was emphasized as long ago as 1923, when Elton (1971: v) suggested that 'The revolutionary idea as it works itself out ... through the history of France from 1789 to 1871 ... is the key not only to modern France but to modern Europe'. Writing soon after the Russian Revolution, Elton (1971: v) recognized the crucial point that at the heart of the revolutionary idea is a revolutionary method. The conscious development of such revolutionary practice was a key feature of 19th century European politics, reaching its culmination in the writings of Marx and Engels (1975a) towards the end of the century, and in the events surrounding the Russian Revolution of 1917. However, the word revolution itself was in wide circulation long before the end of the 18th century.

The idea of a revolution

The 20th century use of the word 'revolution' by social and political scientists has come a long way from its original meaning in medieval times, when it was used mainly in an astronomical sense to refer to the orbit of celestial bodies. Central to such a conceptualization was the idea of return to a particular point, or the completion of a full circuit or cycle. It was not until the 15th century that the word 'revolution' began to be used to refer to great changes in political affairs, and it was not really until the 17th century that this use became particularly widespread. Speck (1988: 1) thus comments that 'In the seventeenth century the word "revolution" did not have the significance which it has acquired since 1789. When Englishmen used it to describe the events of 1688 and 1689 they did not mean by it the violent overthrow of authority, nor the transfer of power from one class to another. Rather it was employed in the sense of the revolution of a wheel turning round to a former state'. While many people viewed the accession of William of Orange as a revolution in this sense, returning England to Protestantism and parliamentary power after James II's shift to Catholicism and absolutist rule, this was by no means universal. Hill (1980, 1986), for example, emphasizes that many contemporaries also used the word 'revolution' in a linear sense to refer to dramatic changes in the course of events (see also Israel, 1991).

This 17th century shift in the meaning of the word revolution coincided with the rise of philosophical liberalism and empiricism, expressed particularly in the work of John Locke (1632–1704). It is thus no coincidence that Locke's (1967) *Two Treatises of Government*, in which he argues that a ruling body should be deposed if it offends against natural law, were first published in 1689. Locke's ideas provided one of the foundation stones of the 18th century

Enlightenment, with their practical expression culminating in the dramatic political upheavals in France at the end of the century. For Braudel (1985: 504) 'The whole of the revolutionary ideology of the Enlightenment ... was directed against the privileges of a leisured aristocratic class, defending by contrast, in the name of progress, the active population – including merchants, manufacturers and reforming landowners' (Braudel, 1985: 504).

The French Revolution is generally acknowledged not only as the defining moment of revolution in Europe, but also as one of the key watersheds in European history (see, for example, den Boer, 1993: 65). In one sense it was the culmination of the liberal Enlightenment, but at the same time it presaged the shape that future revolutions would follow. More importantly for the present purposes it provided a focus for contemplation and reflection, which then polarized intellectual thought and practical action. On the one hand there were those, following Marx and Engels, who saw it as the fulfilment of capitalism's overthrow of feudalism, an analysis of which could provide the basis for the eventual replacement of the modern bourgeois state by socialism. On the other hand, there were those in the wake of Edmund Burke (1968), whose *Reflections on the Revolution in France* was first published in 1790, who saw the French Revolution as challenging every element of stable society and ordered rule, and who would thereafter do their utmost to ensure that the basic institutions and structures of capitalism would survive. The French Revolution was not, though, as organized and as coherent a set of events as subsequent analysis has often tended to suggest. Schama (1989: xiv) has thus recently commented that the Revolution does not 'seem any longer to conform, to a grand historical design, preordained by inexorable forces of social change. Instead it seems a thing of contingencies and unforeseen consequences'.

Born in 1818, Marx was heavily influenced by the revolutionary fervour of the 19th century; he was also one of its greatest publicists and proponents. Through his writings, the *idea* of revolution was to be turned into the formal *practice* of revolution. At his burial in Highgate Cemetery, Engels thus described his close friend and accomplice as 'before all else a revolutionist' (Tucker, 1970: 3). Marx (1960, 1963) not only provided detailed commentaries and critiques of the events surrounding him, but he also projected his revolutionary ideas backwards in order to understand the events that had given rise to capitalism. His views on revolution can be traced to his earliest writings, and many of the key arguments of *Capital* (Marx, 1976), first published in 1867, can be found outlined in the *Manifesto of the Communist Party* (Marx and Engels, 1975a), which was finalized in 1848, the year of the major revolutionary uprisings in France, Italy and Germany. The *Manifesto* asserts that 'the modern bourgeoisie is itself the product of a long course of development, of a series of revolutions in the modes of production and of exchange' (Marx and Engels, 1975a: 43). Capitalism, for Marx, was revolutionary, in that it was the product of the resolution of the internal contradictions of feudalism. Subsequent Marxist writers have thus generally referred to the period when Cromwell and the Parliamentary forces overthrew the Crown in England and then ruled in the Commonwealth from 1648 to 1660 as the English Revolution, seeing in it the violent seizure of power by the bourgeoisie (Hill, 1980; Speck, 1988). In this light, too, the French revolutions of the late 18th and 19th centuries have been interpreted as attempts by the bourgeoisie to replace the increasingly repressed contradictions of the feudal and absolutist state.

For Turok (1980: 5), the central idea of the Marxist theory of revolution 'is that a social formation contains within itself certain limits on its development beyond which it must either transform itself or it will burst asunder through revolution'. Revolution is above all else, in theory, a transfer of state power from one class to another. Tilly (1993: 9) likewise suggests that a revolution is 'a forcible transfer of power over a state in the course of which at least two distinct blocs of contenders make incompatible claims to control the state, and some significant portion of the population subject to the state's jurisdiction acquiesces in the claims of each bloc'. In attempting to identify the key characteristics of revolutions, Turok (1980) has highlighted eight features which they have in common: a rapid and abrupt change of political system; the collapse or overthrow of the government; violence and/or illegality on the part of revolutionists; insurrection; mass mobilization around a revolutionary party or movement; aspirations towards ideals such as liberty, equality, fraternity and national liberation; a high degree of dedication to the cause by the revolutionists; and a developed strategy and tactics. To these, though, must be added two further features. First, revolutions are essentially internal changes of political system and government within a specific state. While they can be fomented by external agency, they must have the sustained support of the majority of the population if they are to achieve lasting success. Revolutions are thus a type of civil war. However, second, they differ from many civil wars in that they involve a fundamental change in the power relations between classes. While civil wars can include conflicts between different ethnic groups or sections of a ruling élite, revolutions replace such élites. Moreover, to be termed revolutions, they must also be successful, if only for a relatively short period of time. Another key feature of revolutions emphasized by Turok (1980) is that they are usually associated with a high degree of violence. The replacement of the old élite is thus usually a bloody affair, involving widespread cruelty, torture and death for those in all groups in society, including both children and the elderly. Invariably, such actions are a direct response to the violence meted out by the previous ruling élite before its fall.

Above all, revolutions arise from a failure in the legitimation apparatus of government (Habermas, 1976). In this light, the events of the late 1980s and early 1990s in the former Soviet Union and eastern Europe can be seen quite clearly as a failure of the minority ruling élite of the Soviet apparatus to legitimate its control of power. Ironically, though, this brings us back to the earlier astronomical definition of revolution: capitalism has come full circle and shown itself able to replace the communist mode of production that was meant, according to Marxist theory, to have been its successor (Habermas, 1990, 1994b).

Political protest within the feudal state

Before the 17th and 18th centuries, there were numerous attempts by groups of people within Europe to revolt or rebel against authority. However, none of these achieved the status and success of being termed revolutions. As Braudel (1985: 495) has so eloquently noted, 'To rebel was "to spit in the sky": the *jacquerie* of 1358 in the Île-de-France; the English Peasants' Revolt of 1381; the *Bauenkrieg* of 1525; the salt-tax rebellion by the communes of the Guyenne in 1548; the violent Bolotnikov rising in Russia at the beginning of the seventeenth century; the Dosza insurrection in Hungary in 1614; the great peasant war which shook the kingdom of Naples in 1647 – all these furious outbursts regularly failed'. In a primarily agrarian society, Braudel (1985) suggests that one of the main reasons for this failure was that labour was dispersed, not only in its scattered agrarian form, but also in its urban expression, where the workforce was divided into small competing units. Moreover, he also notes that Europe's ruling élites, while crushing peasant revolt and rebellion, usually did respond to sufficient of their demands to defuse subsequent protest. Thus, while the peasant uprising of 1358 in the Île de France provoked by the burden of taxation and corvées was crushed by the nobility, it did secure the liberty of the peasantry in the Paris region, and although the Guyenne rising was crushed, the salt tax that had been one of its causes was abolished (Braudel, 1985).

Bercé (1987) emphasizes another important factor limiting the success of medieval revolts. This was that their causes and energies were usually derived from local situations. In his words 'The local community was the fundamental bond, the first resort in cases of confrontation, and the most potent source of outbursts of collective violence' (Bercé, 1987: viii). The state was thus usually able to contain local unrest, by suppressing it with outsiders from other parts of its polity. Central to an understanding of the emergence of more widespread revolt and revolution is thus the relationship between the central state and local communities. Under feudalism, there was effectively a fragmentation of the powers of the state: the Crown granted out its rights in land in return for services, initially mainly military service (Bloch, 1965). It was therefore difficult for the basic producers of society, the unfree villeins, to unite and rise up against the central state. Their first focus of protest was their individual local lord, and in an age when communication was difficult and slow they had little contact with peasants in other parts of the state. In any revolt, however, the lords with their wider spheres of contact and influence could call upon their fellow knights to help to suppress the unrest. Furthermore, throughout the medieval period, the basic order of society was continually reinforced by the preaching of the Church. Again in Braudel's (1985: 493) words, 'The state was there to preserve inequality, the cornerstone of the social order. Culture and its spokesmen were generally on hand to preach resignation to one's lot, obedience and good behaviour'. Religion served to legitimate the inequalities of the social order (see Chapter 4).

During the 15th century, however, the central state increased its power throughout much of Europe, and in so doing, it 'shattered all previous formations and institutions' (Braudel, 1985: 515). In trying to enforce a monopoly on the use of force, in tightening its control on the economy, and in maximizing its legitimation from religious ideology, it provided the context for increasingly widespread outbreaks of violence from the 16th century onwards. Two classic examples of these tendencies can be seen in the Peasants' War in Germany in 1525 and in the responses of the peasantry in France and Spain to increased taxation and repression resulting from the Thirty Years' War (1618–48) (Bercé, 1987). Both examples nevertheless ended in defeat for the peasantry and a reassertion of the power of the central government.

Seventeenth century revolution

The examples of revolt and rebellion so far cited were not revolutionary in that they failed to achieve three things: they did not overthrow the central state, the peasantry failed to gain power and transform the basic social order, and they were local rather than national in character. During the 17th century, however, the power of the old social order was successfully challenged for the first time, and fundamental political and economic changes associated with the replacement of feudalism by capitalist relations of production occurred in England and the Netherlands. Central to these changes were the awakening of new ideas concerning the organization of society and the relationships between ruler and ruled that emerged following the Reformation and which reached their fruition in the Enlightenment of the 18th century.

Feudalism was never uniform in its expression across Europe; its collapse was likewise highly variable in pace and character. While capitalist relations of production emerged quite rapidly in the Netherlands and England, they were much slower to develop in Iberia and France (Aston and Philpin, 1985; Wallerstein, 1974, 1980). It is this very slowness, and unwillingness of the French ruling élite to adapt, that can be seen as having been one of the key factors giving rise to the violence and rapidity of the social and political changes associated with the revolutions of 1789 and thereafter in France.

During the first half of the 16th century the Low Countries formed part of the extensive Habsburg Empire. However, in 1556 Phillip II acceded to the Spanish throne, and sought to reunite the Catholic world under Spanish leadership. In seeking to impose Spanish religious and political control over the Low Countries, Phillip II provoked considerable unrest. The precise causes of the subsequent Dutch revolt have been much debated, but a conjuncture of four key elements occurring at a particular place and time was crucial to the events which followed: the spread of Calvinism, new ideas about the rights of representative assemblies, republicanism, and political freedom (Kossman, 1991). Although the initial revolt was bloodily repressed and it took almost a century before the independence of the Netherlands was eventually recognized by Spain at the Peace of The Hague in 1648, the emergence of new systems of economic production alongside demands for religious and political reform in the late 16th century heralded the dawn of a new era.

The final emergence of the Netherlands as an independent republic coincided with Civil War in England. As in the Low Countries, the English Civil War was provoked by a combination of antagonism towards absolutist tendencies of the Crown and the growing strength of Calvinism. Increasing conflict between the Crown and Parliament, leading to Charles I governing without recourse to Parliament between 1629 and 1640, coincided with the rise of Puritanism, which emphasized individualism and hostility towards Catholicism. Closely associated with these political and ideological conflicts, though, was the increasing economic affluence of both gentry and townspeople, based on trade and on the emergence of new systems of agrarian production. Hostility developed into armed conflict in 1642, culminating in the execution of Charles I in 1649 and the rule of the Commonwealth until 1660.

The comprehensive character of the changes in mid-17th century England and the Netherlands have led many observers to describe them as revolutions. Bercé (1987: 99) has thus emphasized that 'In the two prime cases of England and Holland, what was at stake was the total conquest of power, even if the full extent of the objective only surfaced little by little, by force of circumstances'. In the English case, moreover, the argument is strengthened by those such as Hill (1961, 1972) who see the Civil War as the final resolution of the conflicts within feudalism and the emergence of the bourgeoisie as the ruling élite. The crucial economic significance of these political and ideological changes was that they provided the opportunity for a transformation of institutional structures which were to benefit those wishing to gain individual profit from speculative investment (Wallerstein, 1979, 1980). Indeed, as Hill (1972: 11) has stressed, 'the long-term consequences of the Revolution were all to the advantage of the gentry and merchants, not of the lower fifty percent of the population'.

The collapse of the Commonwealth, and the Restoration of Charles II in 1660, led to the re-emergence of a royal court in England. However, the gains of the Civil War for the rule of Parliament were not entirely lost. As Hill (1961: 222) has commented, 'The Restoration of 1660 was a restoration of the united class whom Parliament represented, even more than of the King'. Subsequently, the attempts of Charles II (1660–85) to impose French-style absolutist rule, and James II's (1685–88) reassertion of Catholicism, led to renewed tensions between Parliament and the Crown. These provided the context for an Anglo-Dutch Protestant political and religious alliance, which saw the Dutch Prince William of Orange (Charles I's grandson, and married to Mary, James II's elder daughter) replace James II as King of England in the so-called Glorious Revolution of 1688 (Miller, 1983). Although bloodless, the events of 1688–89 can be seen as revolutionary in two senses. On the one hand, they reflected the old medieval meaning of revolution, in that they returned the country to the political rule of Parliament (Speck, 1988); on the other, they were revolutionary in that they represented the culmination of the tumultuous events set in motion by the Civil War, finally replacing absolutism by a constitutional monarchy. Furthermore, the Glorious Revolution had a profound influence on revolutionary movements elsewhere in Europe and the colonies. In Europe, as Israel (1991: 31) has stressed, the Anglo-Dutch alliance 'created a real balance of power', whereas previously 'the continent had been virtually prostrate before the political and military supremacy of France'. It also highlighted fundamental questions about the extent to which religious differences should be allowed to influence international politics, and in France it raised the spectre of a parallel revolution to throw off the tyranny of Louis XIV.

Revolution in the 19th century

The flowering of the Enlightenment during the 18th century, with its emphasis on reason, religious tolerance, freedom of thought, and rationality, was closely associated with the new social, economic and political order of the emergent capitalist states of England and the Netherlands. Moreover, it provided an increasing focus for criticism of the absolutist regime of Louis XV (1715–74), as expressed for example in the works of Montesquieu, Voltaire and Rousseau. In France, given the decadence of the Ancien Régime, the parlous state of the country's finances, the continued social disadvantages of the peasantry, and foreign policy failures, it is remarkable that the absolutist state was not challenged by a major revolution before the end of the 18th century. To be sure there were anti-seigneurial uprisings, and opposition to the imposition of various taxes, but the overthrow of the old regime in France occurred almost a century and a half after it did in England.

The American War of Independence (1775–83) provided the spark that ignited revolution across Europe in the 19th century. With the overthrow of English rule, a new type of constitution was established in the United States of America, based on personal liberty, popular sovereignty and political equality. Despite the internal problems of the French regime during the 18th century, it still remained one of Europe's main political forces. By the 1780s, though, the French political system was in a state of crisis, and revolutionary unrest expressed itself in the siege of the Bastille in 1789, the declaration of the Republic in 1792, the execution of Louis XVI in 1793, and the violent bloodshed of 1793–94 (see the case study) (Soboul, 1988; Doyle, 1989; Schama, 1989; Williams, 1989; Townson, 1990).

While narrative histories of the French Revolution have sought to impose some historical logic on the events that took place in the late 18th and early 19th centuries, it still remains difficult to interpret the particular pattern of occurrences that made up the Revolution. Current opinion tends to impute causative power to the local as against the national, and to the individual as against the structure (Schama, 1989). Moreover, Townson (1990: 5) asserts that 'The idea that the Revolution was a class war of bourgeoisie versus aristocrats has had to be discarded, as has the idea that the Revolution led to the dominance of the capitalist bourgeoisie and the development of capitalism'. Nevertheless, by the time Napoleon was crowned as Emperor in 1804, France was a very different place from that which it had been in 1788, and while the bourgeoisie had undoubtedly benefited, so too had the mass of the peasantry (Townson, 1990).

Revolutions in France and Europe, c.1750–c.1900

Pamela Pilbeam

The 1789 Revolution, despite its worldwide reverberations, was an avoidable accident of political ineptitude. Three strands interwove to produce a revolutionary situation. The government of France, the most populous and wealthy state in Europe, had a cashflow problem as a consequence of expensive wars and the absence of a simple, uniform system of collecting revenue. Second, plans for reform polarized opinion. Enlightened writers, such as Montesquieu, suggested that the complexities of the modern state demanded uniform, centralized, rational systems. Successive royal ministers since before the death of Louis XV in 1774 had oscillated between inventing new taxes and attempting radical reform of the very complex privatized fiscal system. Their projects stiffened the determination of the wealthy to have a say in both the size of their contribution and how it was spent. France was in principle centralized, but in reality it was a mass of corporate interests. The resistance of the most powerful of these, the court of appeal, the Paris *parlement*, scuppered royal reform. An attempt in 1787 to consult an Assembly of Notables led to their public defiance of the crown and the decision to call an Estates-General, an elected body which had not met since 1614. Finally an economic crisis, rumbling on since the mid-1770s, with successive poor harvests and financial, commercial and industrial problems, created the unrest which made the argument among the rich open revolt.

When the Estates-General met in May 1789, Louis XVI soon found that his hopes that the assembled clerics, nobles and Third Estate (basically middle classes) would vote for a new system of taxation and gratefully disappear, were ill-judged. They declared themselves a National Assembly and set to work to write a constitution which would make France a limited monarchy (see Table 7.1). Reform escalated into revolution, fed by discontented artisans in conflict with the government over economic problems, with the successful assault on the Bastille fortress, 14 July 1789. Revolution did not stop with the completion of the written constitution in September 1791, because the Assembly's confiscation of Catholic Church lands (the Church was the biggest landowner in the country) and determination to make the clergy paid state servants, stimulated opposition. This counter-revolution produced prolonged civil war in western and southern France. Many nobles and eventually the king openly resisted plans for change, and finally the declaration of war on the Holy Roman Empire in April 1792 crystallized conflicts and led to the declaration of a republic and the trial and execution of the king.

Between 1793 and 1815 the French slithered from the democratic constitution of the Convention (though far from democratic practice of the Committee of Public Safety and Robespierre), to the narrow oligarchy of the Directory, and finally to the open dictatorship of the most successful and best-publicized general, Napoleon Bonaparte, in 1799. The revolutionaries may have written a Declaration of the Rights of Man and Citizen, abolished feudal dues and proclaimed liberty and equality, but their attempts to settle France's political future were a failure as were their projects for religion and education. On the other hand, and ironically given their hopes for decentralization, the revolutionaries set the model for the modern centralized state, with new, uniform institutions, administrative, judicial, fiscal and financial, and codes of law.

French ideas and institutions were given a European dimension by the wars which lasted until Napoleon's defeat in 1815. France acquired a European empire larger than any since Charlemagne. Large areas were absorbed into France, others ruled by Napoleon's friends or relatives, and yet others made vassal states. French institutions were adopted by new local ruling élites and often aspects were retained, including the administrative system and codes of law, after Napoleon's defeat.

In 1815 Europe was superficially restored to pre-revolutionary systems and rulers, but there were repeated outbreaks of revolution in the early 1820s, 1830–35, 1847–51 and 1870–71 (Table 7.1). Why? French instability was a factor, with repeated conflict between rival ruling interests and

continued

a lack of confidence in the ability of systems to endure. However, it was the fragility of economic rather than political structures which did most to create upheaval.

Fundamentally, until the 1860s and 1870s, when Russia and America began to send in surplus grain, the growing European population could not feed itself in times of naturally recurring harvest failure. Political upheaval coincided with prolonged dearth every ten years or so. Starvation was restricted to specific areas, such as Ireland, but many endured shortage, high prices and fear. Rural and urban areas experienced bread riots, forced grain sales and demonstrations against indirect taxes, particularly on wine, tobacco and salt. The downside of the abolition of feudal institutions was that associated common land was gobbled up, depriving the poor of the cushion of common ownership of land for grazing and timber for fuel and construction, a problem fundamental for the poor from Scotland to Sicily.

The industrial sector also suffered a dual crisis, depriving artisans of their independence. It was not a matter of being pushed into a factory; factory workers were not rebels – they had jobs. Expensive new machines, like the Jacquard silk loom (1808), which cost 1000 francs, could be operated in artisan workshops, but could only be acquired by borrowing from silk merchants, who thus became proto-capitalists. There were repeated protests over the use of cheaper labour, women, children and foreigners. In the cotton industry, spinning-machine factories developed more than a generation before parallel technical change in weaving. A brief period of prosperity for independent weavers was sharply followed by the

Table 7.1 Revolutions in France and Europe, *c.*1780–1900: chronological table.

1788	Sieyès, *What is the Third Estate?*
1789	Declaration of the Rights of Man and Citizen (August)
1792	Declaration of War of Liberation of the Peoples of Europe by French State
1792–1814	Expansion of France to Empire of 80 million
1814–15	Defeat of Napoleon; Treaties of Paris
1814	Louis XVIII constitutional king
1814	German Confederation founded under Austrian rule
1820–22	Liberal revolts in Italian states; later in Greece
1827–32	Economic crisis; harvest failures; commercial/industrial recession
1830	July Revolution in France; Louis-Philippe, duke of Orleans, king; revolts in Italian, German, Belgian, Polish lands
1831	Silk workers rising in Lyon (November)
1834	Silk workers rising in Lyon (April)
1839	Blanqui conspiracy in France (May)
1845–48	Economic crisis; harvest failures; commercial/industrial recession
1848	Revolution in France, also Italy, Germany, Austria (February)
	Second Republic in France
	Republic in Venice and Rome; constitutional monarchy in Piedmont
	Frankfurt Assembly writes German constitution
	Radical monarchy in Prussia
	Louis-Napoleon elected president (December)
1849	Frankfurt Assembly closed; Prussian monarchy creates constitutional monarchy
	Italy: constitutional monarchy in Piedmont survives
1850s	Prussia: Frederick-William in conflict with liberals in *landtag*
1851	French: *coup d'état* Louis-Napoleon; popular risings in southern French
1852	French: Louis-Napoleon emperor
1855	Russia: Alexander II, reforming tsar
1858	William regent in Prussia replaces Frederick-William
1859–60	War: France and Piedmont against Austria
	Risings in Italian states
1860	United Italy: unitary state; Victor Emmanuel king
1861	Russia: abolition of serfdom
1862	Bismarck chief minister in Prussia; *landtag* ignored
1866	War: Austria against Prussia; Austria excluded from Germany
1867	North German Confederation under Prussia
1870–71	War: North German Confederation against France; Louis-Napoleon defeated
	Third Republic in France
	Peace of Frankfurt – France loses Alsace-Lorraine
	Paris Commune (March–May)
1905	Revolution in Russia

continued

development of factory weaving. Textile factories employed less-skilled workers, particularly women and children. Cyclical financial and accompanying commercial crises brought growing numbers of bankruptcies and industrial setbacks, with unemployment, underemployment, wage cuts and short-time working.

Urban artisans reacted most radically, but the problems were countrywide; outside a tiny number of leading cities like Paris, industry was growing fastest where labour was cheapest, in rural areas. Town workers were likely to be more literate, reading and writing newspapers and hearing socialist and other radical reformers. They organized themselves into mutual aid and other societies to try to defend their livelihoods. Towns, especially capital cities, were thus the centres of revolt. They were also the hub of increasingly centralized government and most rapid economic growth.

Nineteenth century revolutions were thus above all products of geography. The populations of capital cities grew rapidly, without much expansion of living space. Increasing proportions of inhabitants were migrants, pushed into miserable lodgings in the old artisan quarters and living on the margins. Barricades were easy to erect and defend in the narrow streets (think of the Marais district of central Paris) and hard to defeat. Central artisan districts were cheek by jowl with the heart of government and the press – a volatile cocktail indeed! The consequent problems of public order were not solved until after the 1848 revolutions (Table 7.1).

What changed after 1848 was first that the construction of railroads allowed governments to reinforce both food supplies and garrisons quickly, instead of allowing soldiers to become dispirited and desert to the other side of the barricades – often initially tempted more by food, drink and women than revolution! Within a generation, food imports took the edge off repeated shortages. Urban rebuilding, ostensibly to accommodate railways, allowed troops to manouevre and made barricade construction and defence redundant. Armies were professionalized to make them more reliable, and civil militias, a legacy of 1789 who often had a taste for the wrong side of a barricade, were gradually abandoned.

The fate of the final revolutionary outburst of the century, the Paris Commune (March–May 1871), illustrates the success of governments in learning to control upheaval. The rising, wrongly revered for many years as the first socialist revolution, was a protest, redolent with memories of earlier revolutions, against the uncaring exiled central government by Parisians who had endured a five-month siege by the Prussians, only to find the government agreeing to a Prussian victory march through the undefeated capital. The violent final week of fighting in which 25,000 communards perished, many summarily shot on capture, revealed an army resolute in the defence of the centralized state.

The 1789 Revolution was important in focusing radical thinking and a revolutionary tradition, which is not wholly gone today, but in the 19th century specific economic and social problems were fundamental to insurrection – but revolutions were not the 'class war' beloved by Marx.

Further reading

Doyle, W. (1989) *Oxford History of the French Revolution*, Oxford University Press, Oxford

Pilbeam, P.M. (1994) *The 1830 Revolution in France*, Macmillan, Basingstoke

Pilbeam, P.M. (1995) *Themes in European History 1780–1830*, Routledge, London

Tombs, R. (1996) *France 1814–1914*, Longman, Harlow

But not all of the peasantry were content with the events of 1789–93, and the violent uprisings in the Vendée between 1793 and 1833, in which perhaps as many as 600,000 people were killed (Bercé, 1987), were an expression of the opposition to the new political and economic system that the French Revolution had initiated. The immediate cause of the Vendée revolt was opposition to the state's new conscription policy, but it was also royalist and Catholic in origin. Moreover, it was fundamentally opposed to the economic balance of power possessed by the towns. As Bercé (1987: 205) has described, 'Starting as marches on towns or as resistance to urban power, the Vendée wars were the bloodiest illustration of the irreducible opposition between town and country ... If the suspicious and envious hostility of the peasantry towards townspeople is a commonplace, the crushing of the rebellious provinces was the result of an opposing fury – the middle classes' hatred of the peasantry – which espoused and worsened the totalitarian logic of the revolutionary ideology'.

Just as those who lived through the English Civil War had to wait some 40 years before the events of the Glorious Revolution, so too was the first outburst of the French Revolution followed by a period of further 'revolutionary' change in France, including the July revolution of 1830, which brought Louis-Philippe to the throne (Pilbeam, 1991) and the European-wide uprisings of 1848–51 (Sperber, 1994). While these events, together with the Paris Commune of 1871, have often been seen by Marxists as inexorable stages in the process by which the working classes were to overthrow the bourgeoisie, this view is now no longer so readily accepted (see Pilbeam, 1991).

In February 1848, Louis-Philippe was in turn overthrown during the revolution commencing in Paris, which initiated a year of revolutionary uprisings in Germany, Prussia, Austria and Italy. While the precipitating factors of the 1848 revolutions were the harvest failures of 1845–46 and the recession of 1847, these events were also influenced by the longer-term rising taxation burdens imposed by European governments on their populations, and the emergence of political opposition movements following the French Revolution of 1789 (Sperber, 1994). Across Europe, with the exception of Britain and Russia, riots and demonstrations took place throughout 1848, involving the erection of barricades, land occupations, strikes, boycotts of seigneurial obligations, and violent assaults on those seen as being in positions of power. The prime aims of these political movements were to achieve the final overthrow of remaining absolutist and feudal institutions, to guarantee basic civil rights, to create representative institutions, and to recognize equality before the law (Sperber, 1994).

However, none of the attempted revolutions of 1848 in the long term achieved the dramatic overthrow of political regime that they had intended. In part, this reflected dissension within the ranks of the revolutionaries, and fear among the revolutionary leadership, many of whom were middle-class, of exactly what they were unleashing. In part, too, it reflected the lessons learnt by moderates and conservatives in the sixty years since the French Revolution of 1789, for all of the attempted revolutions of 1848 were eventually thwarted by armies that remained loyal to the king or emperor. Nevertheless, in seeking to retain their overall power, European rulers did accede to some of the revolutionaries' demands. Thus serfdom was indeed abolished in central and eastern Europe, and constitutional governments were also eventually introduced, although, as in Austria, often not until the 1860s.

Russian revolutions

In Russia, the Tsar's army had been mobilized against any possible outbreaks of revolution during the 1848 uprisings elsewhere in Europe, and during the remainder of the 19th century the government ruthlessly sought to prevent any such subversive activity. However, it failed fully to suppress the emerging revolutionary movement, and the crisis of the 1914–18 war provided the opportunity for the Bolsheviks to seize power in 1917. While Alexander I (1801–25) at the beginning of the 19th century had sought to introduce liberal reforms in Russia, the Decembrist uprising of 1825 showed that these had not gone far enough to satisfy the demands of sections of the nobility and intellectuals who had increasingly been influenced by German idealism and French social reform movements. The ruthless suppression of the Decembrists and the introduction of a strong internal security system under Nicholas I (1825–55) prevented further serious outbreaks of unrest, but at the cost of increasing social and political tension. Alexander II (1855–81) on acceding to the throne therefore sought to introduce further reforms, including the abolition of serfdom in 1861 and local government reforms in the 1860s and 1870s. However, these failed to alleviate the situation, and widespread peasant poverty associated with a rising industrial workforce created the conditions in which underground opposition movements could flourish. Many of these built on the anarchist ideals of individual autonomy propounded by Bakunin (1814–76), and the populist movement increasingly called for violent revolutionary action (see case study). Following Alexander II's murder, the ever more absolutist and autocratic styles of government adopted by Alexander III (1881–94) and Nicholas II (1894–1917) exacerbated the already highly volatile situation.

By the end of the 19th century a number of oppositional groupings had emerged in Russia, despite tight official censorship and the prohibition of formal political activity. Thus the first Socialist Democrat Workers' Party congress was held in Minsk in 1898, and following the arrest of most of the delegates immediately thereafter, the key revolutionary activists were deported to Siberia, whence many fled abroad. In 1902, the Social Revolutionary Party was founded as a specifically revolutionary party espousing terrorist violence (Wood, 1986; Williams, 1987), and then in 1903 the second congress of the Russian Social Democrat Party was held in Brussels and London. This led to a split in the party between the Mensheviks led by Julius Tsederbaum Martov, who believed in 'a long period of bourgeois liberal government before socialism could be established' (Williams, 1987: 103) and the Bolsheviks, headed by Vladimir Ilyich Ulyanov (Lenin), who advocated an immediate proletarian revolution led by a small professional party.

The émigré revolutionaries, though, played little part in the first Russian revolution of 1905. The war with Japan, which started in 1904, precipitated unrest, culminating in the shooting of demonstrators in St Petersburg in January 1905, known as Bloody Sunday. This was followed by a general strike, revolts and mutinies across Russia, which were violently suppressed. The Tsar was forced to make concessions, promising a fuller constitution with an elected assembly, the Duma, and initiating agrarian reforms. The outbreak of the 1914–18 war then brought Russia into alliance with Britain and France against Germany and Austria-Hungary, causing enormous internal strains within the Russian economy. By 1917, rapid inflation, food shortages and continuing war losses had led to considerable disillusionment with the government, and rising political unrest. A demonstration in memory of Bloody Sunday was followed by the arrest of Bolsheviks and further protests, but the events of the ten days between 23 February and 4 March caught most people by surprise. In Williams' (1987: 8) words, 'Ten days of popular demonstrations, political manoeuvring and army mutiny developed imperceptibly into a revolution which no one expected, planned or controlled' (see also Katkov, 1967).

The abdication of the Tsar and the establishment of a provisional government under Prince Lvov on 2 March initiated a period of great uncertainty (Acton, 1990). Following Lenin's return from exile in Switzerland on 3 April, he strongly argued against cooperation with the provisional government, advocating instead the nationalization of all land, the cessation of the war, and that all power should be given to the workers' committees, the soviets. Although Lenin's ideas were initially regarded as a minority view by most socialists, the failures of the provisional government over the summer provided the context for the Bolsheviks to seize power in October (Luxembourg, 1961). Rapidly, under Lenin's leadership, the Bolsheviks were able to consolidate their power, in part as a result of the rigorous organization of their party mechanisms, but also through their ruthless application of a policy of terror and execution. Opposition by White Russian military forces loyal to the Tsar was defeated by the Red Army during the Civil War between 1918 and 1921.

Meanwhile, in March 1919, some 50 delegates from various parts of Europe, as well as from China, Korea and Persia, meeting in Moscow set up the Third Communist International, to encourage international support for the beleaguered communist government, and as a first step towards Lenin's aspiration of the global victory of communism (Dukes, 1979). By the time of the Comintern's second congress in July 1920, around 200 delegates from 35 countries attended. The key outcome of this congress was the delegates' acceptance of Lenin's 21 conditions for membership, which emphasized iron discipline, adherence to the concept of the dictatorship of the proletariat, and the need for communist parties to model themselves on the Russian exemplar. The congress also denounced all reformists as class enemies, and advocated insurgency wherever and whenever possible, particularly in the colonies of the imperial powers. The idea of revolution, in the 19th century largely a European concept, was now to be put into practice across the world.

European revolution in the 20th century

In the aftermath of the devastation caused in Europe by the 1914–18 war, reaction to the communist revolution in Russia was pronounced. The war itself had caused fundamental rifts within European socialism, although Lenin and the Bolsheviks had laid a clear claim to international leadership through the success of their revolution. The possibilities of revolution spreading to other countries struck fear into the minds of most European governments.

The clock-making Jura: cradle of anarchism[1]

Antoine S. Bailly

To destroy the state and the capitalist economy

In order to understand the relationship between anarchism and the Jura mountains in Switzerland, it is first necessary to grasp the essential meaning of anarchism. It is then possible to understand its appropriateness to a milieu in which the economy was based on the artisan production of clocks in the middle of the 19th century.

The underlying idea of anarchism is the flourishing of the free person, and in order to achieve this objective it fights against all forms of constraint: those of the state, of religion, of the army, of capitalism and of private property. To gain this end, anarchism proposes a global federative organization based on the commune, an organization very similar to that of the clock-making milieu.

If this movement had its origins in the anti-state doctrines of Bakunin, it was well-known in the Jura, where syndical and political meetings were held as early as 1869, only shortly after the foundation of the first International Workers' Association in London in 1864, and its congresses held in Geneva in 1866 and Lausanne in 1867. The reciprocal influence between the anarchist ideology and an artisan milieu, conscious of its own interests in the face of capitalism and of the industrial revolution, explains the rapid rise of the movement in the Jura. The history of anarchism must thus be situated in a precise socio-economic context, namely that of the social relations of production of the clock-making milieu in the Jura mountains.

The Jura, 'cradle of anarchism'

The Jura clock maker, battered in the middle of the 19th century by various conjunctural crises, was very attentive to internationalist discourses, particularly because this region of Switzerland welcomed numerous political exiles thanks to its intellectual freedom of expression. In this context, the role of the Reclus brothers, of Guesde, of Kropotkin and of Montels must not be forgotten, alongside that of the Jura syndicalists in the growing strength of anarchism.

To relate the socio-economic situation of the Jura to this new ideological system, it is useful to recall the basic organization of the artisan clock-making industry. This was characterized in 1865 by domestic labour in small workshops concentrated in establishments belonging to the workers and divided into precise tasks. One can speak of a threefold division: specialization and technical division of labour; freedom of labour in the framework of corporations; and individual ownership of the means of production. It is therefore scarcely surprising that the industrial revolution, with its factories belonging to proprietors, was perceived as a threat to the corporative regime and a potential crisis for the artisan way of life. The inhabitants of the Jura, who were highly cultured thanks to their freedom, were very conscious of their originality, particularly their individualistic view of the world. The ideas of resistance to capitalism, to the state, and to factory discipline, thus found a very welcome milieu which favoured the development of anarcho-syndicalism. In this context, the analogy between the mountain people of the clock-making milieu and the farmers of the south of Spain, exploited at the end of the 19th century by the large landowners, and preoccupied with the modernization of the means of production, should be noted. The anarchist influence found its most favourable milieu in the Jura and in Andalucía.

The strikes in the summer of 1869, to obtain a reduction in the working day to less than 12 hours and to conserve the 'Holy Monday' day off in order to recover from the exhaustion of Sunday, without any lowering of salaries, confirmed the use of a bottom-up corporative organization in the fight against the emergence of an increasingly capitalist clock industry. This 'Holy Monday' was also called 'Blue Monday' to illustrate the role of absinthe (the blue fairy) in Jura society. The consumption of too much absinthe on Sunday gave people such a headache that they were unable to work the next day, and the phrase 'Holy Monday'

[1] Translation by Tim Unwin

continued

was used to compare it with the 'Holy Sunday'. The defence of labour freedom in the face of mechanization rapidly enhanced the activity of the anarchist workers' movement; the fear of degradation in their social relations of production explains the political commitment of numerous clock makers in favour of a free society in which each would be able to flourish and likewise the collectivity which enabled the labour of each to be enriched.

Birth and breakdown of the Jura Federation

The birth of the Jura Federation on 12 November 1871, in direct opposition to the centralization of the International Workers' Association established by Karl Marx, provided originality to the movement and attracted numerous socialist revolutionaries. In this way, anarchism developed from its corporatist conception towards a desire for the destruction of the state; strikes were thus used to contest the authorities rather than for the defence of the profession. Out of this there emerged in numerous written texts a conception of a destructive and freeing anarchism. The collectivity of the mountains suckled a truly anarchist mystique under the influence of Bakunin. With Brousse, escaped from the commune, and the geographer Reclus, there developed a will to replace authoritarian bourgeois society, founded on the profit-seeking motive, with a fraternal and collective society of cooperation.

Between 1870 and 1920 an intense and original political life reigned in the Jura mountains, established on the idea of governments decentralized to the level of the commune, where goods were held in common. The European federalists found there a source of reflection, but the dream of a perfect society, without an economic base adapted to the world of change, was utopian. The mountain people, poorly armed for the political fight, particularly against capitalism, were battered by the crisis and restructuring of the clock-making industry. For Lenin, on a visit to the Jura in March 1917, the disappearance of the Jura Federation was linked to its inability to seize hold of the government, the banks, the army, and the communication routes. The political fight had changed its face at the end of the 19th century, just as had the social relations of production. The end of the artisan clock maker was also the end of Jura anarchism.

Further reading

Nussbaum, J.M. (1951) Histoire d'un mouvement ouvrier, *Cahiers Suisses*, **3–4**, 83–99

Vuilleumier, M. (1988) *Horlogers de l'Anarchisme*, Payot, Lausanne

After their successes in Russia, communist parties across Europe sought to incite revolution. Germany was widely seen as being the most likely place for success, given the impact of the war and the condition of the proletariat. However, attempted uprisings by the German communists were suppressed by force, and the emerging Weimar regime was able to resist violent opposition from both the left and the right (Dukes, 1979). In part, this failure reflected the influential role in the German Communist Party of Rosa Luxembourg (1961), who had been opposed to Lenin's centralist views, as well as his and Trotsky's willingness to use violence and terror. Organizational weaknesses in the German Communist Party, and its lack of support through the country as a whole, meant that the *Freikorps*, a newly levied force loyal to the Social Democrat Party, was able to suppress the attempted revolution. Elsewhere in Europe, in Italy, France, the Netherlands, Scandinavia and Iberia, smaller-scale revolutionary uprisings were likewise unsuccessful (Dukes, 1979). However, growing nationalist sentiments, an appeal to religious legitimation, and a fear of communism, led to an increasingly violent right-wing backlash. This was reflected in the rise to power in Italy of the National Fascist Party formed in 1921, the growing strength in Germany of the National Socialist German Workers' Party (the Nazis) during the 1930s, and the emergence in Spain and Portugal of similar right-wing governments during the 1930s (Woolf, 1968).

In Spain, the victory of the Left Republicans with their socialist allies in the 1931 elections led to the establishment of a republic supported by both the liberal bourgeoisie and the workers (Browne, 1983). Although the right regained power in 1933, their rule was short lived, and a combined socialist, communist and republican Popular Front was re-elected in 1936. The subsequent attempted labour reforms by the Socialists 'necessarily implied a redistribution of wealth ... (and) constituted a challenge to the existing balance of social and economic power' (Preston, 1994: 1). Opposition to these reforms, fuelled by claims that the leaders of the Republic were anti-religious, and supported by army officers who were furious at attempts to do away with their privileged position, provided the context within which Francisco Franco led a military uprising in 1936, thus initiating the bloody civil war that was to last until 1939 (Trotsky, 1973; Carr, 1982; Browne, 1983; Preston, 1984). The Spanish Civil War is of crucial significance for a study of revolution in Europe for two main reasons. First, it represented a right-wing, fascist reaction to a generally peaceful acquisition of power by the left. Although the Republic was castigated by Trotsky (1973), for example, as being insufficiently revolutionary, it had nevertheless brought a left-wing coalition to power by relatively peaceful methods. It thus showed that socialism, if not necessarily communism, did not always have to gain power by violent revolution. Second, though, the left ultimately failed, in that the counter-revolutionary forces of the Falange and Franco, militarily supported by Germany and Italy, were to achieve victory after three years of bitter fighting. Moreover, this victory was gained despite support for the Republic from the Soviet Union and the involvement of some 60,000 overseas volunteers who formed the International Brigades (Hemingway, 1955).

European fascism was to suffer a severe setback in the wake of German and Italian defeat in the 1939–45 war and the establishment of liberal democracies throughout most of western Europe. However, in Spain and Portugal, both neutral during the war, right-wing regimes remained in government until the 1970s. In Spain, despite increasing social unrest in the late 1960s, Franco retained power until his death in 1975, when his nominated successor Juan Carlos was peacefully crowned as king. In Portugal, the Corporatist New State proclaimed by António de Oliveira Salazar in 1933 was characterized by extensive state management of the economy, tight restrictions on labour movements, prohibition of political parties, strict censorship and a secret political police force. Here a backward rural economy, increasing difficulty in fighting independence movements in its colonies in the 1960s, and growing internal dissent, provided the background against

Figure 7.1 Political wall painting encouraging support for the MRPP (*Movimento Reorganizativo do Partido do Proletariado*), Portalegre, Alentejo, Portugal (*source:* Tim Unwin, 20 July 1981).

which the Armed Forces Movement, supported by the underground Communist and Socialist parties, seized power in a bloodless revolution on 25 April 1974 (Figueiredo, 1975; Mailer, 1977; Graham and Makler, 1979; Unwin, 1987) (Figure 7.1). Following an abortive counter-revolution, the first free elections for almost fifty years were held in Portugal in 1975. The Socialist Party received the largest number of votes, and despite a further attempted coup by the far-left in November of that year, the Socialist Party was confirmed in power in the 1976 Assembly elections.

The future of revolution

The 1939–45 war provided the opportunity for Soviet forces to gain political control of eastern Europe and for communist parties to be established throughout the region without recourse to revolution. Subsequent fears of communism, particularly in the United States of America but also among its European allies in the North Atlantic Treaty Organization, led to a massive counter-revolutionary movement designed to limit the influence of Soviet-led international communism. It was only in 1989, with the collapse of the Soviet economy and system of political repression and legitimation, that a new wave of revolution occurred (Glenny, 1993). In a matter of months, the communist governments of eastern Europe built on the Stalinist single-party model were rapidly overthrown by populist movements which sought to turn them into so-called liberal democracies (Callinicos, 1991). These revolutions, though, were fundamentally peaceful occurrences. Despite the ever-present fear of violence, the political and economic changes that have occurred in Russia and most of the countries of eastern Europe have so far been significant by the lack of bloodshed associated with them. The horrors of 'ethnic cleansing' in what was Yugoslavia, and the warfare practised by the independence movements of the southern republics of the former Soviet Union, are noticeable exceptions to this, but these expressions of violence have resulted primarily from ethnic and religious differences and are not strictly revolutionary in the sense of class-based political change. The summary execution of the Ceauçescus in Romania was also exceptional, and it remains remarkable that such immediate acts of vengeance were not more widespread between 1989 and 1991.

These recent revolutions in eastern Europe, though, are by no means over. Early claims that they represented the unabashed victory of economic and political liberalism, and that they were an 'end of history' (Fukuyama, 1992), have not only been refuted on theoretical grounds (Callinicos, 1991; Habermas, 1994a) but can also be judged as premature on empirical grounds. While their early stages were indeed associated with the twin rhetorics of 'democracy' and a 'free market', the failures of most initial post-1989 governments and the return to power of many former communists emphasizes that it is too early to pass such sweeping judgement (Gowan, 1995). Capitalism has shown remarkable resilience in helping to transform post-capitalist socialism (Habermas, 1990), but the contradictions of capitalist relations of production as they are introduced into many post-communist states have yet to be fully experienced.

In conclusion, it is possible to draw out five main themes from this chapter. First, political revolutions need not be associated with any one type of economic system. Revolutions have thus occurred against political authority associated with feudal, capitalist and communist economic systems. Second, revolutions are generally the result of the failure of the ruling élite to legitimate its actions, and particularly its use of repression and violence. The basic cause of most European revolutions has been opposition towards absolutist rule, vested in the hands of a few people who have lost the confidence of the majority of the population. Third, the origins of the revolutionary idea can be seen to have lain in the intellectual movements of the 16th and 17th centuries, culminating in the Enlightenment of the 18th century. The Reformation, humanist interpretations of individual freedom, and a new search for appropriate forms of government, all precipitated the end of feudal regimes in which kings saw their position and authority as having been ordained by divine right. The collapse of religious legitimation for such political systems was crucial to their demise. Fourth, revolutions have usually taken place within a national context. While there have been limited European-wide attempted uprisings, as in 1848, and despite the Comintern's activities in the 20th century, all successful revolutions have so far been dependent on nationally focused interests, and have been concerned with the replacement of one form of national government by another. Since 1945 the increased emergence of Europe-wide political organizations, and particularly the creation of a

Figure 7.2 Statue of Lenin, now relegated to a corner of the castle courtyard, Narva, Estonia (*source:* Tim Unwin, 27 June 1995).

European Parliament (see Chapter 9), have completely restructured the context in which any future revolutionary activity may take place. It thus seems likely that if future European political revolutions take place, they will be directed against a centralized European bureaucracy that fails to legitimate its activities, rather than against the much-reduced control of power by national governments.

Finally, this survey of the emergence of a revolutionary idea in Europe has also highlighted four key practical attributes necessary for the successful implementation of revolution. First, as Lenin saw so clearly (Figure 7.2), it is necessary for there to be a determined and powerful group of people with a clearly planned agenda, who are able to promote revolution and who, perhaps more importantly, are also able to seize the opportunities provided by the serendipitous way in which events during a revolution unfold. Second, it is necessary for significant sections of the military to be subverted to the revolutionary cause. Most counter-revolutions have thus succeeded in situations where the military have, for whatever reason, remained loyal to the regime in power. Third, it is crucial for the revolutionary party to gain the support of the mass of the population for its programme of change, and fourth, those promoting revolution must be prepared to use violence and bloodshed in their actions, particularly against regimes where institutionalized systems of repression have been most strongly enforced.

Further reading

Bercé, Y-M. (1987) *Revolt and Revolution in Early Modern Europe: an Essay on the History of Political Violence*, Manchester University Press, Manchester

Callinicos, A. (1991) *The Revenge of History: Marxism and the East European Revolutions*, Polity, Cambridge

Habermas, J. (1990) What does Socialism mean today? The rectifying revolution and the need for new thinking on the Left, *New Left Review*, **183**, 3–21

Israel, J.I. (ed.) (1991) *The Anglo-Dutch Moment: Essays on the Glorious Revolution and its World Impact*, Cambridge University Press, Cambridge

Tilly, C. (1993) *European Revolutions*, 1492–1992, Blackwell, Oxford

CHAPTER 8 Political parties

MARK BLACKSELL

Democracy and the role of political parties

Europe is essentially a continent ruled by states with democratically elected governments. In any such society, political parties are the oil that enables the political process to function smoothly. Constitutional provision for public representation and elections may be crucial for democracy, but they are largely meaningless in the absence of organized political parties. Indeed, with particular reference to Europe, Dalton (1988: 127) has described political parties as 'the primary institutions of representative democracy'. Citizens see parties as an agency through which they can identify with government, parties largely manage the process of elections, and governments rely on parties to sustain them in power. In societies new to democracy, such as many of those in central and eastern Europe, the creation or re-establishment of multiple political parties to replace the tyranny of the one-party state is one of the most urgent aspects of the process of social reconstruction (Klingemann and Fuchs, 1995).

Political parties, therefore, need to reflect all the variety of the cleavages in society, while at the same time simplifying and consolidating them, so that a degree of order is allowed to emerge. Indeed, cynics have argued that for this very reason they must be an inherently reactionary force, forever vainly struggling to sustain the *status quo* against the forces of change. In *Phineas Redux*, his great novel about politics and government in 19th century Britain, written when modern democratic government was just beginning to become established, Anthony Trollope (1983: 34–5) claimed that:

> It is the necessary nature of a political party in this country to avoid, as long as it can be avoided, the consideration of any question which involves a great change. There is a consciousness on the minds of leading politicians that the pressure from behind, forcing upon them great measures, drives them almost quicker than they can go, so that it becomes a necessity with them to resist rather than to aid the pressure which will certainly be at last effective by its own strength. The best carriage horses are those which can most steadily hold back against the coach as it trundles down the hill.

This somewhat jaundiced assessment is for the most part an exaggeration; whatever their intrinsic propensity to follow rather than lead, political parties in Europe have been at the forefront in securing peaceful rather than revolutionary change in the continent. Where they have been weak, as in the short-lived Weimar Republic in Germany between 1919 and 1933, or in the former Yugoslavia after the death of Josip Broz (Marshal Tito) in 1980, totalitarianism has tended to flourish; where they have been strong, as in most of western Europe in the second half of the 20th century, they have generally been a creative and positive force, underpinning a climate of political stability.

Only a minority of the plethora of political parties in Europe ever expects to be the party of government. While all may dream about the possibility, most see their prime role as representing different shades of opinion and thus influencing the direction of government policy. Many fail even to do this, acting as no more than a safety valve for specific groups and interests, although as societies become ever better educated and sophisticated, such minority parties have developed increasingly effective means of making their voices heard and ensuring that their agendas are incorporated into government policy.

The influence of political parties is by no means always viewed by governments as benign and their banning is a regular, if not common, occurrence. The right-wing Socialist Reich Party (SRP) and the Communist Party (KPD) have been banned in the Federal Republic of Germany since 1952 and 1956 respectively, because they are viewed as hostile to the constitution. In Northern Ireland, Sinn Féin, which campaigns for a reunified and independent Ireland, has been banned in the past and is still subject to restriction, because of its advocacy of the option of the use of force of arms in achieving its political ends. Similar restrictions have been imposed by the Spanish government on ETA, the Basque nationalist party (see Basque case study in Chapter 6), and by the French government on the FLNC, the Corsican independence party, and there are many other examples. Outlawing political parties is, however, no panacea. If a banned party actually commands substantial public support, it can become a vehicle for revolution and the complete overthrow of the established political order. The 1917 Revolution saw the Communist Party in Russia move directly from illegality to power and the consequences of that lesson have not been lost in Europe (see Chapter 7). In recent years, the force of public opinion has rarely been ignored and, where outlawed political parties are steadily gaining support, ways and means have almost always been found to modify policy and incorporate them into the political mainstream.

The growth and change of political parties

Since the late 19th century European political parties have gone through a succession of developmental stages, which has seen them change from being the preserve of a wealthy landed élite to vehicles for representing mass political opinion (Bartolini and Mair, 1990). Normally, this progression is seen in terms of a three-stage model (Panebianco, 1988), though recently it has been argued that a putative fourth stage, the cartel party, is beginning to emerge (Katz and Mair, 1992). At one level the changes have been driven by the general evolution of society. However, at another level they mirror closely the gradual extension of the political suffrage from wealthy property-owning men, first to the adult male population as a whole, then to women, and most recently to young adults between 18 and 21 years of age. In each country in Europe the details of the timetable for these extensions have been different. Women in Switzerland only received the vote in 1971, and young adults in the UK in 1970. But now universal suffrage includes all adults over 18 years of age in virtually every European country, although there are still variations in the grounds for exclusion for reasons such as mental illness and the serving of gaol terms.

In the first stage, during the late 19th and early 20th centuries, cadre parties were the norm. Their membership was restricted to the wealthy landed male élites who enjoyed the privilege of suffrage and essentially they existed to protect, and promote, the interests of their parliamentary members. In the wake of the social and political upheavals after World War I, however, these cadre parties rapidly began to be replaced by mass parties. The widespread extensions of national electorates meant that the cadres they represented had lost their monopoly on political power and they had to be replaced by more broadly based groupings.

Most famously, the new mass parties, characterizing the second stage, represented the working class, but not exclusively so. Other groups, such as the churches and small farmers, also began to organize politically on the basis of the newly extended franchises, but in all cases they were distinguished from the old cadre parties by the breadth and size of their membership. The large number of members also meant that when representatives of the mass parties were elected to political positions, they were very much delegates of a particular factional interest, rather than trustees for a narrow stratum of civil society as a whole. The very different focus that this engendered has been particularly important since the 1930s, as these mass parties gradually became the parties of government.

The cadre parties did not just meekly fade away in the face of the emergence of mass parties. They responded by actively enlarging and broadening the scope of their membership, and actually derived some advantage from the fact that they were less encumbered by particular sectional interests. Indeed, the growing social and geographical mobility that the mass parties once in government had done so much to encourage, even tended to undermine the fundamental basis of their support. The justification for working-class solidarity became steadily less obvious; the general secularization of society reduced the hold of the church; and the gathering flight from agriculture and rural areas made the small-farmer parties less and less relevant to national politics.

In order to retain their influence and guarantee their finances, the mass parties in their turn had to broaden their appeal. As a result, in the third stage, they became in effect brokers between civil society and the state, run by teams of leaders prepared to compromise their distinctive ideologies in order to attract as large a membership as possible and win electoral support. Crucially, the representatives in government of these so-called 'catch-all' parties (Kirchheimer, 1966) were required to act less as delegates with specific mandates from those who elected them in favour of demonstrating loyalty to the party, so as to sustain it in power.

The fourth stage is not so clear-cut and is more disputed. It is argued in some quarters that the catch-all parties are progressively becoming cartel parties, exercising a monopoly on power amongst themselves. They alternate power through regular periodic elections, but combine to exclude any competitors. They are able to do this because of their virtual monopoly over state funding, and the control they exercise over the media, both of which derive from their existing size and status. It is certainly true that all the larger political parties in Europe are regularly the parties of government and the only significant new party to infiltrate their ranks in the liberal democracies of western Europe is the *Forza Italia*-led coalition, which briefly governed Italy between 1994 and 1996 (see case study of Italy later in this chapter).

The above sequence is essentially a description of events in 'western' Europe over the past century. Most of the countries in southern, central and eastern Europe have endured long, in some cases continuous periods of one-party rule under totalitarian dictatorships of either the fascist right or the communist left. As the 20th century draws to a close, these totalitarian regimes have mostly been replaced by multi-party democracies, in southern Europe during the 1970s, and elsewhere after the collapse of the Soviet Union in 1989. It should, though, be stressed at this juncture that there is no intention of implying that all countries, especially in eastern and central Europe, will move smoothly into a western European mould (Pridham and Lewis, 1996).

Political systems and political parties

The power exercised by parties depends heavily on the nature of the prevailing political system in a country. In Europe these systems are varied and have been subject to widespread and substantial recent change (Dalton, 1991; Duverger, 1964). As a result, there are important variations in the way in which political parties exercise power and in the role they play in civic society.

At its most effective, party government is deeply rooted in all levels of society and party politics are virtually synonymous with the politics of civic society as a whole. The pervasive influence of the Communist Party throughout most of central and eastern Europe in the second half of the 20th century is the classic manifestation of this model, but by no means the only one. In Portugal and Spain right-wing dictatorships held sway for extended periods with almost identical results, while in Greece a briefer period of totalitarian rule between 1967 and 1974 underlined how easily one-party hegemony can become established.

For the bulk of western Europe, however, multi-party democracy has been the norm and this form of government is now becoming established throughout eastern, central and southern Europe as well. Nevertheless, the differences in political culture between independent European states mean that there are significant variations in the role of parties and the way in which they operate. As well as the gradual extension of the franchise from male élites to the whole population, rules on voting and the way in which politicians are actually elected are also important considerations.

Voting may be an important civic duty, but in most countries the choice as to whether or not to exercise that right is left to the individual. Only in Belgium, Italy and Luxembourg is voting compulsory, and the Netherlands only abandoned compulsory voting in 1971. Not surprisingly, these countries have amongst the highest levels of voter participation, consistently running at above 80% of the electorate, but there are many other countries, such as Iceland and Sweden, where voting is not compulsory and which have similar levels. The missing voters in those countries where it is compulsory to vote are explained by spoiled ballot papers, as many as 20% of the electorate opting to choose none of the candidates on offer. At the other end of the scale, the proportion of people voting in some countries is well below 70%. At the most recent national elections the turnout was 66% in France (1993), 67% in Ireland (1992) and Portugal (1991), and 45% in Switzerland (1991).

Taking western Europe as a whole, roughly 25% of the electorate does not bother to vote and that proportion has been growing steadily since the 1960s. In addition, in a multi-party system it is quite possible for the majority party to register only a minority of the total votes cast, so that any claims that they represent the whole of the population are based, at best, on the assumption that the majority is prepared to acquiesce to the result of the election itself.

This is a situation that poses serious questions for the way in which a democratic system functions. In some countries, and the UK is the prime example, representatives are chosen on the basis of a simple majority, with the winner taking all and with a high proportion of power concentrated in the central government. The majority party enjoys virtual monopoly, with opposition parties only able to influence events locally, and at the margins. In other countries, such as Germany and Switzerland, both of which are federations, there is a clear constitutional separation of political powers between the *Länder* and *cantons* respectively. Not only does each level have certain powers devolved to it , but it is also quite possible for different parties to be in control at each level. Thus in Germany foreign policy and defence are clearly federal matters, but education and most social services are matters for the *Länder*, which have the tax-raising powers to fund their programmes independent of central government.

A clear separation of powers between the different levels of government is one stratagem for guarding against single-party monopolies, but the electoral system itself can also be used to this end in a wide variety of ways. For instance, in some countries, there are no separate constituencies and voters choose several preferences from a list of candidates. This method is used in the Netherlands and is favoured because it helps to guarantee a wider range of party representation. Elsewhere, only some of the representatives are chosen on a constituency basis. In Germany, for example, half are selected for single-member constituencies, the other half are chosen from predetermined, ranked, party lists, with any party receiving 5% or more of the total votes cast being guaranteed a commensurate proportion of seats in parliament. One of the main advantages of this system is that it ensures greater representation for minority parties. It is also in practice much more likely to lead to coalitions, with more than one party in government.

Another option, adopted for example in Ireland, is the single transferable vote system, which requires voters to rank all the candidates in the constituency. If their first-choice candidate is not elected, then the vote is transferred to the second choice, and so on until one candidate has at least 50% of the votes cast. The advantage of the system is that it ensures that candidates have at least some support from as large a proportion of the electorate as possible.

Despite such complex solutions to try and ensure effective and fair representation, there is strong evidence of a growing disaffection with the political process. Not only are participation rates in elections steadily declining, but there is growing evidence of votes being cast for 'challenger' parties (Mackie, 1995). People are voting in increasing numbers for candidates who do not represent any of the established political parties and who, therefore, have little chance of gaining power. These are essentially protest votes, born out of a frustration that their particular concerns are not being addressed by the mainstream.

The best-known and most successful of these parties is the Greens, who have established a firm presence widely throughout Europe. In some countries, notably Germany, they have also gained significant political power both at the central and, more especially, at the regional levels. However, in the longer term the most important contribution of new social movements, such as the anti-nuclear movement in western Europe in the 1970s and 1980s, is to radicalize the established order of political parties, rather than to create new ones (Rohrschneider, 1993). Thus, in Germany the anti-nuclear agenda was partially but quite successfully hijacked by the socialist SPD Party at the expense of the Green Party, which had sought to claim it as its own. Similarly, in almost all the countries of the former eastern Europe, the Greens played a leading role in paving the way for the overthrow of communism and its replacement by democratic governments. Subsequently, however, they have signally failed to achieve political power in the new order (Frankland, 1995).

The durability of political parties, or more precisely their underlying organization, is well illustrated by the fate of many of the former, and apparently much reviled Communist parties in what is now central Europe. In Bulgaria, the Communist party was renamed the Bulgarian Socialist Party with a new commitment to multi-party elections. It swept to power in the first democratic elections in 1990 with 53% of the seats in parliament (Koulov, 1995). In Poland, after a period of rejection, the renamed former Communist party narrowly became the party of government after the 1996 elections, while in Hungary the Communists in the guise of the Workers' Party made substantial gains in the east of the country in the 1994 elections, a region that had benefited from substantial industrial investment in the Communist era and which had subsequently experienced a relative decline in the early 1990s (Dingsdale and Kovacs, 1996). In all these cases it is notable that the overthrow of one-party totalitarianism by no means signalled the demise of the party.

Everywhere the former Communist parties were able to capitalize on their formidable grassroots organizations to recast themselves as respectable political forces. Equally, many of the new democratic movements which had initially driven the revolutions forward have found it very hard to translate their initial success into a permanent political power base.

Regionalism and party allegiance

The most arresting feature of political parties in Europe is the extent to which they are tied to national jurisdictions. Every country has its own collection of parties, the majority of which have at best very weak links beyond national boundaries. Equally, the realities of power mean that unless parties aspire to a national mandate they will be consigned either to opposition, or to a relatively minor supporting role in government. Nevertheless, regionalism in party politics is important for two reasons. On the one hand, it is frequently the case that party popularity is uneven across a national jurisdiction and thus, although a party may have formed the government, it is perceived as representing particular regional interests. On the other, parties often are formed specifically to represent and promote a regional minority.

The effects of uneven national support for political parties in Italy, especially after the electoral reform that preceded the 1994 elections, have been analysed fascinatingly by Agnew (1996b). He shows that only the socialist PDS (*Partito Democratico della Sinistra*) and the newly formed *Forza Italia* were truly national parties. All of the other significant players depended heavily on regional power bases, which tainted their credentials as guardians of the national interest. The most obviously regional of these minority parties is the Northern League (*Lega Nord*), which was formed in 1991 by amalgamating a number of small parties in the north of Italy, but the National Alliance (*Alleanza Nationale*) also has a clear regional focus, being heavily concentrated in the *Mezzogiorno* to the south of Rome. The astonishing success of *Forza Italia* is 1994 was based largely on the way its leader, Silvio Berlusconi, managed to forge workable electoral alliances with both the Northern League and the National Alliance, thus bolstering support in those areas where it was weakest (see case study).

Elsewhere the evidence for a pronounced regional element in voting patterns is more contested. After the third successive Conservative election victory in the UK in 1987, there were claims that there was a growing electoral divide between north and south, mirroring a similar perceived economic divide (Johnston *et al.*, 1988). The Conservatives' national victory was based on their strong support in the south and east of England, whereas Labour's support was strongest in the north of England, Scotland and Wales. Although the superficial geography of the electoral results appeared to endorse the two-nation thesis, the results were very controversial and were repeatedly challenged by academics from disciplines other than geography. The gist of the counter-argument was that social variables other than regional location accounted for all but a negligible proportion of the variation and that the latter was therefore unimportant in understanding voter behaviour (Johnston, 1987a, b; McAllister, 1987a, b; Savage, 1987; Johnston and Pattie, 1987). The extended debate did not convince either side of the validity of the other's arguments, but both the major UK political parties nevertheless continue to struggle to make inroads into each other's 'territory'. Labour remains relatively weak in the south and east of England outside London, while the Conservatives are finding it hard to retain a credible foothold in Wales and Scotland.

The regional effect is undeniable amongst many of those parties founded specifically to promote the interests of minorities. The populations of all states include an amalgam of different interests, and inevitably there are always those who feel left out or betrayed by the national consensus. One of the most obvious and popular ways of voicing such discontent is to form a political party whose main platform is to promote these interests. Parties of this ilk exist in many European countries, but in very few instances have they achieved much national significance, other than to act as a focus for decentralizing political power. In a few extreme cases, such as the partition of Ireland in 1922, this has led to the redrawing of the political map, but in the majority of cases central governments have succeeded in neutralizing secessionist pressures by offering a sop to the demands for more decentralization.

It is difficult to encompass the full gamut of regional minorities in Europe, even those that are sufficiently organized to have a political party to represent their interests (Stephens, 1978). The main reason for this is the enormous variation in the internal organization of the central states themselves. At

Italian politics: anomaly or extreme example of contemporary European experience?

John Agnew

The system of *partitocrazia* (coalitional politics revolving around the Christian Democratic Party distributing state resources to affiliated groups and receiving electoral support in return) that had been progressively institutionalized in Italy since the end of the Second World War came under serious challenge in the early 1990s. The end of the Cold War undermined the relevance of the Communist/anti-Communist divide that ran through Italian politics and that had excluded the Italian Communist Party, electorally the second largest party, from national office. Businesses and citizens, particularly in those parts of northern Italy where small firms had created a flourishing export-based economy, increasingly resented the corruption and political favouritism endemic to the Italian fiscal system. The Mafia of Sicily, and organized crime more generally, were popularly seen as depending on political patronage and protection, particularly from Christian Democrat and Socialist politicians. Government economic intervention through fiscal and regional policies was viewed by influential groups, such as taxpayers and big business, as inefficient and based on political rather than economic criteria: rewarding potential voters more than increasing national economic growth. These challenges are often interpreted as showing how different Italy had become from the rest of Europe, not least among large sections of Italian public opinion. Only in Italy was there a blockage preventing alternation in government between the main progressive and centre-right political parties. Only in Italy was there systematic corruption of society by parties dividing up the spoils of government. The Italian system was perhaps unique in Europe in the degree to which political parties had penetrated and corrupted society. It was based, however, on political divisions and historic roots common to other parts of Europe.

In the first place, the Cold War division of Europe ran right through Italy. With the largest Communist Party in Western Europe, even though its dominant factions were frequently critical of the policies of the former Soviet Union, Italy was divided ideologically to an extreme not seen anywhere else. Issues such as membership in NATO, the role of local government, and the social position of the Catholic Church (in the country where it is headquartered), were read through the ideological prism of support for or opposition to the two largest parties, the Christian Democrats and the Communists. Second, the Italian electoral system, orthodox proportional representation (PR) (Deputies and Senators elected by preference ballots from party lists in multi-member constituencies), encouraged both electoral representation for small parties with few votes (discouraging the accumulation of votes around parties that might govern by themselves without entering into coalitions that required dividing up government offices and resources between the parties) and preference voting between candidates (encouraging exchange or patronage voting in which voting blocs could be 'delivered' to specified candidates by organizations such as, most notoriously, the Mafia). This system was a reaction to the excesses of the fascist regime that it replaced after the Second World War, and was designed to provide representation in the Italian parliament of even the smallest organized political minorities. Even though other countries, such as the Federal Republic of Germany and Austria, have modified PR in various ways, they have remained committed to it in general because of past experience with systems that suppressed a diversity of political viewpoints. Third, and finally, only in an aggregate sense were the Christian Democrat and Communist parties truly national parties. Though there was a period from 1963 to 1976 during which they shared an increasing proportion of the total national vote in an expanded number of localities, by and large the Italian political parties had strong local and regional levels of support. Indeed, two regions of Italy were identified strongly with one or other of the two main parties: the north-east (*la zona bianca* or white zone) with the Christian Democrats and the centre (*la zona rossa* or red zone) with the Communists. Elsewhere parties might achieve local dominance (as with the Christian Democrats in, say, western Sicily or the city of Rome), share power (as with the Christian Democrats and Socialists in Lombardy in the 1980s), or engage in vigorous

continued

competition (as with the Christian Democrats and Communists in the Marche region). But this geographical pattern to party competition is not singular to Italy. England and France show remarkable geographical patterns of support for different political parties. Again, Italy is only an extreme version, as a result perhaps of the recency of national unification (coming about finally only in 1870), the history of anti-clericalism (particularly strong in regions where the Church was a landowner or allied with the landowners), and major economic disparities between regions (particularly between the more prosperous north and the poorer south).

Since 1992, the system that prevailed for so long and produced frequent changes of government, with ministers recycled from one ministry to another depending on the balance of power between coalition partners, and that gave rise to the division of control over state institutions (such as the state holding companies, banks and state television) between the parties, including the opposition since the 1970s, has largely collapsed. The slow ending of the Cold War in the late 1980s had already produced a major reformulation of the Communist Party. To this was joined the evidence of massive systematic political corruption exposed in the *mani pulite* ('clean hands') investigation begun in Milan in 1992. This brought about the indictment and disgrace of a large number of politicians (of whom the most famous is Bettino Craxi, the Socialist former Prime Minister of Italy, who is now a fugitive living in Tunisia). The major parties of government – the Christian Democrats and the Socialists – quickly disintegrated. The successor parties – the Popular Party and two others in the case of the Christian Democrats, and the Democratic Party of the Left and Refounded Communism in the case of the Communists – had now to compete, however, in a political arena with two new (one 'instant') parties and one reworked party. The reworked party, the National Alliance, is the renamed former neo-fascist party which models itself (at least publicly) after the British Conservative Party. The new parties are the Northern League (a party committed to a federalist Italy, with most of its support in northern Lombardy and the rural north-east) and Go Italy (*Forza Italia*), the centre-right party invented almost overnight in early 1994 by the media mogul Silvio Berlusconi. The March 1994 election was conducted under a new electoral system designed to encourage pre-election pacts between parties yet maintain local and minority representation (75% of seats are allocated in majoritarian single-member constituencies to candidates from coalitions of parties and 25% go to party lists in multi-member constituencies). In this election, Berlusconi's grouping of himself, the National Alliance and the Northern League won an outright majority of seats in both the Chamber of Deputies and the Senate over the 'progressive' and centre groupings. But this coalition proved extremely fragile. The Northern League's federalist and libertarian leanings coexisted uneasily with the statism of the National Alliance. The Northern League's anti-southern rhetoric went poorly with the heavy dependence of the National Alliance on southern voters. The Northern League withdrew from the government in December 1994. After two governments formed by 'technocrats' without strong party affiliations, the April 1996 elections produced a victory for a new 'progressive' alliance formed around the main inheritors of the two major parties in the old system, the Democratic Party of the Left and the Popular Party. This time the League abandoned pre-election coalitional politics to go it alone. This proved fatal to Berlusconi's Pole of Liberty alliance between himself and the National Alliance. The progressive alliance (known as *L'Ulivo*, the Olive Tree, to symbolize its bringing together of former political adversaries) is a loose congeries of former Communists, former Christian Democrats, so-called technocrats (mainly economists) and various notables (in particular the hero of the Milan corruption investigations, Antonio Di Pietro, who has since resigned). Though its leader and Prime Minister, Romano Prodi, the former head of the massive state holding company IRI, sees his coalition as having the potential to last for a full five-year term, a genuine 'first' in post-1945 Italian history, he depends in the Chamber of Deputies on the support of either Refounded Communism (likely to block any cutbacks in welfare legislation) or, failing that, the Northern League. Prodi's skill as a coalition builder will be put sorely to the test.

continued

But perhaps a 'normal' division between progressive and conservative blocs which can alternate in office is finally emerging in Italy. Only time will tell. If the Northern League has its way, however, a new Italy will also be in the making in which the powers of national office will be secondary to those of regional and Europe-wide tiers of governance. In a deliberate provocation, Umberto Bossi, leader of the Northern League, declared the independence of Padania (his new name for northern Italy) in September 1996. Already, some of the 'national' parties have begun to enunciate policies that point in a federalist direction. After finally experiencing conditions congenial for national political alternation between Left and Right, Italy may be reverting to the status of that 'mere geographical expression' those sceptical of its capacity for political unity have always claimed it to be.

Further reading

Agnew, J. (1995) The rhetoric of regionalism: the Northern League and Italian politics, 1983–94, *Transactions of the Institute of British Geographers*, **20**, 156–72

Agnew, J. (1996) Mapping politics: how context counts in electoral geography, *Political Geography*, **15**, 129–46

Agnew, J. (1997) The dramaturgy of horizons: geographical scale in the 'Reconstruction of Italy' by the new Italian political parties, 1992–95, *Political Geography*, **16**(2), 99–121

Stille, A. (1996) Italy: the convulsions of normalcy, *New York Review of Books*, **43** (6 June), 42–6

one extreme, in Switzerland decentralization to the *canton* level is such that the national government has only limited power and the main rivalries are between the *cantons* themselves. At the other, in the UK for example, where there is heavy emphasis on central decision making, regional dissatisfaction is very much more pronounced.

Despite these difficulties, Figure 8.1 attempts to depict those areas where regional parties have gained significant political power in Europe. In a few cases they have managed to forge an alliance with a national party and have created a genuine national power base within the existing constitution. The outstanding example of this is the alliance between the Christian Socialist Union (CSU) in Bavaria and the national CDU. Elsewhere such alliances, as was the case for the Northern League in Italy, have proved to be very fragile and have ultimately led to renewed and even stronger calls for secession and independence.

More commonly, regional parties do not even aspire to a direct role in government, preferring to act more as pressure groups for change. The question then is whether they choose to do so from inside or outside the political system. In some cases, such as ETA in the Basque country of northern Spain, and Sinn Féin in Ireland, the choice has usually been to operate outside the established order, but more often such parties have decided to work from within (Dahl, 1966).

In Belgium, a country chronically divided by both religion and language, both the Dutch-speaking Flemish majority in the north and the French-speaking Walloon minority in the south have their own nationalist parties. In the early years of the 20th century, Flemish separatism was much more strident in the face of French domination of the education system and the civil service, but since the 1950s they have used their majority steadily to turn the tables. In terms of both the economy and society, the Flemish north is now very much in the ascendancy and its regional party, the *Vlaamse Volksunie*, is in relative abeyance. By contrast, the Walloons in the south are represented by several regional parties, all of which have, or have had, elected members in the national parliament in Brussels. The most important is the *Mouvement Populaire Wallon*, founded in 1961, but it has subsequently been joined by the *Parti Wallon des Travailleurs*, the *Front Commun Wallon*, the *Front Démocratique des Francophones*, and the *Rénovation Wallonne* (Lorwin, 1966).

In the UK, regional parties, representing the nationalist aspirations of Scotland, Wales and Northern Ireland respectively, all have an established presence. The most radical is the SNP in Scotland, which is unequivocal about its ultimate goal of independence, though in the interim it is happy to support any moves towards decentralization within the union. Plaid Cymru in Wales seems less certain

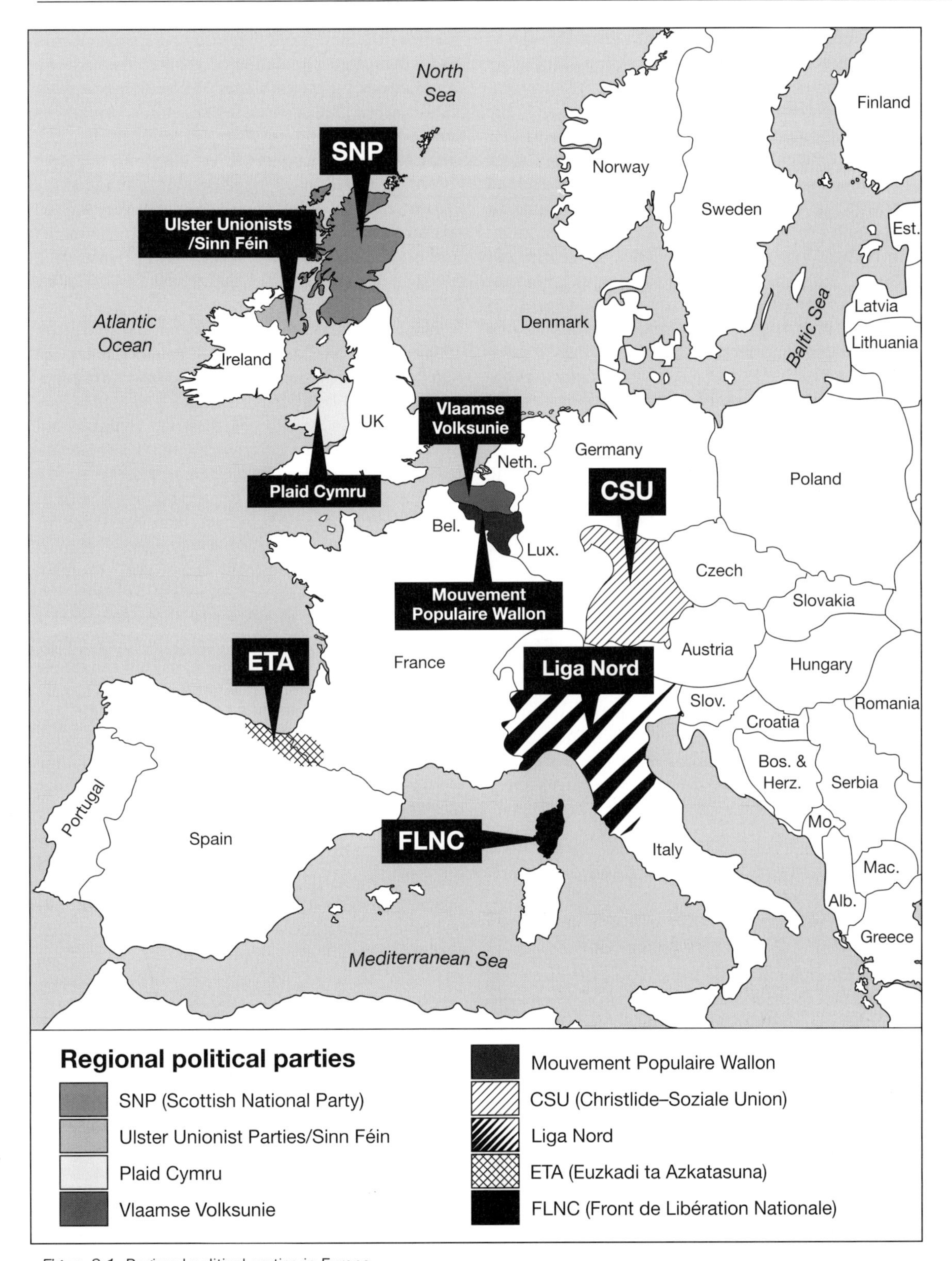

Figure 8.1 Regional political parties in Europe.

about full independence, but would certainly wish to see more decentralization of decision making from the national parliament in London to a Welsh assembly. In Northern Ireland, on the other hand, the largest regional parties, the Official Unionist Party and the Democratic Unionist Party, campaign to see the union strengthened against the all-Ireland nationalism that has already engulfed the south of the country.

The position of minority parties in eastern and central Europe is substantially different from that of their counterparts in western Europe, because of the underlying fragility of the national consensus in many of the newly independent states. In two cases this has already led to secession and fragmentation, and further political reorganization cannot be ruled out elsewhere in the future. In 1993 Czechoslovakia split along ethnic and party lines into the Czech Republic and Slovakia (see case study); and the painful process of redrawing the political map of what was Yugoslavia is still far from complete. Nationalist parties in all the former republics have seized control in the so far independent states of Slovenia, Croatia, Macedonia, and the Serb Republic of Yugoslavia, which incorporates both the former republics of Serbia and Montenegro, as well as Vojvodina and Kosovo. In the remaining republic of Bosnia-Herzegovina the outcome is still unclear, although a three-way partition, into a Republika Serbska, a Greater Croatia, and a small Muslim state seems to be the most likely eventual outcome. Needless to say, forging multi-party democracy out of these nationalist victories remains a huge task.

Beyond nationalism

Given the dynamism of economic integration in Europe, not to mention cooperation in defence matters and social affairs, it is rather surprising that party politics have remained so uncompromisingly national. Nevertheless, there are broader trends and slowly they are becoming a more significant element in the political landscape. The touchstone for this has been the European Parliament, which for all its constitutional weakness is gradually growing in importance (Bardi, 1996). Partly, this is simply a reflection of the growth in the membership of the European Union itself. The original six signatories had a combined population of 190 million, but with 15 countries it now embraces 365 million people. It also, however, reflects a change of status. Initially its members were appointed by member states, but since 1979 four successive direct elections, the most recent in 1994, have considerably enhanced its stature, even though its powers are still very limited in comparison with those of national parliaments and there are still widespread complaints of a 'democratic deficit' at the heart of the EU (Lodge, 1991, 1996).

The last election for the European Parliament in 1994 returned 567 MEPs from 12 member states, the number from each depending on the size of its population. Since then a further 59 have been added as a result of the accession of Austria, Finland and Sweden in 1995. The latter were initially appointees, but they were replaced by elected representatives. In Sweden this happened in September 1995, and in Finland in October 1996. As yet, there is no date for the Austrian elections, but the government is committed to holding them before the end of 1997 (Morgan, 1995; Westlake, 1994).

Since it was first elected in 1979 an increasingly clear-cut party structure has begun to emerge in the European Parliament, which has built on loose groupings of left-wing and right-wing political parties that already existed in Europe. On the left, Social Democratic parties operate under a wide variety of names in 22 European countries and they meet as a group regularly ahead of every EU summit meeting (Bell and Shaw, 1994; Cook, 1996). The main focus on the right is Christian Democracy, with the European Union of Christian Democrats, founded originally in 1947, being relaunched in its present form in 1965. Its members are national politicians from across Europe, united by Roman Catholicism and the ideology of the democratic right (Caliagli *et al.*, 1995; Hanley, 1994). They are not as all-embracing as the Social Democratic alliance, although they do have increasingly close links with other right-wing parties, such as the Conservative Party in the UK. Indeed, in many respects Christian Democratic parties have evolved into classic catch-all parties to meet the changing whims of their electorates.

The Social Democrats, in the form of the Party of European Socialists (PES), and the Christian Democrat and Conservative alliance, in the form of

The environmental agenda in the Czech Republic

Andrew Tickle

Although the independent Czech Republic only came into existence on 1 January 1993, the lands of Bohemia and Moravia have a much longer history of environmental concern. In parallel with other European and North American romantic and aesthetic movements, interest in nature and landscape preservation began in the late 19th century and continued in the first Czechoslovak Republic (1918–38). However, it was only after the Communist 'coup' of 1948 that state environmental legislation and policy began to be formulated in the areas of nature conservation, human health, soils, air, water and forests. The laws themselves were comparatively satisfactory but were subject to very poor implementation, primarily due to distortions inherent in the state socialist planning system, where the pursuit of centrally set quotas and norms dominated production in all sectors likely to affect the environment (industry, agriculture and forestry).

This approach resulted in a severely degraded environment, seen most acutely in industrial areas (North Bohemia and Ostrava), large cities (notably Prague) and rural areas affected by agricultural intensification (Carter and Turnock, 1993). A significant secondary effect was political: Marxism–Leninism stated that state ownership of the means of production would result in a better environment than that typical of capitalist modes. The public were clearly not convinced and thus the 'environment', as one of the few openly discussed social problems, became a significant delegitimating force against Party rule and rhetoric.

Environment and democratic transition

In the 1980s, as environmental degradation became more obvious, some conservation activists became more politically engaged, publishing articles critical of the state and encouraging open debates on controversial issues. This positive deviation against state-induced societal conformity meant that the environmental movement became an attractive haven for more politically minded individuals, who often brought with them close links to more formal opposition groups such as Charter 77. Thus the environmental agenda was significantly suffused through alternative political networks as state power began to crumble in the late 1980s (Tickle and Welsh, 1998).

In the November 1989 'velvet revolution', key individuals from the environmental movement assumed central roles in the political negotiations and ensured a prominent place for enhanced environmental protection in the ensuing political dialogue. This was not solely a value-based decision: it closely mirrored public opinion, which viewed the environment as the most readily identified major problem. Most party manifestos for the first elections (June 1990) also included environmental initiatives, confirming the legitimating power of the issue in the new political climate. Formal politicization of the environmental agenda also occurred with the formation of green parties. However, their role in the Czech Republic since 1990 has been consistently marginal, due initially to a lack of credible personnel and coherent ideas (the major personalities of the pre-1989 environmental movement instead joined mainstream political parties) and, more recently, to waning interest in environmental issues (Jehlicka and Kostelecky, 1995).

The first government (1990–92)

This period was characterized by the building of new institutional capacity, in terms of both novel environmental ministries (at the federal and republic level) and an intense focus on reformulating environmental laws and policies, often derived explicitly from European Union frameworks. Prominent players in this work were the former environmental dissidents, who ensured that environmental policy was to be far more than an area of pragmatic political opportunism. Indeed the high-minded policy aspirations of such players laid the initial seeds of their own downfall with increasingly frequent clashes with the liberal economists

continued

who dominated the new government (exemplified by Vacláv Klaus, the then Finance Minister).

Despite these difficulties, a new and comprehensive body of environmental legislation entered the statute book, including new Czech laws on waste management (1991), air protection (1991), environmental impact assessment (1992) and an overarching federal environmental law (1992), all of which laid a firm basis for future environmental improvements. In addition, significant efforts were made to make the country a significant player in transnational environmental politics, reflecting the philosophy of the global connectivity of environmental issues – abetted by President Havel's desire to cultivate an international diplomatic role. One formal outcome of this international focus was the first pan-European ministerial summit on the environment (at Dobříš in the Czech Republic, June 1991) which gave rise to the 'Environment for Europe' process, now taken on by the European Commission and the European Environmental Agency. In a time of burgeoning international environmental forums and agendas, the success of this initiative has been remarkably significant.

During this time the country also began to be a recipient of international environmental aid packages, the most significant being the environmental components of the PHARE programme of the European Union (EU), whose overall aim is to encourage the changes necessary to build a market-oriented economy and to promote private enterprise. Without tight environmental legislation, however, these aims would be viewed in eastern Europe as being inconsistent with long-term environmental improvements. Furthermore, setting aside any possible negative environmental consequences of foreign 'aid', the involvement of the EU means that environmental policy developments may also become secondary to the goals of free markets and wider European integration.

The second government (1992–96)

The fracture of the post-1989 umbrella ruling group – Civic Forum – into a spectrum of parties contesting the 1992 elections meant that political appointments were now dependent on coalition agreements. This, together with the electoral dominance of Klaus's Civic Democratic Party and his personal antagonism to environmental thinking, ensured the ousting of environmental specialists from government, to be replaced by standard political appointees. This essentially resulted in a severe reduction in environmental influence within government itself, paralleled by a growth in the influence of the free-market paradigm of minimal state interference in environmental matters.

Within six months of the elections, the federal state split, with a unitary Czech Republic coming into existence on 1 January 1993. It is currently difficult to gauge the impact of this major political development in environmental terms, apart from the obvious clarity it now brings to the framework of environmental law, which was previously a complex mix of federal and republic legislation.

In the previous period (1990–92), environmental non-governmental organizations (NGOs), despite being rather weak entities, had been close to government. This relationship was now at an end and the increasing influence of a well-organized industrial lobby (particularly in the chemicals and construction sectors) was confirmed, in both parliament and government. However, the environmental lobby still retained influence, strengthened by the return of former ministers and their advisers, an increasing professionalism in political lobbying, and increasing niche specialization, whereby specific NGOs began focusing on single issues such as recycling, forest protection and energy efficiency.

Important environmental legislation was still passing through Parliament, though the number of acts passed in 1993, 1994 and 1995 together (eight) was less than that in 1991 alone (nine). Some acts were influenced strongly by NGO input, for example the 1993 Ozone Act by the Rainbow Movement (*Hnutí Duha*) and Children of the Earth (*Děti Země*). In the first four months of 1996 the prospect of elections appears to have caused a flurry of proposed environmental bills (six), which may have been a manoeuvre designed to dispel perceptions of a poor governmental record on the environment, underscored by persistent political attacks from almost all opposition groupings (usually championing the merits of state versus market controls).

continued

Future prospects

The general outlook for the environmental agenda in the Czech Republic is not good. Policy development in the area remains a peripheral interest of the government, though the new balance of power from June 1996 – the former coalition now forming a minority government, faced by a strong parliamentary opposition that has been critical of the government's previous environmental record – may mean a revival of the environment's political role. This seems unlikely given that the political agenda is still dominated by the typical issues of macroeconomic progress, wages and cost of living, although opinion polls still suggest that a large proportion of the population continues to view the environment as a major issue. However, as the Czech Republic progresses towards EU membership, the need to harmonize environmental regulations prior to entry (the *acquis communitaire*) may force a re-prioritization.

An additional concern is the attitude of the government (and parliament in general) to civic environmental initiatives. These activities are not encouraged by such institutions, whose view of democracy rests primarily on the freedom of the market and the plurality of political parties. This has given rise to the ironic development that the space for post-communist civil society (and the environmental lobby within it) is narrowing rather than widening. Given that a central paradigm of sustainable development, to which the country is nominally committed, is the involvement of all societal actors, this does not augur well for the future.

Further reading

Carter, F.W. and Turnock, D. (eds) (1993) *Environmental Problems in Eastern Europe*, Routledge, London

Jehlicka, P. and Kostelecky, T. (1995) Czechoslovakia: Greens in a post-Communist society, in Richardson, D. and Rootes, C. (eds) *The Green Challenge: the Development of Green Parties in Europe*, Routledge, London, 208–31

Tickle, A. and Welsh, I. (eds) (1998) *Environment and Society in Transition: Central and Eastern Europe*, Addison Wesley Longman, Harlow (forthcoming)

the European Peoples' Party (EPP), have always dominated the European Parliament. The PES, with 198 seats after the 1994 elections, is narrowly the largest group (Table 8.1), but it is entirely dependent on the support of minor parties for a majority, even with the addition of 19 seats as a result of the accession of Austria, Finland and Sweden. There are seven other significant groupings, the largest of which is the centrist LDR, the Liberal Democratic Reformist group, with 43 seats, but none seems likely to be able to attract sufficient support to be able to dent the dominance of the PES and EPP. The smaller groupings also suffer from their patchy regional representation. Both the two major groupings include MEPs from all the member states of the EU, thus enabling them to claim a breadth of support that none of the others can match.

The growing number of MEPs cannot disguise the underlying weakness of their political groupings, which stems in turn from the fundamentally limited powers of the European Parliament itself. As a result the groupings remain no more than that, with no evidence that they are likely to develop into fully fledged political parties, capable of challenging the hegemony of their national counterparts. It is a situation that seems bound to persist so long as the ultimate political role of the EU remains unresolved.

Conclusion

Political parties and the multi-party system are essential features of western capitalist democracy, steadily growing in stature as this model of society has ever more strongly taken root in the course of the 20th century (Rueschemeyer *et al.*, 1992). They are, therefore, of central importance throughout a Europe that is emerging from the shock of the sudden and rapid demise of the Soviet Union in 1989. In every country, they provide the means by which the multiplicity of popular views can be expressed in government, and

Table 8.1 Political group membership after the European Parliament elections in June 1994.

	B	Dk	F	G	Gr	Irl	I	L	Nl	P	S	UK	Total
PES	6	3	15	40	10	1	18	2	8	10	22	63	198
EPP	7	3	13	47	9	4	12	2	10	1	30	19	157
LDR	6	5	1	0	0	1	7	1	10	8	2	2	43
EUL	0	0	7	0	4	0	5	0	0	3	9	0	28
FE	0	0	0	0	0	0	27	0	0	0	0	0	27
EDA	0	0	14	0	2	7	0	0	0	3	0	0	26
Greens	2	1	0	12	0	2	4	1	1	0	0	0	23
ERA	1	0	13	0	0	0	2	0	0	0	1	2	19
EDN	0	4	13	0	0	0	0	0	2	0	0	0	19
Ind	3	0	11	0	0	0	12	0	0	0	0	1	27
Total	25	16	87	99	25	15	87	6	31	25	64	87	567

Abbreviations of political groups:
PES – Party of European Socialists; EPP – European People's Party (Christian Democrats and Conservatives); LDR – Liberal and Democratic Reformist; EUL – European United Left (including former Communists); FE – Forza Europa (made up entirely of Forza Italia MEPs); EDA – European Democratic Alliance (Gaullist and Fianna Fáil); Greens – Group of the Greens; ERA – European Radical Alliance (left of centre and regionalists); EDN – Europe des Nations (De Villiers' list and some other right-wing anti-EU members); Ind – non-attached (including National Front in France and some right-wing Italians).

are the main guarantors of freedom of speech. It also needs to be remembered, however, that in their present form they are very recent creations; adapting with remarkable facility to the steady broadening of the electoral franchise, which has been one of the key social achievements in Europe's recent history. They are nevertheless entirely dependent on popular support for their success and survival. As Schmitt and Holmberg (1995) have conclusively demonstrated, public indifference will not only sound the death knell for political parties, but also poses a serious threat to the very existence of the multi-party state.

Further reading

Bartolini, S. and Mair, P. (1990) *Identity, Competition, and Electoral Availability. The Stabilization of European Electorates 1885–1985*, Cambridge University Press, Cambridge

Lorwin, V.R. (1966) Belgium: religion, class, and language in national politics, in Dahl, R.A. (ed.) *Political Oppositions in Western Democracies*, Yale University Press, New Haven and London, 147–87

Mackie, T. (1995) Parties and elections, in Hayward, J. and Page E.C. (eds) *Governing the New Europe*, Polity Press, Cambridge, 166–95

Morgan, R. (1995) *The Times Guide to the European Parliament 1994*, The Times, London

CHAPTER 9 The European Union: cumulative and uneven integration

ALLAN M. WILLIAMS

The making of the European Union

The European Union (EU) was conceived as an institutional framework for European integration in the 1950s, in conditions which were very different to those of the 1990s (Table 9.1). Over time, this institutional framework for the European space has been both widened (through enlargement) and deepened (through increasing transnationalization of regulatory activity). The process of cumulative integration has, though, been uneven with tensions between differing levels and political and economic arrangements.

The deliberations in the 1950s, which led to establishment of the European Economic Community, were influenced by three main concerns (Williams, 1994): first, the legacy of World War Two, and particularly the need to create an international regulatory framework which both facilitated and regulated Germany's economic renascence; second, a need to create a more effective political voice for Europe in the international arena amidst the tensions of the Cold War, but without creating a single European collective defence and security space; and third, although in the mid-1950s Europe was in the middle of its longest and most sustained economic boom of the 20th century based on mass consumption and Fordism, there were already concerns about its global competitiveness, especially in relation to the power of American transnational capital (Aldcroft, 1980).

There were both political and economic goals for the establishment of the European Communities. One aim was the creation of a third global political force, but earlier, futile attempts to establish a European Defence Community and a European Political Community had already exposed the constraints on such an ambition. In the absence of effective collective political mechanisms, therefore, the European Communities constituted 'a sleeping giant' on the world political stage, and were effectively the USA's junior partner in NATO. Greater stress was placed on the second goal, reinforcement of the logic of capital accumulation via a series of neo-liberal economic reforms to reduce barriers to international competition. Given the prevailing western European social democratic model of state regulation, the limitations of neo-liberalism were recognized, and were tempered by common policies for key sectoral interests. But, in line with the neo-liberal emphasis, such policies were considered to be short-term measures for facilitating market adjustments (Hodges, 1981: 44), as in the case of Social Fund intervention in the labour market (Mazey, 1989). Even the Common Agricultural Policy was presented (if not necessarily conceived) as a mechanism for short-term market adjustment rather than as the enduring and structural feature of agricultural production which it became.

By the 1990s the European Union was operating in a radically different global context. Easing of the Cold War signalled the end of its unquestioned role as the cosy ally of the USA and, at the same time, there were more localized disputes in and around the margins of Europe. The then British Foreign Secretary, Douglas Hurd, summed this up with the phrase that 'the Cold War was unfriendly but stable' (Budd, 1993: 6). With increasing isolationist tendencies in the USA, and a rising tide of nationalism (especially in the Balkans), there were pressures for Europe to adopt a more collective and autonomous role in defence and security matters.

The long economic boom had also been replaced by increasing uncertainty after the early 1970s as the limitations of Fordism were exposed, as American hegemony weakened, and as global competition intensified. There was particular concern about 'Eurosclerosis', Europe's comparative weakness in international high technology trading. Global rather than European competitiveness was the urgent issue, and the response was more neo-liberal reforms

Table 9.1 Key dates in the establishment of the European Union.

1952	Creation of the European Coal and Steel Community (ECSC) by Belgium, France, Federal Republic of Germany, Italy, Luxembourg and the Netherlands (Treaty of Paris, 18 April 1951)
1957, March 25	Two Treaties of Rome, creating the European Economic Community (EEC) and the European Atomic Energy Community (EURATOM)
1960	Creation of the European Free Trade Association (EFTA) by Austria, Denmark, Norway, Portugal, Sweden, Switzerland and the UK
1967, July 1	Creation of the European Community (EC) out of the ECSC, the EEC and EURATOM (Merger Treaty, 8 April 1965)
1972, January	Summit agreeing to enlargement through accession of Denmark, Ireland, Norway and the UK
1973, January	Accession of Denmark, Ireland and the UK to the EC
1979	Introduction of the European Currency Unit (ECU)
1979, June	First direct elections to the European Parliament
1981	Accession of Greece to the EC
1986	Accession of Portugal and Spain to the EC
1986, February 26	Single European Act to come into force on 1 July 1987
1987, July	Ratification of the single amendment, changing the EC's voting structure to a weighted majority on a range of issues
1991, October	Agreement between the EC and the EFTA to establish a European Economic Area
1991, December 11	Maastricht Treaties agreeing to common foreign and security policy, and establishment of a European Central Bank and single currency in 1999
1993, January 1	Removal of all international economic barriers within the EC
1993, November 1	Foundation of the European Union (EU) as successor to the EC
1994, January 1	Establishment of the European Monetary Institute (EMI)
1995, January	Accession of Austria, Finland and Sweden to the EU

through the Single Market programme. The Economic and Monetary Union (EMU) project represents a further extension of neo-liberalism, as the convergence criteria dictate sharp reductions in public sector deficits and debts, requiring reduced – or at least changing forms of – state interventionism.

There has also been a greater need since the early 1970s to confront structural social and economic inequalities, especially unemployment and social exclusion. Reduced economic growth after 1974 has been associated with increasing inequalities (Glyn and Miliband, 1994). Whereas unemployment rates in the EU had been below those in the USA in the 1970s, and they had risen sharply in both during the early 1980s, they subsequently fell in the USA while remaining persistently high in the EU. In 1994 unemployment in the EU exceeded 10%, but it was 6% in the USA, and in Japan, despite recent economic uncertainties, it was only 4%. There were also differences in job-generating capacities; employment had been relatively static in the EU over the long period 1960–92, but had almost doubled in North America and had increased by approximately one-half in Japan (Lee, 1995: 1581).

Europe's comparative economic weakness was characterized by inherent sectoral and spatial inequalities. For example, there was an average 5.5 percentage point differential in unemployment between the member states in the period 1964–70. By 1971–80 this had become a 7 percentage point differential, and by 1981–90 had widened to a 16 point differential between Spain and Luxembourg. There was also a root change in regional inequalities after the early 1970s as the previous tendency to convergence gave way to divergence. Dunford (1994: 95) argues that, in part, this is because '... a neo-liberal programme of market integration and an intensification of competition for investment creates a zero-sum gain in which the gains of the winners are often at the expense of the losers'. In response to these pervading inequalities, at all spatial scales, employment generation and amelioration of social exclusion have become major issues for the EU.

Over the last forty years, there have been considerable changes in the institutional framework for European space provided by the EU, partly because of these global economic and political shifts, but also in response to the different and often contradictory domestic political interests of the member states (see Table 9.1). The resulting changes can be characterized in terms of Balassa's (1961) seven stages of economic integration: a free trade area, a customs union, a common market, an economic union, a monetary union, an economic and monetary union, and full economic union (total economic unification, implying a high level of political union). In broad terms, the EU in the 1960s was an imperfect free trade area, customs union and common market which, by the 1990s, had become significantly more complete in these respects while also being committed to economic and monetary union. Progressive deepening of economic integration has created a distinctive economic space by way of trade diversion and trade creation, as well as by increasing the freedom of movement of labour and capital. The Single Market programme and the Treaty on European Union have been particularly important in this respect.

Changes in the political constitution of the EU have tended to lag behind economic integration. From its inception, the EU has always been a distinctive international body politically because the powers of the member states were counterbalanced – though unevenly and modestly – by those of the Commission, the European Parliament and the European Court of Justice (Pinder, 1991). However, it has essentially been an intergovernmental body, particularly since the 1966 Luxembourg compromise asserted the principle of unanimity (or the right to veto) in decision making by the member states.

The key moments in the laggardly and uneven construction of political integration, other than the signing of the Treaty of Rome in 1957, have been 1972, 1986 and 1991. The 1972 Summit was a particular high point as it marked two events: completion of the first widening of integration via enlargement (see case study on the northern expansion), and a public commitment to achieve fuller economic and monetary union by the end of that decade, including completion of the single market, introduction of a common currency and a central European bank. However, in what I have referred to elsewhere as 'bright openings and bitter endings' (Williams, 1994: 57), the late 1970s and early 1980s were actually characterized by minimal integration. In particular, the limitations of intergovernmentalism were exposed in a Europe which was increasingly subject to uncertain economic growth. Political integration was decisively renewed only in 1986, with the passing of the Single European Act, followed by the Treaty on European Union in 1991. These increased the scope for majority voting in the Council of Ministers, while transferring limited powers to other EU institutions, as well as establishing new goals for the EU in defence and security matters.

Kirchner (1992) argues that, as a result of these political shifts, decision making in the EU in the early 1990s could be characterized as a cross between intergovernmentalism and cooperative federalism. Progress in recasting the European political space has been uneven, crabwise and subject to shifting coalitions of interests. Most significant redistributions of power from the intergovernmentalist Council of Ministers to potentially federalist EU institutions have been achieved only in response to particular 'moments of crisis'. A notable example was how the need to implement the 300 measures of the Single Market programme led to the Single European Act, which introduced majority voting on a range of issues. The strongest lever in bringing about such change has been the tension between uneven economic and political integration. This was used to advantage by Jacques Delors in the so-called 'Russian dolls' strategy to advance the cause of economic and political union. The prime example of this was the argument that increased economic integration in the 1980s has created a 'democratic deficit', the 'logical' remedy to which was greater political union.

Having outlined some of the background to the changing nature of the EU, the remainder of this chapter focuses on more recent features of its evolution. The next two sections, respectively, examine the renewal of integrationist tendencies in the 1980s and the 1990s, and demonstrate the limitations of greater union in the latter decade. The final section considers some of the significant challenges facing the EU at the end of the 20th century.

The 1990s: European crises and reconstruction

The Single Market and the Single European Act: a twin strategy for integration

There have been two main, and interlinked, battlegrounds in the contested progress to economic and political union in the 1980s and the early 1990s: the Single Market (SM) programme and the Maastricht Summit. The SM programme was a response to two particular pressures: the crisis of European global competitiveness, and the need to relieve the near-paralysis of decision making. In essence, the SM is a territory without internal frontiers, in which there is guaranteed free movement of goods, services, people and capital. The economic advantages of a single market lie in increasing economies of scale, specialization, agglomeration effects and industrial investment,

The northern expansion

Roger Bivand

Beyond Europe

Only two Nordic countries have a heritage of continuous statehood of more than a century. For this reason among others, the Treaty of Rome has not necessarily been seen as building a bulwark against continental conflict by increasing mutual dependence, but rather as constituting a threat to local democracy and national sovereignty. Of the four larger countries, Denmark and Sweden have by turns dominated Finland and Norway. Norway was Danish from shortly after the Black Death until 1814, when it was granted to Sweden. It broke free from Sweden in 1905, perhaps because of diplomatic pressure from the neighbouring major power, Russia. Finland was won by Russia from Sweden during the Napoleonic Wars, and achieved independence in 1917, an independence wounded by a brief but bloody civil war between Reds and Whites (Isachsen, 1968). Denmark joined the European Community over 20 years ago; Norway applied to join at the same time, but a referendum in 1972 rejected membership. Sweden and Finland were unable to consider membership until the collapse of the Warsaw Pact. These three countries, together with Austria, drew EFTA into the Single Market in the European Economic Area (EEA) framework early in the 1990s, but before the EEA had begun to function, all of them had applied for full membership in the EU. All except Norway joined the Union in 1995; the 1994 no-vote in Norway was only a little less strong than 22 years earlier. Denmark and the southern, urbanized regions of Finland, Sweden and Norway are not very different from other regions of north-west Europe already in the Union. The specific difference implicit in the northern expansion becomes apparent in geographical terms when we turn to the vast inland wilderness, both beyond the Arctic Circle and stretching west and east of the Gulf of Bothnia (Figure 9.1).

There are three ways of viewing the northern areas: as a string of coastal settlements facing the North Atlantic and the Arctic, principally along the Norwegian and Russian coasts; as the colonization of the interior wilderness from the Gulf of Bothnia in today's Finland and Sweden (Bylund, 1960); and using the geographical name still found in many atlases – Lappland (*Sápmi* or *Sábmi*). While Sami languages and Finnish belong to the same non-Indo-European family, the peoples do not share a common history (see Chapter 2). The Sami, with similar peoples in the Russian Arctic, are indigenous, and have experienced incursions from without which have threatened and often dispossessed them. Nils Gaup's 1987 Oscar-nominated film *Ofelas* (The Pathfinder) is a contemporary rendering of Sami oral traditions about such raids. Gaup invented a 'language' for the film bandits which could not be identified as specifically Russian, Finnish, Swedish, Danish or Norwegian, not just to avoid losing a potential market but also to symbolize the historical and continuing conflict. It is conservatively estimated that over 40,000 Sami now live in Norway, about 20,000 in Sweden, and several thousand each in Finland and Russia.

In addition to the Sami, we see homesteaders, trappers and fur traders, miners, missionaries both Lutheran and Orthodox, soldiers, fishermen and tax collectors. Finnish migrants moved anticlockwise around the Gulf of Bothnia, settling in Northern Norway and in Sweden. Much of this enormous area was subject, until 1800, to multiple taxation by Russian, Swedish and Danish–Norwegian authorities, and some communities were not finally split by national borders until 1826. 'Pomor' trade between Norwegian and Russian coastal settlements developed vigorously over the next century, reducing dependence on Bergen as the monopoly entrepôt (Sjøholt, 1990). These contacts were terminated in 1917 but have revived since 1989.

The demarcation of the border between Sweden and Russian Finland took place under somewhat chaotic circumstances, with surveyors unsure about the river they were supposed to follow. It was many years later that prospecting for metal ores showed that national boundaries did matter in the wilderness, and that natural resources could become a cause of war; Nazi Germany raced in 1940 to secure access to Swedish iron ore, which its enemies wished

continued

Figure 9.1 The northern expansion: Objective 6 designated areas (*source*: World Vector Shorelines and CIA World Data Bank II as distributed in GMT: Generic Mapping Tools at http://www.soest.hawaii.edu/soest/gmt.html. Draughtsman – Kjell Helge Sjøstrøm).

continued

to interdict. France and Britain had themselves been half-heartedly planning to cross through neutral Sweden to join the Finns in standing against the Soviet invasion at the same time. The end of World War II saw scorched earth tactics used in both northern Norway and northern Finland; Norwegian Kirkenes was captured by the advancing Red Army, after most of the northern counties' inhabitants had been forced southwards by the retreating Germans. Similarly, Rovaniemi in Finland was completely reconstructed after wartime destruction, partly by architect Alvar Aalto; the Soviet Union annexed Karelia, Viborg and other territories as spoils of war. Sweden emerged from the war with a mission to forge ahead in building a centralized welfare state, adopted only in part by its neighbours as time passed.

In the post-war period, all three national governments saw a clear need to modernize the far north, to bring it under greater state control. They all saw the strategic importance of the wilderness, open to assault from the east. Murmansk, with its 450,000 inhabitants, is almost ten times larger than other cities in the region, and lies just 120 km from the Norwegian border. Over and above vulnerability to the testing of nuclear bombs at Novaya Zemlya (1955–90: 132 devices with a yield of approximately 470 megatons, 94% of the total Soviet test yield), the Kola peninsula has arguably the largest concentration of nuclear weapons and reactors in the world. In addition to the former Soviet Northern Fleet, operating more than 60 nuclear submarines, the peninsula is home to civil nuclear-powered vessels, dumps of waste from these vessels and other uses, a large nuclear power plant, and many other inadequately secured radiation sources (Bellona, 1996).

The regional contexts and attitudes to the EU

The three regional contexts have conditioned attitudes to the EU, as indeed they have influenced relationships with national governments. Since the war, a second tier of elected local government has been put in place at the provincial level, overlaying the small-scale municipal units instituted in the 1800s. Many municipalities are not only tiny in population size, but also huge in area (Guovdageainnu–Kautokeino in Norway extends over 9687 km^2, Kiruna in Sweden 19,447 km^2, Inari in Finland 15,251 km^2) (Figure 9.1); they are only exceptionally concentrated in single centres. Attempts to rationalize settlement and municipality structures have been more successful in Sweden than in Norway, using coercion through the location of health, education and social services. The modernizers deployed a range of instruments covering many policy fields which have rendered the northern municipalities and provinces dependent on transfers from the central budget. Despite this, distrust with regard to centralizing tendencies remains strong, except in larger industrial and administrative centres, which feel that they have influence among national élites, mostly through Labour Party and Trades Union channels.

During the last twenty years, few central government policies for the northern regions have been successful. Starting from the coast, fisheries regulation and overfishing have combined to ruin both coastal fisheries and most fish-processing industries; whaling has largely been given up because of international protests. The Sami lost their battle against the damming of the Alta River for hydropower in Norway; substantial overgrazing is affecting the reindeer-carrying capacity of their heartlands; land rights remain unsecured. Most state-owned mines and smelters have been either closed or divested, private ones closed or restructured. Restructuring has also hit the forestry and wood products industries. The public sector is now the largest employer in these areas, providing 35–40% of total employment. Confidence in solutions proposed by central political élites is not strong, although a survey from Swedish Norrbotten shows that people feel very dependent on the central state (Johansson, 1995). Consequently, when the central state élites, using the same kinds of modernizing language, recommended voting to accept the terms negotiated with the EU, voters had to assess whether they were going to be let down once again.

continued

Membership, Objective 6, and a future in the EU

The only majority for EU membership across the northern areas was at Rovaniemi in Finland; the no-vote was stronger in Norway and Sweden than in Finland, and stronger overall for smaller municipalities and lower population densities. Nationally, the results were Finland yes-vote 56.9%, Sweden yes-vote 52.2% and Norway no-vote 52.3% (Arter, 1995; Hansen, 1996; Murphy and Huderi-Ely, 1996; Myklebost and Gläßer, 1996; Petterson *et al.*, 1996; Sogner and Archer, 1995). Although Norway's no-vote has attracted most attention, it would be wrong to discount the overall northern no-vote in terms of its potential impact on coherence within the Union. Norway's no-vote was furthermore conditioned by long-standing territorial cleavages and sentiments, with adherence to specific perceptions of national sovereignty paramount.

In negotiation, the three Nordic countries sought to address the fears of people in the north by winning acceptance for support from structural funds over and above the objectives which already existed. Some of the northern communities have low GVA per capita levels, but not in general low enough for Objective 1 support to be made available. Consequently, the new Objective 6 was proposed, to include municipalities or parts of municipalities with a population density of under 8 per km^2 (Figure 9.1). The designated areas of Finland and Sweden have an area of over 450,000 km^2, and a population of under 1.3 million, yielding an overall population density of under 3 per km^2. In addition, Norwegian negotiators, at the instigation of the Norwegian elected Sameting, opened the door for a Swedish protocol to its accession treaty related to the rights of the Sami, in particular to their exclusive right to herd reindeer in the traditional Sami area. Sami politicians have reported that contacts with Brussels have been developing positively, and that indeed more understanding for cultural diversity is shown in the Commission than they are used to in Stockholm (Labba, 1996). The Swedish Sameting is busy working on Measure 5 of the Swedish SPD for Objective 6: 'Sami development. Measures aimed at preserving and developing Sami culture and the traditional activity of reindeer husbandry'. They are fascinated to note that Brussels is willing to apply the subsidiarity principle to the elected Sameting in policy areas where both Sweden and Norway have been dragging their feet. There is optimism that this may be extended to land rights.

In addition to Objective 6, INTERREG II programmes are being considered, partly to supplement two existing institutions. The Nordkalott Committee was set up by the Nordic Council in 1967 to group northern provinces of Norway, Sweden and Finland; the same provinces would like to continue or add to this within the INTERREG framework. Following the dissolution of the Soviet Union, Russia became an active partner for the three Nordic countries, and cooperation was established for the Barents Euro-Arctic Region, including the Nordkalott provinces and three Russian regions. This project is being actively pursued by Finland as part of its aim to become Russia's gateway to the EU.

EU membership is widely seen by people in the far north as a political project of the central élites, removing them even further from influence on their own lives. Only minority groups, such as the Sami, and environmentalists horrified by the nuclear rubbish dump on the Kola peninsula, have displayed initiative in contacts with Brussels, initiative which they feel has been well rewarded. Many Europeans would be very happy to experience Europe's last wilderness, as tourists or students, but few inhabitants seem able to find the motivation to get started on their new European chance, waiting for somebody else to sort out their problems. While this points to atrophy in civil society, the actions of the Sami and environmental groups demonstrate that parts of society are ready and willing.

continued

Further reading

Bylund, E. (1960) Theoretical considerations regarding the distribution of settlement in Inner North Sweden, *Geografiska Annaler*, **42**, 225–31

Hansen, J.C. (1996) Les pays nordiques et l'Union européenne: intégration ou isolement?, *Cahiers de Géographie du Québec*, **40**, 255–65

Isachsen, F. (1968) Norden, in Sømme, A. (ed.) *A Geography of Norden*, Cappelens Forlag, Oslo, 13–19

Murphy, A.B. and Huderi-Ely, A. (1996) The geography of the 1994 Nordic vote on European Union membership, *Professional Geographer*, **48**, 284–97

Sjøholt, P. (1990) Marginality, crisis and the response to crisis: some development issues in the Scandinavian Northlands, Department of Geography, University of Bergen, Bergen (Geografi i Bergen Nr 135)

as well as reduced uncertainty and enhanced competitive pressures. Lord Cockburn's White Paper (Commission of the European Communities, 1985) listed the 300 measures which were required to remove physical, fiscal and technical barriers to the creation of the SM. The technical barriers included subsidies, nationalistic state procurement policies, and non-tariff barriers to trade and competition.

Given the pessimism and minimalism which characterized the EU in the early 1980s, it is remarkable that the SM programme was largely implemented on time, even if decisions on 18 measures were still outstanding at the end of 1992 and some agreements had involved significant compromises of the single market principles. The success was based on a well-crafted strategy which had four key elements: the Cecchini (1988) report, which claimed that the cost of not implementing the SM would be to forgo a 4.5% increase in GDP and a two million increase in employment; publication of a seemingly definitive and achievable target list of 300 measures; the urgency generated by setting a deadline of 1 January 1993 for introduction of the SM; and the Single European Act, which instituted majority voting on most of the measures required. In this context, Ross's (1991: 49) comment that the EU was 'constructed out of the materials of capitalist crisis by gifted politicians' seems particularly apposite.

In practice, much of the debate about the SM was ill-founded, including the scale of the economic gains predicted by Cecchini. However, the significance of the Single Market programme is not in doubt, if only because of four particular implications:

1 It was a catalyst for a wider debate on Europe's global competitiveness.
2 The introduction of increased majority voting by means of the Single European Act provided an important precedent for constraining national sovereignty.
3 The neo-liberal vision of the European economy was extended by reductions in national sovereignty in respect of state subsidies, some areas of taxation, mergers and acquisitions, and public procurement.
4 The SM contributed to what Padoa-Schioppa (1988) termed the inconsistent quartet: free trade, capital mobility, exchange rate fixity and national monetary sovereignty. While the SM liberalized trade and capital movements, most member states were locked into virtual exchange rate fixity by membership of the Exchange Rate Mechanism. In these circumstances, national monetary sovereignty was not sustainable, and – arguably – monetary union had become inevitable, as well as necessary for fulfilment of the neo-liberal economic vision of Europe.

The next stage – the Treaty on European Union – can be seen as an inevitable outcome of the uneven nature of the integration instituted by the SM programme and the Single European Act.

Maastricht: rebalancing the political and the economic

Even in the early stages of the SM programme, many proponents of greater political and economic union were looking beyond this to further integration. In particular, the 1988 Delors Report mapped out a route to Economic and Monetary Union (EMU) as the logical extension of the SM. The member states agreed to convene an Intergovernmental Conference (IGC) to consider the economic issues. Later, on Chancellor Kohl's insistence, it was agreed to convene, simultaneously, a second IGC to address the manifest democratic deficit in the EU. The resulting Treaty on European Union, although dogged by political compromises, involved three main 'pillars', a Social Agreement and limited reform of decision making. In some ways it can be seen as a further extension of the neo-liberal project via EMU, together with necessary reforms to realign political mechanisms with the new economic realities.

The *First Pillar* represented an extension of existing treaties in order to widen EU competencies to include new areas such as consumer protection and culture. The most important element was the commitment to EMU. There was a particular logic in the timing of this agreement, for Krugman (1990) argues that, with higher levels of integration, the costs of EMU decline while its benefits increase. Artis (1994) summarizes the costs and benefits. The main benefits are elimination of transaction costs; reduced uncertainty for company location decision making; reduced risk premiums in real interest rates as a result of reduced uncertainty, thereby encouraging investment; reduced market segmentation; and giving Europe a stronger collective voice in world economic management, conceivably as one of the potential G3, alongside the USA and Japan. The main cost would be having to forgo using nominal exchange rates as an economic shock absorber, although their efficiency is questionable anyway given the possibility of wage–price reactions as a result of price increases in imported goods, following devaluation.

A short and demanding timetable was agreed for the introduction of EMU. Stage One (by 1 July 1990) involved currency linkage via the Exchange Rate Mechanism, and progress towards economic convergence. This was deemed largely to have been accomplished. Stage Two (to be achieved by 1 January 1994) required creation of the European Monetary Institute, stronger convergence, and agreement by December 1996 on the initial membership of and timetable for EMU. In Stage Three (to begin in January 1999 – unless otherwise specified), a European System of Central Banks and a European Central Bank are to be created, the ECU can be used as a common currency in addition to national currencies, and procedures for controlling budget deficits are to be strengthened. Both the UK and Denmark secured partial opt-outs from the timetable, contributing to what was known as the 'variable geometry' of the structure of the Union.

The *Second Pillar* involved commitment to a common foreign and security policy including, in the longer term, the framing of a common defence policy. The background to this was growing US insistence that the EU should play a greater role in international affairs. Given the sensitivity of extending EU competence in this sphere, at the costs of national sovereignty, the Treaty was only able to set out broad aims for the second pillar. The detailing of objectives remained the preserve of the intergovernmentalist European Council, and so required only minimal transfers of powers to the Union.

Soon afterwards, at the Lisbon Summit, the Council specified the key factors which created sufficient common interest to merit collective EU action: issues which involve geographically proximate areas, impact on the Union's political and economic stability, and/or threaten its security. However, the second pillar had very weak foundations. First, and perhaps most importantly, it was established on an intergovernmental basis outside the main EU institutions, notably the Commission, which was probably the only body with sufficient resources to have made it effective (Mortimer, 1995: 20).

The second weakness is that, instead of bringing this pillar inside the Union and its institutions, primary responsibility for the Common Foreign and Security Policy (CFSP) was allocated to the Western European Union (WEU). First established in 1954, following collapse of the proposed European Defence Community, the WEU had been 'in hibernation' for most of the subsequent thirty years (Duke, 1996: 167). Its role lacked clarity, particularly in respect of other European security bodies such as NATO and the Organization for Security and Cooperation in Europe. A third weakness is that membership of the EU and the WEU were not coincidental: Ireland, Denmark and, until 1992, Greece, were not members of the latter. Fourth, the WEU lacked an effective military capacity and depended on NATO infrastructure. Not surprisingly, in the absence of an effective means of pursuing a collective CFSP, most Member States have opted to pursue specifically national strategies.

The background to the *Third Pillar* is that removal of barriers to the circulation of goods, services and capital in the SM also removes barriers to cross-border crime, illegal immigration, fraud and drug running. The third pillar is an attempt to introduce effective collective EU action into this arena. Of equal importance in understanding this pillar is the backdrop of racism, unemployment, and fears of what, emotionally, is termed the 'invasion' of a poorly defended Fortress Europe by illegal migrants. The Treaty proposed a twin-track approach: the European Council would decide on common visa requirements, while cooperation would be developed in areas of common interest, such as asylum policy. Perhaps the single most important provision of the third pillar was to bring asylum within the competence of the EU.

While the Treaty on European Union addressed some of the decision-making and foreign policy concerns of the EU, it had little direct significance for the structural social problems of the Union, which were fundamentally distributional questions. The main exception was the Social Agreement appended to the Treaty, from which the UK had opted out. Recognizing the social polarization (between capital and labour, and between labour in different countries) which was inherent in the neo-liberal project, the agreement set out a three-tier model of competency in respect of labour markets: issues on which majority voting was permissible, such as working conditions; issues which required unanimous voting, such as social security provision and protection of redundant workers; and matters over which the EU had no competence, such as strikes and pay. It was essentially an enabling rather than a prescriptive agreement, and has brought about no more than limited harmonization of some aspects of labour market regulation.

The Treaty on European Union, which had been conceived of as necessary to address the problems of uneven integration following the SM programme, was itself 'an unstable compromise' (Newman, 1993: 13). This was explicitly recognized in the agreement that the treaty would be reviewed at an IGC in 1996. While apparently wide-ranging, the Treaty is flawed by British and Danish opt-outs of key policy areas, and by the lack of prescription particularly in respect of the second and third pillars. Its importance, therefore, lies not so much in providing a blueprint for economic and political union, but in representing a symbolically important step towards this goal.

Post Maastricht: intentions and realities

The Treaty on European Union was one of the high points in the process of cumulative European integration. Thereafter, the recession of the early 1990s undermined popular confidence in the value of further integration, cast doubt on the anticipated economic benefits of union, and caused governments to become more defensive in the face of electoral discontent and rising unemployment. Ratification of the treaty proved to be problematic. A referendum to approve the treaty was narrowly passed in France, while it took a second referendum and further opt-out clauses before being accepted in Denmark. The UK parliament only ratified the treaty after a long and tense conflict. One consequence was that the 'inevitable logic of integration' was increasingly questioned in the cold light of reality in the 1990s.

First pillar: the price of EMU

To overcome German reservations about EMU, tough qualifying criteria for membership were established, based on Germany's traditional macroeconomic concerns: exchange rate stability, low inflation and interest rates, and low levels of public debt and budget deficits. These were defined in relation to average or best performances in the EU as a whole. The key point is that they required rigorous public sector financial management. The challenge of these stringent convergence criteria was intensified by the recession of the early 1990s. There was also intense speculative pressure against several European currencies, partly because the high costs of German unification were forcing up interest rates across Europe. As a result, in the autumn of September 1992 the UK and Italy were forced out of the ERM, and Portugal and Spain devalued their currencies.

Despite the initial difficulties, most of the convergence criteria became achievable for most of the member states in the course of the 1990s. The most difficult remaining obstacles were levels of public sector debts and deficits. Governments have used both creative accounting and real deficit-cutting measures to achieve these targets. For example, there have been massive privatizations in Spain, Italy and France to raise revenues and reduce deficits, even if such measures, of necessity, make a one-off financial contribution. EMU, therefore, has reinforced neo-liberal policies which were being introduced in the name of increasing global competitiveness.

In practice, there is some room for flexibility in interpreting both the convergence criteria and the effectiveness of governments in achieving these (Artis, 1994: 365). Member states can be accepted into the EMU, even if they do not meet all the criteria, if other governments believe that they are at least converging. Therefore, the ultimate timing and the initial membership of EMU will depend as much on political will as on economic fundamentals. Germany will play a critical role in this, given its traditionally strong monetary and fiscal disciplines. Equally important is the alliance between France and Germany; without this, the EMU project would already have been abandoned in the face of the social and political costs faced by some member states.

Overall, progress towards EMU seems to be all but inevitable. Equally inevitably, its membership and impact will be highly uneven. What, then, will be the consequences of creating a new and more tightly integrated European economic space? In some senses, meeting the convergence criteria will be a relatively easy one-off event compared to the difficulties of meeting tight annual fiscal targets. This will require that the weaker members of the EMU either will have to implement further cuts in public expenditure and will have to receive increased structural aid from the EU (increasingly problematic as all member states have to hold down their public sector deficits), or will have to increase flexibility in labour markets and prices, given that competitive devaluations will no longer be possible. Whichever strategy is followed, there is a clear implication that some social groups, in particular countries and regions, may have to pay a high price for EMU. Furthermore, the EMU macro-region will also throw its shadow over neighbouring territories; non-members will be pressured into tight public sector expenditure curbs in order to remain competitive with the new 'economic heartland' of Europe.

Second pillar: the Balkans and the WEU

The ending of the Cold War was expected to realize a peace dividend for western Europe. Instead, Europe faced a rising tide of nationalism in countries without the mediating frameworks of democratic traditions. Almost simultaneously with the discussions at Maastricht, the first major challenge to the second pillar was brewing in the Balkans, and the crisis in the former Yugoslavia would soon confront the EU's common foreign policy ambitions (Williams, 1996: 216). The member states clearly had shared interests in a collective response to this crisis: it was characterized by proximity, threatened the security and stability of at least one member state (Greece), disrupted trade flows and generated massive refugee movements. It was the first major challenge for the EU and there was a comprehensive failure to formulate and implement a collective response to it.

There were at least three major failures in the EU's collective response. First, due to Germany's insistence on early recognition of Croatia and Slovenia, there was a failure to work out a convincing set of principles for recognizing new states and to be sensitive to the history of the region and to the existence of complex territorial claims. Second, the EU failed to broker a peace in Bosnia, and lacked the resources for either peace making or peace keeping. Third, the decisive European political role was played not by the EU, but by three member states acting in their individual capacities within the five-member contact group: the USA, Russia, France, Germany and the UK (Carter *et al.*, 1995; Williams, 1996).

There were two main reasons for the failure of the WEU. First, it lacked the resources, the command structure and the experience to operate 'out of area'. Only NATO, and in reality the USA, had such military means. Second, the member states were insufficiently committed to the WEU, either because of traditions of non-intervention or, as in the cases of France and the UK, because of their insistence on playing an independent security and foreign policy role. On the evidence of the crisis in the ex-Yugoslavia, then, the EU and the WEU fall far short of constituting a geopolitical institution independent of the established military powers (Smith, 1996: 14).

Third pillar: a failure of limited ambitions

There has been only limited progress in implementing the third pillar concerned with Europe's internal security. There have been achievements, notably an extradition convention, and a Europol Drugs Unit. However, there has also been a failure to agree a common approach to policing external borders, common visa rules, and common approaches to asylum and immigration. Such failures are not unexpected, because of the prevalence of intergovernmental decision making in this area. For example, in 1996 the UK government blocked a joint initiative to fight racism on the grounds that this required changes to UK race relations laws.

The starkest failure of collective action has been in respect of asylum. The background to this was a large increase in refugees in the 1980s at a time of rising unemployment, racism and electoral support for the extreme right. The crisis in the ex-Yugoslavia generated large-scale flows of refugees, as did ethnic nationalism elsewhere in central and eastern Europe. Given proximity, and the long history of labour migration from Yugoslavia to the EU, it was hardly surprising that large numbers of refugees fled to the latter. By late 1992, 550,000 people from the ex-Yugoslavia had become refugees abroad and 220,000 of them had sought asylum in Germany (Williams, 1996).

Not only were the demands for asylum – and the associated financial burdens – distributed unevenly across the EU, but there was also the contradiction of unequal barriers of entry to a Union where there was relatively free movement, particularly following the Schengen agreement. The logical and minimalist response to this was harmonization of processing of asylum applications across the EU. This was a very limited objective and failure to achieve this indicates the fragility of the third pillar. It could be argued that the scale of the refugee problem explains why governments retreated into nationalistic approaches to asylum, but the real crisis was a lack of collective political will rather than the scale of the refugee movements (Robinson, 1995).

Into the 21st century

The European Union in the mid-1990s lies very much in the shadow of Maastricht, or more accurately a failure to harness many of the enabling powers of the Treaty on European Union. Much of the debate in the 1990s about the economic and political spaces of the EU has been concerned with remedying these lacunae. The politics of Maastricht, of the EMU and of the 1996 IGC have been concerned with the regional balance of power and of how to balance the different interests and powers of the UK, France and Germany. In contrast, the debate about the economic future of Europe has long since shifted from Europeanization to globalization and the need to counterbalance the competitive challenge of the Pacific Rim. Debates on European governance can appear parochial when set against the global challenges facing the European economy.

Recession has made the task of reconstructing the European space, and repositioning Europe in the global economy and geopolitics, even more challenging; between 1990 and 1995 the average annual economic growth rate in the EU was only 1.5%, whilst the unemployment rate increased from 8.2% to over 11%. There is also the challenge of how to create a new institutional framework which will facilitate the reintegration of the western and eastern European economies, as well as recognizing the aspirations of the southern applicants for membership: Turkey (see case study), Malta and Cyprus. Progress on these issues is not made any easier by increased support for right-wing, nationalistic and anti-EU political factions across Europe.

One of the difficulties for the EU is the tension between different levels of integration in its economic, political and social spaces (Williams, 1994). In particular, there is a relatively high degree of economic integration, and this will be widened by pre-accession agreements with the eastern European membership candidates, and deepened by EMU. But there is far less political integration where most key decisions remain subject to the cumbersome procedures of intergovernmentalism. There is a compelling logic to bring the EU's economic and political spaces into better alignment if an increasingly Europeanized economy is to be more effectively and democratically governed. The 1996 IGC – or 'Maastricht Two' – provided an opportunity to effect such a realignment, for above all it was about making 'Maastricht One' work and utilizing its potential.

The key issues at the IGC were how to proceed with EMU, and with implementing the objectives of the second and third pillars of the Treaty on European Union. To some extent this is a question about the competencies of different EU institutions and subsidiarity. This is a contested area, reflecting fundamentally different visions of the future of Europe. The Commission favours more majority voting in order to make the second and third pillars effective, and there are also arguments that the European Parliament should be strengthened to remedy the democratic deficit. The member states have very different agendas. Germany generally favours increased economic and political union, including a single European space with guaranteed freedom of movement, a joint approach to asylum and immigration, cooperation against terrorism, and merger of the WEU with the EU. It also accepts increased majority voting as being necessary for efficient functioning of the Community. In contrast, the UK's goal is containment, particularly with respect to extension of majority voting or the powers of the European Parliament.

Turkey and the EU

Jesús del Río Luelmo

Turkey's history as a state is deeply linked to the westernization tendencies launched by its founder, Mustafa Kemal Atatürk, in the first half of the 20th century (Kinross, 1964; Kazancigil and Özbudun, 1981). Indeed, ever since, the neighbouring European bloc has became a model – almost an obsession – for Turkey's political and economic élites. The first significant step towards formal integration into the main Western institutions was its membership of NATO in 1952. Subsequently, the Association Agreement signed with the then European Economic Community in 1963 created a tangible framework for channelling this relationship, and this persists until the present moment. Full membership of the EEC was not necessarily the final goal of this agreement, but its possibility was contemplated in Article 28. In the particular geopolitical circumstances of the Cold War, this led to Euro-centrist views dominating Turkey's foreign policy. Turkey's efforts in this direction, despite variable degrees of success, have been remarkably persistent (Forum Europe, 1992 and 1994). However, at the very moment when the European Union is seeking deeper forms of integration, the issue of Turkey's possible accession seems to have become marginalized in the EU's enlargement agenda. The particular conjuncture of internal and external developments, together with the complexity of the relationship, are at the root of this partial failure of integration.

Turkey's 1987 application for full membership caused surprise in both the domestic and international spheres. The Commission's opinion on it seemed designed simply to delay the final decision. But the world situation changed dramatically in the following years. The signing of the Treaty on European Union, the end of the Cold War, the break-up of the Soviet Union and the Gulf War made the way to accession extremely hard, if not impossible, for Turkey. Hence, there has been a search for an alternative to integration which meets the needs of both Turkey and the EU. The possibility of Turkish integration in the EU is of immense importance. For the EU it has brought into question the very concept of Europe, and has created serious friction in otherwise relatively smooth relations. In addition, Turkey's complex socio-economic and political nature, its internal contradictions and contrasts, have further complicated the issue and have to be taken into account when assessing the relationship with Europe.

An uneasy relationship

One of the keys to interpreting the character of the institutional framework for the relations between Turkey and the EU is the complexity of the international political situation in the early 1960s. First, the rivalry with Greece played a central role in Turkey's decision to follow its neighbour's example and apply for a special relationship with the EU. No less important, the full implications of the process of European integration, so recently initiated by the Treaty of Rome, were not well understood at that time. Therefore, Turkey's application did not receive due consideration by the EEC, which was under pressures of security from the Cold War political environment.

After the signing of the 1963 Ankara Agreement (Evin and Denton, 1990), Turkey began to take the necessary steps to become a full member of the EEC. There were, however, considerable obstacles to the implementation of the agreement. Turkey's economic performance, although characterized by impressive growth during these years, averaging an annual 6.5% between 1963 and 1978 (Balkir and Williams, 1993), has been below that required to fulfil the schedules fixed by the agreement. Besides these internal economic difficulties, other circumstances prevented the deepening of the relationship. There have been three military interventions in Turkey (1960, 1971 and 1980), which have severely disrupted the country's political life. Regardless of their nature and development, these coups had the immediate effect of distancing Turkey from the Western bloc in general, and in particular they slowed down the process of its integration into the European Community. This in turn contributed to delays in the association timetable and to a widening of the social and economic gap rather than the

continued

supposed convergence. From an external point of view, the already delicate relations with Greece suffered a severe blow with the Cyprus crisis of 1974. Turkey's occupation of the northern part of that country placed it in a weak position in the international political arena.

Despite these difficulties, Turkey did advance its westernization strategy during this period (Zentrum für Turkeistudien, 1989): it found a place in the international order, continuous European support for its modernization, and substantial economic improvement (Sen, 1993). These led Turkey to apply officially for membership in 1987 (Eralp *et al.*, 1994).

Turkey's changing relationships with the EU

The relationship between Turkey and the EU has been particularly troublesome. Turkey's economic performance has been highly uneven, and this has hampered both market reforms and trade liberalization (Krueger and Aktan, 1992; Gumpel, 1992). Without any doubt, a *de facto* integration within the EU in economic terms has been the main economic feature of the association between Turkey and Europe. The establishment of strong economic links with EU member states, which account for almost 50% of Turkey's foreign trade, together with its sustained economic growth averaging 5% per annum during the 1980s (OECD, 1995a), are sufficient by themselves for Turkey to be considered as an important candidate for membership (Table 9.2; Figure 9.2). However, other imbalances have threatened to overwhelm these positive aspects: regional differences, economic instability, sectoral imbalances, international migration, democracy and human rights, and the role of Islam.

Questions have arisen about the sincerity and depth of Turkey's democratization process, a prerequisite for any deepening of relations with the EU. The omnipresence of the military has always been a negative factor in any such assessment (Dodd, 1983). Its role as warden of the republican order, and hence its interventions, have provided a significant obstacle to its inclusion in the Western bloc. The presence of political Islam does not improve this view. The recent rise of Islam in politics is often perceived as a threat in the West. To what extent such ideologies accept the pluralist democratic system still remains to be seen, for Erbakan, chairman of the *Refah Partisi* (Welfare Party), with Islamic views, was appointed Prime Minister in 1996. This has happened for the first time since the foundation of the Republic, with Turkey being the only secular Muslim state.

Islam as the creator of a distinctive culture is perceived as one of the main obstacles to a smoother relationship between Turkey and the EU. A certain sense of cultural shock is a consequence of deeply different political, social, economic and cultural traditions (Manisali, 1988; Heper *et al.*, 1993). This might be a sign of the shallowness of Turkey's westernization process. The lack of integration of Turkish workers in Germany is an example of the complexity of this issue. This labour force, now no longer needed, struggles to find its identity between Germany and Turkey (Martin, 1991).

Turkey's poor human rights record does not offer better impressions (HRFT, 1994). Besides general human rights problems, the position of the Kurds seems to be far from resolution, and reveals some of the state's political weaknesses. Kurdish contested sovereignty has proved to be the main domestic concern for the Turkish authorities, partially due to the international echo that the problem has recently produced.

The strategic dimension has always been considered as decisive in the establishment of any association between Turkey and Europe. Turkey's importance to Western defence was a key factor during the Cold War period, particularly for US interests, as reflected in the Truman Doctrine. The end of the Cold War has not changed this importance, giving Turkey the chance both to search for new scenarios and to offer a model for other states in the region, thanks to its secular political system and its close ties with the West.

However negative this overview might appear, the picture would not be complete without taking into account the advantages of a closer relationship. Besides the strategic dimension, there are other considerations. From an economic point of view, Turkey offers the EU not only a large and

continued

Table 9.2 Turkey: GNP and economic growth, 1980–94.

	1980	1981	1982	1983	1984	1985	1986
Nominal GNP per capita ($)	1.556	1.574	1.387	1.280	1.234	1.353	1.487
GDP by origin (%):							
Agriculture	24.2	22.6	22.7	21.6	20.3	19.4	18.8
Industry	20.5	21.5	21.9	22.4	23.1	23.6	25.0
Services	55.4	55.9	55.4	56.0	56.6	57.0	56.2
Real GNP growth (%):	–2.8	4.8	3.1	4.2	7.1	4.3	6.8
Agriculture	1.3	–1.8	3.3	–0.8	0.6	–0.3	3.6
Industry	–3.6	9.9	5.1	6.7	10.5	6.5	13.1
Services	–4.1	5.8	2.3	5.3	8.2	5.0	5.2

	1987	1988	1989	1990	1991	1992	1993	1994
Nominal GNP per capita ($)	1.671	1.748	2.005	2.682	2.620	2.707	3.004	2.193
GDP by origin (%):								
Agriculture	17.2	18.3	16.6	17.0	14.7	14.3	14.8	15.7
Industry	24.9	25.1	25.9	25.0	25.2	25.0	26.9	27.0
Services	57.9	56.7	57.5	58.0	60.1	60.7	58.3	57.3
Real GNP growth (%):		9.8	1.5	1.6	–0.4	6.4	8.1	–6.0
Agriculture	0.4	8.0	–7.7	0.3	–0.3	4.0	–1.3	–0.3
Industry	9.2	2.1	4.9	–2.8	2.9	6.3	8.2	–5.7
Services	13.2	–0.8	3.2	6.6	0.4	6.1	9.8	–9.2

GNP is in 000s.
Source: State Institute of Statistics, 1996.

relatively wealthy market, but also an open gate to economic penetration to the East. In relation to the process of European integration, it is possible to state that Turkey's full membership would be an efficient way of establishing a counterbalance to the EU's new northern orientation since the 1995 enlargement. From a cultural point of view, it would contribute to a richer, more diverse society.

To be, or not to be a member? Towards a special relationship

While none of the obstacles outlined above can explain by itself the opposition within the EU to Turkey's full membership, in combination they seem to offer a pessimistic panorama when considering the possibilities for future integration. Strategic interest has been the key to the somewhat artificial formal relationship that has been developed, based on the institutional need for cooperation rather than on a real will to integrate. From this assessment arises the need to search for a balance between Turkey's aim of seeking full membership, and the EU's interest in keeping Turkey outside the EU but not at too great a distance. The question, then, is not whether Turkey will gain access to the EU but, instead, what kind of relation will be established between the two parties. The complexity and peculiarities which define this case seem to open a door to the development of a special relationship. The Customs Union has probably already achieved the limit of the present tendency towards full membership. At the same time, the sudden post-1989 incorporation of other peripheral areas of the continent is widening the debate about the nature of post-Maastricht Europe, and raises the question of whether it is necessary to pursue new forms of integration.

The opportunity for Turkey to secure full EU membership seems to have passed. The southern enlargement in the 1980s was a missed opportunity, particularly in the context of Greece's accession.

continued

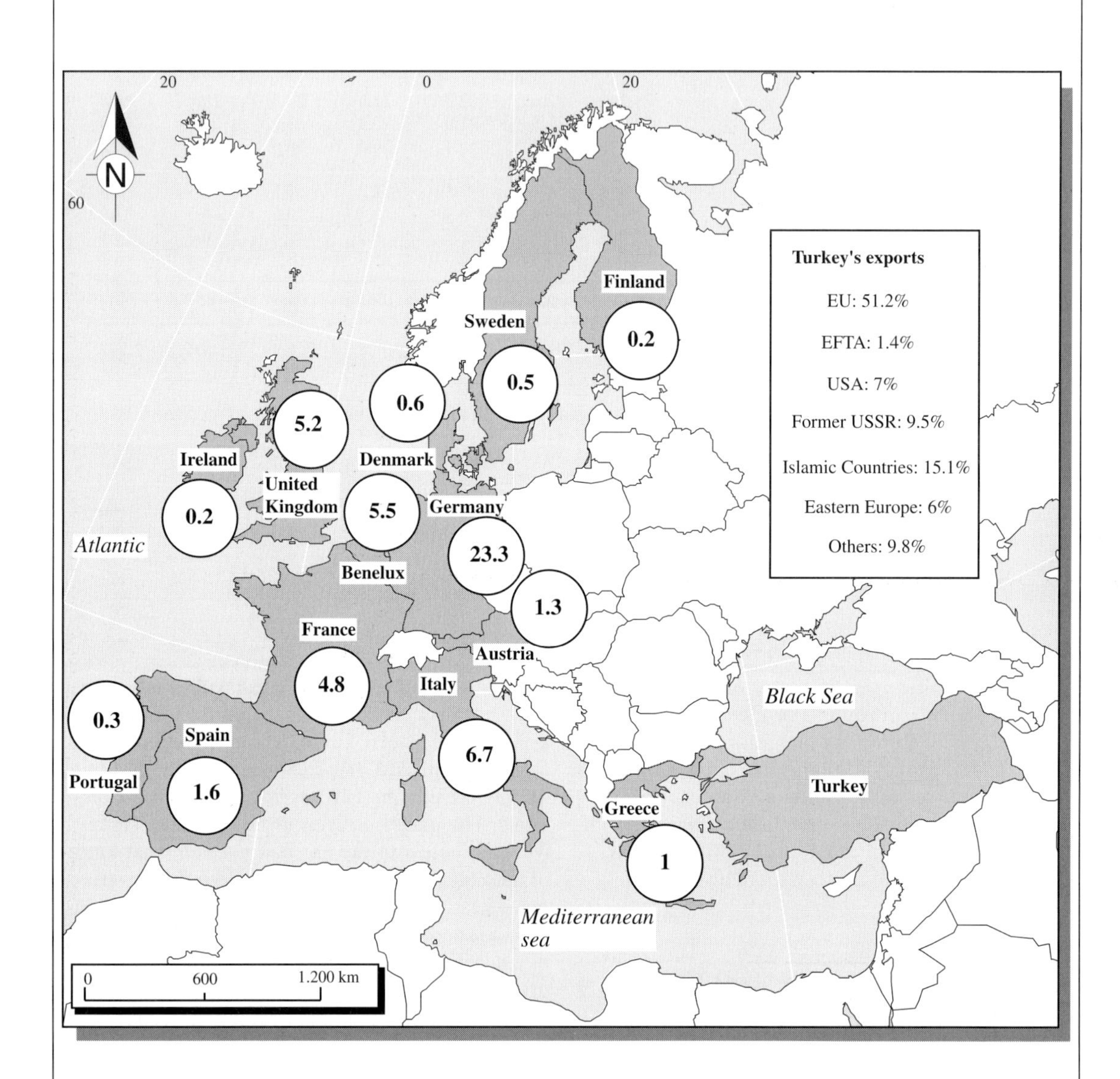

Figure 9.2 (a) Turkey's exports in 1995 (%) (*source*: State Planning Organization, Ankara, unpublished data).

continued

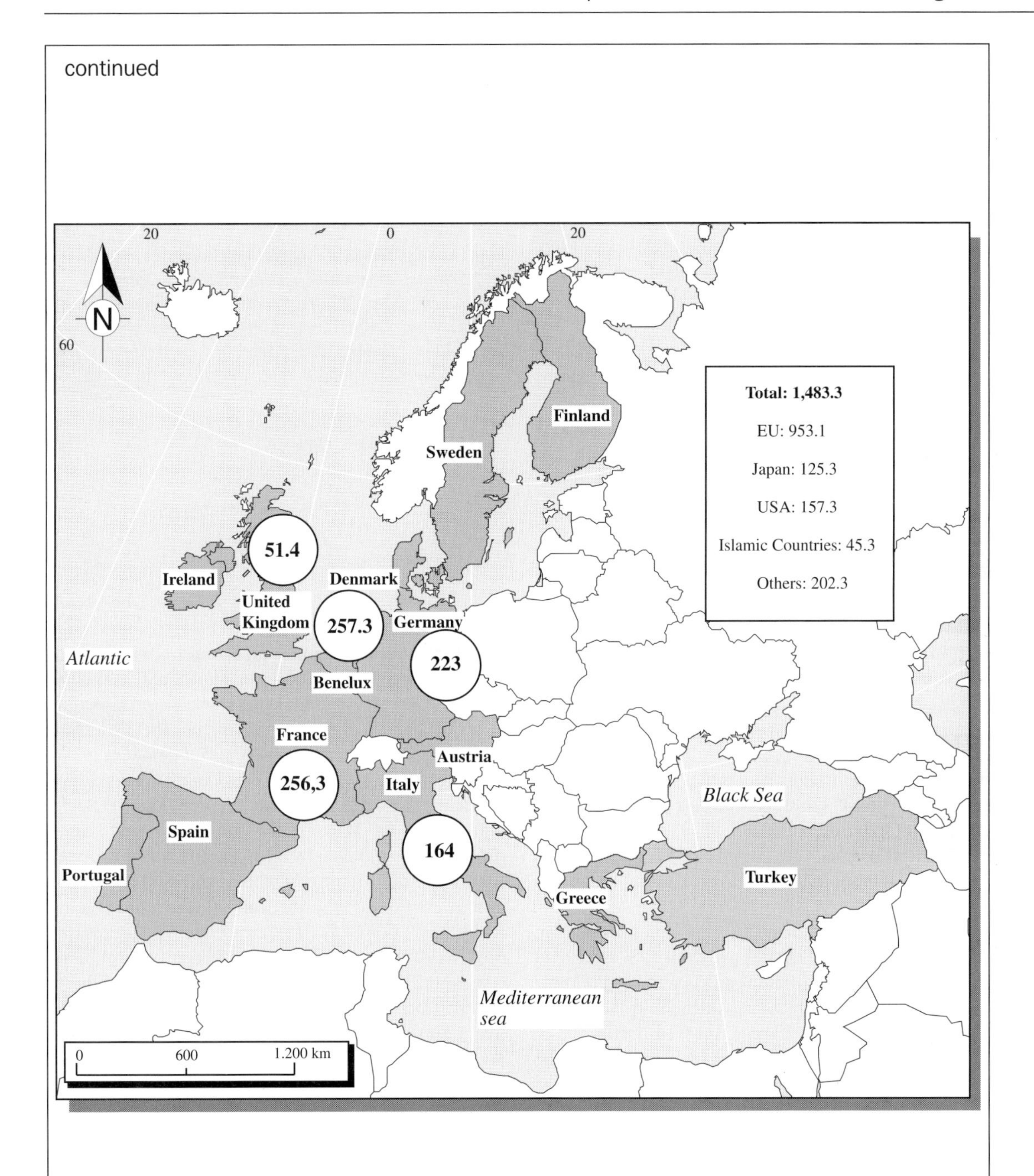

Figure 9.2 (b) Turkey's sources of foreign direct investment in 1994 in $ millions (data from SPO, August 1996).

continued

Maastricht, with its implications for a deeper integration, and the opening up of eastern Europe seem to have drawn a line under Turkey's application. The changing geopolitical context of the 1990s also suggests that Turkey's economic, political, cultural and strategic pivotal role could be negatively affected by a very clear pro-Western option in its foreign policy. Indeed, the recently opened Turkic Republics, and the Black Sea Cooperation Scheme, offer interesting alternatives to European integration. Continuous rejection by the EU is having a negative effect on Turkey's internal stability and its Western orientation. Any 'final' rejection will have to be handled very carefully and needs to be accompanied by clear alternative political and economic strategies.

Further reading

Eralp, A. *et al.* (eds) (1994) *The Political and Socioeconomic Transformation of Turkey*, Praeger, London

Evin, A. and Denton, G. (eds) (1990) *Turkey and the European Community*, Leske and Budrich, Opladen

Gumpel, W. (ed.) (1992) *Turkey and the European Community: An Assessment*, Verlag, München

Heper, M. *et al.* (eds) (1993) *Turkey and the West. Changing Political and Cultural Identities*, I.B. Tauris, London

Sen, F. (ed.) (1993) *Turkey and the European Community*, Leske and Budrich, Opladen

Recent experience has shown that a redistribution of power between EU institutions is a prerequisite for meaningful operationalization of the second pillar and for further enlargement. However, any deepening of integration resulting from the 1996 IGC is also likely to increase the obstacles to enlargement. As Smith (1996: 6) writes, 'At one and the same time, the EU has appeared as a model of democratic and economic stability to be pursued by the new or newly democratic countries of Europe, and as a symbol of how far they have to go to reach the promised land'.

Finally, we need to return to the point stressed in the introduction to this chapter: the EU was created in the 1950s in very different political and economic conditions to those prevailing in the mid/late 1990s. The 1980s and 1990s have witnessed a number of attempts to reform the political institutions and the markets of the EU in pursuit of its founding goals: to create a genuine single market in line with a neo-liberal vision, and to give Europe a collective voice in geopolitics. However, the economic backdrop to, and social fabric of, the EU have also changed fundamentally. Perhaps the most salient feature of late 20th century Europe is increasing social polarization, related to the repositioning of Europe in the global division of labour.

In this context, the crisis of how to maintain a social democratic welfare state model will become increasingly a European rather than a national issue. Lipietz (1996: 371) summarizes the key contradiction: 'The social crisis of the EU stems from the dichotomy of social norms which are still "social democratic" on the one hand, and ultra-liberal institutions which condemn these norms on the other'. While the Social Agreement in the Treaty on European Union provided a modest reinforcement of European social democratic norms, this is outweighed by the macro-economic implications of the convergence requirements of EMU. Lipietz (1996:371) argues that the loss of macro-economic sovereignty implied by convergence implies that countries end up seeking to 'export' their unemployment by being more competitive than their neighbours. It can also be argued that membership of the EMU is likely either to demand increased labour market flexibility from the participants or to result in further exposing their structural weaknesses, especially in terms of unemployment. In any event, the greatest future challenge for the EU may be neither reform of its political institutions nor enlargement, but how to reconcile a neo-liberal strategy for increased economic global competitiveness with the social and political tensions that this seems certain to generate.

Further reading

Artis, M.J. and Lee, N. (eds) (1994) *The Economics of the European Union,* Oxford University Press, Oxford

Lipietz, A. (1996) Social Europe: the post-Maastricht challenge, *Review of International Political Economy*, **3**, 369–79

Smith, M. (1996) The European Union and a changing Europe: establishing the boundaries of order, *Journal of Common Market Studies*, **34**, 5–28

Williams, A.M. (1994) *The European Community: the Contradictions of Integration*, 2nd edition, Blackwell, Oxford

PART III Economic production and exchange

CHAPTER 10 Agricultural production

BRIAN ILBERY

The context of change

Europe has long been characterized by a diversity of agricultural production systems, both within and between western and eastern regions. Major structural changes have occurred in the types and intensity of agricultural production since 1950. In western Europe, the modernization and industrialization of agriculture reached a peak in the late 1970s and early 1980s. Since then emphasis has been placed on reducing agricultural output and farming in more environmentally beneficial ways. In eastern Europe, agricultural industrialization also occurred, but it was different in nature from that found further west. A second major restructuring of agriculture has taken place since 1990, this time based on a return to private farming.

The main driving force affecting agricultural change in both western and eastern Europe has been government policy. Nevertheless, it has evolved quite differently in the two regions, regulating agrarian production in contrasting ways. The reasons for government intervention have varied significantly. In western Europe, farm income levels have long been of serious concern, and a major objective of agricultural policy has therefore been to ensure a fair standard of living for the agricultural community. Indeed, the Common Agricultural Policy (CAP) has played an important role in providing support for farm families. However, the shift towards a post-productivist farming system (Shucksmith, 1993) since the mid-1980s has led to a radical rethink of the CAP and financial support for farm families. The problem is that policy changes could have major implications for the economic and social well-being of many farming communities, especially those in 'peripheral' agricultural areas. In eastern Europe, the adoption of socialist principles after 1945 led to large-scale government intervention in agriculture, at an unprecedented level compared with western Europe. Socialist ideology meant the attempted elimination of private ownership in agriculture and the incorporation of farming into industry, the perceived backbone to economic development. Agriculture was to conform to centralized five-year plans and state regulation. However, the demise of state socialism in 1989–90 led to a movement towards market economies and the reintroduction of private farming. This has not been an easy transition and many problems still confront agriculture in most east European countries.

Government policy and agricultural restructuring

Two major phases of restructuring characterize the development of agriculture in both western and eastern Europe since the Second World War. In western Europe, the first phase of agricultural *productivism* lasted until the mid-1980s. This was typified by an almost uninterrupted modernization and industrialization of agriculture, with an emphasis on raising farm output through high levels of government support (Ilbery, 1990a). The second phase of *post-productivism* has developed since the mid-1980s and is characterized by the integration of agriculture within broader rural economic and environmental objectives. Although the nature of post-productivist agriculture is as yet little understood, various attributes of what has been termed the *post-productivist transition* (PPT) can be listed (Lowe *et al.*, 1993; Shucksmith, 1993; Ilbery and Bowler, 1998). These include a reduction in food output, an emphasis on food quality rather than quantity, the progressive withdrawal of state subsidies for agriculture, the growing environmental regulation of agriculture, and the creation of a more sustainable agricultural system. In eastern Europe, the post-war continuation of land reform and resettlement, where small strips and blocks of land were redistributed to the peasantry, was seen as the first step to more radical changes in agriculture and an attempt to lessen the opposition to centralized industrial policies (Warriner, 1969; Rugg, 1985). Through a process of *collectivization*, farmers were encouraged to cooperate and form larger enterprises which were effectively controlled by the state. This phase lasted until 1990, since when it has been replaced by the *privatization* and/or *reprivatization* of farming, where the aims are to recreate landed property rights and to develop an efficient market system of agricultural production. Thus, while western Europe is currently attempting to control the volume of agricultural production, eastern Europe is trying to introduce measures which will increase output.

Policy development in western Europe

The basis of agricultural policy in western Europe has been a system of *price support* for different products. This helped to achieve the twin goals of increasing both agricultural productivity and farm incomes. In the European Union (EU), for example, the CAP fixed *guaranteed prices* for a range of agricultural products. These were set well above world market prices and applied to all farmers within the EU. The system worked through *intervention buying*, whereby intervention agencies purchased any surplus production in the marketplace in order to keep prices artificially high. Such high guaranteed prices had the effect of encouraging farmers to produce as much food as possible, since there was no upper limit on the amount of production eligible for guaranteed prices. Both efficient and inefficient farmers were thus encouraged to remain in farming and the consequence was the overproduction of many agricultural products, leading to the infamous 'lakes' and 'mountains' of food. Such surplus production had to be either stored, exported outside the Union at subsidized prices, or destroyed.

Geographically, the price support system of the CAP favoured 'northern' products such as milk, beef and cereals over 'southern' products like wine, vegetables and fruit. As the northern areas of the EU also tended to have larger farm size structures, agricultural production became increasingly concentrated in these 'core' areas (Bowler, 1985). This led to increasing regional disparities in farm incomes and thus to the designation in 1975 of a number of Less Favoured Areas (LFAs). Within these 'peripheral' areas, an annual compensatory allowance was available to farmers to cover the increased costs of farming; this was payable as a grant per hectare of land or per head of livestock. However, such areas cannot compete with agricultural production in 'core' areas, and it became clear that the problem of low farm incomes in marginal agricultural areas could not be solved simply through production grants and subsidies.

By the mid-1980s, it was recognized that the established model of agricultural productivism, whereby farm incomes can be maintained by increasing output at high guaranteed prices, had to be dismantled. As the Commission of the European Communities (1986: 8) stated, 'European agriculture has to accept economic realities and learn to produce for the market, to adapt to commercial demands, and to continue to modernise'. A series of measures designed to control agricultural production in the EU were introduced in the mid-1980s. These included guaranteed thresholds for cereals, sunflower seeds, processed fruit and raisins, and compulsory quotas in the milk sector (Ilbery, 1992). Also introduced were such environmental measures as Environmentally Sensitive Areas (ESAs) and a farm woodland scheme, and extra funding for diversification into tourism and crafts in LFAs. A major effort to control output in the cereals sector came in 1988 with the introduction of a voluntary arable set-aside scheme, where farmers were compensated for taking at least 20% of their arable land out of production altogether for five years.

However, by the late 1980s it was clear that production control measures were not solving the major problems of EU agriculture. Consequently, the European Commission recast the objectives of policy reform in agriculture to make producers more subject to market controls. This was translated into action through the MacSharry proposals (1991) and reforms of the CAP (1992). MacSharry proposed substantial cuts in support prices for cereals, with smaller cuts for beef, milk and butter. He also advocated controls on sheep farming, extensions to the original set-aside policy, early retirement for farmers, and a system of *income aid* to compensate farmers with small and medium farms for their projected loss of income (Robinson and Ilbery, 1993).

The final CAP reform package in 1992 was quite complex, but the main measures are summarized in

Table 10.1. Guaranteed prices for cereals were cut by 29% over three years and, to compensate, *all* farmers receive income aid in the form of area-based direct payments. However, these payments are available only if farmers set aside a proportion of their arable land on a rotational basis over six years. Non-rotational set-aside is also possible. Effectively, set-aside is now compulsory if arable farmers wish to qualify for income aid, except for those producing less than 92 tonnes. In the beef and sheep sectors, a system of quotas was introduced to limit the number of livestock on which farmers can receive payments (Wynne, 1994); milk quotas were also retained. The overall objective of the CAP reforms, therefore, is to decouple the link between farm incomes and the volume of food produced, through a movement away from high support prices for food towards direct income aid for farmers. Such aid can be in the form of compensation for lower guaranteed prices and/or payments for 'farming the countryside'.

Indeed, the CAP reforms also require member states to implement a package of 'accompanying measures' relating to the afforestation of farmland, early retirement, and an agri-environmental programme. The latter seeks to limit agricultural production by encouraging farmers (voluntarily) to adopt more extensive and less polluting farming practices (see case study about Germany). Farmers are to be financially compensated for loss of income and 50% of the eligible expenditure is borne by the EU budget. Each member state had to submit and have approved an agri-environmental package by the end of 1993. Unfortunately, many agri-environmental schemes are restricted to 'designated areas' and are 'bolted on' to an economically driven policy (Robinson and Ilbery, 1993).

Table 10.1 Reforms of the Common Agricultural Policy, 1992: key points.

1	29% cut in cereal support over three years
2	Compensation for compulsory rotational set-aside (specified % of arable land)
3	Income aid on remaining arable land for farmers setting aside specified % of arable land
4	Removal of co-responsibility levies
5	Farmers allowed to withdraw from original (voluntary) set-aside scheme
6	New non-rotational set-aside to be introduced
7	Quotas in the beef, sheep and milk sectors
8	Additional financial aid for environmental protection, forestry and early retirement (accompanying measures)

Policy development in eastern Europe

The main objective of post-war agricultural policy in eastern Europe was to increase self-sufficiency, especially through the development of grain production. This was to be achieved by adopting the Soviet model of *collectivization*, whereby farmers were encouraged to cooperate and form larger enterprises. Collectivization was dominated by two types of enterprise: state farms and agricultural cooperatives (collectives). The former were owned by the east European states and financed by central authorities. They were established mainly from privately owned large estates in areas of low population density. Their objective was to act as models of management and experimental technology, often in relation to specialist types of farming. The latter were, in theory, controlled by the collective members, among whom the profits were shared (on the basis of the volume of work contributed). In reality, the collectives were controlled by central authorities, which made major decisions on production levels, input and output prices and agricultural subsidies, and took a quota of the profits and produce. Cooperative members were allocated a *household plot*, of around 0.5 ha, for their own private production of vegetables, fruit and livestock. Both state farms and collectives were incorporated into the totally planned economies of eastern Europe and thus guided by five-year plans.

Despite considerable resistance to this new state system of agriculture, notably in Poland, the former Yugoslavia, Czechoslovakia and Hungary, different versions of collectivization had been established as the basic form of land occupation in eastern Europe by the mid-1960s. In Bulgaria, for example, over 1 million private farms had been replaced by 1000 large-scale collectives, each over 4000 ha, by 1957 (Cousens, 1967). Collectivization was more readily accepted in the less developed east European economies, which lacked the capital to modernize in the 1950s. However, by the end of the 1960s it was already clear that the centralized system of control was constraining the growth of agricultural production. Individual farmers had been deprived of their land and thus lost the incentive to work hard.

Between the late 1960s and 1989, three waves of attempted reform affected east European agriculture (Csaki, 1990). The first wave advocated such policy measures as decentralization, increased independence for the collectives and greater use of economic incentives to encourage production. Most countries

Agri-environmental issues in Germany

Geoff A. Wilson

Germany occupies a special role in Europe with regard to agriculture. Not only is it the most populous and one of the largest countries, with 80 million people, but it has also recently undergone dramatic political and socio-cultural changes as a result of German reunification in 1990 (Jones, 1994). Germany, therefore, provides an interesting case study with which to evaluate agri-environmental issues in a changing Europe.

Germany has a complex agri-environmental situation. The country includes a wide variety of different landscapes, ranging from high mountainous areas in the south, to flat and marshy areas in the north. The result is a diversity of agricultural land uses, including extensive pastoralism on high mountain meadows, intensive arable areas on fertile loess soils in central Germany, and highly intensive specialized horticultural land use including, for example, viticulture or hop production in the south and south-west.

Like most other countries in the European Union (EU), agricultural production in Germany increased dramatically after the Second World War. Intensification brought with it increased use of fertilizers, pesticides and herbicides, as well as the destruction of many semi-natural habitats on farms, the pollution of groundwater and rivers, and soil erosion and degradation (Knickel, 1990). In descending order of priority, agricultural mismanagement has affected biotopes and habitats, groundwater, land and soils, surface waters and food quality (Höll and Von Meyer, 1996). Yet, compared with many other European countries, Germany was relatively quick to respond to the degradation of the countryside. Germany has been a crucial partner in the initiation of a variety of EU regulations aimed at controlling environmentally destructive agricultural management practices.

In this regard, Germany stands out from other countries in terms of its agri-environmental policies in three major ways. First, its federal political structure means that implementation of policies rests largely with the individual regions (*Länder*) rather than with the national government. This leaves substantive decision-making power to individual regions in terms of interpreting EU and national regulations, resulting in a set of agri-environmental policies that vary considerably from *Bundesland* to *Bundesland* (Höll and Von Meyer, 1996). As a result, implementation of agri-environmental policies in Germany is different from that in more centralized European countries, such as France and the UK, and is more similar to countries with regionalized policy implementation structures, such as Spain (Hoggart *et al.*, 1995; Whitby, 1996). In environmental terms, the federal structure of German agri-environmental policy making has both advantages and disadvantages. On the one hand, individual *Länder* are able to cater for specific environmental needs in their respective countryside and may, therefore, be more willing to implement small schemes targeting specific habitats or groups of farmers (Wilson, 1994). On the other hand, a holistic approach towards countryside conservation is often precluded, as the boundaries of scheme areas often coincide with the boundaries of regions. Comprehensive protection of similar habitats across regions may, therefore, be impeded. An overall result of Germany's federal structure is that it has the most complex set of agri-environmental schemes in Europe. Overall, 91 schemes are offered in the 16 *Länder*, often overlapping in aims and objectives, and relying on a relatively large bureaucracy to implement and monitor this vast array of schemes (Wilson, 1994, 1995).

Second, the environmental movement developed early in Germany as a political force (*Die Grünen*), and substantive pressure was exerted from the 1970s onwards by German environmental pressure groups to establish schemes for the protection of the countryside and to switch from environmentally damaging input-intensive farming to organic farming. Indeed, Germany has one of the highest rates of organic farming in the EU (Höll and Von Meyer, 1996). These developments are also partly linked to the strong 'alternative' movement in Germany, which has forced both policy-makers and farmers to reassess their policies and environmental practices. As a result, Germany initiated its own regionally funded agri-environmental schemes comparatively early. For example, pro-

continued

grammes were established in the 1970s for the protection of wildlife strips along fields and for the protection of orchards (Wilson, 1994). This has meant that by the time funding became available at EU level based on European Commission regulations (797/85, 2078/92), the West German agri-environmental programme was already well under way. Nonetheless, a recent analysis of schemes across Germany has shown that the amount of financing available for agri-environmental schemes varies considerably from region to region, and that former East German *Länder*, in particular, are only just beginning to implement large-scale and long-term schemes (Wilson, 1994, 1995). Indeed, in former East Germany little regard was paid to the environment during the pre-1990 communist era (Wilson, 1996), and much work now remains to be done with regard to the implementation of agri-environmental policy making and countryside protection.

Third, several regions, all within the west, offer some of the most comprehensive agri-environmental schemes in Europe in terms of available scheme finance and possible uptake by farmers. On the one hand, this suggests a conservation-oriented attitude among many policy-makers in German regions, but on the other it also reflects the drastic need to curtail environmentally destructive management practices in some of the worst-affected areas. Schemes such as the Bavarian Kulturlandschaftsprogramm (KULAP) and the MEKA programme in Baden-Württemberg (both in the south of Germany) command budgets that are well in excess of many national agri-environmental schemes in the EU. A closer look at the latter programme offers some interesting insights into the approach by one of Germany's regions to tackle the problem of environmental degradation in the countryside.

The MEKA programme in Baden-Württemberg (south-west Germany) is a relatively recent scheme (Figure 10.1). Initiated in 1992, it was entirely financed by the region for two years, until the EU agreed to pay half of the costs (since 1994) through the accompanying measures of the Common Agricultural Policy based on the MacSharry reforms. MEKA is special in two respects. First, it operates on a unique points system that allows almost any landholder (only very small holdings of under 2 ha are excluded) to enter parts or all of their land into a variety of land management agreements. Second, 55,000 farmers (65% of those eligible) are participating in the scheme (Wilson, 1995). This large number of participants indicates the vast scale of the scheme, and the large financial backing necessary to support continuous payments on these farms (yearly budget well in excess of £200 million). A survey in one part of Baden-Württemberg showed that most farmers were happy with the scheme, and that MEKA seemed to be particularly well suited to farms located in water protection areas (i.e. relatively strict regulations on farm inputs were already in place before the scheme started) (Wilson, 1995). This study also revealed that MEKA for the first time offered opportunities for sustainable management of rare juniper moorlands, and that farmers welcomed additional payments for the maintenance of these threatened semi-natural habitats. Most important, however, MEKA also has important socio-economic impacts, ensuring the survival of many economically marginal farms, which obtain up to 40% of their annual income through the programme.

To conclude, the German agri-environmental situation reveals two important issues with regard to countryside protection in a changing Europe. First, it highlights the increasing importance that many countries are attaching to agri-environmental schemes aimed at rewarding farmers for the continuation, or introduction, of environmentally friendly farming practices. Second, it also shows that many European countries are following rather individualistic paths in terms of interpretation of what 'countryside conservation' should entail, and what schemes should be implemented to alleviate existing problems.

Despite doubts over whether other countries have the financial resources that Germany has available to support such agri-environmental issues, there may be lessons to be learnt from the German situation. For example, comprehensive schemes that have been well received by farmers, such as the MEKA programme, could provide a blueprint for other European regions, particularly where at the moment only relatively small schemes in geographically restricted areas are available. Other European countries could also learn from the incremental agri-environmental policy-making

continued

Figure 10.1 The MEKA scheme in Baden-Württemberg (adapted from G.A. Wilson, 1995, German agri-environmental schemes II: the MEKA programme in Baden-Wüttemberg, *Journal of Rural Studies* **11**(2), 149–59, copyright 1995, with permission from Elsevier Science Ltd, The Boulevard, Langford Lane, Kidlington, Oxford OX5 1GB, UK).

continued

Figure 10.1 The MEKA scheme in Baden-Württemberg.

Total number of participants:	55 000	
Total number of eligible holdings:	85 000	
Participation rate:	65%	
	Area, ha ('000)	% of eligible area
Extensive pasture management	479	27
Specific pasture management prescriptions	140	23
Steep pasture management	64	11
Maintenance of traditional orchards	61	10
Abondonment of wheat growth-regulators	74	35
Arable extensification (reduced planting density)	64	12
Conversion arable to pasture	207	25

Source: adapted from Wilson (1995).

process in Germany, which often involves not only 'official' environmental managers, such as agricultural officials, but also consultations with other environmental actors such as environmental NGOs or farmers' organizations.

Environmental degradation is still continuing in some parts of the German countryside, and the recency of many agri-environmental schemes makes it difficult to assess their overall environmental impact at this stage. What is increasingly clear, however, is that German agri-environmental schemes have at least reduced environmental degradation in the countryside. They have also heightened awareness among the general public that such schemes need to be put in place if the European countryside is to be protected from further intensification and environmental mismanagement.

Further reading

Hoggart, K., Buller, H. and Black, R. (1995) *Rural Europe – Identity and Change*, Edward Arnold, London

Whitby, M. (ed.) (1996) *The European Environment and CAP Reform: Policies and Prospects for Conservation*, CAB International, Wallingford

Wilson, G.A. (1994) German agri-environmental schemes I – a preliminary review, *Journal of Rural Studies*, **10**(1), 27–45

Wilson, G.A. (1995) German agri-environmental schemes II – the MEKA programme in Baden-Württemberg, *Journal of Rural Studies*, **11**(2), 149–59

Wilson, O.J. (1996) Emerging patterns of restructured farm businesses in Eastern Germany, *GeoJournal*, **38**(2), 157–60

adopted some measures which did weaken central control and foster decentralization. However, political factors remained dominant and many reform ideas were never implemented. Real changes did occur in Hungary, where the socialist orthodoxy was tempered by partial devolution of management decisions to the level of the individual enterprise and a more open approach to the development of small-scale agricultural production on the household plots (Symes, 1993). Whilst still conforming to state plans and production targets, Hungarian collectives could sell their products as they wished; they were even offered incentive schemes by the government. The household plots flourished and their economic significance became inversely related to their size. Not surprisingly, agriculture developed more rapidly in Hungary than in any other east European country.

Economic crises in the late 1970s, both internally and globally, led to a second wave of reforms, very much along the lines of the first wave. These again proved ineffective in most areas and it was not until the third wave of reforms in the mid-1980s that radical change was demanded. Csaki (1990) outlined how the reform package aimed to improve the efficiency and quality of agricultural production, move towards a price policy which reflected the real costs of production, increase the role of financial incentives, give greater decision-making freedom to the cooperatives, widen the possibilities of private agricultural production, and increase the significance of the household plots. This increased flexibility, though, was unevenly applied. For example, while both Hungary and Czechoslovakia actively pursued price reforms, financial inducements and private agriculture, East

Germany and Romania experienced very limited decentralization and decision-making freedom. The state continued to price agricultural products at low levels and to keep the cost of essential inputs high. Not surprisingly, many collectives entered the new agricultural era in 1989–90 heavily indebted.

The year 1989 saw the demise of state socialism in eastern Europe and the search for a new agricultural structure based on *privatization* and a *market economy*. This search is clearly evident in all countries, although the rate of progress has varied and been slowest in Bulgaria and Romania, where Communist parties have remained fairly stable (Repassy and Symes, 1993; Agocs and Agocs, 1994; Kopeva *et al.*, 1994; OECD, 1994a, b, 1995b). The privatization of agriculture will continue into the next century, but the move towards a market economy has included the following dimensions:

1. The immediate end to centralized planning and traditional production subsidies.
2. The creation of a free market in agricultural production, where both producer and consumer prices are freed from state control.
3. The privatization and/or reprivatization of land and the development of independent farming.

Whilst many common problems confronted east European agriculture as it moved towards a market economy, including a lack of capital for modernization of the food supply system and the breakdown of trading links with the former Soviet Union, most attention has been focused on land reform and the re-creation of individual property rights. In the longer term, the privatization of agricultural land is inevitable. However, in the shorter term there has been considerable debate about whether land should be privatized or reprivatized. The latter involves returning land which was collectivized (restitution) to former farmers and their heirs. This would inevitably lead to an agricultural structure based on large numbers of very small holdings, which would not be able to compete in increasingly global markets. The former involves the distribution of cooperatively held land (collectives) to members according to their contribution to the cooperative. Members have the right to pool this land and to farm it together in a reorganized and privatized cooperative. Such a privatization process helps to ensure larger farm size structures and greater continuity of land occupancy, as well as slowing down the inevitable transformation of agricultural structures. In reality, different forms of privatization and reprivatization have been introduced in eastern Europe and it has proved difficult to develop an agricultural land market (Kopeva *et al.*, 1994; Wilson, 1996).

Changing patterns of agricultural production

The different phases of agricultural restructuring in Europe have been associated with different patterns of agricultural production. European agriculture was 'industrialized' in contrasting ways in the 1960s and 1970s, since when both western and eastern regions have developed separately as they make the transition to post-productivism and private farming respectively.

The modernization and industrialization of agriculture

Agricultural production systems were modernized and industrialized throughout Europe before the end of the 1970s. A detailed account of these processes in western Europe has been provided by Bowler (1985, 1986). He suggested that agricultural modernization was characterized by three structural dimensions: intensification, specialization and concentration. *Intensification* can be measured in terms of either increasing farm inputs (e.g. fertilizers, herbicides) or increasing farm outputs (e.g. production of crop or livestock products) per hectare of land. All western European countries were characterized by agricultural intensification between 1950 and the mid-1980s (Jansen and Hetson, 1991). This was particularly noticeable in south-east England and central France, areas associated with large farm structures and arable farming. However, by the mid-1980s there were early signs of a trend towards agricultural extensification in parts of Greece, northern and western France, Wales, Northern Ireland and northern England (Bowler and Ilbery, 1997).

Specialization refers to the proportion of the total output of a farm, region or country accounted for by a particular product. Regions in western Europe have tended to increase their specialization in the products for which they have a comparative economic advantage, thus creating a pattern of increasing regional specialization (Bowler, 1985). Bowler (1986) demonstrated how two-thirds of the census regions in the EU-9 became more specialized in agricultural

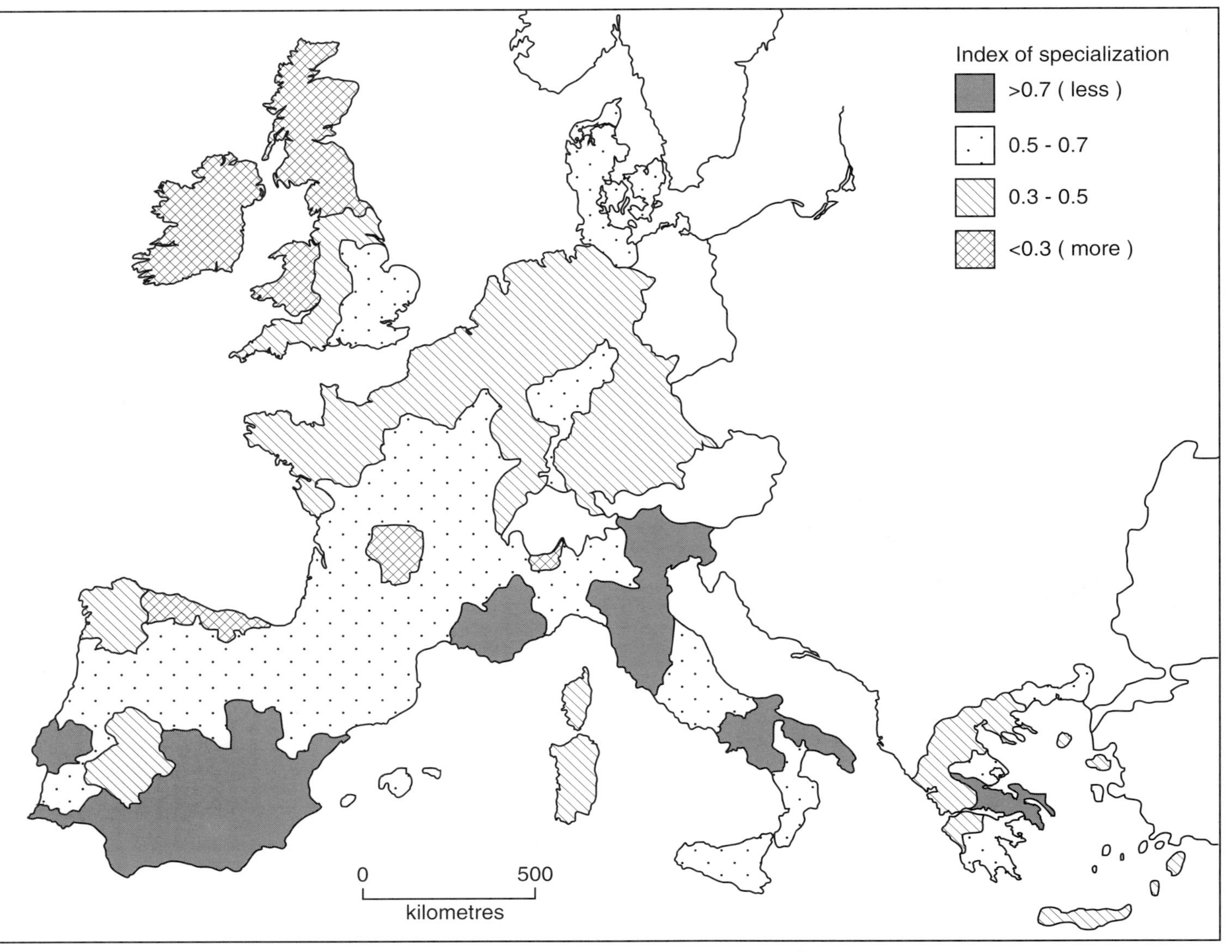

Figure 10.2 Regional specialization in EU agriculture, 1987 (*source*: © INRA Paris 1997, *CAP and the Regions*, Laurent, C. & Bowler, I. (eds), map from I. Bowler and B. Ilbery).

production between 1964 and 1977, with north-western regions associated with grass-based beef and milk cows, interior lowlands in France specializing in field crops and cereals, interior upland and mountain zones based on beef and sheep, southern coastal fringes characterized by wine and intensive fruit crops, and the Netherlands and Belgium specializing in intensive (often factory-based) livestock. By the end of the productivist period, patterns of regional specialization were quite pronounced (Figure 10.2). Using an index of entropy based on five major land uses in the EU in 1987 (cereals, permanent pasture, fresh vegetables, industrial crops and vines), Bowler and Ilbery (1997) demonstrated high levels of agricultural specialization in northern, mountainous and peripheral regions, where grazing pasture dominates the land use structure (Figure 10.3). These reflect well-known differences between the dairy regions of the Netherlands, the sheep farming regions of Scotland, and the beef cattle farming regions of west Ireland. In contrast, lower levels of specialization are found within the Mediterranean area of the EU, including southern Spain, south-east France, parts of central and southern Italy, and central Greece; these are explained by the traditional combinations of vegetables, cereals and vines in Mediterranean farming systems (Figure 10.4).

Concentration describes the increasing proportion of either productive resources (especially labour and capital) or farm production (outputs) located in a smaller number of regions. In terms of outputs, for example, Bowler (1985, 1986) has shown how the production of wheat, potatoes, milk and oilseeds within the EU became increasingly concentrated in Denmark, Ireland, the UK and West Germany, while the production of fresh fruit, pork, eggs and mutton became more concentrated in France, Belgium, the Netherlands and Italy. At the farm level, concentration has led to a decline in the number of small farms and the concentration of land into larger farms.

In addition to these three dimensions, a fourth dimension – the absorption of agriculture into the wider *food supply system* – led to the industrialization of agriculture. This process was further developed in some regions than in others. For example, the development of large-scale farms with close economic links with agro-food companies and food retailers encouraged 'agribusinesses' in such regions as East Anglia (UK), the southern Netherlands, the Paris Basin (France) and Emilia Romagna (Italy). These agribusinesses offer production contracts to those farmers within relatively close proximity to processing plants who have larger farms and particular types

Figure 10.3 Dairy cattle grazing near Jever, northern Germany (*source*: Tim Unwin, 6 August 1985).

Figure 10.4 Wheat, vines and olives growing near Lake Trasimeno, Umbria, Italy (*source:* Tim Unwin, 12 July 1987).

of farming, such as vegetables, fruit, pigs and poultry. This further perpetuates the spatially uneven development of agriculture within western Europe, with particularly adverse consequences for farmers in peripheral (lagging) regions. Concentration also continues throughout the food chain, with mergers between different food processors and between processors and food retailers, thus exacerbating the continued spatial differentiation of agriculture.

In eastern Europe, three main features of agricultural modernization need to be emphasized. First, *collectivization* did not lead to a unified agricultural production system. Its uneven application created large differences in farm structures and farming systems. For example, Poland and the former Yugoslavia remained dominated by small owner-occupied farms. The average farm size in Poland was less than 4 ha (Morgan, 1992) and private farming accounted for nearly 80% of the land and 75% of agricultural production during the 1980s. As structural measures were not introduced, farming remained backward and inefficient. In contrast, Bulgaria and Albania had well over 80% of their land in state farms and cooperatives. These large spatial differences in farm business types were still apparent in the late 1980s (Table 10.2).

The main objective of collectivization was to increase self-sufficiency in basic foodstuffs. Thus emphasis was placed on grain production, at the expense of livestock husbandry. Those countries with higher rates of collectivization became dominated by grain farming, with high applications of fertilizers to

Table 10.2 Farm business types in eastern Europe, 1988.

	Agricultural area ('000 ha)	State farms (%)	Cooperative farms (%)	Private use (%)
Bulgaria	6 162	89.9	89.9	10.1
Czechoslovakia	6 765	30.3	63.6	6.1
German Democratic Republic	6 182	7.7	82.5	9.8
Hungary	6 497	14.9	70.9	14.2
Poland	18 742	18.5	3.6	77.9
Romania	15 094	90.5	90.5	9.5

Source: (Csaki (1990: 1238).

increase yields. In the drive to increase output, the costs of production often exceeded purchase prices, by as much as 60% in Czechoslovakia (Repassy and Symes, 1993). It was not until the 1970s that attempts were made to develop livestock enterprises and these were quite separate from arable activities. Indeed, crop and livestock farming often became separated into different units and regions; as a result, livestock farming was usually dependent on imported concentrated feed rather than on locally produced fodder crops.

Second, the 1960–80 period was characterized by processes of *rationalization* and *integration* within collectives (Figure 10.5). Two trends were discernible. First, fewer and larger farm units were formed through horizontal integration, whereby collectives joined together to form agro-industrial complexes (APKs). The theory behind this was that the incorporation of groups of villages into one APK would permit greater specialization and economies of scale. Second, greater links within the food chain were sought through vertical integration, whereby collectives and factories fused together to form giant industrial–agricultural complexes (PAKs). Effectively, industrial and marketing institutions were imposed on agriculture, creating a true industrialization of agriculture at the regional scale (Troughton, 1986). Unfortunately, most regions tended to practise the same type of farming, creating bottlenecks in the food-processing industries and preventing regional specialization in agriculture and the exploitation of the comparative economic advantage of different regions.

Third, the *household plots* (auxiliary farming) allocated to cooperative members became a vital source of certain foodstuffs, including fruit, potatoes, milk, eggs and pork. For example, over 50% of the total production of potatoes (in Bulgaria, Hungary and Romania), fruit (Czechoslovakia, Hungary and the GDR), eggs (Bulgaria, Hungary and Romania), pork (Hungary) and milk (Romania) came from these small units. They became a vital part of the agricultural economy and were especially encouraged in Hungary to complement the APKs. Indeed, increasing links were developed between small (private) and large (collective) farms in particular areas of Hungary; by 1989, the 12% of agricultural land devoted to private farming accounted for a remarkable 37% of total agricultural production (Agocs and Agocs, 1994). This left collectives and state farms occupying 88% of the land and producing 63% of the food. Over 90% of rural households and 20% of urban households in Hungary had access to house-

Figure 10.5 Collective farm at Reinholderode, German Democratic Republic (*source:* Tim Unwin, 7 April 1990).

hold plots. The situation was very similar to the development of pluriactivity in western Europe in that the household plots provided additional income for people working on state farms/collectives and in non-agricultural sectors. Overall, therefore, a polarized agricultural structure developed in many parts of eastern Europe, dominated on the one hand by the large socialist sector (cereals and industrial crops) and on the other hand by small private enterprises (fruit, vegetables, eggs and pork).

The transition to post-productivism in western Europe

The CAP has been reformed since the mid-1980s in response to agricultural overproduction, the high costs of price support and the environmental disbenefits of productivist agriculture (Figure 10.6). Farmers are adjusting to this post-productivist transition (PPT) in various ways, termed *pathways of farm business development* by Bowler (1992). Land diversion (set-aside) and pluriactivity have emerged as two important elements of the PPT. The former involves paying farmers for taking arable land out of production for a specified period of time. Initially introduced on a voluntary basis in 1988, set-aside became compulsory in 1993 and, by 1995, over 7 million ha of arable land had been set aside. The voluntary scheme was characterized by both national and regional variations in patterns of uptake. At the national level, Italy and the former West Germany accounted for 60% of the total (Table 10.3), but the response was minimal from many countries. Regionally, set-aside proved more attractive in marginal rather than core cereal-growing areas, 'where yields are lower and cereals are a smaller component of the agricultural system' (Briggs and Kerrell, 1992: 94). This finding was confirmed by Jones (1991) and Jones *et al.*, (1993) in West Germany, and by Ilbery (1990b) in England. Jones *et al.* (1993) have also noted the tendency for German farmers to use voluntary set-aside payments to restructure their farm businesses.

Compulsory set-aside in 1993 led to a major increase in the amount of 'retired' arable land in the EU, to 4.6 million ha. In particular, France had progressed from contributing less than 14% of set-aside land in 1992 to being the main 'actor', with over one-third of the EU total (Table 10.3). One consequence of the compulsory scheme has been the better targeting of 'core' cereal areas, even though this has been on a rotational basis with few environmental and

Figure 10.6 Posters against reform of the CAP, near Nevers, France (*source:* Tim Unwin, 14 August 1992).

Table 10.3 Voluntary and compulsory set-aside in the European Union.

	1988–92[a] Area (ha)	% of EU total	1993–94[b] Area (ha)	% of EU total
Belgium	880	0.05	19 000	0.41
Denmark	12 813	0.74	208 000	4.52
Germany	479 260	27.78	1 050 000	22.80
Greece	713	0.04	15 000	0.33
Spain	103 169	5.98	875 000	19.00
France	235 492	13.65	1 578 000	34.27
Italy	721 847	41.82	195 000	4.23
Ireland	3 452	0.20	26 000	0.56
Luxembourg	91	0.01	2 000	0.04
Netherlands	15 373	0.89	8 000	0.17
Portugal	Exempt		61 000	1.32
United Kingdom	152 700	8.85	568 000	12.33
Total	1 725 790	100.00	4 605 000	100.00

[a]Voluntary set-aside.
[b]Compulsory set-aside.

Source: DG VI, European Commission.

landscape benefits. Indeed, EU countries have been encouraged to consider long-term set-aside as part of their agri-environmental package. The United Kingdom, for example, has introduced a 20-year Habitat Improvement Scheme for the constructive use of set-aside land. Overall, both cereal production and stockpiles of grain in the EU have fallen since 1991, together with a corresponding fall in the use of fertilizers and pesticides.

Defined as the generation of income from on-farm and/or off-farm sources in addition to that obtained from primary agriculture, *pluriactivity* has assumed unaccustomed significance among post-productivist agricultural policies, especially in peripheral (lagging) regions. The incidence of pluriactivity has undoubtedly increased in western Europe, although it varies considerably between regions (Fuller, 1990). Off-farm pluriactivity dominates and over one-third of farm households obtain more than 50% of their income in this way. Geographically, patterns of pluriactivity reflect the interaction of various 'external' and 'internal' factors. Among the former are included regional economic conditions and such social and cultural factors as the increased participation of women in the labour force and differences in language (Efstratoglou-Todoulou, 1990; Edmond and Crabtree, 1993; Bateman and Ray, 1994). Internal factors include farm size, income, levels of indebtedness, the age and education of the 'adopters', and stage in the family life cycle (Ilbery and Bowler, 1993; Bowler *et al.*, 1996).

The return to private farming in eastern Europe

East European agriculture was in a state of crisis in 1989. Production was inefficient, collectives were run mainly by the elderly, production costs and indebtedness were increasing, and export markets were falling; there was a lack of technical equipment and capital investment in agriculture and a failure to respond to changing consumer demands for more meat, fruit and vegetables. Food shortages increased and both Bulgaria and Romania became recipients of food aid. Yet, agriculture still occupied a key position in the economies of eastern Europe, accounting for high levels of employment and gross domestic product. So, after years of being relegated to a position of low priority, agriculture suddenly occupied a status of high priority. Unfortunately, this led to quickly devised plans for land privatization before developing a full policy concerning the types and levels of agricultural production needed (Unwin, 1994).

The re-creation of individual property rights has proceeded in slightly different ways in east European countries. However, as Symes (1993: 296) suggests, it has 'not yet proved a catalyst for dynamic structural change' and, indeed, radical reform has often been rejected by former collective members. In Bulgaria, for example, reprivatization was introduced quickly, but 90% of the land has since been returned to cooperative use (Kopeva *et al.*, 1994). Similarly in Hungary, 90% of 'farmers' have reaffirmed their status with reorganized forms of cooperatives (Symes, 1993). The same kind of development has been reported in eastern Germany (Bergmann, 1992) and Czechoslovakia (Hundeckova and Lostak, 1992). Although the dominance of (reorganized) cooperatives is delaying the inevitable privatization process, their continued presence is vital for agricultural production during the transition period (see case study).

There are good reasons why former collective members do not wish to farm privately. These range from the threatened loss of household plots and pensions to the lack of qualifications and experience to run family farms, and the wish not to risk uncertainty and incur heavy debts. Members prefer the security and assured standard of living associated with agricultural cooperatives and have developed a preference for working cooperatively. There is also the awareness of being unable to compete with the 'larger farms' in western Europe, where their future markets are likely to lie. So, members prefer to leave

The transformation of agricultural cooperatives in Slovakia

Eva H. Karasek

The 1989 'Velvet Revolution' changed the direction of Slovakia's economy. Private ownership became legal and was encouraged by waves of privatization, when state-owned enterprises changed owners and individuals started their own businesses. This also opened a range of new opportunities for the agricultural sector. However, the change failed to fulfil its initial promises. The transformation of agricultural cooperatives can be characterized rather by what it was not. It was not well organized, not approached with adequate know-how, and not regulated by sound legislation. This case study describes the original intentions and the actual outcomes of the restructuring of agricultural cooperatives. The first section provides a brief description of Slovak agriculture in the past, and this is followed by an account of its transformation in the 1990s. The final section illustrates regional differences in the outcomes of this transformation.

Cooperatives before the transformation

Before the breakdown of the centrally planned economy, agricultural production was organized in cooperatives. These were founded by an aggressive socialization process between 1955 and 1960, which involved persuasion or direct forcing of landowners to give their land and agricultural properties to the state, and join the cooperative. By the 1980s, 95% of the total of 7 million hectares of land was owned by the socialized sector, and only 5% was in private hands. Landowners became the members and the employees of the cooperatives, which were the sole source of employment for the area, employing generations of entire families in smaller towns and villages and providing incomes equal to those of urban workers.

State organs such as the Ministry of Agriculture and Nutrition and the Communist Party Congresses determined *what* the cooperatives would produce and *how* they would do so in the form of a central plan. They also determined *whom* the cooperatives could supply. As a result, most cooperatives were highly specialized, focusing on the production of a limited range of crops, and there was also a division of labour among the cooperatives. Given that the state was making all of the decisions, the role of the cooperatives' managers was restricted to the management of the cooperatives' operations to ensure that the central plan and its directives were fully executed.

The restructuring of cooperatives

The goal of transformation was to separate cooperatives from the state in terms of decision making, finance and commerce. This was supposed to happen by redividing the fixed assets (land and properties) of the cooperative among the members from whom the land was taken, according to their contribution during the socialization process. The members were also supposed to receive a sum according to the number of years worked for the cooperative. The members could become the cooperatives' shareholders or, if they chose to leave the cooperative, private farmers. The cooperative would then have to register as an autonomous business entity of some form (a limited liability company) and continue functioning with the remainder of the land and fixed assets.

What actually happened was very different from the above sequence. One barrier to a smooth transformation process was the inconsistency of legislation with the conditions it was supposed to regulate. The taboos of the past 40 years, such as 'ownership' and 'profit', became the way of life for which the majority of the cooperatives' members lacked the appropriate mindset. They did not understand what 'owning things' meant or why they should claim their land. Special education and legal advice was provided by newly found consulting companies, which were charging high rates. Not all cooperatives could therefore afford such a service. Given the 40 years of authoritarian management style, however, managers found this both useless and threatening to their positions. Many cooperative members were vulnerable to being

continued

misled and to the opportunistic behaviour of managers. Most managers also lacked the appropriate managerial know-how. Therefore, they not only mismanaged the process, but also fell victim to the speculation of more resourceful colleagues who managed to realize their own business ambitions using the cooperatives' fixed assets, with the help of others' ignorance.

Despite these inelegancies, the transformation process did have some results. New agribusinesses have been established by managers and members of cooperatives, by uniting their land and resources. By 1994 there were already 950 agricultural small and medium-sized enterprises, from which 930 dealt with direct agricultural activities. The operations of agribusinesses, however, are not problem-free. Due to the lack of markets, agribusinesses are still loss making, and are forced to rely on the private funds of their owners. Difficult access to capital is another serious barrier to the development of agricultural businesses. While the philosophy of transformation emphasizes equality of ownership forms, the same philosophy makes it impossible for the new businesses to borrow capital. Banks are reluctant to lend money because the new businesses have nothing to back the loans with.

The gaps in legislation also allow cooperatives to withhold the claimed land and finances for an unlimited period. Most agribusinesses are still in various court procedures to get their shares, but by the time cooperatives settle their 'priority' debts towards the state and the banks, there is little left to share.

Another outcome of the transformation process has been the rise of private farming, founded by those cooperative members who left the cooperative and started a business on their land. In 1995 there were 17,000 registered private farmers, farming on 230,000 hectares of land. Their survival, however, is even more complicated than that of other agricultural businesses. As well as the above financial problems, they are also facing problems of the discriminatory attitudes of the Slovak Land Fund and the Land Reform Law. Farmers are forced to rent land from the Fund until they receive their own from cooperatives. At the same time, however, the Fund rents land preferentially to those businesses which have a longer agricultural tradition. Therefore, farmers are forced to rent land from individual landowners, who are demanding 40–50% of the farmers' income in turn. If farmers finally receive their land from the cooperatives or the Fund, they get it bit-by-bit at different locations. Since the Land Reform Law forbids the uniting of these bits into a manageable whole, farmers also have to face higher production costs because of the larger distances involved.

Despite these constraints, the performance of farmers is still impressive. By 1995 they dominated the market with a 75% share in potatoes, 50% in eggs, 20% in meat and about 15% in grains.

Regional differences

The different historical backgrounds, agricultural conditions and traditions in Slovakia's main regions have played an important role in the transition process. Both the process itself and its outcomes were surprisingly different in the west, the centre and the east of the country.

Attracted by the ideal conditions for agriculture, large landowners settled in the Danube valley (West Slovakia) at the time of the Austro-Hungarian monarchy. Because these landowners owned large areas, the socialized cooperatives were much larger than in other regions of Slovakia. During the transformation process most cooperatives restructured successfully, despite the fact that many members claimed their lands for purposes of private farming. Agribusinesses came into existence, after compensating the private farmers and those who separated from the cooperative, by the registration of the cooperative as a limited liability company. Their activities are well organized. Farmers and agribusinesses formed various forums, which provide them with business services, such as market information and legal advice, and also serve as a 'meeting place' for them to exchange information and views and to form business contacts. Management skills and foreign capital are also more readily available in the west due to the region's closeness to the Austrian and Hungarian markets.

continued

The landscape further towards the north, centre and east of Slovakia is characterized by mountains, hills and smaller valleys with conditions mostly unfavourable for agriculture. Therefore, agricultural production has traditionally had a minor role in these areas, apart from wineries, animal husbandry and forestry. During the socialization process cooperatives were formed by uniting small family farms in various villages into one bigger unit. Based on the demands of their members, after 1989 the cooperatives broke up and divided the land between the villages which were united in the late 1950s. The result was many small cooperatives which managed to survive only because none of the members claimed their land and financial compensations. Some of them, however, are in the process of liquidation, due to incompetent management. Agribusinesses were formed differently than in the west. The largest ones were the result of the united effort of some opportunistic managers and some members who took out from the cooperative everything that was of any real value, and then left the cooperative to its fate. These agribusinesses are more short-term oriented, producing on an *ad hoc* basis, depending on what they can sell in a given season. Management skills are still underdeveloped, and foreign investors are generally discouraged by the 'money hungry' culture of most businessmen.

Further reading

OECD (1996) *Agricultural Policies, Markets and Trade in Transition Economies: Monitoring and Evaluation 1996*, OECD, Paris

Swinnen, J.F.M. (ed.) (1994) *Policy and Institutional Reform in Central European Agriculture*, Avebury, Aldershot

their land in a cooperative in return for a lifetime annuity rather than to return to private farming. This, in part, reflects the elderly demographic structure of many agricultural cooperatives. However, the offspring of cooperative members may wish in due course to exercise their right to farm privately.

By 1994, the majority of agricultural land was still being farmed by reorganized cooperatives, even though family farms may have become numerically dominant. This was the situation reported by Wilson (1996) for eastern Germany, where the emerging farm business structure consists of family farms, farm partnerships and reorganized cooperatives/companies (Table 10.4). Although family farms accounted for 71% of all farm businesses in 1994, their average size was very small in comparison with the size of registered cooperatives and joint stock companies.

Whatever farm business structure is emerging, there is no doubt that the transition to private farming has created great uncertainty and had a negative impact on agricultural production. The situation has not been helped by governments leaping into privatization too quickly and withdrawing subsidies for agriculture (Agocs and Agocs, 1994). Free markets do not mean that state help for agriculture is not required, yet the Hungarian government, for example, reduced subsidies by 41% between 1988 and 1991. This resulted in lower production, higher consumer prices and imported foodstuffs. Only belatedly did it respond with increased subsidies, by which time much damage had been done (Agocs and Agocs, 1994). Clearly, it will take a long time to develop a commercially successful system of private farming in eastern Europe.

Table 10.4 Farm business structures in eastern Germany, 1994.

Farm type (legal form)	Number of holdings	Average size (ha)
Family farms full-time	6 355	152
Family farms part-time	5 999	25
Farm partnerships	1 977	402
Registered cooperatives	1 333	1 441
Joint stock companies	1 976	790
Total	17 640	305

Source: Wilson (1996: 159).

Conclusion

This chapter has provided some insights into the nature of agricultural restructuring in Europe since 1950 and its uneven impact on agricultural production.

Although all types of European agriculture have been affected by restructuring processes, there continues to be a diversity of production systems. Three key points emerge. First, aided by contrasting types of government intervention, agricultural restructuring has been proceeding in opposite directions in western and eastern Europe. The transition from productivism to post-productivism in western Europe, for example, is associated with greater economic and especially environmental (re)regulation of agriculture. In contrast, the transition to private farming in eastern Europe has been accompanied by government deregulation and the movement towards a freer market for agricultural production.

Second, although productivism and collectivization created contrasting patterns of agricultural production in Europe, both led to the 'industrialization' of farming systems. This was accomplished in different ways and whilst regional specialization and integration into the wider food supply system were encouraged in western Europe, they were not successfully achieved in eastern Europe. However, both systems did lead to a polarized structure of farm businesses; in the west this consisted of large-scale agribusinesses and smaller-scale family farms, whilst in the east it comprised large collectives and very small household plots.

Third, the different pathways of farm business development associated with the PPT in western Europe are encouraging even greater spatial differentiation in agricultural production, especially between 'core' and 'peripheral' areas. This is already apparent with set-aside and pluriactivity and highlights the need to consider agriculture as just one element in an evolving and localized rural policy for western Europe, in which non-agricultural policies are required to stimulate rural development. In eastern Europe, this situation has not yet been reached and the debate about the future of rural areas is dominated by the development of landed property rights and the privatization of agriculture. However, eastern Europe too will have to consider policies for environmental protection and rural diversification in the longer term.

Further reading

Bowler, I.R. (1985) *Agriculture under the Common Agricultural Policy: a Geography*, Manchester University Press, Manchester

Commission of the European Communities (1986) *Europe's Common Agricultural Policy*, Commission of the European Communities, Brussels

Ilbery, B.W. (ed.) (1998) *The Geography of Rural Change*, Longman, London

Laurent, C. and Bowler, I. (eds) (1997) *CAP and the Regions: Building a Multidisciplinary Framework for the Analysis of the EU Agricultural Space*, INRA, Versailles

Robinson, G. and Ilbery, B.W. (1993) Reforming the CAP: beyond MacSharry, in Gilg, A. (ed.) *Progress in Rural Policy and Planning*, **3**, 197–207

CHAPTER 11

Industrial restructuring in Europe: recent tendencies in the organization and geography of production

RAY HUDSON

Introduction

The last two decades have seen an increasing concern to reorganize the production process within Europe and to experiment with alternatives to mass production in both its western and eastern European forms. These restructuring processes have taken place in the context of profound changes in the geopolitical map of Europe and in the character of contemporary capitalism. Perhaps the most significant of the latter changes relates to hotly contested claims as to an epochal shift from one dominant growth model to another and related claims about the rise of a global economy (see Amin, 1995; Amin and Thrift, 1994). Globalization reflects, *inter alia*, decisions by national states to change the international regulatory framework through participating in institutions such as the IMF, the World Bank, the World Trade Organization (formerly GATT) and the Multi-Fibre Agreement. Globalization is partly linked to perceived limits to national state regulation, associated with the demise of a Fordist regime of accumulation. State involvement did not abolish crisis tendencies within the capitalist mode of production, but rather transformed and internalized them within the state. In due course, they erupted as crises of the state itself and its mode of crisis management. Paralleling this crisis of state management and regulation of the economy in the capitalist West, there was the more profound crisis and collapse of the state socialist economies of the East.

Recognition of the limits to state capacities stimulated a search in western Europe (and the USA) for new neo-liberal macro-scale regulatory models that accepted national states' limited powers to counter global market forces. These in turn were exported to eastern Europe as a form of 'shock therapy' which helped insert these nation states into the wider global economy and redefined the map of locational possibilities for production within Europe. These various neo-liberal and 'post-Fordist' regulatory experiments led to the advanced capitalist countries of the West becoming characterized by a greater degree of socio-spatial inequality, while 'shock therapy' created new inequalities in the East. These changing intra-national and global conditions in turn defined the parameters within which corporate and industrial restructuring took place within Europe.

The most far-reaching geopolitical change within Europe has been the redefinition of the relation between East and West into one between eastern and western Europe. Opening up significant swathes of territory denied to capital for decades has major implications for geographies of production within this enlarged European space, although the full repercussions have yet to be revealed. It has created new opportunities for some companies and places; conversely, it has generated a serious threat to others. As well as these changes, however, there were others underway in western Europe, originating in the 1950s with the creation of the EEC and EFTA (see Chapter 9). The next two decades witnessed an ongoing redefinition of the boundaries between the two parts of western Europe and the deepening as well as widening of the EC as it became the EU. The EU simultaneously encourages globalization and is a site of resistance to it, and this is reflected in tensions between its industrial policies seeking to promote globally competitive companies (through support for R&D, a permissive attitude to intra-EU acquisitions and mergers, especially in key high-tech sectors) and those seeking to promote social and spatial cohesion and equity within the EU. This tension has had an important influence on corporate restructuring and patterns of territorially uneven development (Ramsay, 1990). In addition, with the prospect of the Single European Market, inward investment into the EU became an increasingly significant influence on the organizational and territorial pattern of production.

In summary, these macro-scale changes – globally, within western Europe and between East and West – set the context in which industrial restructuring and the search for new viable models of production organization at the micro-scale has occurred. In the following sections, these alternatives are explored more fully. First, mass production and its geographies are briefly outlined. Second, various alternative forms of high-volume production (HVP) are illustrated. Third, their geographies within Europe are considered, and fourth, the strands of continuity between pre-existing and new forms of HVP are considered. Fifth, claims and evidence about the growth of small firms and their territorial concentration in industrial districts within Europe are examined, and finally, some conclusions are drawn about the implications for future geographies of production and restructuring tendencies within Europe.

Mass production in Europe: western 'Fordist' and eastern state socialist variants

Mass production in Europe was, and still is, characterized by a high degree of vertical integration within the firm, *a fortiori* in the combinats of eastern Europe. It is also characterized by a particular social division of labour in production, regulated politically in the state socialist economies and primarily regulated via prices and market relations between customers and suppliers in the capitalist ones. Typically, capitalist companies pursued dual (or multiple) sourcing strategies to minimize the risks of problems with their suppliers. Even so, regulating relations between companies in this way requires large and expensive buffer stocks, giving rise to the characterization of production as 'just-in-case'. The uncertainties of the eastern 'command economies' generated similar uncertainties, leading to the hoarding of parts and materials. One implication of extensive buffer stocks is that problems of quality are concealed and then dealt with at the end of the line, via fault rectification sections. Despite extensive stocks, problems of maintaining an adequate balance in component flows along the line remain, which interrupt its smooth flow.

Figure 11.1 Automobile assembly, Opel, at Bochum in the Ruhrgebiet: a 'classic' Fordist mass-production plant established in the 1960s in the Ruhr, following colliery closures there (*source:* Ray Hudson).

These systems of mass production are marked by a deep technical division of labour within companies and their production units. There is a sharp distinction between occupations requiring mental and manual labour, informed by Taylorist views of scientific management. In addition, they are marked by a strong vertical hierarchy of control; individual workers are restricted to single, specialized – often deskilled – tasks (Braverman, 1974). This can lead to problems – for employers and managers – of lack of workers' motivation and innovation for it fails to capture the knowledge that workers develop through doing their job. One consequence of alienation is a tendency for the production line to go down for long periods, with a direct impact on productivity. Capitalist companies typically produce the same components or products in different plants to minimize the risks of labour unrest disrupting production.

These forms of mass production also have characteristic geographies. In eastern Europe these were determined politically, both intra-nationally and internationally within the COMECON bloc, with particular areas assigned roles in spatial divisions of labour (for example, see Smith, 1996; Swain, 1996). In western Europe such geographies were determined by an interplay of corporate decisions and state policies. Initially production was concentrated in those major industrial conurbations in which mass production of consumer goods had sprung up in the decades between the 1920s and 1960s. As this approach to production reached its limits in the 1950s and 1960s, companies sought new 'spatial fixes' to preserve the viability of the old production model. Thus new spatial divisions of labour emerged, first intra-nationally (Figure 11.1), then internationally (Hudson, 1988). While key strategic decision making and R&D remained in the old core areas, various peripheral locations (previously unindustrialized agricultural regions and deindustrialized former workshops of the world) became the location of unskilled and/or deskilled work in component production and/or routine assembly. Companies were persuaded to locate such functions there by the attractions of abundant available labour, along with often generous financial incentives in the form of grants and loans.

Mass production increasingly became crisis prone in western Europe from the late 1960s, for three main reasons: contradictions internal to this model of production organization, changes in the international division of labour, and deep systemic crises in the political economy of capitalism. While there was some counterbalancing service sector growth in the core areas, this was generally not the case in the peripheries. It seemed as if concerns about the vulnerability of the branch plant economy were justified (Hudson, 1995a). In the east this form of production collapsed spectacularly with the political changes of the post-1989 era. Such forms of production nevertheless remain within Europe, although they are certainly much less prevalent than a decade ago and coexist alongside new forms of HVP, which are considered in the next section.

Big is still beautiful – experimenting with new forms of high-volume production in Europe

The last decade or so has seen a surge in acquisition activity by non-EU transnationals in Europe, along with enhanced cross-border acquisitions within the EU, stimulated by the imminent Single European Market (Hamill, 1993). There has also been an increasing prevalence of strategic alliances between these bigger companies in search of economies of scale and scope as part of strategies to remain competitive. These have been particularly prominent in industries characterized by high entry costs, a continuing importance of economies of scale, rapidly changing technologies and substantial operating risks (such as IT, biotechnologies, automobiles and new materials industries; Dicken and Oberg, 1996) (see case study on the Hungarian motor vehicle industry).

Such alliances and the heightened centralization of capital have often been a key precondition for a more 'flexible' (from the point of view of capital) reorganization, restructuring and relocation of production. Such changes in corporate anatomy have therefore often been a prelude to experiments with new ways of producing in great quantities within Europe, including just-in-time (Sayer and Walker, 1992), lean production (Womack *et al.*, 1990), dynamic flexibility (Coriat, 1991), flexible automation (Veltz, 1991) and mass customization (Pine, 1993). Perhaps their key common characteristic is a concern to combine the benefits of economies of scope, small-batch craft production and a greater flexibility in responding to consumer demand with those of economies of scale. This is most sharply exemplified by mass customization, with its goal of batch sizes of one – that is, uniquely customized commodities. This emphasis on product differentiation is not, however, simply a response to more fragmented consumer demands but a part of strategies to enhance competitiveness by segmenting markets in new ways. While HVP methods are in many respects more flexible than just-in-case production, their specific requirements introduce their own rigidities into the organization of production within and between firms (Sayer and Walker, 1992).

Moreover, these *are* forms of *high-volume* production, variations around the basic mass-production theme. Consequently, there are strict limits, deriving from the material and social requirements of commodity production, to the range of industries and products in which these new HVP approaches *could* be applied. A company such as Dell may provide PCs and workstations on a mass-customized basis; it is difficult to see how soap powder or frozen pizza could be profitable if produced in this way. Moreover, these 'new' HVP approaches in practice combine elements of 'old' methods of mass production with some new production concepts and practices.

Equally, while these new HVP approaches share common characteristics, there are significant differences between them. Companies face a choice in deciding the what, how and where of production. Furthermore, the economic viability of such approaches assumes (albeit usually only tacitly) that certain macro-economic conditions will be fulfilled. The choice of a particular HVP approach is linked to these labour and product market conditions. Companies have moved away from very high levels of automation to more labour-intensive high-volume production technologies within Europe in the 1990s. Lower fixed-capital intensity offers greater scope for combining flexibility with profitability as aggregate levels of demand have declined and/or as labour-market conditions have changed (see Hudson and Schamp, 1995).

The introduction of HVP has redefined the social relations of production, between companies and between employers and employees. HVP incorporates producing 'just-in-time' and close relations between customers and suppliers – for example, between manufacturers and their suppliers, or between retailers and their suppliers (Crewe and Davenport, 1992). Relations between companies are sometimes based

Transforming the motor vehicle industry in Hungary, 1989–95

David Sadler and Adam Swain

This case study focuses on industrial reorganization in central and eastern Europe in the period after 1989, drawing on the motor vehicle industry for illustrative evidence. This sector figured centrally in the wave of foreign direct investment (FDI) which has accompanied political and economic reform in this region over the last decade. Manufacturers quickly established the capacity to assemble over one million vehicles annually there, as old patterns of production were subsumed in the dash for the market.

In the first round of inward investment (in the period up to the end of 1993), $4bn was committed to the reconstruction of the auto industry in east Germany, and a further $1.9bn elsewhere, including $900m in Hungary and $800m in the Czech Republic. The leading investors were Volkswagen and General Motors. In a second round, Fiat and Daewoo were most evident, as they established production bases in Poland, Romania and Uzbekistan. This case study focuses on a country which was amongst those most prominent in the earlier round – Hungary – because some of the impacts of FDI are more clearly evident over this longer time-span. The account which follows first describes the main investment projects and their interaction with local and European markets and with existing producers. Subsequent sections deal with the ways in which manufacturers began to develop linkages with local suppliers, and the systems of labour relations which were established in the new factories.

The main investments and their interaction with existing producers

Hungary played a distinctive role in the motor vehicle industry under the former centrally planned system, producing trucks and buses but not passenger cars, although it did produce substantial volumes of components for assembly by other car manufacturers elsewhere. There were three main state-owned firms in the sector. Ikarus produced buses, Rába supplied Ikarus with axles and engines, and Csepel supplied Ikarus with transmissions and steering mechanisms. Rába and Csepel both also produced trucks. Together, these three firms formed an integrated motor vehicle complex, although – in contrast to the situation found in most centrally planned economies – they also depended on a number of other, separately structured (but still state-owned) components suppliers.

The Hungarian government actively sought FDI in the auto industry from the late 1980s onwards, offering a five-year tax holiday and at least a 60% allowance for a further five years. From the point of view of inward investors, one of the main initial attractions was market access, particularly in a context where there were high expectations of new sales growth. As the full extent of economic collapse in central and eastern Europe in the early 1990s became increasingly evident, however, the emphasis shifted slightly and some firms increasingly viewed their bases in the east as low-cost production locations for export to the rest of Europe.

There were four major projects in this period. Suzuki established a car assembly plant at Esztergom (north of Budapest on the Slovakian border), General Motors an engine assembly and small-scale car assembly operation at Szentgotthárd (in the west of the country, on the Austrian border), Ford a components plant at Székesferhérvár (south-west of Budapest), and VW-Audi an engine assembly facility at Győr. All of these were relatively substantial undertakings. For instance, Suzuki aimed to produce 60,000 vehicles annually by 1997, whilst GM developed the capacity to produce 400,000 engines annually. The Ford and VW-Audi plants were wholly owned by their parent companies. Suzuki established a joint venture (in which it held an initial 40% stake) with a consortium of local (state-owned) components firms, Autokonszern. GM held a 67% stake in a joint venture with Rába. These joint ventures were one reflection of the Hungarian state's desire to use FDI projects as a tool of industrial policy (although subsequently Suzuki increased its holding to 78%, and GM assumed full control of its operations in Hungary in 1995).

continued

The context for state policy was, however, set by the impact of the 'shock therapy' of the market, which saw existing firms come perilously close to failure. Ikarus, for example, lost practically all its traditional export market in the former Soviet Union virtually overnight, and output slumped from 8000 buses in 1990 to 3000 in 1992. Such 'old' firms were kept afloat only by virtue of government guarantees, in recognition of their continued strategic role in the national economy.

Localization and component procurement: the case of Suzuki

The attempt by government to use key firms as generators of change was particularly evident with respect to Suzuki. The Hungarian government held great expectations for this project, anticipating that some 18,000 jobs might be created elsewhere in the economy through the purchase of local components, in addition to the 1300 employed directly at the plant. Suzuki also had an interest in utilizing local suppliers, for Hungary's association agreement with the European Union meant that at 60% local content the vehicles could be exported there duty-free. 'Local' in this context meant derived from European sources, even though Suzuki's initial negotiations with the Hungarian government had envisaged 70% local content derived from within Hungary alone.

There were, however, formidable barriers in the way of increased procurement within Hungary. The scale of the assembly operation meant that there was little incentive for Suzuki's Japanese suppliers (or any other firm, for that matter) to invest directly in Hungary purely to supply Suzuki there, so emphasis was placed on local (Hungarian-owned) firms. Suzuki identified 129 such firms which were interested in becoming suppliers, but only 25 of these (subsequently reduced to 15) were deemed by Suzuki to have any realistic chance of becoming a supplier, and only then on condition of investment in Japanese technology. The first to be offered a contract was a firm which had previously supplied seats to Ikarus. As production began in 1992, the local content figure of 35% included 21% derived from the assembly operation itself.

To meet the local content target Suzuki encouraged its suppliers in Japan to cooperate with Hungarian firms, and arranged lines of credit for these firms to purchase technology under licence. One year after the start-up of production, local content reached 52%, of which 26% came from Hungarian firms, 21% from the assembly operation, and 5% from firms in the EU. The Hungarian supplier base comprised 32 firms, although just 13 of these actually produced components. Significant obstacles in the way of greater sales by Hungarian firms included financing and technological problems, plus disagreement over contractual terms. Prices were based on the cost structure for equivalent parts in Japan plus an agreed profit margin; there was little scope for covering indirect costs (such as healthcare and social facilities) that had previously been carried by the state-owned enterprises.

There was also a clear price for Hungarian firms to pay for the opportunity of doing business with Suzuki: external dependency. Whilst the assembler was a leading force behind the establishment of the Hungarian Association of Automotive Component Companies, and did on occasion offer financial assistance towards the cost of technical licences, most such licences precluded Hungarian firms from supplying other Suzuki plants outside Hungary. Quality control was devolved to the Japanese firm from which the Hungarian supplier purchased technology, with a clause in all agreements that if local quality was deemed inadequate by Suzuki, the Japanese partner would fly in replacement components at its expense. Sales to customers in western Europe were also prohibited by the terms of the agreements, although sales to other assemblers in eastern Europe were permitted.

Thus the scope for the further development of Hungarian suppliers remained very severely delimited. The prospect of re-creating an integrated auto industry complex (as envisaged at one time by the government) was also a remote one, although the country remained highly attractive for externally owned firms to locate in order to produce components for a wider European market. In these ways, the auto industry in Hungary was radically transformed.

continued

New forms of labour recruitment and organization

The other major characteristic feature of these FDI projects was their pattern of labour recruitment and organization – more important in the long term even than prevailing low labour costs (wage levels were around one-tenth of those in western Germany in the early 1990s). From the point of view of the companies concerned, the opportunity to experiment with new labour processes was an important factor in their search for alternative models of production organization. Prevailing conditions of high unemployment set a context within which employers could begin to develop new norms of workplace behaviour.

The first stage of this process lay in highly selective recruitment procedures, made possible by the sheer volume of applications received, and necessary by a requirement to break with previous practices. Thus, for example, the first 120 employees at Ford's components plant were selected from 7500 applications by means of a complex, multi-stage process in which attitude rather than technical ability was a primary consideration. Once recruited, considerable stress was placed upon the socialization of employees to new productivity and performance targets, and to group- or team-centred structures of organization and communication. Teamworking prevailed and these teams had wide-ranging responsibilities. At Ford, for instance, each work team was expected to measure its own performance against three variables: throughput, inventory and operating expense. Suzuki sent its first recruits to Japan for training in working practices and group values. The results of these processes were mixed. Some Hungarian employees of Suzuki, dissatisfied with their long working hours in Japan, staged a short token strike, and there was relatively high labour turnover at Esztergom in the early stages. On the other hand, managers at Ford and GM reported that they were extremely satisfied with the degree of flexibility and resultant high productivity achieved. Such outcomes were studied very carefully by these companies, with an eye to imitation elsewhere in Europe.

Conclusions

During the early 1990s, inward investment played a decisive role in the reshaping of the motor vehicle sector in Hungary, as the country offered an opportunity for firms to experiment with alternative models of production organization. The Hungarian government welcomed and encouraged such changes, seeing FDI as a means to an end of economic transformation, and hoping to use key firms as 'dynamics of change' around which new industrial structures might emerge. The results were less than expected, at least in the short term, partly because of the way in which new investors created conditions of external dependency amongst their supply base. On the other hand, workplace conditions were fundamentally redefined as companies took advantage of the opportunity to recruit, socialize and motivate their employees in wholly new ways. By the mid-1990s, the Hungarian motor vehicle sector was radically altered as the country adapted to a new role in a different spatial division of labour from that which had been created under central planning.

Further reading

Sadler, D. and Swain, A. (1994) State and market in eastern Europe: regional development and workplace implications of direct foreign investment in the automobile industry in Hungary, *Transactions of the Institute of British Geographers*, **19**, 387–403

Sadler, D., Swain, A. and Hudson, R. (1993) The automobile industry and eastern Europe: new production strategies or old solutions?, *Area*, **25**, 339–49

on trust and linked with single-supplier deals. An important consequence of producing 'just-in-time' is that there are minimal buffer stocks. Quality has to be 'built in' from the outset. Furthermore, there are minimal problems of balancing flows of parts on the line since components are ordered and delivered only in response to orders from customers (either other companies or final consumers), or from workers further up the line. If, however, such imbalances do arise, production quickly grinds to a halt.

New forms of relations between companies presuppose new forms of capital–labour relations. Forms of HVP necessarily require workers to perform a wider range of tasks than on the Fordist line; in that sense the technical division of labour is not so deeply inscribed. There is considerable debate as to the implications of this change for those carrying out the tasks of production. Some (for instance, see Wickens, 1986; Cooke, 1995) see workers as multi-skilled, empowered and more creatively involved in the production process, and place great stress on teamwork. Companies (and their intellectual supporters) emphasize that these new forms of work provide better jobs, recombining the mental and the manual which Taylorism had torn asunder. Others dispute this (Garrahan and Stewart, 1992; Beynon, 1995). For them, jobs in these factories are no better than those on the old mass-production lines of earlier years. What is involved is not multi-skilling but multi-tasking, a search for new ways of intensifying the labour process. The production line keeps running continuously with the aim of minimizing the number of workers needed for a given line speed. These alternative interpretations stress not the qualitative differences from the old Taylorist model, but rather the continuities between the old and the new. Indeed, from the point of view of labour, the new HVP approaches may involve greater intensification of work and stress than before.

In order to introduce new ways of working, there have been corresponding changes involving greater selectivity in recruitment, putting more emphasis on appropriate attitudes, making sure companies hire the 'right' people on the labour market, workers who will accept and adapt to new ways of working. At this point links to broader macro-economic and geographical contexts become crucial – for high unemployment, spatially concentrated, provides the context in which the texts about employment practices can be rewritten within Europe.

These new HVP approaches both require and permit the shattering of old forms of trades unionism and the institutions of labour, and their recasting in new moulds. These may be non-union but may as easily be one-union, as trades unions have often been willing to trade off sole bargaining rights for various 'sweetheart' deals to combat their own falling memberships. They are grounded in a very different conception of capital–labour relations to the previously dominant one as the already asymmetrical power relations between capital and labour swung sharply in favour of the former.

There is a further and venomous sting in the tail from the point of view of labour. For the new models of production are undeniably predicated on there being *no* return to 'full employment'. Such labour-market conditions would destroy at a stroke companies' capacity to be so selective in deciding which individuals to employ. Companies are acutely vulnerable to interruptions to production in approaches built around vertical disintegration, just-in-time principles and minimal stock levels. This was sharply emphasized by the strike in October 1991 at Renault's engine and gearbox plant at Cleon, which soon ground production to a halt in many of Renault's other plants, for which it was the sole JIT supplier. Companies therefore have to strive to ensure that individual workers and organized labour remain cooperative and compliant, to reform the institutions of organized labour and redefine the culture of capitalism in Europe in a way that is consistent with their interests.

New geographies of high-volume production in Europe?

In the post-Fordist era, companies are seeking new routes to competitiveness in an increasingly stretched out and 'mobile' economy, with accelerating flows of capital and commodities as time–space compression reaches new levels. Capital is becoming even more (hyper)mobile but, seemingly paradoxically, place has become more significant to it as globalization has proceeded apace. Exploitation of place-specific concentrations of people in search of work, often reproduced in social and cultural settings that shift the costs of labour power onto local societies and states, has assumed an enhanced significance (Figure 11.2).

Figure 11.2 Decentralization of labour-intensive production in traditional sectors to the southern periphery of Europe: shoe production, Komotini, north-east Greece (*source:* Ray Hudson).

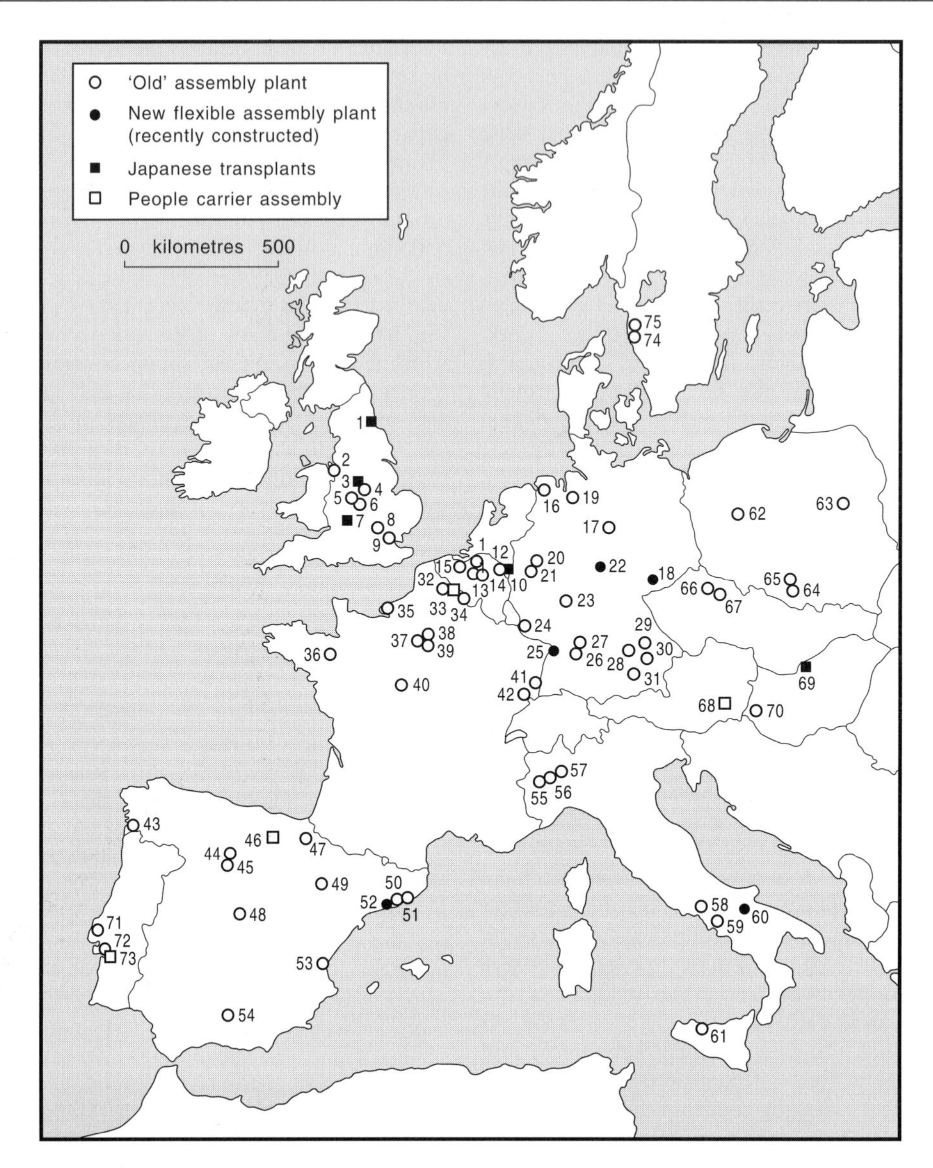

Figure 11.3 Automobile production facilities in Europe in the early 1990s (adapted from R. Hudson and E.W. Schamp, 1995, *Towards a New Map of Automobile Manufacturing in Europe?*, pp. 239–41, copyright 1995, with permission from Springer-Verlag GmbH & Co. KG, Tiergartenstr. 17, D-69121 Heidelberg, Germany).

One consequence of introducing just-in-time approaches initially seemed to be that they would lead to a regional reconcentration of production. Such interpretations were heavily influenced by the clustering of Japanese component companies around the new 'transplants' of the Japanese automobile companies in the USA (Mair *et al.*, 1988). Inward investment from Japan into Europe quickly revealed, however, that there is no precise correlation between just-in-time and in-one-place. For *some* components in the car industry, there *is* undoubted evidence of in-one-place 'synchronous production' (Figure 11.3). These are typically high-bulk, low-value components, with frequent deliveries and changes of specification (for example, see Ferrão and Vale, 1995). Furthermore, in *some* locations, especially where

Figure 11.3 **Key**:

Number	Place	Firm	Type (see key)
United Kingdom			
1	Sunderland	Nissan	■
2	Halewood	Ford	○
3	Burnaston	Toyota	■
4	Ryton	PSA	○
5	Longbridge	Rover	○
6	Cowley	Rover	○
7	Swindon	Honda	■
8	Luton	GM	○
9	Dagenham	Ford	○
Netherlands			
10	Born	Volvo/Mitsubishi	■
Belgium			
11	Antwerp	GM	○
12	Genk	Ford	○
13	Brussels	VW	○
14	Brussels	Renault	○
15	Ghent	Volvo	○
Germany			
16	Emden	VW	○
17	Wolfsburg	VW	○
18	Mosel/Zwickau	VW	●
19	Bremen	Mercedes-Benz	○
20	Bochum	Opel/GM	○
21	Köln	Ford	○
22	Eisenach	Opel/GM	●
23	Rüsselheim/Frankfurt	Opel/GM	○
24	Saarlouis	Ford	○
25	Rastatt	Mercedes-Benz	○
26	Sindelfingen	Mercedes-Benz	○
27	Neckarsulm	Audi	○
28	Ingolstadt	Audi	○
29	Regensburg	BMW	○
30	Dingolfing	BMW	○
31	München	BMW	○
France			
32	Douai	Renault	○
33	Valenciennes	Fiat/PSA	□
34	Maugeuge	Renault	○
35	Sandouville	Renault	○
36	Rennes	PSA	○
37	Flins/Paris	Renault	○
38	Paris/Poissy	PSA	○
39	Paris/Aulnay	PSA	○
40	Romorantin	PSA	○
France (continued)			
41	Mulhouse	PSA	○
42	Sochaux	PSA	○
Spain			
43	Vigo	PSA	○
44	Palencia	Renault	○
45	Valladolid	Renault	○
46	Vitoria	Mercedes-Benz	□
47	Pamplona	VW	○
48	Madrid	PSA	○
49	Zaragoza	Opel/GM	○
50	Barcelona	Nissan	○
51	Barcelona	Seat	○
52	Martorell/Barcelona	Seat	●
53	Valencia	Ford	○
54	Jaen	Suzuki	○
Italy			
55	Turin/Rivalta	Fiat	○
56	Turin/Mirafiori	Fiat	○
57	Arese	Fiat	○
58	Cassino	Fiat	○
59	Pomigliano	Fiat	○
60	Melfi	Fiat	●
61	Termini Imerese	Fiat	○
Poland			
62	Poznań	VW	○
63	Warzaw	Opel	○
64	Bielsko Biala	Fiat	○
65	Tychy	Fiat	○
Czech Republic			
66	Mlada Boleslaw	Skoda	○
67	Vrchlabi	Skoda	○
Austria			
68	Graz	Chrysler	□
Hungary			
69	Esztergom	Suzuki	■
70	Szentigotthárd	GM	○
Portugal			
71	Azambuja	GM	○
72	Setubal	Renault	○
73	Setubal	Ford/VW	□
Sweden			
74	Götborg	Volvo	○
75	Trollhättan	Saab/GM	○

transport infrastructure is very poor, as in the former German Democratic Republic, there is of necessity a greater degree of spatial clustering. Even so, this is often 'pseudo-just-in-time', with buffer stocks held in warehouses near to assembly plants. There is thus no firm correlation between just-in-time and in-one-place in what is typically seen as the prime exemplar industry. It therefore seems reasonable to conclude that this will also be the case in other sectors in which just-in-time approaches can be implemented in Europe. Indeed, more generally the key issue is predictability and regularity of deliveries. The crucial variable is 'temporal certainty' rather than spatial proximity *per se*. In such cases, the vehicles used in transport function as mobile warehouses in which inventories are held.

Industrial restructuring in the Lisbon Metropolitan Area: towards a new map of production?

Mário Vale

Introduction

The Lisbon and Tagus Valley NUT II level region is one of the most favoured economic regions in Portugal. Situated within it is the Lisbon Metropolitan Area (LMA), the densest urban area in the country with important economic and administrative functions (Gaspar, 1995). From an economic point of view, the LMA is also one of the most competitive regions as a result of its close capital–labour relationships, which give rise to higher than average productivity levels.

Changes occurring during the 1980s in the organization of production helped to prepare the region to face stiffer competition as a result of the Single European Market. To a great extent, the slump in heavy industry, particularly on the Setúbal Peninsula in the south of the LMA, opened up the way for far-reaching restructuring of production. Industrial restructuring in the steel foundries and the chemical industry as well as ship-building and naval repairs gave rise to a sharp increase in unemployment in the area, which in turn had serious repercussions for the labour market.

With the most difficult phase now left behind, the LMA has begun to show signs of economic growth, characterized by contained unemployment, an improvement in workers' qualifications, new investments in production and the development of production infrastructure. This case study presents and discusses the changes in production in the LMA towards the end of the 1980s and at the beginning of the 1990s within the context of globalization and increasing competition in the Single Market. The region's relative success may be explained in terms of its role as a beacon pointing to a much wider process that will lead to a redesigning of Europe's production map. However, restructuring industry and seeking new models of production means that the less favoured regions are constantly having to face new challenges. These are usually placed beyond their reach unless they obtain the explicit back-up of policies aimed at sustainable development.

Production change and job loss

In 1992, the Lisbon and Tagus Valley NUT II region provided about 29% of the jobs, and generated more than 51% of the domestic industrial Gross Added Value (GAV), in national manufacturing activity. This high level of productivity was far greater than that found in other regions. The productive fabric in the LMA shows evidence of an influential Fordist heritage. Large labour- and capital-intensive industrial plants have been set up where scale economies have been an essential strategic vector (Ferrão, 1987). At the same time, an improvement has been noted in the standard of job qualification, and there has been an upsurge in the labour market. As an outcome of the depression of the 1970s, the region spearheaded the process of industrial restruc-turing, and this had a profound effect on unemployment, on production organization and on the downsizing of firms. One of the main results of the slump lay in the way the region turned to specialized production in sectors which had been deeply affected by the international recession (namely shipyards, steel foundries, metalworks and chemical plants) due to depressed foreign markets and stiffer worldwide competition (Gaspar *et al.*, 1996). From the end of the 1970s until the mid-1980s, the region was plunged into economic and social turmoil with an extremely high unemployment rate, particularly on the Setúbal Peninsula, where the percentage of jobless in 1986, at 20.1%, was twice the national average of 10.3%.

After the mid-1980s, the LMA's haul to economic recovery was begun against the backdrop of Portugal's full membership of the European Union. In effect, the structural funds helped to set up development programmes involving enterprises, infrastructures and occupational training. At the same time, and despite the fact that the process of industrial rationalization on a European scale had phased out some types of jobs, foreign industrial enterprises were increasingly interested in the LMA due to new needs arising with the Single Market. There was a marked upswing of inward

continued

investment, therefore, particularly on the Setúbal Peninsula, where large production plants had been set up in the electronics and automobile industries.

However, not enough new jobs were opened up between 1983 and 1993 to attenuate the levels of industrial unemployment, and there was a 9% drop in the number of jobs. Table 11.1 illustrates that the well-established industries, such as the steel foundries and the chemical industry, were by far the most seriously affected sectors in terms of job loss as a result of the restructuring process. The decline of industrial activity in the LMA has sparked off a discussion about deindustrialization of the region. However, job loss is not synonymous with deindustrialization, as the region continues to have a high volume of production. Baptista (1989) believes that job loss does not mean that deindustrialization is in progress but, rather, that industrial restructuring has allowed firms to increase their levels of productivity and safeguard their competitiveness in the European market. On the one hand, industrial restructuring has been characterized by rationalization of all those industries formerly dependent on obsolete technology. On the other, it has meant relocating industrial plants in other regions while maintaining their supervising headquarters in the LMA.

Spatially, it is evident that Lisbon municipality has embarked upon a deindustrialization process. Figure 11.4 thus shows that traditional industrialized municipalities, such as Almada, Barreiro and Vila Franca de Xira, are declining in favour of the new production centres. The most dynamic of these are Palmela on the Setúbal Peninsula, and Sintra along the northern part of the LMA. In these two municipalities, inward investment has acted as a powerful lever for economic development.

The importance of being European: from crisis to economic success?

The LMA's economic recovery continues to make progress, although the greatest thrust of manufacturing activity is confined to the Setúbal Peninsula. When seen within a Portuguese context, the Setúbal Peninsula is endowed with unique characteristics owing to its sea-going access to the northern European markets, a young and green labour force, high qualification and education levels, and regional wage levels below the EU average.

Two main lines of action were followed in order to enhance development in the region. One was based on carrying out a development programme on the Setúbal Peninsula that was largely financed by European structural funds, while the other attracted foreign investment direct backed up by industrial policies at home. The role of the structural funds has been crucial in allowing the regional economy to develop. On the one hand, the conditions for applying investments have been improved; on the other, it has been possible to channel direct funding into some of the foreign-owned industrial enterprises.

Table 11.1 Manufacturing employment change in the Lisbon Metropolitan Area.

	Employment (1993)	% of employment (1993)	Employment change, 1988–93 (%)
Food, beverages and tobacco	26 252	14.98	3
Textiles, clothing and leather	18 247	10.41	–21
Wood and cork products	8 685	4.96	–2
Paper, pulp, printing and publishing	20 749	11.84	20
Chemicals	22 662	12.93	–29
Non-metal products	10 254	5.85	–3
Iron and steel	4 416	2.52	–40
Metal products, electronics and transport Equipment	62 027	35.40	–4
Other industries	1 947	1.11	
Total	175 239	100.00	–9

Source: MESS (1989, 1993).

continued

AutoEuropa is a practical example of what this regional economic recovery strategy has achieved. New production strategies have led transnational corporations (TNCs) to establish European production networks, reduce risks and share costs (Dicken *et al.*, 1995). In the case of *AutoEuropa*, which involved a strategic alliance between Ford and Volkswagen, these guidelines are only too obvious. The two manufacturers set up production on a new kind of motor vehicle – the Multi-Purpose Vehicle – with the aim of penetrating a segment of the market where, at least in Europe, there was little experience to be had.

The decision to locate the plant on the Setúbal Peninsula in Palmela reflected the changes which had already been made to the map of European industrial production. Effectively speaking, the manufacturers chose a region where they were

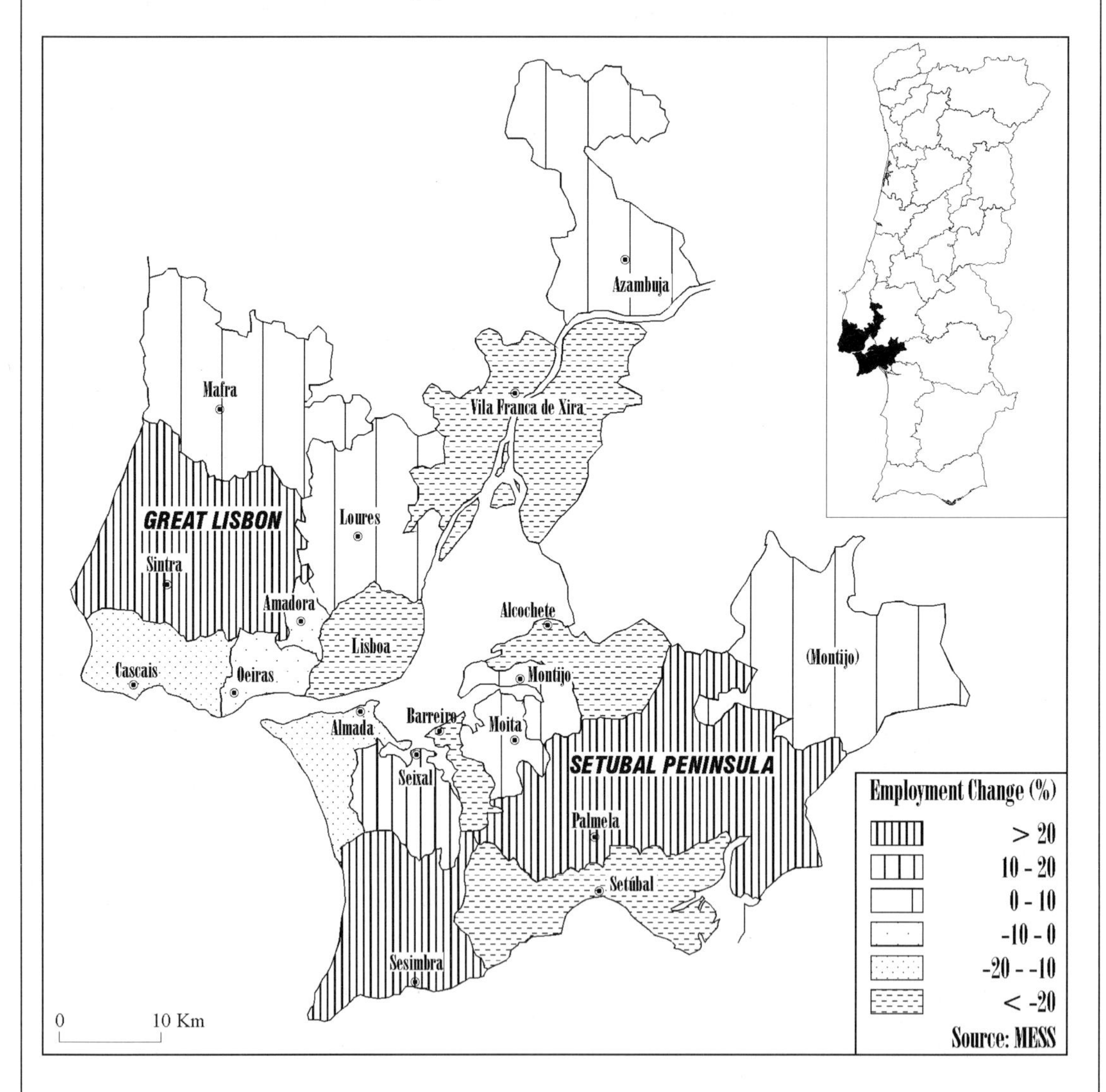

Figure 11.4 Manufacturing employment change in the Lisbon Metropolitan Area, 1988–93 (author's copyright, with thanks to Joaquim Seixas for cartographic assistance).

continued

likely to obtain generous funding, mainly from structural fund coffers, good sea-going access to the northern European markets, and inexhaustible supplies of cheap labour-power earning wages that were very much lower than in other parts of Europe. If, at first glance, these conditions appear not to provide evidence of a new production strategy, as they seemingly run along typically Fordist-type principles regarding investments, deeper analysis shows that some of the guidelines do indeed illustrate a new strategy based on production organization at a European-wide scale (Ferrão and Vale, 1995; Pike and Vale, 1996). *AutoEuropa* has thus involved huge capital investments from elsewhere in Europe, it is producing a sophisticated vehicle, has forced Portuguese firms to become more competitive in the European market, and is an example of the way in which TNCs increase local embeddedness (Dicken *et al.*, 1994).

In conclusion, changing the map of industrial production in Europe would seem to be a reality whereby some of the peripheral areas have been allotted an important role. Nevertheless, the less favoured regions offering a greater likelihood of economic development are mostly those which may be regarded as the most highly favoured in terms of the country itself. This seems to be the case with the LMA.

Acknowledgements

This case study builds on research supported by the Junta Nacional de Investigação Científica e Tecnológica, project PCSH/C/GEO/715/93.

Further reading

Baptista, A.M. (1989) Perspectivas de desenvolvimento económico da Área Metropolitana de Lisboa, *Sociedade e Território*, **10/11**, 43–8

Ferrão, J. and Vale, M. (1995) Multi-purpose vehicles, a new opportunity for the periphery? Lessons from the Ford/VW project (Portugal), in Hudson, R. and Schamp, E. (eds) *Towards a New Map of Automobile Manufacturing in Europe? New Production Concepts and Spatial Restructuring*, Springer, London, 195–217

Gaspar, J. (1995) Lisbon Metropolitan Area: structure, function and urban policies, in Fonseca, L. (ed.) *Lisboa: Abordagens Geográficas*, CEG, EPRU 42, Lisboa, 81–104

There is no 'obvious' geography to other new HVP methods in the way that there initially seemed to be with just-in-time. As a concern with economies of scale remains central in many industries, the introduction of lean production almost certainly means fewer factories. Nevertheless, *where* these factories might be sited remains an open question, subject to locations satisfying labour market and other production requirements. From one point of view, one might expect heavily automated assembly plants, requiring very high levels of fixed capital investment, to be drawn to 'core' regions, near to the main markets of the EU. On the other hand, the availability of substantial regional policy grants in peripheral locations, plus the ready availability of labour there with no previous history of the industry in question, facilitating continuous shift working and new working practices, may make *these* attractive (see case study of Lisbon). For instance, placing new Fiat plants in the Mezzogiorno in the 1990s could be seen as supporting such an interpretation (see Conti and Enrietti, 1995). Equally, the locational strategies integral to the penetration of capital into eastern Europe would seem, *a fortiori*, to exemplify such an approach, as with General Motors' new plant at Eisenach and Volkswagen's at Mosel (see Schamp,1995; and the earlier case study of the Hungarian motor vehicle industry). Furthermore, labour-intensive component production, organized on classic just-in-case principles, will continue to find the cheap labour peripheries of Europe an attractive destination.

There is a similar indeterminacy associated with the geographies of mass customization, though there are suggestions that these will involve a sophisticated use of spatial differentiation. As the example of companies in Europe such as Dell suggests, there is considerable flexibility over choice of production location, given the possibilities offered by advances in information technologies in communication and production. Furthermore, this concept of mass customization can be linked with ideas about globalization, localization, and 'glocalization' (see, for example, van Tulder and Ruigrok, 1993). Mass customization as a production strategy can thereby be connected with concepts of segmenting markets by territory, culture and taste and then 'regionalizing' production in response to this. Even so, the implications of such an approach for the location of the jobs that it generates within Europe remain ambivalent. In addition, however, in this case the greater inputs of mental labour into production, the greater weight attached to the role of R&D, and the much greater emphasis placed upon after-sales services as part of the product could help pull such production to core rather than peripheral locations.

So how new is the new? The novelty of HVP and its geographies

In assessing the extent to which there have been significant changes in the dominant forms of organization of production in an (allegedly) post-Fordist European economy, it is important to recognize the variety of forms and relations of production that were to be found within the (allegedly) monolithic forms of mass-production organization in east and west. In its western capitalist variant, there was a proliferation of small and medium-sized enterprises (SMEs), 'one-off' and small-batch production by big firms, and widespread evidence of outsourcing and subcontracting alongside a high degree of vertical integration in mass-production sectors. In the eastern state socialist variant, there were several forms of 'informal' production between and around the formal state firms and vertically integrated combinats while mass production existed alongside other forms of small-batch production within the state sector.

Much of the supposedly radical change in forms of HVP was therefore prefigured in the era of mass production. Considerable emphasis has been placed upon the transformation in the role of former mass-production manufacturers and 'first tier' component suppliers to a new role as systems integrators, coordinating the activities of a range of suppliers to which they have subcontracted various production activities. Capital goods producers have operated in this way as 'systems integrators' for decades (Vaughan, 1996). This particular organizational model is not, therefore, new. Often each new order or contract involved the creation of a new web of suppliers and subcontractors for the precise purpose of fulfilling *that* order, and attempts to introduce mass-production principles (for example into shipbuilding) met with at best very partial success. Moreover such companies have 'traditionally' produced to order, just-in-time, and continue to do so, precisely because of the characteristics of the commodity and product markets. This system of production developed temporally alongside mass production of consumer goods. Thus to think of a linear sequence of just-in-case replaced by just-in-time is to miss the point. Rather the transition from just-in-case mass production to new just-in-time approaches to HVP in Europe involves combining production concepts from the system of 'one-off' production of capital goods with those from the more continuous line system of mass production.

Furthermore, the suggestion that the new HVP approaches led to production both just-in-time and in-one-place and a regional reconcentration of the economy is open to question on two rather different fronts. First, the empirical evidence for such a regional clustering within Europe is limited, even in the exemplar automobile industry (Hudson and Schamp, 1995). Second, during the Fordist era there were in fact two rather distinct tendencies reshaping the geography of production within Europe. First, there was decentralization associated with the search for spatial fixes for just-in-case mass production – a search which, moreover, continues. Second, there was a very different tendency towards spatial agglomeration in highly automated continuous-process industries such as bulk chemicals and metals. This form of territorial clustering was and is based on input–output linkages (often between separately owned companies) because of a continuous interchange of products as one plant's or company's outputs become another plant's or company's inputs. This territorial clustering was often associated with a move to new locations dotted around the coasts of Europe as these maritime production complexes increasingly relied upon imported raw materials and exported products (see Lisbon case study). Often they were located in peripheral problem regions in which very generous state financial aid was available to offset the massive fixed capital investment costs involved in establishing such complexes on green field sites (Hudson, 1983; Hudson and Sadler, 1989).

At the same time, it is clearly premature to announce the death of Taylorism. The opening up of eastern Europe to capital investment from the west has again redefined the ways in which Europe's peripheral regions are being incorporated into the production chain. It has created fresh opportunities for corporate use of peripheral spaces to preserve Taylorism and intense competition between peripheral locations in Europe, east and west, for such investment. As well as using the periphery to seek to preserve the old, companies are also using peripheral locations to experiment with new ways of producing without endangering production in their main core plants (Hudson, 1994a, 1997).

So how new is the new? SMEs, networks and the geography of industrial districts

For some, the growing numbers of small firms and, more specifically, locally agglomerated flexible production systems represent a new departure and a progressive response to the crisis of mass production (see Scott, 1988; Storper and Scott, 1989). This, however, ignores the extent to which SMEs have been a persistent and numerically dominant organizational form rather than a sudden arrival on the European corporate scene. Two additional points can be made about the relationships between firm size, plant size and tendencies to produce 'flexibly'. First, there has certainly been a considerable expansion in the number of small firms in parts of Europe. Not least, this is because a switch to small firms and endogenous growth has characterized national and regional development strategies. Such small firms may deploy quite technically sophisticated production methods. Small does not necessarily mean technologically backward or archaic forms of production organization. Nevertheless, many small firms in the East that emerged in the interstices of the former state socialist economies and continue to exist in their ongoing transformation do have precisely these qualities of obsolescence and archaism (see Smith, 1996). Second, many big firms produce 'flexibly'. The pursuit of economies of scope is not necessarily inconsistent with the pursuit of economies of scale. Conversely, many small firms *do* produce 'rigidly', locked into either subcontracting arrangements, whereby they produce to the qualities and quantities specified by their customers at the controlling retailing end of the production filiere or, alternatively, into the routine production of standardized components for those manufacturing companies further up the filiere which form their markets. Much of the growth of both new small firms and new service sector activities, especially those denoted as producer services, has been precisely linked to the new subcontracting and outsourcing strategies of major companies. Many small firms therefore exist in the niches which are unattractive to bigger companies, are often subservient to them and established as a direct consequence of their strategies of concentrating on core competencies, and are engaged in fierce price competition with one another. The market position of small firms is typically based on profoundly asymmetrical power relations between companies in the production filiere, characterized by fierce competition between SMEs for the markets which bigger companies provide.

Industrial districts are characterized as relatively self-contained, product-specialized regional economies of linked small firms, which form internationally competitive nodes in a global economy. Other industries, not necessarily based around small manufacturing firms, are also claimed to be organized as industrial districts; for example, the international financial services sector focused on the City of London (see Amin and Thrift, 1994; and also the second case study in Chapter 12). Such a view sits uneasily with the notion of territorially agglomerated networks of SMEs, however, since the City of London contains some of the most important controlling agents of the global economy.

Most small manufacturing firms in Europe are clearly *not* therefore embedded into the social and technical production structures of industrial districts. Such forms of regional industrial structure are rare and will remain so. Where they do arise is in circumstances in which there has allegedly been a significant and irreversible growth in consumer sovereignty, alongside marked volatility and shortened product life cycles, which requires that production be organized on a flexible basis. Mass production cannot meet consumer demands for better quality and more differentiated goods, with increasingly reduced life cycles. Consequently, satisfaction of market demands necessitates decentralized coordination and control, a horizontal division of labour between independent but interlinked producers, numerical and task flexibility among workers, who at the same time are required to display greater ingenuity and innovatory 'on-the-job' capacities, and elimination of time and wastage in delivery and supply. Such forms of production organization are claimed to be particularly evident in industries in which volatility and product innovation in niche markets is most pronounced. The new market conditions allegedly require a radical transformation of the production system towards flexible intra-firm and inter-firm arrangements which simultaneously combine economies of agglomeration, scope and versatility.

Moreover, it is argued that such a transformation implies a spatial reconcentration of the different agents involved in a production filiere. Such agglomeration offers a series of (Alfred) Marshallian benefits, upon which a system of vertically disintegrated and 'knowledge'-based production can draw. This includes reductions in transaction and transport costs; the production of a local pool of expertise and know-how; a culture of labour flexibility and cooperation, grounded in associative social relations of trust and dense localized social interaction within civil society (see Cooke, 1995; Storper, 1995); and the growth of a local infrastructure of specialized services, distribution networks and supply structures. The resultant geography of production becomes organized around a set of internally cohesive regional economies which each grow in a virtuous circle of self-reproducing and self-regulating territorially based social relationships.

Over the last couple of decades, heavily influenced by Bagnasco's (1977) seminal study revealing the existence of a 'Third Italy', empirical studies have been carried out that are claimed to lend credence to the notion that industrial districts are (re)emerging in parts of Europe. In some cases high technology and certain sorts of micro-electronics production are seen as pivotal (Isaksen, 1994). Commonly cited examples include the M4 corridor, the M25 ring and the area around Cambridge in the UK, and Grenoble in France. Baden-Württemberg in Germany represents a different case, in which the emphasis is more on the applications of new production technologies within existing networks of inter-firm relations, in which large firms are typically prominent. A third example is constituted by various sorts of areas specializing in niche-market consumer goods such as designer clothing and craft products, such as central and north-east Italy and parts of Jutland in western Denmark (see Dunford and Hudson, 1996).

Two types of questions arise from such studies. First, do the specifications of process accurately reflect the social relationships that underpin these industrial districts? It is clear that the social processes constitutive of spatial agglomeration differ in very important ways (for instance between Baden-Württemberg and Emilia-Romagna), so that there is a danger of classifying together common spatial forms produced through very different sorts of social process. As Garofoli (1986) points out, one can identify at least three very different types of territorially based forms of production organization within the Third Italy:

1 Specialized productive areas, which are monosectoral, export-oriented and characterized by weak inter-firm linkages.
2 Local productive systems, which are also monosectoral and export-oriented but with strong intra-firm connections.
3 Territorially integrated systems areas characterized by multisectoral production through diversified networks of local firms operating within a supportive institutional environment and in a climate of cooperation between the main social actors.

This points to important differences in localized social structures, relations and production conditions *within* the exemplar region of the Third Italy itself. There are clearly grave dangers in indiscriminately grouping together European regions with very different social relations of production into an undifferentiated category of 'new industrial spaces' (as Scott, 1988, does).

Second, to what extent are such organizational–territorial forms themselves spatially and temporally specific (see Hudson, 1989)? There is growing evidence of firms within industrial districts increasingly internationalizing their production strategies via subcontracting and/or direct investment to exploit the spaces of eastern Europe recently reopened to capital (Dunford and Hudson, 1996). Furthermore, the range of commodities that can be produced via strategies of flexible production directed at niche markets is extremely limited. There are powerful systemic pressures within capitalism which lead to a strong tendency to high-volume (if not always Fordist-style mass) production and consumption.

Conclusions: a complex mosaic of new and old forms of production and their geographies

The range of options open to industrial companies is now wider than it has ever been. The possible combinations of choices of location, both inside and outside Europe, production technologies and products vary with forms, types and sizes of companies. Overall, however, it is unprecedented. On the other hand, new constraints, such as those of ecology, may be becoming more important (Hudson, 1995b). How, then, will the geography of production in Europe evolve in the foreseeable future?

The continuities with the past will be at least as strong as the breaks with it. The suggestion of a future of generalized regional regeneration around flexibly specialized ensembles of firms linked into industrial districts is fundamentally misconceived. For there always have been sectors and areas within capitalism within which small-batch production for niche markets has been present. But there are strong systemic pressures within capitalism towards high-volume production as a consequence of competitive pressures, both between capital and labour (to force down wage costs) and between companies (over the distribution of surplus value). The industrial geography of Europe will continue to be dominated by major transnationals (many controlled from outside Europe) and by high-volume production and consumption.

The future industrial map of Europe will equally be one of growing concentration of corporate power, R&D, and advanced manufacturing producing the technically sophisticated new products in existing 'core' regions. There is some evidence of the pressures of time-compression in product and process development forcing firms in advanced or growth sectors to rely on spatial propinquity as a key element in their competitive strategies. However, the suggestion that Europe will experience a generalized reconstruction of regional economies around clusters of HVP industries producing just-in-time and in-one-place is no more tenable than that of a future of industrial districts of networked small firms. Indeed, the European industrial economy remains one that is spaced out, and although the specifics of the spatial patterns have altered (and will continue to do so), there are strong threads of continuity with the spatial divisions of labour of mass production in both East and West.

Further reading

Benko, G. and Dunford, M. (eds) (1991) *Industrial Change and Regional Development: the Transformation of New Industrial Spaces*, Belhaven, London

Dicken, P. and Oberg, S. (1996) The global context: Europe in a world of dynamic economic and population change, *European Urban and Regional Studies*, **3**(2), 101–20

Hudson, R. (1994) New production concepts, new production geographies? Reflections on changes in the automobile industry, *Transactions of the Institute of British Geographers*, NS **19**, 331–45

Hudson, R. and Schamp, E.W. (eds) (1995) *Towards a New Map of Automobile Manufacturing in Europe? New Production Concepts and Spatial Restructuring*, Springer, Berlin

Humbert, M. (ed.) (1991) *The Impact of Globalization on Europe's Firms and Industries*, Pinter, London.

CHAPTER 12 European financial systems

JANE S. POLLARD

Introduction

Financial services occupy a pivotal position within the European Union's (EU) programme of economic and political integration. Indeed some would argue that financial integration is *the* key step in the Single Market Programme of 1992 (Masera and Portes, 1991) for three main reasons. First, financial integration affects the performance of all agents in the EU economy. As Strange (1986) points out, we are all, involuntarily, engaged in the international financial system. The profits of businesses large and small, the opening or closing of firms, the affordability of mortgages and all kinds of consumer loans are tied to movements in exchange rates and interest rates. Second, financial integration raises critical economic and political questions concerning which regulatory regime will come to dominate the way that financial firms will operate in Europe. Regulatory arrangements differ across Europe and affect the competitiveness of financial industries. Banking markets in the UK, Denmark, the Netherlands and Germany, for example, are relatively open, in contrast to the more closed markets of Spain, Portugal and Greece (Dixon, 1991). The importance attached to firms, individuals and governments having confidence in the financial system is such that all the players in the European market have a stake in the regulatory standards of everyone else. Third, financial integration and monetary union have profound implications for the geopolitical project which is driving European integration, namely the strengthening of Europe's position *vis-à-vis* the United States and Japan (Cutler *et al.*, 1989).

This chapter explores three key aspects of Europe's financial integration. The first section considers the context for further financial integration in Europe, focusing on how changes in *global* financial markets are shaping the restructuring of finance in the EU. The second section focuses in greater detail on the ongoing restructuring of financial services and the moves towards a single currency in the EU. By way of conclusion, the third section describes how financial restructuring is likely to affect prevailing regional inequalities in Europe, particularly in the context of the changing map of Europe and the possible enlargement of the EU to include central and eastern European countries.

European finance in a global economy

European financial integration is a complex phenomenon involving not only changes in the regulatory systems of EU members, but also a series of changes in global financial markets (Llewellyn, 1992). The move towards European financial integration has a long history. Underlying this diverse, and problem-laden, history of attempted monetary cooperation in Europe is the desire to protect European economies from the costs – economic, social and political – of financial instability and speculation. Full monetary union means different countries agreeing to adopt a single currency. In the European case, this means the replacement of the deutschmark, the lira, the franc – and the currencies of other countries which join the single currency – with a new currency, the 'euro'. A single European currency, it is argued, could deliver all manner of benefits for EU member states. It would eliminate foreign transaction costs for firms and individuals (Krugman, 1990), it could spread German-style price stability across Europe, and perhaps most importantly, it would remove currency uncertainty. The removal of such uncertainty about exchange rate movements would not only boost investment and trade, but also underpin Europe's single market in goods and services. A single currency could also lead to a reduction in the market discrimination which, for example, makes cars more expensive in Britain than in France, as quoting prices in one currency makes price differences transparent. Added to this, a single currency would impose a new framework of incentives and constraints on member states' budgets, enhancing financial coordination and discipline in Europe (Emerson *et al.*, 1992). Finally, a single currency could strengthen the 'single voice' of the EU *vis-à-vis* the United States and Japan.

Many of these arguments in favour of a single currency dwell on the economic and technical aspects of money, its importance in facilitating trade, and its ability to express the relative value of different commodities in different countries (see Dyson, 1994). Yet money clearly has cultural, political and psychological dimensions to it, and some of these attributes are invoked in arguments against a single European currency. The pound sterling, the French franc and the German deutschmark are expressions of statehood and national identity; they have long histories and have acquired symbolic as well as economic importance. A single European currency is similarly bound up with political and cultural baggage. If EU member states adopt a single currency, they not only lose their national currencies, but also the ability to use exchange rate policy as an instrument to help adjust to changing economic conditions. To explain the push towards a single European currency, we need to appreciate first how national systems of financial regulation in Europe have been, and still are, inextricably bound up with international financial affairs, and second, how this international financial order has changed since the 1960s.

The collapse of the international monetary order

Leyshon and Thrift (1995: 111) argue that 'the process of European financial consolidation may be seen as a regional solution to the collapse of the global monetary order'. The global monetary order in question here was the Bretton Woods system of fixed exchange rates, which anchored a period of relative financial stability and impressive economic growth in the period after 1945 (Grahl and Teague, 1990; Block, 1977). The Bretton Woods agreement of 1944 signalled the passing of the baton from Britain to the United States in international monetary affairs, and established the dollar as the world's reserve currency. With European currencies pegged to the dollar – which was in turn convertible into gold – European economic development became bound up with the monetary and fiscal priorities of the United States. The exchange rate stability generated by the Bretton Woods system in the 1950s and 1960s (Figure 12.1) was such that monetary union within Europe was a minor issue, and one deemed secondary to issues of trade and industry (Grahl and Teague, 1990). What concerns us here is not why the Bretton Woods system unravelled in the 1960s and early 1970s – there are many treatments of this elsewhere (Brett, 1983; Block, 1977; Evans, 1988; Leyshon and Thrift, 1995) – but rather its effects on European finance, which are difficult to overstate.

There were five key repercussions of the collapse of the Bretton Woods system. The first was an increase in international financial flows and much greater volatility in financial markets (Thrift, 1996; Harvey, 1989; Gardener and Molyneux, 1990) (Figure 12.1). This was problematic for three main reasons. First, a number of European currencies were subjected to speculative assaults which led to devaluations of sterling (1967) and the French franc (1969). Moreover, waning confidence in the US dollar pushed investors to look elsewhere and, given the strength of the West German economy, to buy deutschmarks. This oft-since repeated pattern of currency movements has turned the deutschmark into an object of speculation, pushing up its value relative to other European currencies. Second, such exchange rate instability makes investment decisions more hazardous and imposes all manner of adjustment costs on firms and governments exposed to exchange rate fluctuations. A third problem for the EU countries concerned the effects of this exchange rate instability on their attempts at economic integration. Typical of this were the difficulties it created for their attempts to run a customs union and set common prices for agricultural produce (Grahl and Teague, 1990). Exchange rate fluctuations meant that uniform external tariffs could not be applied; currency devaluations had the effect of reducing tariffs on agricultural produce entering EU countries. Exchange rate volatility thus became a fundamental problem for European states in the early 1970s.

Second, the collapse of the Bretton Woods system acknowledged that credit provision had become an international activity and one which compromised the regulatory abilities of nation-states. The growth of multinational corporations, the rapid growth of trade, and the development of the comparatively unregulated Eurodollar markets (offshore credit

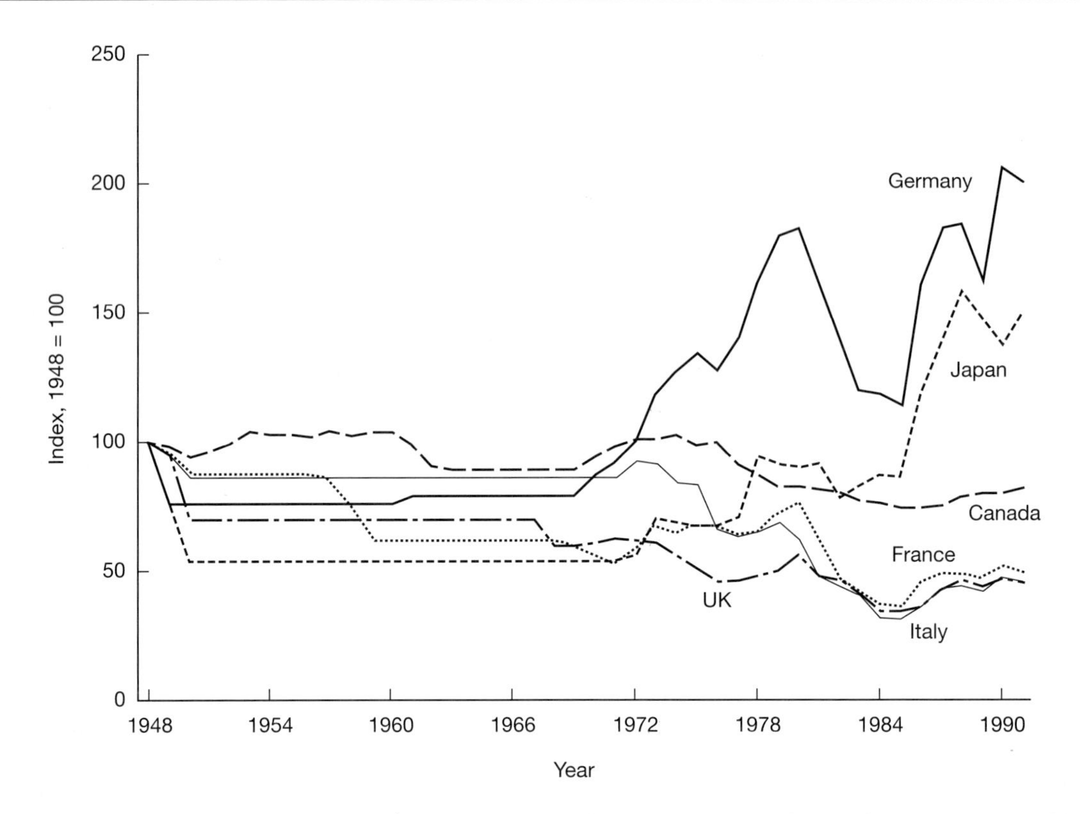

Figure 12.1 Exchange rates in terms of US$, 1948–91 (*source*: Chart 7–1, *Economic Report of the President*, copyright US Government, 1993, reproduced by permission).

money markets beyond the regulatory reach of the US Federal Reserve) in London signalled the ability of international financial capital to evade financial regulation designed at the national level in the Bretton Woods era. Through the 1950s and 1960s there was, in essence, a gradual *deterritorialization* of credit creation (Evans, 1988; Strange, 1988; Thrift and Leyshon, 1994) as increasingly international and private markets operated outside the regulatory boundaries of nation-states. This internationalization of credit creation has reduced the ability of nation-states to regulate their financial affairs. Related to this, the coordination of financial regulation is now a *global* affair (Llewellyn, 1992) and changes in regulation in EU member states reflect this reality. In the aftermath of the Less Developed Countries' debt crisis of the 1980s, for example, the Bank for International Settlements (BIS) introduced capital adequacy standards for banks in all major industrial countries, not just EU member states.

Third, just as the collapse of the Bretton Woods system of fixed exchange rates ushered in a greater volatility in exchange rate markets, so it marked the start of a period of intense innovation in international financial markets, as firms looked for ways to insure against risk. Trading in foreign exchange has boomed since the early 1970s, spawning a multitude of new financial instruments, and markets, to allow firms to spread risk. Piggy-backed on foreign exchange markets now are markets in foreign exchange futures and options – so-called 'derivatives' in that their value is derived from movements in the price of other assets – designed to spread risk. These markets are highly speculative, a feature of what Susan Strange (1986) has termed 'Casino Capitalism'. As financial markets have become more volatile, they have become more innovative and complex.

Fourth, alongside the growth of international private banking in largely unregulated offshore markets, there have been moves to deregulate and reregulate financial markets (Moran, 1991; Moran and Prosser, 1994) by breaking down barriers between different specialisms and encouraging innovation and risk taking. With the ascendance of neo-liberal, supply-side economics in Britain and the United States in the 1980s, the 'market' took on new significance. These regulatory changes have affected financial firms and markets in several ways. Generally, banks and other financial firms have been given greater freedom to determine where and how they do business. At a national scale, aided and abetted by new telecommunications technologies, protected domestic markets have been opened up to foreign competition. Within countries, we are seeing greater competition between different market segments, for example insurance and banking, which is blurring the boundaries between different financial specialisms. In essence, the international financial system has changed from a system managed by nation-states (though under the hegemony of the USA since 1944) to one in which private investors and the machinations of financial markets in London, New York and Tokyo have much greater sway.

Finally, in the period since the collapse of the Bretton Woods agreement, the European map has changed profoundly. Since the end of the 1980s we have witnessed the collapse of communism and the break-up of the Soviet Union and its sphere of influence, Czechoslovakia has been divided and Croatia, Slovenia and Bosnia-Herzegovina now mark the remains of the former Yugoslavia. The former German Democratic Republic has entered into political and monetary union with the former Federal Republic of Germany, resulting in a *de facto* enlargement of the EU. Hungary, Poland and other central and eastern European countries have similarly embarked on market-based reforms and sought to improve their relations with western European countries. A number of these countries, including Bulgaria, the Czech Republic, Estonia, Latvia, Hungary, Poland, Romania, Slovakia and Slovenia, have recently applied for EU membership, and are being considered for what would be the fifth enlargement of the EU since its inception (as the 'European Community') in 1957 (see Chapter 9).

Since the late 1960s, then, the international financial system has changed profoundly. The state-centric Bretton Woods system has been replaced by a system operated, but not controlled, by a series of public and private groups. Central banks and currency speculators now operate in a global, privatized credit system, in a proliferating series of interacting markets. The international financial system has changed to become much more dynamic, complex, volatile and innovative. This is the global financial context in which European economies now find themselves.

The European search for monetary stability: from currency unions to a single European currency

The collapse of the Bretton Woods system, between 1968 and 1973, ushered in an era of floating exchange rates, and posed problems for European states because 'the uncertainty that rules in the financial world spills over not only onto individual lives but into the fortunes of governments and of countries – and sooner or later into the relations between states' (Strange, 1986: 3). This uncertainty was the impetus for European economies to start discussing fuller economic and monetary union, with a view to creating a European zone of currency stability.

In 1970, the six original members of the European Community (France, West Germany, Netherlands, Belgium, Italy and Luxembourg) adopted the Werner report, which called for full economic and monetary union (EMU) by 1980. While this was an unrealistic timetable, the Werner report recognized that something had to be done to curb currency fluctuations within Europe. In 1972, the EU member states and Britain and Ireland joined what was to become known as the 'snake', the 'thin wriggling shape traced by EC currencies as they moved up and down against the Dollar' (Grahl and Teague, 1990: 107). The strength of the West German economy was such that the snake was essentially a deutschmark zone in which European economies attempted to keep their currencies stable. The strength of the deutschmark, however, imposed severe adjustment costs on EU countries. Countries with weaker currencies had to intervene in foreign exchange markets and push up their interest rates, dampening domestic growth, in order to maintain parity with the deutschmark. Such an arrangement was inevitably unstable; the French left the snake three times between 1972 and 1978, and only the Netherlands and Denmark managed to stay in it with West Germany for its duration.

The next European attempt to move towards monetary integration, the birth of the European Monetary System (EMS) in 1978, acknowledged the need for a new international monetary system not based around the gold–dollar system and the institutions of the Bretton Woods era. The EMS also recognized the uneven burden of adjustment imposed on European countries by the snake. In the EMS, all countries were to be involved in maintaining monetary stability, not just those with the weakest currencies relative to the deutschmark. Under the snake, each country had responsibility for maintaining currency in fixed parity with the deutschmark; in the EMS, however, exchange rate parities were calibrated around a basket of EU currencies termed European Currency Units (ECU). At the heart of EMS was the exchange rate mechanism (ERM), a series of bilateral exchange rates set between countries, with a permissible band for fluctuations. Defence of the parities involved collective responsibility, with central banks with strong currencies lending to those with weaker currencies (Grahl and Teague, 1990). Originally, a supranational European Monetary Fund (EMF) was to be established to oversee monetary policy and to support the ERM of the EMS. Yet, in a forerunner to contemporary debates, such an institution was vetoed by the Bundesbank, fearful of a loss of control over the German economy. In essence, 'the failure to back the ecu with an appropriate monetary institution meant that while the ecu was the *de jure* lead currency in the EMS, the *de facto* lead currency was the deutschmark' (Leyshon and Thrift, 1995: 132).

European economies participating in the EMS were thus subject to what Grahl and Teague (1990: 126) term 'a double hegemony'. On the one hand, their exchange rates were controlled directly by movements in the deutschmark. Indirectly, however, European currency movements were still linked to the fortunes of the US dollar. As such, the EMS was largely a defensive measure, an effort to insulate European economies from the volatility of the US-dominated international financial system (Dornbusch, 1991; Grahl and Teague, 1990), and not an attempt to establish the ECU or a European Monetary Fund to seriously threaten the US dollar as the world's reserve currency. Nevertheless, the EMS did deliver greater monetary stability to European economies, particularly through the mid- to late 1980s. Between 1984 and 1989, for example, there were just 12 currency realignments within the ERM, compared with 27 between 1979 and 1983 (Emerson *et al.*, 1992). The EMS helped to stabilize exchange rates and provided a framework for counter-inflationary policies in Europe (Artis, 1994).

Since the introduction of the EMS, issues of monetary integration in Europe have become further bound up with broader features of economic and political integration. The EU has promoted further economic integration, harmonizing taxes and subsidies, in order to try to create a 'level playing field' for member states. In terms of financial services, the three major elements of a level playing field are freedom of capital movements between member states, freedom of location in member states, and the removal of barriers to cross-border trade. Through the 1980s and 1990s, a number of important policy movements have furthered this agenda.

The first such movement came about in the form of the Cockfield White Paper of 1985 (CEC, 1985), which listed a series of measures – to be adopted before 1992 – to ensure the free movement of individuals, goods, capital and services in Europe. In so doing, however, Cockfield attacked the prevailing strategy, which insisted that no internal market could be created until regulatory arrangements had been harmonized between member states. Instead, Cockfield established an agreed *minimum* set of harmonized regulations and encouraged mutual recognition of member states' national regulatory principles. A second important feature of the Cockfield White Paper was its emphasis on using home country – and not host country – requirements to regulate firms; British firms, for example, were to be supervised under British regulations regardless of where they were doing business in the EU. These principles were enacted in the 1986 Single European Act, which set out the broad parameters for the completion of the internal market by 1992.

The Cockfield White Paper was also the impetus for further legislation concerning financial services. The Second Banking Directive, issued in 1988 and in effect from 1 January 1993, established the provision of 'a single passport' for banks in the EU. Granted by an institution's home country, the passport permits firms to open branches and engage in a variety of defined permissible activities in EU countries without seeking authorization from their host country (Table 12.1). In similar manner, securities firms also work with a single passport (Investment Services Directive), although there is potential for confusion in that the home state is supposed to enforce regulation and deal with conflicts of interest, while the host state is responsible for other business rules, for example, advertising and marketing, investor compensation arrangements and so forth (Dale, 1992). There are similar moves afoot to create a single European market for life and non-life insurance, although major differences in national regulation remain (Llewellyn, 1992).

Table 12.1 Banking services mutually recognized in the Second Banking Directive.

1	Deposit taking and other forms of borrowing
2	Lending
3	Financial leasing
4	Money transmission services
5	Issuing and administering means of payment (credit cards, travellers cheques and bankers drafts)
6	Guarantees and commitments
7	Trading for own account or customers' account in: (a) money market instruments (b) foreign exchange (c) financial futures and options (d) exchange and interest rate instruments (e) securities
8	Participation in share issues and services related to such issues
9	Money broking
10	Portfolio management and advice
11	Safekeeping of securities
12	Credit reference services
13	Safe custody services

Source: CEC (1989).

The Delors Report and a single European currency

Another major advance on the road to economic and monetary union was the report of the Delors Committee in 1989. As momentum for further integration gathered pace in the late 1980s, the committee was established to consider what actions were needed to facilitate full economic and monetary union within the EU. Many of its recommendations were enshrined in the Treaty on European Union (TEU), which was agreed at the inter-governmental conference (IGC) in Maastricht in 1991 (see Chapter 9). The Delors report established three stages that member countries had to go through in order for the EU to achieve full economic and monetary union (Table 12.2). At the Maastricht conference in 1991, timetables were established for each of these three stages, a set of criteria for membership and transition between the stages were specified (Table 12.3), and a constitution for the new European Central Bank was decided.

During stage one, deemed to have commenced in July 1990, all capital controls were to be phased out and EU members were to work towards economic convergence, as measured by interest rates, inflation and the size of their respective budget deficits. One of the most controversial aspects of the Delors report was the idea that participating countries would not only agree to forgo some independence in fiscal and monetary policy as a *result* of monetary union, but that en route *towards* such a union they should also be required to make hard political choices concerning public expenditure, inflation and unemployment in order to meet the convergence criteria specified at Maastricht. In stage two of convergence, the European System of Central Banks (ESCB) should be established and margins of currency fluctuation narrowed further. Finally, and for those countries which have fulfilled the convergence criteria, in stage three, exchange rates would become irrevocably locked and lead to full monetary union – the adoption of the euro – with an independent ESCB lending policy and starting to manage foreign currency reserves.

Progress through stages one and two, however, has presented different member states with all kinds of difficulties. Three issues in particular have reared their head. First, since the signing of the Maastricht Treaty, Europe has fallen into recession, experiencing GDP growth of only 1.1% in the community in 1992 (Williams, 1994). During the recession, public sector debt rose across the member states as tax revenues declined and social security spending increased. Second, and related to this, there have been the problems of meeting the convergence criteria laid down at

Table 12.2 The three stages of Monetary Union.

Stages of Monetary Union	Timing	Actions
Stage 1	Up to January 1994	1 Closer monetary coordination 2 Phasing out of capital controls
Stage 2	1 January 1994	1 Establishment of European System of Central Banks 2 Currency realignments only in exceptional circumstances 3 Council of Ministers can recommend changes in national government budgets
Stage 3	1 January 1999 at latest	1 Participants' currencies are irrevocably fixed and replaced by a common currency

Table 12.3 Convergence criteria for Economic and Monetary Union.

Rates of inflation	A country's inflation rate should not exceed by more than 1.5% that of the three best performing countries (over a period of one year prior to examination)
Interest rates	A country's average long-term nominal interest rate should not have exceeded that of the best three performers by more than 2% (over the same period of time)
ERM stability	A country should have been in its ERM normal band 'without tension' for 2 years
Fiscal (1)	Ratio of budget deficit to GDP should not exceed 3%
Fiscal (2)	Stock of outstanding government debt to GDP should not exceed 60%

Source: derived from Artis (1994).

Maastricht; governments have been forced to bring in austerity packages to cut their deficits. In France, public service workers and students took to the streets at the end of 1995 to protest against the reform of the social security system (Anon., 1995). Even Germany, Europe's largest economy and the one assumed to lead the way into monetary union, is having problems as its growth in GDP has slowed and its budget deficit crept past the 3% mark. Most EU members, including Germany and France, have yet to meet all the convergence criteria for monetary union laid out at Maastricht (Molyneux *et al.*, 1996; Barber, 1996).

This questioning of the path to monetary union was also fuelled by the events of September 1992, when speculation in foreign exchange markets wrought havoc in the European ERM. Sterling and the Italian lira were forced to withdraw from the ERM, and other currencies were devalued against the deutschmark; only the Dutch guilder and the Irish pound emerged unscathed (Bladen-Hovell, 1994). In August 1993, finance ministers, recognizing the weakness of the ERM, decided to relax the rules for membership, by widening the bands of permitted currency speculation from ± 2.5% to ± 15%. The crisis in the ERM in 1992 and 1993 highlights the contradictory nature of monetary integration. As Grahl and Teague (1990) have argued, it is impossible to reconcile fixed exchange rates betwen national currencies, independent monetary policy in the countries concerned *and* full capital mobility between the countries. This is so because freely mobile capital will seek out the highest rate of return (interest rate). In a fixed rate system, then, national differences in interest rates are unsustainable as countries with lower rates of interest would suffer an outflow of funds until eventually either the fixed exchange rate, the capital mobility or monetary policy has to change. With the liberalizing of capital flows in the EU, in preparation for full monetary union, it has become easier for capital to move into and out of EU states. As a result, speculators have been able to push money into the more inflationary states in the system (for example Italy and Spain) in anticipation of an interest rate rise (Leyshon, 1993). In pushing ahead with monetary union, the European Commission was 'seeking to draw the Community closer together, for fear that, unless matters were moved along, the ERM would collapse under the weight of its internal contradictions' (Leyshon, 1993: 1555).

These difficulties point to a third, and seemingly intractable, problem affecting the introduction of a single currency, a problem which afflicts all attempts to effect currency stability, namely the marked levels of uneven development and regional inequality in Europe (Dunford, 1994). The German economy, for example, is much stronger – in terms of per capita GDP and purchasing power – than the countries that joined the EU in 1981 and 1986, Portugal, Spain and Greece. In a European context, these countries have relatively low per capita income and high unemployment and are disproportionately dependent on agriculture (Cole and Cole, 1993; Williams, 1994). Moreover, within member states, there are significant interregional, inter-urban and intra-urban variations in economic and social well-being. Within the EU, then, there are very marked variations in regional and national economic performance and, consequently, variations in capital flows, rates of inflation and unemployment. Traditionally, movements in exchange rates have been a mechanism for compensating for such variation; governments have used changes in their exchange rate as a 'shock absorber'. The economic and political strains of the reunification of Germany, for example, were reflected in interest and exchange rates across Europe and precipitated the 1992 crisis in the ERM. In December 1991, the Bundesbank raised interest rates because of concern about the costs of reunification and the possibility of inflation in Germany. Other member states were forced to follow suit, sacrificing their domestic priorities, until the tensions forced the devaluations of 1992. After monetary union, European governments will have to rely on changes in wages and prices – and not revaluation or devaluation – to adjust to changes in global economic conditions.

This raises two questions. First, what will be the priorities of a European Central Bank? Specifically, will it serve as an instrument *of*, or rather a control *on*, governments (Grahl and Teague, 1990)? The Delors Committee, influenced by neo-liberal doctrine, favoured the latter view. Yet, German unification saw the autonomy of the Bundesbank being overruled, and the strains of unification are now prompting the Germans to stress the importance of economic convergence before entry into European monetary union (Anon., 1996). Will European governments have such control of a European Central Bank to help them manage economic cycles? Second, and related, what are the likely regional implications of greater financial integration: who will be the winners and losers of further financial integration?

Implications of financial restructuring in Europe

Underpinning moves to further economic integration are a very orthodox, neo-liberal set of beliefs, namely that reducing barriers to trade and movement and harmonizing markets will allow more efficient firms to increase their share of the European market. The logic of the Cecchini Report (1988) assumes that as barriers to trade and investment come down, cheaper imported goods and services will substitute for local products and resources will be thus released into other activities, generating greater export earnings (Dunford and Perrons, 1994). Much of the rhetoric about liberalizing capital flows and establishing minimum regulatory requirements is couched in terms of 'levelling the playing field', and of harmonizing the costs of financial services across Europe around a lower average (Cecchini, 1988; Leyshon and Thrift, 1992). For O'Brien (1992: 1) this signals 'the end of Geography', in that liberalizing capital flows and making the price of capital more similar in different countries leads to 'a state of economic development where geographical location no longer matters'. Yet, in the short term, such a levelling is highly unlikely given the marked regional inequalities between and within EU member states. In the longer term, it is likely that prevailing patterns of uneven development will be reworked, and possibly accentuated, with some countries and regions benefiting from the restructuring of financial services, and others losing out. In essence, greater integration does not signal the 'end of geography', rather a reconfiguration of the prevailing regional inequalities in Europe.

Initially, though, it is important not to overstate the likely implications of changes in European financial services, for a number of reasons. First, as the case study illustrates with respect to Hungary and the Czech Republic, European countries have developed very different regional financial structures. EU member states, for all their regulatory harmonization, are still very different creatures; there remain considerable differences in their competitive, organizational and tax structures (Molyneux *et al.*, 1996). While it is true that some financial markets – such as wholesale banking – have been international in nature for a long time, other markets, like retail banking, have been and still are organized nationally (Llewellyn, 1992).

Second, the effects of 'levelling the playing field' for financial services in the EU will not appear immediately, if at all. There is still a great deal of diversity in national payment systems, for example, and a single currency will not immediately facilitate more efficient cross-border payments. It takes time for banks and other financial firms to restructure to take advantage of opportunities to expand into other member states.

Third, and related, consumers may prefer to purchase their financial products from domestic firms they know, rather than new EU competitors. While it is true that technological changes are enabling information about customers to be assembled at a distance, rather than face to face, established indigenous institutions may still retain considerable advantages over their overseas rivals, not only in gathering information about customers but also in having more regular contact with customers and building up their brand name and reputation with potential customers. So economic, political and cultural differences between member states will probably remain as barriers to entry into different financial markets, particularly retail markets, which have traditionally been organized on national lines.

What, then, are the likely impacts of this restructuring of European financial services? We can think of the impacts at urban, regional and international scales. Financial services are concentrated in comparatively favoured regions, both within and between countries, in Europe. Thus, the areas around London, Paris, Frankfurt (Figure 12.4), Brussels and Amsterdam are centres of financial activity and employment (Begg, 1995). This existing structure of

The development of regional banks in Hungary and the Czech Republic

Britta Klagge

The development of financial markets in eastern Europe is an important part of the transformation from planned to market economies. Banks play a central role in this process, as banking functions such as payment, clearance and the provision of finance are at the heart of a capitalist market economy and thus are important factors for economic development and growth (Caprio and Levine, 1994; Dittus, 1994). The banking systems in eastern Europe, however, are passing through major transformation processes themselves, in the course of which strong centralization is given up and market forces become more important.

From a regional perspective, regional banks (banks that are not headquartered in the capital city or main financial centre of a country) are especially important in a market economy. They are often more ready to provide credit to small companies and entrepreneurs in their respective region than branches of centrally located banks, and can thus be important contributors to more balanced regional development (Chick and Dow, 1988; Hartmann, 1977; Klagge, 1995). This case study compares the development of regional banks in Hungary and the Czech Republic, two of the most advanced transformation economies in eastern Europe. Although the two countries are the same size (roughly 10 million people), their regional structures have always been markedly different, and the development of their regional banks differs accordingly.

Regional structures in historical perspective

Both Hungary and the Czech Republic were part of the Habsburg monarchy, centred on the main capital of Vienna until 1918. Hungary managed to attain some form of independence in 1867 and from then on Budapest served as capital for the Hungarian part of the Austro-Hungarian 'double monarchy'. The Hungarian half of the empire included not only Hungary as it has existed since 1918, but also many of the neighbouring countries and regions, such as Slovakia, Croatia, and parts of Romania. After 1867, Budapest developed to be not only an administrative but also an important economic and financial centre, and soon it was the second largest city in the empire. At the same time Prague slipped to third place as surrounding Bohemia and Moravia, which were parts of the Austrian half of the empire, were governed from Vienna. In 1918, Prague became capital of the newly established state of Czechoslovakia, which split into Slovakia and the Czech Republic at the beginning of 1993.

Apart from the different functions of Budapest and Prague before 1918, there were major differences in the economic structure and functions of the respective parts of the empire. While Bohemia and Moravia were the industrial 'heartland' of the empire, Hungary served as the main provider of agricultural products. Industrial production as well as services were strongly concentrated in Budapest, while the rest of the area was rather underdeveloped and showed a strong west–east disparity. Industry in Bohemia and Moravia was distributed more evenly and regional disparities were much smaller.

These historical regional structures still exist today (compare Figures 12.2 and 12.3). Budapest maintains the position of a primate city within today's Hungary. Its 2 million inhabitants represent almost 20% of the population, while the second and third largest cities, Debrecen and Miskolc, have only 218,000 and 190,000 inhabitants respectively. Prague has a population of only 1.2 million, or 12% of the Czech population, and the next largest cities are Brno and Ostrava with 390,000 and 326,000 inhabitants respectively. Thus strong centralization and large regional disparities in Hungary and a rather decentralized regional structure in the Czech Republic were preserved during the socialist era, despite efforts to support regional development outside the Budapest region and especially in the eastern part of the country.

continued

Transformation of the banking system and the development of regional banks

During the socialist era economic control was centralized and market forces were abandoned in all eastern European countries. This was especially true for the banking system. The existing banks were socialized after World War II, and a so-called one-tier banking system was established. In this system, the national bank performed both the functions of a central bank and commercial banking activities. There also existed a few specialized state financial institutions, but they were controlled by the national bank. The national bank in turn had to comply with the economic plans of the central planning bureaucracy and the requirements of the socialist government; credit lines, for example, were determined and planned by central institutions of the 'command economy' rather than as a result of market mechanisms and yield considerations. Thus economic and banking control were concentrated in the capital cities of Budapest and Prague, with the rest of the country being served by branch networks.

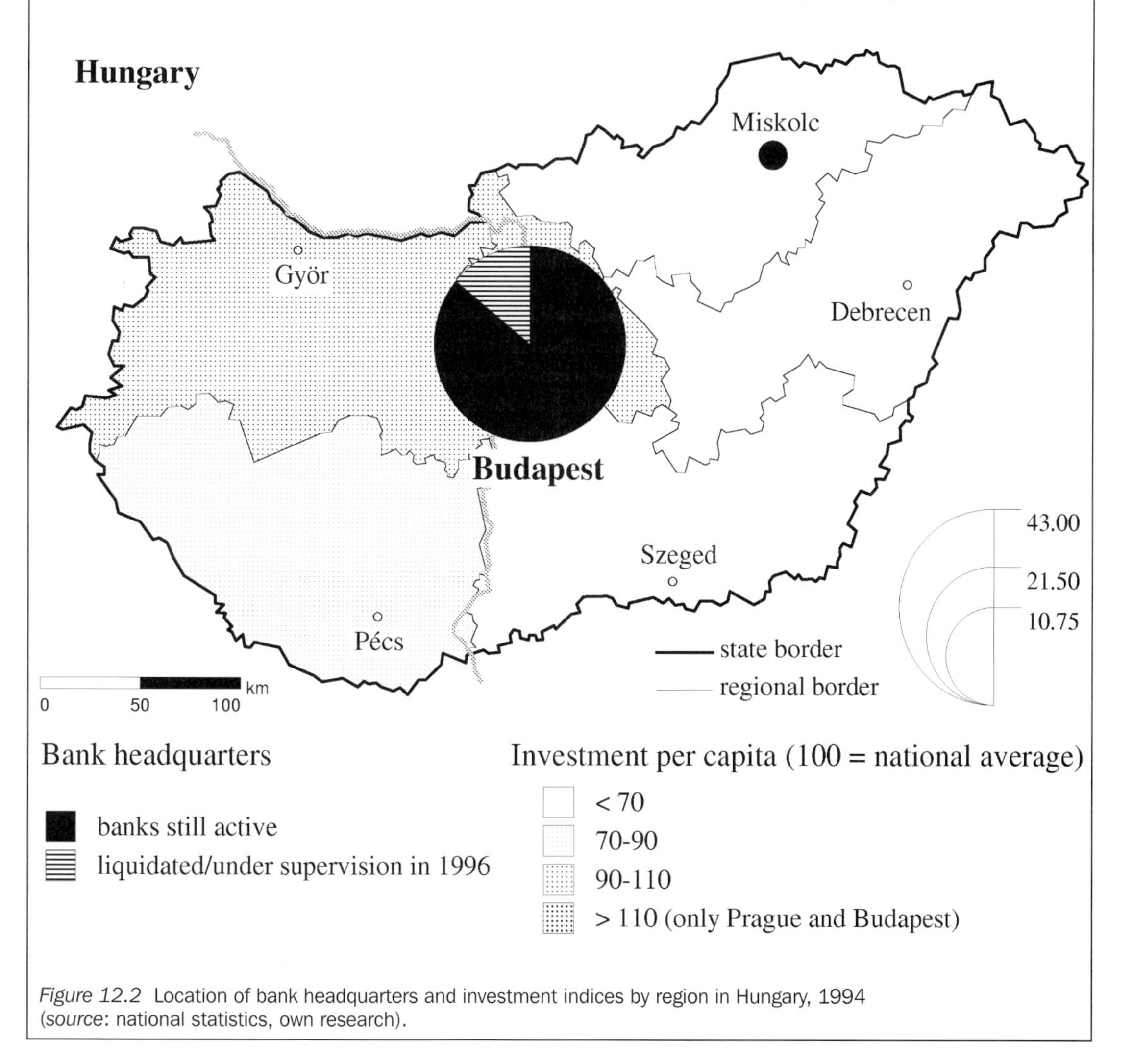

Figure 12.2 Location of bank headquarters and investment indices by region in Hungary, 1994 (*source*: national statistics, own research).

continued

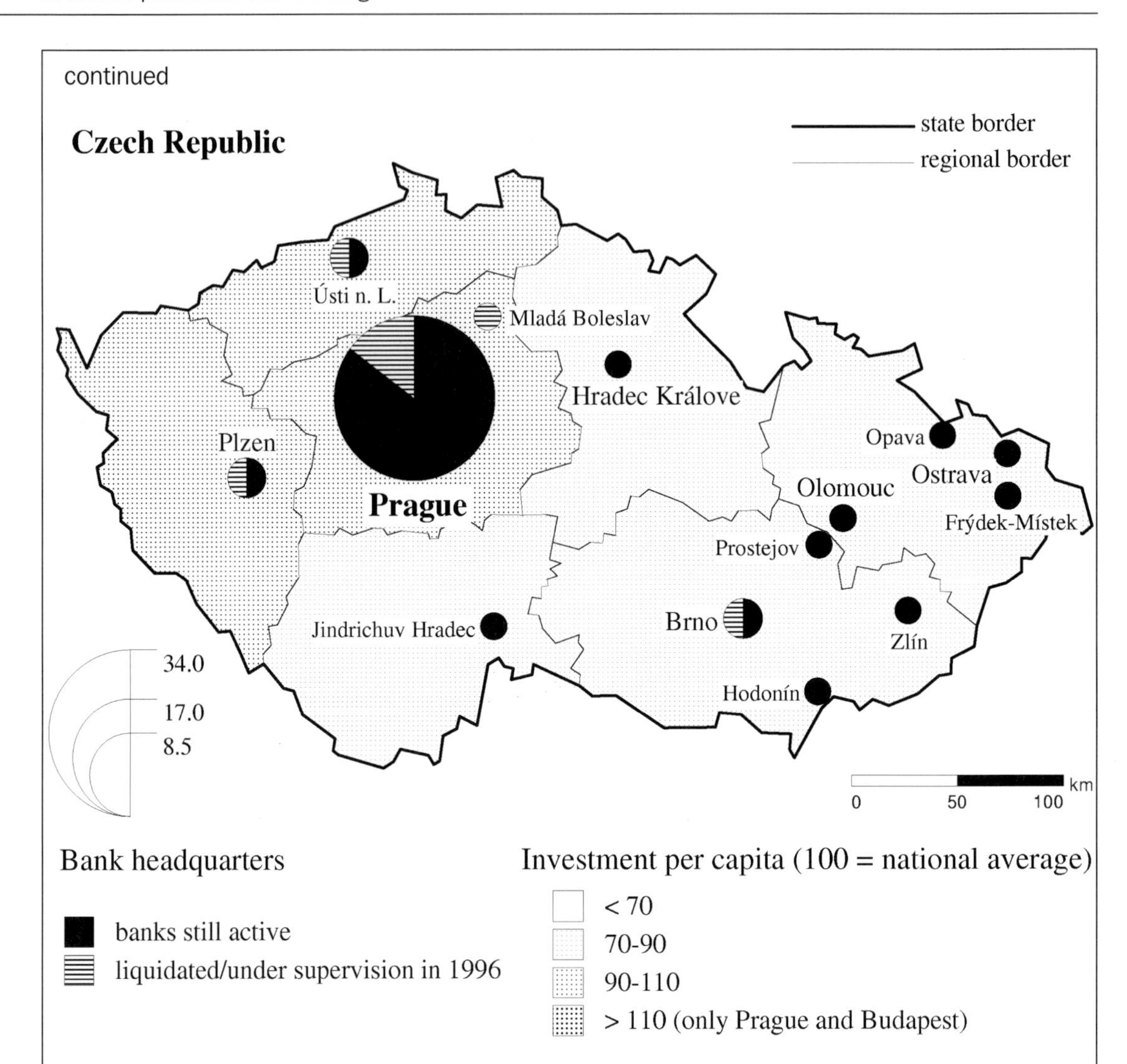

Figure 12.3 Location of bank headquarters and investment indices by region in the Czech Republic, 1994 (*source*: national statistics, own research).

In Hungary, efforts to decentralize the banking system and to introduce competition again started very early. Hungary has a long tradition of step-by-step reforms and liberalization, dating back to 1968. New kinds of financial institutions which could operate somewhat independently from the national bank were allowed in the early 1980s. In 1987 a two-tier banking system – separating the central from the commercial banking functions – was established again after 40 years of socialist rule. From then on, it was gradually adapted to western banking standards, especially after 1989–90. In the Czech Republic – or Czechoslovakia at that time – the first official reform considerations were voiced in the very late 1980s. A two-tier banking system was established immediately in 1990 and from then on transformation proceeded quite rapidly.

With the establishment of a two-tier banking system the commercial banking functions controlled by the national bank were transferred to a number of banks, which also took over most of the branches. All of these banks were headquartered in the respective capital city and were designed to operate nationwide (in Czechoslovakia, though,

continued

there was a regional separation between Slovakia and the Czech Republic). The central banking functions remained with the national banks, which from that time had all the rights and responsibilities of a central bank in a market economy.

Apart from the banks that took over the commercial banking functions from the national banks, new banks were established by domestic as well as foreign investors. Table 12.4 shows the number of banks headquartered and operating in Hungary and the Czech Republic from 1988 to 1994 (including branches of foreign banks). Despite their different developments, by the end of 1994 the two countries had roughly the same number of banks (including branches but excluding building societies and local savings cooperatives), 50 in the Czech Republic and 44 in Hungary. A special feature in Hungary is the approximately 260 savings cooperatives, which were already operating under socialist rule. Each of them is very small and they mainly operate in rural towns, where they were and still are the main providers of banking services. These services, however, only comprise very basic functions such as payment and clearance services for private customers and small enterprises; their credit business is economically not very significant due to the fact that they are undercapitalized and lacking in know-how.

Table 12.4 also gives an indication of the significance of regional banks. While in 1994 there was only one such bank in Budapest, there were 16 regional banks in the Czech Republic, representing more than 30% of all banks. The only regional bank in Hungary is headquartered in Miskolc (Figure 12.2); it was founded by the state and some of the surrounding communities in order to support regional development in one of the most depressed regions in Hungary. As opposed to the absence of 'normal' regional banks in Hungary, there is at least one and often more than one regional bank headquarters in each of the seven regions of the Czech Republic (Figure 12.3). Generally these are rather small, although most of

Table 12.4 Number of banks in Hungary and the Czech Republic, 1988–94 (year-end figures).

	1988	1990	1992	1994
Hungary (total)	23	30	37	44
Hungary (headquartered outside Budapest)	0	0	0	1
Czech Republic (total)	–	9	40[a]	50
Czech Republic (headquartered outside Prague)	–	1	10[a]	16

[a]Beginning of 1993, after separation from Slovakia.

Sources: National banks, own research.

them have branch networks that cover the surrounding areas. While some of them have recently been declared bankrupt, their overall importance has not diminished since most of them were taken over by other regional banks.

A comparison of Hungarian and Czech developments shows that existing regional structures not only reflect banking activities and credit policies, but also have an influence on the establishment of new regional banks. In a country with strong regional centres and generally balanced regional development, regional banks are more likely to spring up than in an environment that is traditionally characterized by strong centralization.

Further reading

Bonin J.P. and Székely, I.P. (eds) (1994) *The Development and Reform of Financial Systems in Central and Eastern Europe*, Elgar, Aldershot

Móra M. (1994) Banking in Hungary, *WIIW Members' Information*, **1/1994**, 23–31

Pick, M. and Turek, O. (1993) Banking in the Czech Republic, *WIIW Members' Information*, **12/1993**, 18–24

Saunders, A. and Walter, I. (1992) The reconfiguration of banking and capital markets in eastern Europe, in Siebert, H. (ed.) *The Transformation of Socialist Economies Symposium 1991,* Institut für Weltwirtschaft Kiel, Mohr, Tübingen, 101–30

financial services in different countries is going to affect the impact of the single European market. At an urban and regional scale, established financial centres and their surrounding regions may be the beneficiaries of moves toward the single market, particularly if we see further rounds of mergers – like those between Banco de Bilbao and Banco de Vizcaya and between Banco Hispano Americano and

Banco Central in Spain (Jones, 1990) – which concentrate financial services in established centres like London, Frankfurt, Paris and Madrid (see case study of London). While financial integration may bring with it intensified competition, the long term is likely to see an increase in concentration in sectors like banking and insurance (Dow, 1994). Indeed, Begg (1995) argues that prevailing national hierarchies of financial centres will probably become pan-European in scope, with London, Frankfurt and Paris, for example, paralleling the dominance of New York, Chicago and Los Angeles in North American financial markets.

Mergers and acquisitions activity in financial services, and especially banking, increased in the late 1980s and early 1990s in preparation for the single European market (Molyneux *et al.*, 1996) (Table 12.7). Given the limited number and relatively large size of the major banks in the UK, Germany, the Netherlands, Belgium and France, however, acquisition is an expensive strategy and one unlikely to be popular with regulators. Thus, most majority acquisitions in the banking sector have been domestic affairs. Banks can extend their geographic reach in other ways, however, through cross-border share holdings and joint ventures (Bellanger, 1990; Leyshon and Thrift, 1992). This latter strategy is proving popular with banks and insurance companies outside the EU seeking a foothold in the European market (European Commission, 1994). Again, there may be parallels with North America, in that banks may seek to become 'super-regional' institutions, focusing initially on neighbouring banking markets. Many European banks have entered markets across the EU. Deutsche Bank, for example, opened a regional office in Manchester (1989) and has links in the Netherlands, Portugal, Italy and Spain (Gardener and Molyneux, 1990). Similarly, Abbey National and the Woolwich Building Society have acquired specialist mortgage institutions in France (Lafferty Business Research, 1992). Thus far, much of the cross-border merger and acquisition activity has taken place in southern Europe, with banks in Italy and Spain being targets for northern European banks (Gual and Neven, 1993). Again, this suggests that further integration may lead to a greater concentration of financial firms and expertise in established centres and regions. Related to this merger and acquisitions activity and the increasing competition in financial markets, there has been a spate of job losses and branch closures affecting banking and insurance in the UK, Denmark, the Benelux countries, Germany and Spain (Lafferty Business Research, 1992).

Figure 12.4 The centre of Frankfurt, Germany's financial capital (*source:* Tim Unwin, 11 July 1983).

Always here and there: the City of London

Nigel Thrift

The City of London is one of the very few 'economic' spaces in Europe to have survived for over 300 years. It has achieved this feat by constant change, and yet at its heart the same activity has been consistently important – the exchange of information.

In the Governor's room of the Bank of England, up on the wall above the meeting table, there is a round dial which looks, at first glance, like a clock. In fact, it is a wind direction indicator, which dates from the days when the Directors of the Bank needed to know when the wind changed so that they could hurry down to the wharves and oversee incoming cargo. The indicator signifies the importance of the Port of London in the history of the City over a long period of time. This 'merchant' City apart, there have been other 'Cities' too. There was, for example, a City which was based on manufacturing (Michie, 1992). But these other Cities have all now faded into memory. The Port of London finally collapsed in the 1960s. The other Cities had all but disappeared before that.

The one City that was left was the one whose practical and spiritual centre was the Bank of England. Founded in 1694 by opportunistic merchants in order to fund the war against France and based on the management of the national debt, the Bank, along with later institutions like the Stock Exchange, gradually became a pivot around which a City of money and finance clustered.

Over time, it is this financial City that has become the main occupier of the City of London's space, so much so that it is now often regarded as coincident with it. From the demise of the financial markets of Amsterdam during the Napoleonic Wars up until the beginning of the First World War, the City of London was the world's chief financial centre: London was the world's largest, most liquid and most sophisticated financial market and sterling was the chief currency of international traders (Cain and Hopkins, 1993a, b; Kynaston, 1994, 1996). Bagehot (1873) famously described the City thus: 'the briefest and truest way of describing [nineteenth century] Lombard Street is to say that it is by far the greatest combination of economical power and economical delicacy that the world has ever seen'.

But the First World War marked a decline in the City's fortunes. Though the City was not physically harmed by the war, its financial markets were affected by the uncertainty. After the war, London's plight worsened as the productivity growth of the British economy lagged behind that of countries like the United States. The government responded by overturning the *laissez-faire* policies that had propelled the earlier growth. For example, they retreated steadily from the gold standard and banned UK banks from lending to foreigners. In combination, these actions cut the foreign share of London's capital markets. In turn, this declining foreign activity in the London markets crippled the internationally oriented merchant banks, forcing them to scale down or even halt their main business in foreign lending and trade acceptances.

By the 1950s the City appeared to be in the economic doldrums, in danger of becoming merely a financial backwater. The United States had emerged as the world's leading economic power, while the US dollar had become the chief currency of international trade. Not surprisingly, the British merchant banks had lost much of their international influence, being enlivened only by a few new entrants like Warburgs. Equally, the social structure of the City had become sclerotic, based upon who you knew rather than what you knew. The City's space had also become ossified, consisting of commuters who came from all over London and south-east England, but chiefly from specific areas of the 'Home Counties'. For example, there was a well-defined 'stockbroker belt'. The actual space of the City and its feeder spaces were the domain of men in heavy dark suits who spent their lives engaged in a daily round of travel and work (Courtney and Thompson, 1996).

But since the 1950s the City has changed, almost beyond recognition. The world has come to the City and the City has gone out into the world again, just as in its 19th century heyday. What accounts for this transformation? There are many reasons, but four stand out. First, the floating of exchange rates and the growth of the Eurodollar

continued

market in the 1960s were a signal that the world has increasingly become one of money and finance. Since the 1960s large numbers of new financial instruments have been continuously invented in places like London, often piggy-backing on other financial instruments and thereby creating a system of echoes which mean that money can be traded and capital borrowed and lent farther than ever before. As a result, cities like London, New York and Tokyo can now call on vast reserves of money, and their markets are continuously liquid (Table 12.5).

Second, and partly as a result of the first reason, London has become a vast conglomeration of technical expertise. Its 272,000 workers in financial services in the City include experts on a full range of financial issues and, taken together with their compatriots in New York and Tokyo, they are, in a sense, the world financial system. This is a community that is very different from that of the 1950s. Most importantly of all, it is cosmopolitan because of the arrival, from the 1960s, of large numbers of foreign banks – 257 of the world's largest 1000 banks now have branches in the City of London, more even than in New York – and because, for many of the jobs in the City, the labour market is now international (Thrift, 1994; Leyshon and Thrift, 1997; McDowell, 1997).

Third, London has become reconnected to the rest of the world. It is connected physically through the large numbers of telecommunications networks which are centred on the City. And it is connected socially. The City is not a static place, rather it is a place through which many thousands of people flow every day – to do deals, to seek advice, to gain knowledge or to seek out new business opportunities. Similarly, people who work in the City spend much of their time travelling overseas doing precisely the same things (Table 12.6). In other words, the City is a vast pool of information which is constantly being called on and updated (Thrift and Leyshon, 1994). After all, in the world of money and finance victory goes to the most well informed. In turn, that also makes the City a very sociable place: to obtain information you have to know people and that means building relationships of trust and confidence. So cafés, restaurants, wine bars, meeting rooms and the like are as much a part of the City's daily currency as money.

Fourth, the City is a space where government regulation favours making money. The City is not a 'deregulated' space. Indeed parts of it are now strongly regulated, usually as a result of the discovery of financial fraud. But much regulation is 'light touch' and, combined with the Bank of England's understanding of the City, this makes it a relatively easy place to do business (Moran, 1991).

Table 12.5 Indices of the City of London's world financial pre-eminence: London's share of world business in selected financial markets.

Market	London's share (%)
SEAQ-I (global cross-exchange securities trading)	64
SEAQ (global trading in domestic securities)	6
Equity options	4.7
Equity futures	4.4
Interest rate futures and options	11
Commodity futures and options	15
Foreign exchange	27
Swaps	35
International equity underwriting	65–70
Euro paper	90–95
Euro MTNS	90–95
Eurobond underwriting	60–75
Eurobond trading	75
Bank lending	18
International fund management	81
International mergers and acquisitions	40–50
International insurance	7.5
Ship broking	5.0

Source: London Business School (1995: 2.9).

Table 12.6 City pairs with highest volumes of air passenger traffic involving a European city.

Rank		Volume (thousands of flights in both directions)
1	London/Paris	3 402
2	London/New York	2 276
3	Amsterdam/London	1 775
4	Dublin/London	1 721
5	Frankfurt/London	1 222
6	New York/Paris	1 217
7	London/Los Angeles	1 015
8	Brussels/London	950
9	London/Tokyo	908
10	Frankfurt/New York	831

Source: International Civil Aviation Organization, Civil Aviation Statistics (1993).

continued

So the City of London is a space which is always both here and there. Money, capital, people and information flow into it and out again, providing it with an enormous sphere of influence. This sphere of influence extends over Europe and beyond. Over the last ten years competitor financial centres in Europe like Paris and Frankfurt have attempted to challenge London's hegemony. Paris has had a financial centre growth policy, while Frankfurt has had a *Finanzplatz* policy. But these policies have failed, as symbolized by the events of 1995, when the Germans came to London in force. For example, Deutsche Bank began to move its Frankfurt-based investment bank to London and merge it with its merchant banking subsidiary, Morgan Grenfell. Dresdner Bank purchased the UK merchant bank Kleinwort Benson to build up its London presence and buy skills it could not develop in Frankfurt. Westdeutsche Landesbank's investment banking arm, West Merchant Bank, took over Pamnure Gordon, a London stockbroker; and Commerzbank, the only major German bank without a British securities house, expanded its London office and moved all of its trading of non-deutschmark products from Frankfurt to London.

Of course, London's position in the world financial hierarchy can still be threatened – by financial scandal, by the proposed European Monetary Union, by new and so far unforeseen technological developments which mean that face-to-face communication becomes less important than it currently is. But, for now at least, London reigns supreme in Europe as a financial centre.

Further reading

Courtney, C. and Thompson, P. (1996) *City Lives. The Changing Voices of British Finance*, Methuen, London

Kynaston, D. (1994) *The City of London. Volume 1: A World of its Own 1815–1950*, Chatto & Windus, London

Kynaston, D. (1996). *The City of London. Volume 2: Golden Years 1890–1914*, Chatto & Windus, London,

Leyshon, A. and Thrift, N.J. (1997) *Money/Space. Geographies of Monetary Transformation*, Routledge, London

Michie, R. (1992) *The City of London. Continuity and Change, 1850–1990*, Macmillan, London

More broadly, in the absence of radical, interventionist regional policy or large-scale migration, further economic integration looks set to intensify existing regional inequalities in Europe. For weaker nations and regions greater integration could mean more import penetration and greater competition in markets for financial services and other products. Moreover, as competitive conditions are harmonized across the EU, one of the incentives for financial firms to relocate investment to peripheral regions disappears; firms no longer have to relocate in order to manage differences in exchange rates, taxes, tariffs and so forth (Dunford and Perrons, 1994; Artis, 1994). Capital is highly selective, and financial services investment and expertise is concentrated in a few key regions and metropolises in northern Europe. If further financial integration is synonymous with further specialization, we are likely to see an intensification, and not a narrowing, of prevailing regional inequalities in Europe. Related to this is a further question: how are weaker EU economies to adjust to the greater competitive pressures that come with further integration? In the past the EMS operated 'as a short term stabiliser: when a serious conflict arose over the medium term between a member country's exchange rate and the direction of its other economic policies it has been the exchange rate which gave way' (Grahl and Teague, 1990: 136).

With a single currency, however, wages and prices will have to give way, which does not bode well for workers. While a single currency is designed to impose greater economic discipline on member states – workers know that inflationary wage increases cannot be offset by devaluation, for example – such discipline may be administered in the form of higher levels of unemployment and falling output and generate a deflationary spiral (Dunford and Perrons, 1994).

The applications for EU membership from post-socialist economies in central and eastern Europe add a further dimension to this argument. In the

Table 12.7 Mergers and majority acquisitions in banking, insurance and finance, 1987–93.

Year	National[a]	EU[b]	EU/international[c]	International/EU[d]
1987–88	540	41	121	46
1988–89	704	123	139	80
1989–90	859	206	122	122
1990–91	800	168	106	122
1991–92	683	136	76	73
1992–93	606	122	82	108

[a]Operations between firms in the same country.
[b]Operations between two or more EU member states.
[c]Operations in which EU firm acquires non-EU firm.
[d]Operations in which non-EU firm acquires EU firm.

Source: European Commission (1994), Annex IV.

wake of German reunification, and the difficulties member states are having in fulfilling the Maastricht convergence criteria, EU plans for further expansion may have to be moderated. If the absorption of the former German Democratic Republic, one of the stronger economies of eastern Europe, caused such problems in the ERM, how is the EU to absorb the likes of Hungary, the Czech Republic, Bulgaria, Estonia, Latvia, Poland, Romania, Slovakia and Slovenia (Leyshon, 1993)? What would further expansion of the EU imply for investment in existing EU regions with low productivity and a high dependence on agriculture (Hudson, 1994b)? Are the Italian Mezzogiorno, and parts of Spain, Portugal, Greece and Ireland, to compete for investment with central and eastern European economies? Although regional development programmes are in place, not all regions can emerge as 'winners' in the changing economic bloc that is Europe (Dunford, 1994). How the EU reacts to the changing geopolitical realities in central and eastern Europe returns us to the geopolitical ambitions underpinning the single-market programme: is financial restructuring concerned with increasing competition in the EU, fostering regional economic convergence, and/or promoting successful European 'champions' which can compete globally?

Further financial integration holds out the potential of many benefits for EU member states. Financial stability is one key part of managing the uncertainty and volatility that marks the contemporary global economy. Although a more stable, less costly European financial base is not, on its own, a recipe for regenerating European regions, it is critical in allowing decision makers to plan ahead, protected from the uncertainty and economic damage associated with large swings in exchange rates. Yet financial integration, for all its potential benefits, poses some heady challenges, not least of which is the construction of alternative supranational institutions with a pan-European and not a nationalist, Bundesbank-dominated, vision of Europe's future. This institutional project 'of balancing the interests of "actual-existing" nation-states with the gains that might be realised from ceding power to a supranational monetary authority... has dogged the integration project from the very outset' (Leyshon and Thrift, 1995: 138).

It remains to be seen how the market logic of the Delors Report can hold together the fabric of regional inequalities and tensions on which the single European market in financial services is being sewn. Without significant, interventionist regional policy and a European Central Bank dedicated to democratic, investment-led development strategies, financial integration is likely to exacerbate regional inequalities and push countries to adopt deflationary policies. Such defensive adjustment strategies will, in turn, make the potential benefits of financial integration, not to mention the social cohesion emphasized in the single-market programme, recede further on the horizon.

Further reading

Gillespie, I. (ed.) *Banking 1992: A Eurostudy Special Report*, Eurostudy, London

Grahl, J. and Teague, P. (1990) *1992 The Big Market: The Future of the European Community*, Lawrence and Wishart, London

Grauwe, P. de and Papademos, L. (eds) (1990) *The European Monetary System in the 1990s*, Longman, Harlow

CHAPTER 13 Trade, European integration and territorial cohesion[1]

JEAN PAUL CHARRIÉ

Immediately following the creation of the European Economic Community (EEC), trade between the six founding countries progressed enormously, to such an extent that the EEC rapidly appeared to be an inescapable element of European integration. The British, and the countries with which they maintained privileged commercial links, soon admitted that the dynamism of their economies was fundamentally linked to their membership of this vast continental market.

This quite extraordinary increase in trade might be considered to have been linked to the work of GATT (the General Agreement on Tariffs and Trade), or to the globalization of trade, as much as to the introduction of the fundamental liberties which lie at the heart of the Treaty of Rome. Recent years have shown that the construction of the European economy has acquired a certain autonomy in relation to the general evolution of the global economy. Commercial flows continue to increase at a higher rate than those between other industrialized countries.

The creation of the Single Market, by suppressing all the administrative and technical barriers still hindering the movement of commodities, would allow the commercial flow between states to increase still further and thus introduce new economic and social dynamics. Concealed behind this evident success are some fundamental questions about the future of Europe (see also Chapter 9). The ultimate objective is the transition to a single currency, which would result in the suppression of distortions due to currency variations. However, the rapid expansion of trade is overloading existing infrastructures and creating a demand for new ones, whose design must correspond to a more European, and less national, logic. Will this opening up be beneficial to all the states as a whole and all the regions, or will it in fact aggravate the disparities between the different parts of the European Union (EU)?

The issue of the integration of central and eastern European countries can be included alongside these questions directly concerning the functioning of the EU. Are the 15 existing members in a position to help these other countries succeed in their transformations, and will they find appropriate solutions to bring about their entry into the present system?

Increased commercial integration

From 1983 to 1992 trade within the European Community experienced remarkable dynamism and a greater increase than that of global trade. This is demonstrated by the growing proportion of trade represented by intra-EEC as against other types of trade (Eurostat, 1995) (Table 13.1). Whereas at the beginning of the 1980s the ratio of internal to external

Table 13.1 EU commercial balance (goods) by geographical zones.

	1983	1988	1990	1992
Intra-European 12	54%	59.6%	60.6%	61.9%
Extra-European 12	46%	40.4%	39.4%	38.1%
of which old EFTA	10.1%	11.3%	10.8%	10.2%
Intra-European 12 + EFTA	64.1%	70.9%	71.4%	72.1%

Source: Eurostat (1995).

[1] Translation by Kate Wilson.

trade was relatively close (54% to 46%), the gap between intra-European and external trade widened over the next ten years to stand at 62% to 38% by 1992. This evolution is even more significant if the countries belonging to EFTA (European Free Trade Association) are included, as the intra-European proportion of trade in commodities increased from 64% to 72%.

Such integration is weaker where services are concerned, since in this case the balance of trade is unfavourable to the EU, because dealings with countries outside the EU take it up to 51.7% against 48.3%. Yet this difference was much more pronounced in 1983 (55.7% against 44.3%), and the ratio also changes considerably if the 11.8% of the EFTA countries are included in the EU services balance of trade. In fact 60% of the trade in services took place within the European Economic Zone in 1992.

This increased integration is true of all the countries of the EU and the former EFTA. In 1983, the share of trade between European countries remained less than 50% in Denmark, Italy, the United Kingdom, Finland and Iceland. In 1992, it was more than 50% in all countries, which indicates the acceleration of trade within Europe. The extent of commercial integration, however, depends heavily on the role of the state. Integration is generally very strong, close to or more than 70%, in the small states that participated in the first stages of the construction of Europe and acquired a wider market for their businesses: Belgium, the Netherlands and Luxembourg. But this is also true of the states that greatly depend economically on one of the member countries, notably Portugal and Ireland. Among the great economic powers, only France has brought itself up to nearly 65%, which demonstrates its current deep involvement in the European economic system, after having previously been a protectionist country (Figure 13.1).

It may seem surprising that Germany, in this evaluation of integration on the basis of trade, appears at the same level as the United Kingdom or Denmark; that is, 55% in 1992. It is well known that the United Kingdom continues to favour trade with Commonwealth countries and the United States, and Denmark at first established links with Scandinavian countries, although their entry into the EU will modify this percentage. Where Germany is concerned, this modest percentage expresses the increased degree of globalization in the economy of that country, closely involved in the functioning of the EU but at the same time equally open towards the former eastern countries and very competitive in Asia and North America.

The Europeanization of trade of the three states to join the EEC between 1980 and 1986 should also be noted. If the integration of Greece and Portugal was due to their being small underdeveloped countries, the same is not true of Spain, which now maintains 66% of its trade with the 11 other countries of the EU, compared with 50% ten years earlier.

This acceleration of trade between the 15 countries of the EU is based on specific advantages as well as on the specializations which allow for competition

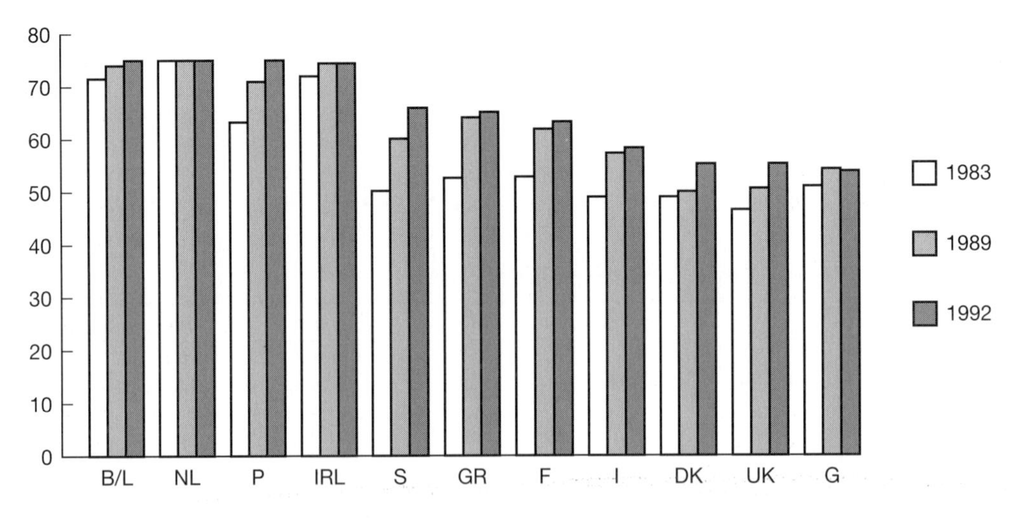

Figure 13.1 The role of the EU in the exports of European countries, 1983, 1989, 1992.

between trading partners. It is not surprising to discover that France is the leading exporter of agricultural products in trade within the Community, achieving 22% of the total, ahead of the Netherlands and Germany. But while France has strengthened its position in gaining 1% over ten years, Germany's and particularly the Netherlands' shares are declining. If Germany occupies first place in the export of manufactured goods, with nearly 30% of the total, ahead of France, Italy and the United Kingdom respectively, this domination is less than it was in 1983. This is due to the progress made by the other states, in a context of increase in industrial production and the flow of trade between the major partners. In fact, the influence of Germany in trade within the Community is declining in the manufacture of vehicles, industrial equipment and general machinery, as well as electrical appliances. However, Germany still maintains at least 30% of its trade with the 11 countries of the EU, or even more if the states of the European Economic Zone (EEZ) are included.

Where the eastern and central European countries are concerned, trade agreements aim to create (bilateral) free trade zones over the course of the next ten years. The agreements are expected to allow a reduction in quantitative restrictions on the majority of goods, and restrictions on industrial products are expected to be abolished completely. The EU is already the leading trading partner of these countries, accounting for more than half their trade, following an unprecedented development in imports and exports up to an annual rate of more than 15%.

A series of favourable factors

This increased integration of the economies of the 15 member states could be considered the result of the implementation of the four fundamental liberties written in the Treaty of Rome and the evidence that this free trade zone remains a significant driving force, affecting all countries. Among these four liberties – free movement of goods, workers, capital and businesses – only the free movement of goods has had real success, although even this is limited by the normative obstacles and tariffs introduced by states to maintain concealed protectionism.

Yet other factors played a part in this evolution, even though the economic crisis made the construction of Europe more difficult, and the growth of global trade was more rapid than that of trade within the Community. Everything points to the fact that there is relative European autonomy in this area and that this autonomy, in relation to the rest of the world, has been strengthened by the introduction of the Single Market, the restructuring of industry, and in a more circumstantial way by the reunification of Germany.

Although the Single Market came into effect on the first day of January 1993, many decisions were effective before this date, and European companies anticipated its introduction. This objective indisputably boosted European integration. The disappearance of customs duties at the end of the 1960s was not accompanied by any other action to suppress the other tariff and non-tariff related obstacles to the free movement of goods. The Single Market proposed the strengthening of the four fundamental liberties in order to achieve the free movement of people, companies, goods and capital and succeed in creating a Europe without borders (Commission Européenne, 1993a).

The Single Market allows a vast market to be created, one of the most extensive in the world, within which businesses can improve their performance while to a great extent being liberated from the constraints imposed by national standards, which often constitute disguised protectionism. The big companies soon realized the advantages that the creation of the Single Market brought. They supported the Commission right from the start, and took measures to organize their production and marketing systems in this new framework. These advantages are related to increased volume of production, economies of scale and access to the previously very protected public market. Yet small and intermediate companies have often shown themselves to be more reticent in the face of this economic globalization, which is weakening them.

The change in behaviour of industrial groups is a good illustration of the effects of this opening of borders. For more than 20 years the big firms acted against each other, more likely to formulate agreements with foreign, especially North American, organizations while carrying out strategies which were in fact hostile to the construction of Europe. The increasing competition of Japanese products in Europe, the reorientation of Japanese politics and the rise of investments in Europe incited European countries to support the aims of the Single Market. The prospect of the Single Market accelerated mergers and takeovers in all branches of industry. Even if it is not appropriate here to present the movements between firms in detail, their effect on European integration must be emphasized.

Let us first take, for example, the cooperation programmes in the field of high technology, as demonstrated by the increased power of the Airbus or Ariane–Espace programme, to which several industrial groups belong. European research programmes such as Eureka or Esprit can also be included. More effective for the development of trade is the necessity for firms to be present or to increase their presence in the new markets. From this results a rapid development of trade, not only of goods but even within the company, due to the fragmentation of the production process between different departments. The automobile industry has made considerable use of this option, manufacturing in huge volumes vehicles whose components may originate from several countries (see Chapter 11).

On a broader scale, the coexistence of imports and exports of the same products is based on the convergence of industrial networks and relations between companies. The majority of the big countries appear to be becoming increasingly non-specialized, that is to say they have a presence in most branches of industry, which signifies that intra-European trade will increasingly reflect trade within each branch. Of course, each country tends to have a specialization, like Italy, whose textile–clothing industry has clients in all the EU countries, and like the countries of the Iberian Peninsula, which still appear to be oriented towards very labour-intensive industry. Yet they all recognize the transfer to high-added-value industry, which should reduce the flows linked to the price of labour, and strengthen those resulting from the internal and European organization of the large companies. This can be illustrated by the situation of the UK, where besides the two big oil companies controlled by British capital, the principal UK export firms are British Aerospace, Ford Motor Co., Rolls-Royce and IBM. These are good examples of participation in European programmes and strategies.

The reunification of Germany in the early 1990s constitutes a supplementary factor favourable to European integration through trade (Deutsche Bundesbank, 1992). In effect, the German trade balance, which was in considerable surplus up to 1990 (up to 35,000 million ECU in 1989), has been declining sharply since 1991. This reversal of the trend is above all a result of the great increase in imports, partly induced by the consumer demands of the population of the former German Democratic Republic. This brisk development in German imports has benefited other European countries, which have increased their sales on average by nearly 30%. This phenomenon has been particularly noticeable with regard to car sales, whereby Spain and France have profited, having despatched in 1992 alone between 350,000 and 400,000 vehicles, whereas the UK and Italy accounted for only 100,000 to 200,000 cars. Even if the effects of reunification wear off, trade between European countries has been influenced by it, and interaction is stronger.

European trade zones and regional inequalities

The bilateral analysis of trade in commodities emphasizes the pivotal role played by Germany in the construction of Europe, and also highlights the importance of the impact of the reunification on the other countries as a whole. In effect, Germany is the leading supplier of all the European countries, whether within the 15 member states or in the EEZ, with the exception of Ireland and Norway. These two exceptions are due to privileged relations that Ireland still maintains with the UK, and similar relations of Norway with Sweden. Germany's position is particularly strong in relation to the small countries which traditionally remain in its sphere of influence, such as most of the Benelux countries, Austria, Switzerland and Denmark. It is also still significant, with approximately 20% of imports, in the cases of France, Italy and Sweden.

European integration cannot be measured solely by the role of intra-Community commerce within overall trade. It also depends on the volume of this trade expressed in thousands of millions of ECU. Only two countries pass the 100,000 million ECU mark with regard to both exports and imports: France with 115,000 million ECU and particularly Germany with 160,000 million ECU. The UK with 80,000 million ECU does not do much better than Belgium or the Netherlands. In addition, Germany and France have comparable weight with respect to their imports and have all the partners of the EU as their clients.

Because of this, the Franco-German connection plays a role increasingly determined by the volume of trade between these two countries, even if the balance has more often than not been in deficit for France. The political advances in the construction of Europe, discussed between these two countries, reflect in a way the degree of trade integration and the obligation to carry through financial and monetary policies based on these strong interrelations. The share of France, foremost customer and supplier of Germany, in Germany's imports and exports comes to 12% of the total trade and to nearly 25% if the Europe of the 12 member states alone is considered. Germany is by far the leading buyer and seller in France with almost 30% of the total amount of trade from the EU (Banque Indosuez, 1990) (Figure 13.2).

Trade zones can be seen to form around the Franco-German axis, resulting from geographical proximity and the long history of economic relations in which both of the EU trade powers participate. Of course, the economic weight of the principal countries – Germany, France, UK, Italy – is such that the latter trade mainly among themselves, but that does not exclude the formation of trade sub-zones within the 15 member states.

The existence of an Alpine Zone is evident, owing to the multiple links maintained by Germany, France and Italy, since Italy is the second supplier and the third or fourth biggest customer depending on the case. The neighbouring countries of Austria and Switzerland are also included. Switzerland, hesitating to enter the European Economic Zone, is Austria's third biggest customer, Italy's fourth and Germany's sixth. This zone must obviously be enlarged when including the central and eastern European countries with which Germany maintains a large share of its trade and is strengthening its position.

The existence of a Nordic Zone is equally evident. It is dominated by Germany and the UK. The strong influence of Germany in the whole of this zone has already been observed. The influence of the UK is equally evident, though more as a customer of the

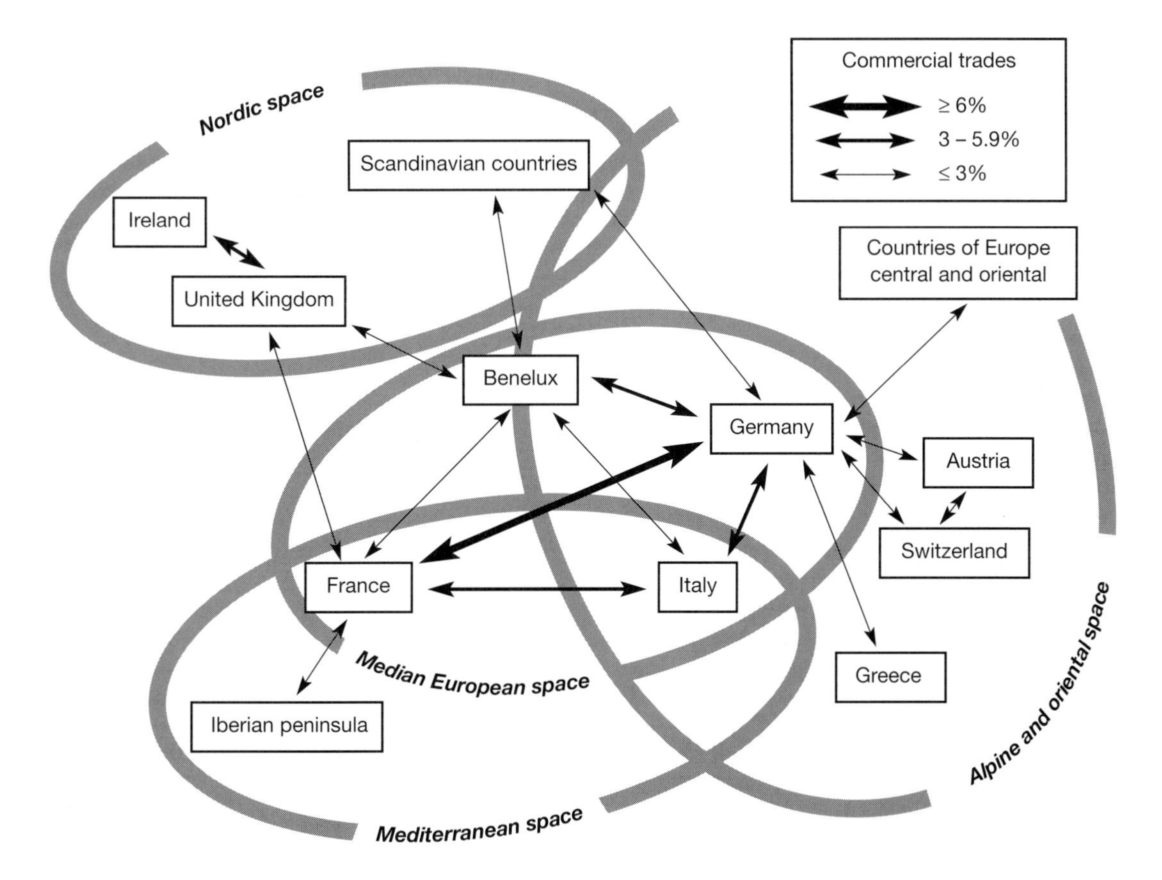

Figure 13.2 Europe's major commercial zones and principal commercial currents.

countries of this zone than as a supplier. The UK is Ireland's and Norway's foremost trading customer, Denmark's and Sweden's second and the Netherlands' and Finland's third.

The Mediterranean Trade Zone is dominated by France and Italy, but the increased integration of the Iberian countries strengthens the influence of this geographical sub-zone of trade. Spain has notched itself up to fifth place in the ranks of France's suppliers and customers, placing itself just below the UK regarding exports to France.

The analysis cannot be conducted solely on the ranking of the state; regional performance must also be taken into account, all the more since the suppression of borders and the rapid expansion of decentralization policy is increasing the political and economic weight of the regions. Within states, just as within trade zones, the regions are not all on an equal footing. This is demonstrated for some large continental states by calculations of rates of export in terms of the relationship between exports and GDP at the end of the 1980s (Bourdon, 1991). Such calculations show the differences in performance of regions and their degree of economic openness towards the rest of the European Union. In a schematic fashion, one can ascertain that the most dynamic and better structured regions have an export rate which is equal to or greater than 30.

All of the regions in the west of Germany have export rates above the European average, but the highest rates are to be found along the Rhine from Rhineland-Westphalia to Switzerland. In France, regional imbalances are much more pronounced. The whole Atlantic area has low levels, while the eastern half appears more open to the outside, with Alsace, Franche-Comté, Nord–Pas-de-Calais and the Midi-Pyrénées having values above the European average. In Germany, as in France, all of the regions interact with the rest of the European Union, but this is not the case in the Mediterranean countries, where certain regions have an export rate approaching zero. While the regions of northern Italy achieve average values of around 30 and more, the central and particularly the southern regions present very mediocre outcomes. This situation is even more pronounced in Spain, where no region achieves a level as high as the average 30. Only Navarre, due to the presence of multinationals within its area, approaches it; Catalonia and Valencia do not even reach 20. All of the other Spanish regions, as in Portugal, have very low values, reaching down to zero in the case of Extremadura.

It can be concluded from these few components that the regions which participate the most in intra-Community trade, with a real effect on the exports of the country, are those which are close to the tertiary and industrial heart of the EU. It is not by chance that in this description based on export levels, there exists a strong domination by 'Rhine Europe'. Inclusion of the Benelux countries and the London Basin would only reinforce such a perception. It is a question of the most industrialized areas, those where businesses are the most dynamic and open to international competition. The relative geographical proximity of these regions – those of north-east Spain are less close – favours trade between companies. This varying regional participation in intra-Community trade also depends on the level of training of company managers and employees. There is an evident correlation between the high average level of training in all the German regions and the low regional differentiation, whereas this is completely the opposite in Portugal and Spain. This retarded development is not only economic, but is also due to the level of training provided for the population.

Commercial flow of goods and means of transport

It would be totally illusory to consider solely the movement of goods within EU countries, as changes in the functions of businesses are accompanied by considerable increase in trade in services. It is evident that some of the needs can be satisfied by the introduction of new telecommunications tools, including those for the enhancement of services (see case study of the role of the World Wide Web). The appearance of agencies dealing with trade, finance and international business law might suggest the elimination of much of their clients' need to travel when dealing with foreign enquiries. Business travel would thus be expected to decline. Yet recent observations show that this is by no means the case, and that business travel continues to increase. For the time being it is difficult quantitatively or qualitatively to evaluate the relative importance of these various changes on the modal distribution of flows or quite simply for the importance of flows. Yet the role played by airports and the research into their connection with the different terrestrial means of transport certainly indicates the trend.

The World Wide Web in Europe and European exchange

Eric F. Berthoud

The World Wide Web (WWW) is a unique European product. It was developed by Tim Berners Lee at CERN (www.cern.ch), Geneva, Switzerland, in the early 1990s. Initially, the only widely available browsers were purely textual, and the only graphical one was Lee's NeXT implementation. Due to the tremendous development of the Web, a need was felt for search engines that attempted to provide a general catalogue for WWW resources. One of the first was designed by Oscar Nierstrasz. The W3 Catalog (iamwww.unibe.ch/~scg/W3catalog/history.htm) ran from 2 September 1993 to 8 November 1996, at the Centre Universitaire d'Informatique (CUI) of the University of Geneva. It now has been superseded by many others, such as Alta Vista and Hot Bot.

According to the Internet Society (www.isoc.org) and Network Wizards (www.nw.com), there is no way to determine how many users are on the net, or the number of Web servers, other than by making guesses and estimates. Nevertheless, counts allow the minimum size of the Internet to be estimated. The worldwide number of hosts was thus approximately 213 in August 1981, over one million in mid-1992, and over 13 million in September 1996 scattered in some 106 countries. The net continues to double in size approximately every 12–15 months.

Although there is not necessarily a correlation between the host domain name and the country where it is located, a survey by the RIPE Network (www.ripe.net) suggests that over 3.3 million hosts were running in Europe in September 1996, among which 85% were in the EU. In the last months of 1996, the average growth was around 5% to 6% every month in Europe as a whole. Some countries had a higher rate. In the middle of 1996, for example, it was 7% in the UK and 8.5% in Russia.

Very early on, publishers perceived the use that they could make of the Web. In Europe, at the end of 1996, more than 300 news magazines and daily newspapers had a server on the Web (200 in the EU and 100 in the other countries), not counting radio and television as well as other specialized media.

Most media using the Internet perceived the Web as a means to get themselves known and to cross the borders of the geographical area in which they are situated. Most daily newspapers express themselves strongly against censorship. For example, the 'Unione Sarda' (http://web.tin.it/UNIONE) wants to show that the Web allows local collectivities to get heard without being controlled by central governments.

Governments and administrations can measure the efficacy of the tool. In her address to the 'Openness and the European Union Institutions' Conference held in Stockholm on 27–28 June 1996, Mary Preston, head of the 'Citizen-oriented Measures' unit of the Secretariat General of the European Commission, declared that: 'Possibly the most exciting development as regard to the dissemination of existing information is the creation of the European Institutions WWW server EUROPA' (www2.echo.lu/legal/stockholm/welcome.html). The site (europa.eu.int/) opened in February 1995, and over 58,000 Internet surfers visited it every day in July 1996. Among the various types of information available are the Intergovernmental Conference database, which groups together documents from working groups, institutions, member states, contributions from universities, trades unions, employers' federations and chambers of commerce. This particular service was consulted by 30,000 people in May 1996. According to Mary Preston, not only must European citizens be kept informed, but they must also be given the opportunity to ask questions and to state their positions. The Web interface is particularly well suited for this practice of democracy.

This objective will only be reached if people understand the hows and whys of the Web. Among the various efforts towards Web literacy, Web for schools (inf.vub.ac.be/wfs/) should be rewarding (see also Report of the Task Force Educational Software and Multimedia, www2.echo.lu/mes/en/report796-toc.html, and the Education and Training Sector of the Telematic Applications Programme, www2.echo.lu/telematics/education/en). If its first objective is to link teachers and pupils over Europe in the production of electronic educational material, its other objective is even more important in the long run. It is meant to set the stage for a concerted effort of

continued

academia, politicians, industry, educational bodies and publishers to accelerate the uptake of the new technology for the benefit of schools and of society as a whole. A budget of 3.5 million ECU has been launched to organize a first experience. It concerns 150 schools (750 teachers) with on-line training material and Europe-wide Web support as well as educational Web resources. The delivery of end results was expected at the end of 1996. The target is to reach 30,000 schools in 1998.

Commerce and finance find the Web a useful platform, allowing them both quick information retrieval and communication facilities, as well as the possibility to read information on a relatively anonymous basis. This last aspect is of importance, since it allows any company to be better informed before deciding on the validity of a contact as well as permitting them to bypass a sometimes complex diplomatic procedure. Typical examples of the use of the Web in commerce include the following:

- Private and public sites offering an exchange forum. Many European chambers of commerce participate in the forum set by IBCC-Net (www.worldchambers.com/gbxhp.html).
- Immediate access to communication with contacts in business around the world is offered by Interactive Global Marketplace™ set by PANGAEA.NET (www.pangea.net). Most of the recent companies to join this site are situated in Slovenia, Romania and other countries of the former eastern bloc.

In finance, many sites offer brokerage services at very low commission rates. If, at the present time, only 1% of transactions realized by these firms actually occur on the Web, the habit may increase quickly. Private Swiss bankers (B. Guisani, *L'hebdo*, 4 July 1996, www.webdo.ch/hebdo/hebdo_1996/hebo_27/adf_27_ndi.html) declare that a Stock Exchange on the Web can be effective before the year 2000.

The Web is also more and more widely used not only to present a company but also to find new workers:

- Job offers on Web pages are increasing. On these pages, users can find job descriptions and information on application procedures, order brochures and application forms, read about development programmes, and read or hear comments of employees. British Telecom was looking for 750 graduates in the winter of 1996, and French National Education published offers for 21,100 teaching positions at the secondary level in 1995.
- Specialized sites like www.job-online.cegos.fr provide information on job offers nationwide.

The Internet is also particularly well adapted to 'infomercial' types of advertisement, where the viewer is given the chance to adapt and to customize a product or a service to their own taste, and even to order it on-line. Most automobile manufacturers are now present on the Web. Some, like Fiat, even ask the user to fill in a questionnaire on the type of car he or she likes to drive and the type of driver they are. Other advertisers offer on-line games and gimmicks, which draw more and more younger people onto their sites. Primarily used in the USA, this technique is now quickly gaining European sites. It also allows a closer profile of potential buyers to be drawn, as advertisements placed on the World Wide Web sites can be tracked in minute detail according to the number of visits, the type of enquiries, even the network address of each visitor.

The fight for the control of standards is located at the planetary level. We can be sure of one fact: whoever controls standards will control the network of the future and will be able to make people pay an extra dime on any computer-linked activity, be it business, communication research or services. Although the European electronics industry is not directly engaged in the battle of browsers, it is active in many other fields, following, as far as the EU is concerned, the lines set by the INFO2000 (I*M-EUROPE, www2.echo.lu) programme of DGXIII of the European Commission. Examples of such activity include the following:

- The new numerical televisions allow Web pages to be read and electronic mail to be used. Philips is already (*Libération*, 23 September 1996) competing with Samsung, Sanyo and Mitsubishi.
- On the mobile side, Nokia (and others) have developed a portable telephone able to connect to any Traveller version of Internet provider.

continued

- The building of the information superhighways (www2.echo.lu/other/national.html) of the future has already been under way for some years. The UK Department of Trade and Industry has launched a programme entitled Developing Broad-band Communication; Finland is spending US$175 million annually; France is spending up to US$10 billion from 1994 to 2000; while groups like Havas (France) (*Le Monde*, 5 October 1996) want to allow 40,000 users to connect to cables with 6 MB/s at a low price (£10/month).
- The multilingual nature of Europe is a challenge. The Multilingual Information Society (MLIS) programme (www2.echo.lu/mlis/en/infnote.html) has been launched by the EU with a budget of 15 million ECU for 1997–99. It will encourage the translation sector and language use in the business environment. A first step is the availability of EURODICAUTOM: the multilingual on-line dictionary of the Translation Services of the EU (www2.echo.lu/edic/). A vision of a Europe in which every citizen has full equal access to information services using her or his own language will help develop expertise and infrastructure in this promising field.

A study on Electronic Publishing Developments (MLIS, www2.echo.lu/elpub/en/infonote.html) predicts that about 1 million multimedia-related jobs will be created in the 15 EU member states within the next ten years. Another member state study on the market volume of Electronic Information Services (Information Market Observatory, www2.echo.lu/impact.imo/ms-study.html) in the European Economic Area (EU + Norway and Iceland) estimates a 1994 market volume of 4139 million ECU (UK 28.4%, Iceland 0.2%). Switzerland and the eastern European countries contribue a further 10%. The demand for content creators and developers is expected to show the highest growth rates (+95% until 2005). I*M FORUM (Information Market Forum: www2.echo.lu/echo/databases/forum/en/foruhome.html) contains details of companies active in the information market.

The Internet and the Web are bound to play an important role in the information society of the future. Probably the most important skills to be applied to cyberspace will be the development of the sense of a shared community of interest. Europe is well placed to face this challenge.

Further reading

Check out the following sites:

Arnaud Dufour, a young author of books on the Internet and cybermarketing (inforge.unil.ch/adufour/welcome.htm)

Hance, Oliver (1996) *Business and Law on the Internet*, McGraw-Hill, New York.

Philippe Quéau, Directeur de la division information et informatique à l'UNESCO (1996), Qui contrôlera la cyberéconomie? *Manière de Voir*, October (web.ina.fr/People/Philippe.QUEAU/ QUEAU.fr.html)

Don Tapscott and his book *The Digital Economy: Promise and Peril in the Age of Networked Intelligence*, McGraw-Hill, New York (www.mtnlake. com/pardigm/presentations.html#de_seminar)

Follow international (and Swiss) news on the Web and read the *Cyberjournal of Whebdo* (www.webdo.ch).

In the same way it is difficult to determine which cities will truly internationalize and benefit, or suffer, from these new trade dynamics. It is known that the majority of large cities work on the concept of the international city and provide themselves with facilities to attract, even to induce, business travellers and service enterprises in transition to stay. Everywhere on the transport networks and at every moment, the passing business traveller must feel perfectly at ease, whether travelling on a brief segment of the network, or staying at a 'node', connected to multiple telecommunication systems and offering all the conveniences of everyday life: upmarket hotels belonging to international chains, *bureaux de change* open 24 hours a day, international schools, in a linguistic context marked by the omnipresence of English.

Despite the increased role of movements of people connected with business, the flow of goods retains a dominant place and their transport, by sea and particularly by land, poses formidable problems owing to the development of long-distance trade, from one end of the EU to the other, according to regional dynamics and company strategies.

Analysis of the tonnage transported in the eight most important EU countries shows that barely one tonne in 20 crosses a border, and that this ratio has remained very stable over time. If this finding is broken down by country, appreciable differences appear. On the one hand, for the Benelux countries, more than one tonne in three goes to or comes from other countries; on the other, for the remaining six countries less than one tonne in ten crosses a border. The dynamics of the great North Sea ports along with the narrow links existing historically between the Benelux countries largely explains these disparities (see case study of Rotterdam).

This weak share of intra-Community trade in relation to the internal trade of each state conceals significant differentials in growth. Globally, internal trade increased at an annual rate of 2.1% between 1982 and 1990, whereas trade between EEC countries increased at a rate of 2.8%. In a general context of increase in land-based trade, with the exception of Ireland, which has lost almost 15 points in the last nine years, the superior dynamism of intra-Community trade within the EU is confirmed by almost all countries. The only ones to go against the rule are Germany, although it has very similar rates of increase in internal and in international trade, and Greece, whose geographical situation is hardly favourable to an explosion in trade with its partners in the Community. Intra-EU trade has increased in France, for example, by 28% in nine years, whereas internal trade increased by only 14%. There is therefore a more rapid increase in trade between EU countries, but in a context of predominance asserted by internal flows (Table 13.2) (Diaz Olivera *et al.*, 1995).

Out of almost 750 million tonnes which move between the 15 countries of the EU, a quarter moves by sea routes and the rest is transported by land. Oil products are conveyed from the principal unloading ports and large refining centres, among which the port of Rotterdam (see case study) is in first place by a long way. Behind navigation, which accounts for more than 40% of movements, come the pipelines (37%), then the navigable routes (17%), the latter being used principally in the Rhine countries due to the existence of a substantial network.

Table 13.2 Inter-community trade according to mode of transport in 1991.

	Road		Sea		Navigable rivers		Rail		Others		Total	
	Mt[a]	%	Mt	%	Mt	%	Mt	%	Mt	%	Mt	%
Agricultural products	35.2	54.4	17.5	27.0	7.0	10.8	5.1	7.8	0	0	64.7	100
Food commodities	43.1	59.7	14.2	19.7	4.7	6.5	2.7	3.7	7.6	10.5	72.2	100
Combustible minerals	4.5	27.3	5.5	33.0	3.1	18.8	3.5	20.9	0	0	16.6	100
Petroleum products	5.4	3.1	72.8	42.3	29.2	17.0	1.5	0.9	63.2	36.7	172.1	100
Ore and scrap metal	8.2	15.0	8.4	15.5	29.8	54.8	5.4	10.0	2.6	4.7	54.4	100
Metallurgical products	24.5	43.7	13.9	24.8	3.3	6.0	14.3	25.4	0	0	56.1	100
Minerals and construction materials	56.5	43.0	21.5	16.4	46.6	35.4	5.9	4.5	1.1	0.8	131.5	100
Fertilizers	5.0	29.2	6.6	38.7	4.0	23.3	1.5	8.7	0	0	17.0	100
Chemical products	41.9	54.7	19.0	24.8	7.3	9.6	5.6	7.3	2.9	3.7	76.6	100
Manufactured goods	54.8	73.9	13.6	18.3	0.3	0.4	4.6	6.2	0.9	1.2	74.2	100
Total	279.1	38.0	193.0	26.2	135.3	18.4	49.9	6.8	78.2	10.6	735.5	100

[a] Mt = million tonnes.
Source: Comext-Trex, L'Europe en Chiffres (1994).

The port of Rotterdam as an engine for European integration

Bart Kuipers

Rotterdam: Germany's largest port

Since 1962, the port of Rotterdam has been known as the largest port in the world or, slightly ironically, as 'Germany's largest port'. In the 1990s, however, it has tended to be referred to more as 'the Gateway to Europe' or 'Mainport Europe'. These new slogans, used particularly to promote the port, are more than just rhetoric. They indicate a transition for the port of Rotterdam from a port oriented towards the trans-shipment of bulk cargo, notably oil, into one oriented towards the trans-shipment of both bulk and containerized cargo. With the rise of the maritime container, new port-related phenomena emerged in the port of Rotterdam, such as European Distribution Centres and Intermodal rail and barge shuttle services. These owed their success to European integration.

A European port network

In the late 1990s, the port of Rotterdam has three main functions: the world's largest port for the trans-shipment of bulk cargo, Europe's largest container port, and a world-scale petrochemical complex. But, despite the rise of the container and logistical innovations, Rotterdam is still Germany's largest port, as well as being the largest port in the world with a total throughput of 294 million tonnes in 1994 and 291 million tonnes in 1996.

From the 1970s onwards, three driving forces were of fundamental importance for European ports, and for the port of Rotterdam in particular: first, the rise of the container; second, European integration; and third, the logistics revolution. The result of these forces is an emerging European port network, replacing the traditional port hinterlands, which were national in character.

The rise of the container industry

The sea-container is the most successful and dominant innovation in the transport system of the last few decades. Without the container, the move towards a global economy would hardly have been possible. Introduced in the ports of the Hamburg–Le Havre range in the late 1960s, container volumes had increased to 5.1 million twenty-feet equivalent units (TEUs) in 1980 and to 12.6 million TEUs in 1994; they are expected to grow to more than 25 million TEUs by 2005. Rotterdam is by far the largest container port in Europe, with a market share of 36%, compared with 22% for Hamburg and 18% for Antwerp.

The container industry uses two concepts that are of great importance in the integration towards a single European transportation system. The first is the 'load centre concept' and the second is the rise of intermodal transport. The prime driving force behind the load centre concept is the search for economies of scale and increasing levels of productivity in container operations. The container industry has been able to find these improvements, both at sea and in the inland transport system, in an integrated European port network.

The load centre concept, also called the 'gateway' or 'mainport concept', and in the air cargo industry the 'hub-and-spoke concept', implies a concentration of deep-sea container traffic at a limited number of large ports and the distribution of containers from these large ports by feeder services, or by rail, road or barge transport, to final inland destinations. The rationale behind this concentration of container cargo is the need to decrease the turnaround time of the large container ships in ports, because of the high operating costs of these ships. From a load centre, containers are distributed to the whole continent, instead of to a limited 'natural hinterland'. Load centres, like the ports of Rotterdam, Hamburg and Antwerp, are therefore competing for the same European hinterland.

As a result of this load centre concept, an export container from a location in Europe has no natural port of export. Underlying the choice of a port of export are factors such as the logistic concepts employed by a logistic service provider, the port-of-call patterns of the large container lines, and the supply of intermodal services. Geographical factors are diminishing strongly in importance. Instead, a European port network is emerging (Figure 13.3).

continued

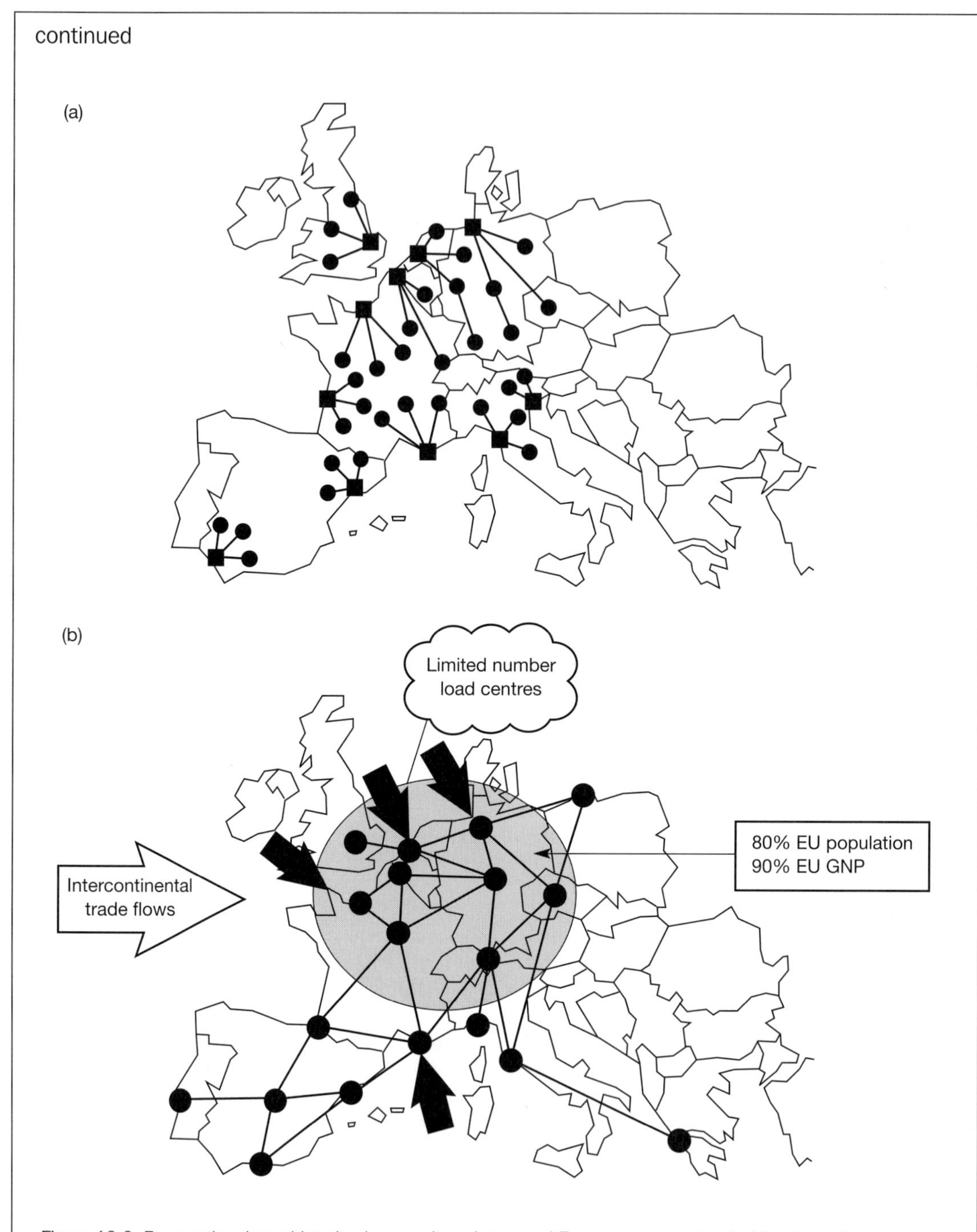

Figure 13.3 From national port hinterlands towards an integrated European port network: (a) prior to European economic integration, ports served mainly national markets; (b) European economic integration, together with load centre and intermodal.

continued

A major factor in favour of this European port network, alongside the load centre concept, is intermodal transport. This is the second consequence of the rise of the container. Although several definitions of intermodal concepts are in use, the combination of unit load, especially containers, with rail, barge or short-sea services, is usually referred to as intermodal transport (see Hayuth, 1987). Intermodal transport has extensively modified the basic hinterland patterns towards a port network on a European scale. The shuttle concept (a rail or barge intermodal service from origin to destination, without stops en route) and emerging trans-European networks, in particular, serve as motors for an integrated European transport network.

In the 1990s, private rail-service providers are organizing rail shuttles in a much more market-oriented fashion. These new and fast-growing shuttle services are usually joint ventures between large container shipping lines and European national railways. This more market-oriented method of operating railway services is also an important factor in the diminishing importance of national railway organizations, towards a number of more integrated services on a European scale. Finally, trans-European networks are the means by which these intermodal services will function on a European, instead of a national, infrastructure. Ports like Rotterdam are important nodes that 'feed' this infrastructure network with cargo.

European economic integration is of vital importance

In addition to the rise of the container, other factors are of crucial importance in the creation of a European port network. European economic integration is one such factor. European shuttle services and trans-European networks could not really exist with the traditional barrier effects of national borders. The free flow of persons and goods within the European Union is the main reason why European intermodal and port networks can now function. After 1993, national strategies aimed at encouraging national ports became subject to strict regulation. Because of the process of European integration, competition between seaports has also had to conform more to the market, leaving much less opportunity for national incentives.

The logistics revolution: speed and rationalization

In the 1980s, managers of industrial enterprises became aware of the contribution of the logistic function in their endeavours to lower operating costs and improve customer service. Two factors are of particular relevance: the adoption of just-in-time (JIT) logistic concepts, and the emergence of the European Distribution Centre. With JIT-like logistic concepts, both lower logistic costs and better customer service are possible (see Chapter 11).

The European distribution centre provides the optimum solution, combining a large-scale, low-cost warehouse function and the possibility of distributing goods – often manufactured in Asia or the Americas – throughout Europe within 24 hours of the order. Before 1993, distribution centres were usually organized at a national scale. A large multinational could have 10 to 15 warehouses in Europe. After 1993, a concentration and rationalization took place, whereby the number of warehouses was reduced to three or even one large central European distribution centre. The Netherlands has proved to be a very attractive location for such European distribution centres, with 49% of all US and 42% of all Japanese European distribution centres located there. Due to logistic innovations, in combination with deregulation in Europe, the continent is now seen as a whole in logistic terms, instead of as a number of separate countries.

Europe without its major port engine?

The port of Rotterdam is one of the main building blocks of a European port network and of the

continued

European economy. However, its importance should not be exaggerated: without the port of Rotterdam, processes such as the formation of an integrated European port network would still continue. But what would be the magnitude of the economic effects on the European economy if a major disaster happened and the port of Rotterdam disappeared? The destruction of the port of Kobe in Japan as a result of an earthquake in 1995, for example, suggests that this is not a purely hypothetical situation.

In an intellectually impressive research study, the Netherlands Economic Institute (NEI) has calculated the effects on the European economy as a whole of a Kobe-like 'sudden death' event for the port of Rotterdam. The costs related to the destruction of the port of Rotterdam for parties outside the Netherlands directly dependent on the port would be 23.7 billion Dutch guilders (in 1995: US$14.7 billion). These opportunity costs are the result of, amongst other things, higher transportation costs related to the use of alternative ports and other, more expensive, transport solutions, as well as loss of sales for companies because the total European port capacity is unable to cope with the large Rotterdam cargo flows and some companies would relocate their business out of Europe.

The costs associated with these adaptations would be much higher for bulk cargo than for containerized cargo. Of the total opportunity costs of 23.7 billion Dutch guilders, 5.1 billion is associated with containerized cargo and 18.6 billion with bulk cargo. Cargo flows destined for Germany would pay, at 10.8 billion Dutch guilders, the highest opportunity costs. Compared with the product value of cargo trans-shipped in the port of Rotterdam, a Rotterdam sudden death would mean, for bulk cargo, extra costs in the order of magnitude of 21% of the value of bulk goods trans-shipped. For containerized cargo, these extra costs are modest: some 3% of the value of the cargo trans-shipped.

For containerized cargo, a highly efficient European port system is operational and port hinterlands are no longer dependent on one port. Alternative transport solutions are available without much extra cost. However, for the large flows of bulk cargo through the port of Rotterdam, this is not the case: the port is to a large extent 'captive' for the industry in the German hinterland using large amounts of bulk cargo. If the port of Rotterdam disappeared as the result of a sudden death, it would be especially the steel industry of the Ruhr region that would be faced with serious health problems.

Further reading

Buck Consultants International (1996) *Seaports and their Hinterland*, National Spatial Planning Agency/ European Commission, The Hague/Brussels

Cooper, J., Browne, M. and Peters, M. (1991) *European Logistics. Markets, Management and Strategy*, Blackwell, Oxford

Hayuth, Y. (1987) *Intermodality: Concept and Practice. Structural Changes in the Ocean Freight Transport Industry*, Lloyd's of London Press, London

NEI (1997) *Het voorwaarts economisch belang van de Rotterdamse haven in Europees perspectief (VEEM II) (The Forward Linkages of the Port of Rotterdam in a European Perspective)*, Rotterdam Municipal Port Management, Rotterdam

van Klink, H.A. (1995) *Towards the Borderless Mainport Rotterdam. An Analysis of Functional, Spatial and Administrative Dynamics in Port Systems*, Tinbergen Institute & Thesis Publishers, Rotterdam

River navigation led the international means of transport up until 1989, and is characterized by its spatial concentration. Whereas road and rail traffic present quite a homogeneous spatial distribution, river trade is strongly concentrated: approximately a third of the tonnage transported by river is concentrated in a single relationship between Germany and the Netherlands, in dealings with the large North Sea ports. Correlatively, two-thirds of the tonnage exchanged between these two countries is transported by river. There is therefore an original bond there, differentiating them from other pairs of countries also having relations dominated by the route. This particularity prevents an excessively heavy concentration of trade on the other land-based infrastructures between the Netherlands and Germany.

Roads are increasingly becoming the most used means of transport: 38% of intra-Community trade in goods chooses road transport, but this share rises to more than 50% for agricultural produce, foodstuffs and chemicals, and to almost 75% for manufactured goods. As goods become more perishable, more fragmented in their distribution and of higher added value, the more roads supplant other means of transport. With all countries put together, the strength of roads internationally is a lot weaker than for internal trade alone, since in 1990 roads moved less than one tonne in two internationally, while in internal transport, nine tonnes out of ten were in a lorry. But the rates of increase qualify this initial perception. In nine years between 1982 and 1990, roads gained 10% of trade internationally, to the detriment of rivers but also of the railways, which now account for less than 9% of trade between the eight countries. Roads overtook river navigation for the first time in 1990.

The distribution and receipt of goods by tonnage participating in intra-Community trade are very heavily concentrated in four countries. Germany and the Netherlands are each concerned with 30% of the tonnage exchanged, Belgium 20% and France 15%, that is, almost 95% of intra-Community trade. Italy represents approximately 5% of intra-Community trade, all forms put together. Because of its insularity and despite its very significant internal trade, the UK has difficulty generating notable external land-based trade. Of course, sea links have multiplied and the opening of the Channel Tunnel provides even more opportunities, but despite a very rapid increase in its influence within the Community during these last years, the UK represented less than 2% of land-based trade at the beginning of the 1990s. With regard to Spain, despite strong dynamism internationally, with an increase of more than 50% during the four years that followed membership of the EU, its influence still remained weak in 1990, with less than 3% of the total intra-EEC trade. As for Portugal and Luxembourg, their respective contributions do not even reach 1%.

EU policy regarding infrastructures and territorial cohesion

The aim of the Single Market and, more generally, the philosophy of the Commission, is to achieve the formation of a vast market of trade, people, goods and services. The suppression of administrative and technical borders should stimulate the economic development of nations and regions and thus find solutions to the problem of unemployment. Such a perception, the value of which can only be verified in a few years time, demands that answers should be given for the retarded development affecting the peripheral regions, and that solutions be found for the congestion problems already occurring in the domain of transport on the busiest routes. The Commission's White Book (Commission Européenne, 1993b) advocates a healthier, more open, more decentralized, more competitive, and more interdependent economy. In fact, the EU has specific aid policies for infrastructures and economic development of the most disadvantaged regions in order to avoid aggravation of disparities.

Out of the three main lines of development proposed by the 1993 White Book, two focus on networks: information networks and trans-European transport and energy networks. Presented as a network 'of arteries in which the economic blood of the European Union will flow' (Commission Européenne, 1993b) this programme consists of proposing cross-border projects in order to facilitate intra-Community and not simply national logic on a continental scale. These trans-European networks are a vital feature for the creation of the Single Market, but also theoretically offer the advantage of reinforcing social and economic cohesion between the richest and poorest regions. They aim to bring the peripheral regions closer to the centre of the EU and allow them to benefit from the trade explosion which is to accompany the success of the Single Market.

From 26 infrastructures contained in the 1993 White Book, just over ten programmes, judged to be priorities following various meetings of the European Council, have been looked at. The majority of these projects concern the high-speed train (TGV), which is a response to the needs of businessmen and not to the traffic of goods, but they are attempting to bring a rail solution, thanks to planned transport, to the blocking-up of the motorways across the Alps (Brenner routes), and again in the crossing of the Channel by tunnel.

These trans-European programmes would have to find Community financing to create border crossings, the profitability of which is unconfirmed, and to succeed in linking the whole of the EU territory together through infrastructures, in order to assure the development of trade. The north–south routes which are to bring the most distant states and regions closer are therefore privileged: the Brenner route, the Rhône Corridor and its extension towards Madrid, and the estuary routeways on the Atlantic coast. Other main routes are to improve east–west traffic, such as the links between London, Paris and Cologne, making full use of the Channel Tunnel, or the new Lyon–Turin link by TGV, or Madrid–Lisbon by motorway.

These priority programmes have suffered huge delays since nothing has yet been decided by the heads of state and governments. Furthermore, they provoke objections from environmental groups when they concern crossing sensitive areas like mountains, or using rivers or developing motorways when rail seems more appropriate. The controversy over the Somport Tunnel in France, which mobilized ecologists from all over Europe, the canalization of Doubs in France to maintain Rhine–Rhône links, or the resistance of the inhabitants of Kent in Great Britain faced with the influx of TGV lines at the opening of the Channel Tunnel, are good illustrations of this.

Despite the willingness displayed by the Commission, does the EU have the will and the means to influence an offer which remains essentially produced, at least regarding its infrastructure dimension, at a national scale? Research into profitability in a difficult budgetary context in fact privileges connections which already experience significant traffic. If the European high-speed rail network projects are considered, a juxtaposition of highly profitable national sections can be distinguished, with the international links playing the role of the poor relations. Whether motorway or high-speed rail networks are considered, the links created first are never cross-border ones. Quite the reverse: these interlinks, because of their low profitability, call for assured political willingness for their promotion and their construction, and the responsibility naturally lies with the states to achieve a high-speed railway line between Lyon and Turin.

The interlinking of the EU by motorways, roads and high-speed railway lines is an essential but not sufficient element for an increase in intra-Community trade and the development of a new economic dynamism. The most dynamic regions are those which profit most from the suppression of borders. The location of Japanese and American foreign investments over the last few years emphasizes these strong regional disparities, while also reinforcing them. These multinationals establish themselves in the areas which are best equipped for conquering new markets or developing new relations between companies. These regions are those of Rhine Europe, at the heart of the EU. Rare are the exceptions like Scotland and its 'Silicon Glen', where computer companies have established themselves successfully, or capitals like Dublin, which offers the advantages of 'free zones'.

In view of the Single Market and more intense competition between regions, owing to the opening of borders, the Commission has managed to expand regional aid so that the most prosperous member states help the most disadvantaged to accelerate their development. The main aim (Objective No. 1) of the Regional Development Fund is to concentrate financial resources in the poorest regions, all situated on the periphery. This action has been complemented by the creation of a cohesion fund, following discussions about the transition to a single currency, from which Greece, Spain, Portugal and Ireland all benefit. This structural action, which accounted for only 20,000 million ECUs in 1993, will reach 30,000 million ECUs in 1999, that is, more than 35% of the EU budget. During the period 1993 to 1999 the most disadvantaged regions will receive almost 100,000 million ECUs (at 1992 rates), 74% under Objective No. 1 and 11% under Objectives 3 and 4. The four poorest countries received 48% of aid in 1992, and are due to receive almost 55% by the end of the current programme in 1999.

It is obviously difficult to judge the efficiency of this policy before the end of this century, but results obtained over the period 1989–93 prompt considerable reservations. Of course, road infrastructures have been improved in Spain and Portugal, and motorways are making previously marginalized areas accessible. But the effects of these financial contributions on economic growth remain extremely modest, even ineffectual in some cases. In effect, the EU's assessment of the regional aid policy between 1989 and 1993 makes clear that encouraging signs have been noted in terms of economic convergence in a majority of regions but 'this process has generally been slow and regions have been affected to different degrees' (Commission Européenne, 1993b). Generally it is insular areas that remain isolated from the growth of the Community – notably Corsica, Sardinia, the Canaries – and the same applies to Greece and Northern Ireland, although there are specific reasons for the latter's situation.

Conclusions

The analysis of intra-Community trade is a good indicator of European integration. This is clearly progressing, since now at least half of state exports, sometimes three-quarters, take place between the member states of the EU, and recent expansions only confirm this trend. Of course, the nature of the trade is becoming ever more complex, as services account for an increasingly large share in relation to the movement of goods, and as trade with eastern Europe is incorporated more closely with that of the west. In the same way, it is necessary to distinguish between proximity trade, between regions very close to each other, and trade related to new business strategies. In the first instance, trade within the cross-border regions of the industrial and tertiary heartland, in other words Rhine Europe, is marked by territorial continuity and the close links of trade in people, goods, business and leisure. In the second instance, trade in business and goods depends on the functioning of companies in an increasingly internationalized economic area. This is illustrated perfectly by trade connected to the relocation of the major car manufacturers, notably in the Iberian Peninsula.

These developments show just as clearly that the EU cannot avoid heading towards further political integration. Whether it be regarding infrastructures increasingly in demand from the continuous growth in trade, or the environment threatened by the transfer of factories and the construction of new transport facilities, action needs to be taken with more of a European and less of a national perspective. European integration cannot occur by way of increased territorial disorganization. Interdependence is necessary to achieve stronger integration. This requires similar levels of development and facilities. Progress towards a Europe with a single currency and an increasingly competitive market cannot be held up by ever greater discussion on territorial development. This is one of the negotiated points during the Intergovernmental Conference which must be completed in 1997 and which must give rise to new perspectives for the EU.

Further reading

Commission Européenne (1993) *Le Marché Unique Européen, Documentation Européenne*, Commission Européenne, Brussels

Commission Européenne (1993) *Livre Blanc: Croissance, Compétitivité, Emploi, Les Défis et les Pistes pour Entrer dans le XXIe Siècle*, Commission Européenne, Brussels

Commission Européenne (1994) *Europe 2000+, Coopération pour l'Aménagement du Territoire Européen*, Commission Européenne, Brussels

CHAPTER 14

Consumption and retailing: sameness and difference

RONAN PADDISON AND ANDREW PADDISON

In Europe the development of consumerism has become critical to the present historical moment. Most accounts of contemporary social structure and change attest to the importance of consumption: in the postmodernized consumer culture the act of consuming, captured in such terms as lifestyle, fashion and taste, functions as a (if not *the*) main source of social differentiation (Crook *et al.*, 1992). Further, most accounts of the rise of the contemporary consumer culture place it firmly within the overarching framework of globalization – the spread of consumerism has become worldwide. Although it began at least two centuries ago among the bourgeoisie of the major European cities, its diffusion has become a defining characteristic of late capitalism. Yet, within contemporary Europe consumerism has been divisive. While in western Europe it has become central to daily life, in eastern Europe its denial helped occasion the dramatic changes of the recent past.

As part of the development of the consumer culture, shopping itself, beyond meeting mundane needs, becomes more than a necessity and attains status as a leisure activity. Similarly the sites through which retailing is conducted become an integral part of everyday experience. Nowhere is this more evident than in cities, often themselves in the vanguard of social change, but most clearly the sites of conspicuous consumption. Physically, it is in the cities that successive rounds of retail innovation have been instrumental in structuring and restructuring urban space. The arcades of the 19th century are paralleled in the 20th century by the profusion of planned shopping centres and the emergence of out-of-town shopping districts, trends which characterize much of western Europe, and which are emergent in some of the larger cities of eastern Europe. Innovation serves the needs of retail capital, adapting to changes in consumer demand, but is itself part of the larger marketing project underpinning the consumer culture in which consumers are encouraged to want more than they need.

Consumption involves more than the utilitarian, encapsulating cultural and social meaning besides its immediate practical significance. To Douglas and Isherwood (1979) such a meaning of consumption was no less true for the pre-industrial as for the industrial society – the consumption of material goods not only has intrinsic value, but also acts as a basis within which social relations are conducted. In the consumer society the centrality of consumption has become closely bound with subjectivities, where the act of consuming becomes more than a means, but an end in itself. Consumption, then, has become linked to desired lifestyles in which, in the shift from modernity to late modernity, its role has become intertwined with the processes of self and group identity; consumption rather than production-based work has become the critical basis for defining who we are (Bocock, 1993).

In present-day Europe, and particularly in the more affluent societies of western Europe, this centrality of consumption to everyday life has become apparent in a variety of ways. Lury (1996) lists some of the major features of modern consumption: the constantly increasingly number and types of consumer goods, the accentuation of style and design, the growth of shopping as a leisure activity, the deepening process of marketization of services previously provided by the state, the growing visibility of sport and leisure activities, the development of novel sites for retail consumption and of consumerist movements, the inexorable penetration of advertising and even the appearance of addictive forms of consuming, encapsulated in the frequently quoted aphorism 'shop till you drop'.

Within Europe, predictably, these trends are unevenly developed, though the seductiveness of consumer capitalism is likely to ensure that the spread of consumerism, already dominant in western Europe, will also progressively envelop much of eastern Europe. Indeed, the traumatic political events following from the 'fall' of the Berlin Wall and the demise of socialism in eastern Europe have been attributed, in part at least, to the inability of communist regimes to match consumerist aspirations. Even if it is a moot point whether the events following late 1989 were to signify the power of the consumerist ideology rather than the failure of socialism as such, the aspirations of East Berliners eager to gaze at the shopping centres of West Berlin is an abiding image. Following liberalization the market shortages endemic to the socialist period throughout much of eastern Europe were to become history, the problem to come being one of affordability rather than of accessibility to western goods.

While within postmodern accounts the act(s) of consumption permeate virtually all aspects of human experience, in this chapter we confine attention to those forms of consumption traded through the marketplace, and specifically by the retail market. This excludes major aspects of the consumption experience, such as the consumer as tourist, whose gaze has had profound spatial and cultural effects in European localities (see Chapter 19). Nevertheless, the retailing–consumer nexus is sufficient to help evaluate how consumption processes are simultaneously associated with outcomes which are linked to homogenization and local difference within present-day Europe. As part of the wider project of marketing, retailing is closely bound to consumerism, and indeed has become instrumental in fostering consumption. This chapter is divided into three parts. The first section examines how the retail market is differently structured within Europe and how in particular its capitalization has evolved. In the second section, the linkage between the changing retail sector and space are explored through an examination of the different forms of consumption that have accompanied retail change. In the final section we examine critically how consumption processes are linked simultaneously to the production of sameness and difference in present-day Europe.

The changing pattern of consumption, and of the network of retail institutions through which consumer goods are traded, affords the opportunity to look at some of the questions currently salient within Europe – that the spread of marketization processes, and the installation in the European Union of the Single Market, through reducing the significance of national boundaries, are leading to increasing homogenization; that, for example, 'Euro-patterns' of consumption are emerging. Such an argument oversimplifies the richness of cultural difference within contemporary Europe – as much as marketization and commodification processes are a force for homogenization, they have had to contend with the renaissance of ethnic difference. Nor are such processes unaffected by the forces of globalization. Consumer cultures need to be located within the wider processes of globalization in which the bonds of national and cultural distinction are fracturing the spread of consumerism in western, and increasingly in ex-socialist, European countries.

Retail structure and change in contemporary Europe

Retailing in Europe is both a diverse activity and one which is undergoing rapid change. Its diversity is amply illustrated by the differences in the structure of the retail market in contrasting European countries. The set of institutions through which retailing is conducted varies considerably in countries as different as Portugal and the United Kingdom. Yet in all countries retailing is a highly dynamic sector, and particularly so in eastern Europe, where much of the sector had stagnated during the socialist period. In countries where the retail sector is already relatively advanced, such dynamism reflects the need of the sector to meet the demands of a more segmented and sophisticated post-industrial consumer market, rather than merely those of mass consumerism of the industrial society. In other countries, notably those in eastern Europe but also including southern European countries such as Spain, such dynamism reflects a process of 'catching up' with retail practices initiated in affluent Europe.

Retail change in Europe is a complex and variable process, but one in which common trends are identifiable. These trends include the increasing concentration and capitalization of the industry, the development of mass retailers represented in both the food and non-food sectors, the simultaneous development of specialist retailing catering for fragmenting consumer markets, the growth of international retailers, and the adoption of new technologies to maximize the efficiency of operation. Based on such trends Tordjman (1995) has attempted to classify national retail systems according to their 'level of development', effectively the extent to which these trends characterize the retailing structure, either as a whole or with regard to particular sectors of it. Four types were defined ranging from the 'traditional' to the 'advanced', Greece and Portugal typifying the former, and Germany and the United Kingdom the latter. Such classifications should be read cautiously; even if they are only interpreted as heuristic devices, the assumption of unilinearity which they incorporate masks the different ways in which retailing is developing nationally and locally in Europe. Further, in practice retail development is much affected by national regulatory regimes – in some countries, such as Italy, the development of large retailers has been hindered by legislation; in others, such as Britain in the 1980s, their development has been fostered by the relaxation of planning controls.

One of the chief dimensions by which retailing has become differentiated is that of concentration. Basic measures of the scale of retailing, as one facet of concentration, are provided by such indicators as the frequency of outlets and the average number of employees per enterprise. Table 14.1 illustrates for the member states of the European Union, together with Norway, Switzerland and Iceland, the considerable differences characterizing the retail market. Barring the exceptions, such as Belgium, in which the relatively large number of small retailers partly reflects the effect of government policies designed to protect independent traders (Merenne-Schoumaker, 1995), the major contrast is between northern and southern Europe. The structure of the retail market in countries such as Greece and Portugal differs radically from that of most of northern Europe, in which retailing, and particularly certain sectors such as food, have effectively become dominated by mass retailers.

These variations reflect differences in the level of retail modernization within which the processes of concentration/capitalization play such a critical role. The development of retail chains (multiple retailers) is another commonly cited signifier here, harnessing the advantage of scale economies and in many European countries capturing an ever larger proportion of trade. Again, as Table 14.2 shows, for the same list of countries, there are marked differences in the sales penetration of multiple retail organizations, though their patterning does not simply replicate the trends of the earlier statistics. While in countries in northern Europe, in particular the United Kingdom, multiple retail organizations tend to account for a high proportion of retail sales, this is by no means always the case. Equally in some southern European countries, multiple retailing has already established a strong commercial footing. In Spain, for example, retail modernization developed rapidly during the 1980s, which decade saw the entry of both domestic and French hypermarkets, the development of out-of-town shopping centres and the decline in the number of (small) independent traders.

The patterning of retail trade concentration varies not only between and within countries, but also between sectors. Food retailing is of particular significance, not only because variations in trade concen-

Table 14.1 Basic scale characteristics – European retailing, 1991.

	No. of retail outlets per 10,000 population	No. of employees per outlet
Portugal	192	2.11
Greece	184	1.9
Italy	171	2.59
Belgium	141	2.14
Spain	134	3.15
Luxembourg	116	5.17
Denmark	100	4.15
France	97	4.52
Sweden	94	6.28
Norway	92	3.8
Ireland	90	4.48
Germany (West)	85	5.35
Switzerland	83	9.06
United Kingdom	81	8.7
Netherlands	80	6.71
Finland	77	5.33
Iceland	67	5.33

Source: adapted from Euromonitor (1996).

Table 14.2 Retail significance of multiple chains, 1991.

	% of all sales accounted for by multiples
United Kingdom	51
Spain	47
Ireland	38
France	38
Austria	37
Netherlands	35
Norway	32
Portugal	25
Switzerland	25
Germany (West)	22
Denmark	19
Finland	17
Sweden	16
Greece	15
Luxembourg	10
Italy	8

Source: adapted from Euromonitor (1996).

tration are pronounced, but because this in turn is linked with different marketing practices. Where retail trade concentration in the food industry is linked with the growing power of retailers in the marketing chain – itself taken as an indicator of modernization – the extent to which multiple retailers have been able to capture trade through 'own brand' selling varies widely. In the United Kingdom, where food retailing is highly concentrated, with well over 60% of all sales being accounted for by just five retail chains, a measure of their strength is reflected in the high penetration of own-brand sales (as a proportion of all food sales), more than five times the rate for Italy and Greece. Their importance in the British case reflects the hyper-competition which has developed between an oligopoly of bulk retailers, in which the selling of own brands has become one route by which to achieve differentiation while maintaining competitive position.

Cultural factors, too, help explain such differences, the tradition of buying food from specialized (usually small) outlets remaining as a strong preference amongst consumers in some countries. To the extent that market forces tending towards concentration may run counter to such cultural preferences, the state may attempt to buttress the retail market. In France recently enacted legislation (1996) has sought to protect the dwindling numbers of *boulangeries*, which evoke a powerful image within French food retailing, by preventing use of the term other than to authorized (usually small independent) traders. The extent to which such resistances to market forces and retail capitalization will be able to maintain culturally specific forms of retail trading in the longer term must be in doubt given the power of market forces, though at the very least the retail market will need to be able to meet regional and local preferences for particular types of foods, which are often both culturally distinct and ingrained (Delamont, 1995).

If the development of retail capitalization in Europe has become associated with a long-run trend towards retail concentration – reflected in the spread of vertically integrated channels of distribution in which the retailer has assumed an increasingly assertive position and through the proliferation of large retail outlets, represented by the profusion of supermarkets, hypermarkets and discount stores – such trends are unevenly expressed. Nowhere has this become more apparent than in the contrasts in the structure of retailing between the affluent, consumerist economies of western Europe as opposed to those of eastern Europe, in which the state had sought to control the operation of retail distribution. Even if such control was by no means total – small-scale private trading was often allowed to persist – the socialist experiment in controlling distribution resulted in different types of retail institution. Since liberalization, as the case study demonstrates for Poland, there has been a proliferation of small-scale retailing, as well as large-scale outlets, in some cases represented by the opening of branch outlets of international retailers, restricted to the major urban centres.

If trends such as concentration signify the convergent nature of retailing in Europe, the development of international retailing is another of its more obvious signifiers. The advent of transnational retailing in Europe is recent, though after its slow growth in the 1960s and 1970s the number of international establishments in the European Union trebled in the 1980s (to 870 outlets), since when it has continued to accelerate. For the larger retailers, food as well as non-food, generalist as well as niche-market, internationalization represents a logical step. Where there are limits to continued national expansion, where regulatory barriers limiting cross-border expansion have been relaxed, or where, as for specialist retailers, international trading is possible because of the development of niche markets having similar tastes, internationalization is a means of achieving continued growth.

In fact, international retailers vary – in Treadgold's (1988) classification of global, transnational to multinational retailers, convergence is most pronounced in the case of the global corporation in which there is very little alteration of the format and marketing across boundaries. IKEA exemplifies the global retailer, with outlets spread widely across the continent, though concentrated in northern Europe (Figure 14.1), catering for distinctive lifestyle demands. Overall, international retailing is more dominant within the consumer markets of northern Europe (Burt, 1995) – not only are most of the cross-boundary retailers headquartered there, particularly in France, the United Kingdom and Germany, but most of their outlets are located in these countries. Nevertheless, as the Eurostat (1993) report *Retailing in the European Single Market* indicated, within the medium term international retailing will have considerably deepened its penetration of the consumer markets of southern Europe.

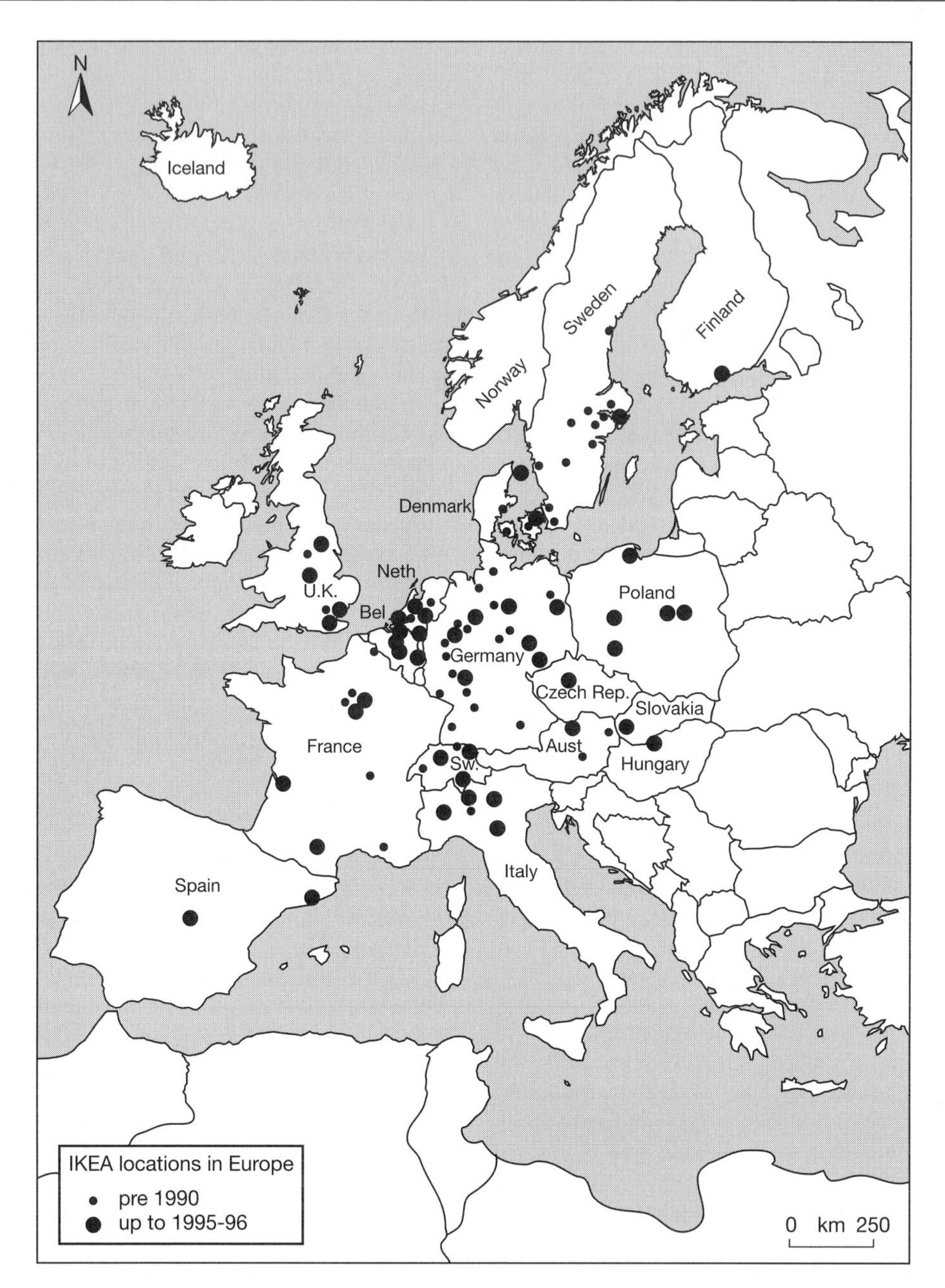

Figure 14.1 Store development, IKEA, before and after 1990 (*source: IKEA Facts and Figures, 1995–96*).

Consumption sites and socio-spatial change

In Europe, as elsewhere, retail innovation has successively redefined the spatiality of consumer opportunities and behaviour through the creation of new consumption sites. Often such changes have had profound implications on urban form, (re)defining its commercial structure and the usage shoppers make of the city. Department stores in the 19th century were among the first of the major retail innovations to redefine the relationships

The restructuring of Polish retailing

Andrzej Werwicki

The socialist heritage of the centrally planned economies was poorly adapted to the requirements of modern economic development and has required considerable technical and managerial adjustment following the collapse of these systems in the late 1980s. The sphere of retailing has been typical of such changes, requiring adaptation to the free market both in legal and in organizational terms.

The processes of restructuring

The first requirement of a transition towards a market economy was the creation of new economic legislation, and in Poland this was introduced as early as 1990, when laws were promulgated concerning property ownership and the abolition of the state monopoly in the wholesale and overseas trades. A privatization law also permitted the development of private retailing on a nationwide scale. Second, a free system of rents as well as new currency regulations were introduced, permitting the free circulation of foreign currency. The shortage of appropriate shop premises, high rents, restricted currency flows, and the general poverty of the population, though, all restricted the development of retailing at this time.

However, in the early 1990s the market was starving because of a general lack of consumption goods. Consequently, between 1990 and 1992 there was an explosion in the number of small, private retailers who offered a wide range of foreign-made products from stalls and boxes located along the main shopping streets of Polish cities, frequently situated opposite the old retail outlets dating from the socialist period, which were almost empty of produce. As well as such stalls there were also numerous hawkers, most of whom were unregistered and paid no taxes. In February 1991, more than 500 dealers were thus recorded along a single main shopping street in Warsaw. 200 of these possessed a permanent kiosk, leaving some 300 as hawkers or with only a movable table as their place of trade.

In 1987, there were some 224,357 retail outlets in the whole of Poland, of which 146,893 were regular shops. Only 22,376 shops (15.2%) were registered as being in private hands (Statistical Office, Poland, 1989). By 1991, the number of private shops had risen to 301,526, representing 96.9% of the total number of shops at that time (310,966) (Statistical Office, Poland, 1992, 1993). These figures emphasize the extremely rapid transformation of the Polish retailing sector.

Two other processes were linked with the advent of privatization: fragmentation of the retail enterprises, and a restructuring of employment. Between 1988 and 1992 the number of shops in Poland more than doubled (Table 14.3), indicating both a process of physical partition of previously larger shops, and also a general decline in the size of shops (Table 14.4). Privatization also meant that many people working in retailing became their own employers. Thus, the total numbers of people working in retailing grew from 1.51 million in 1987 to 2.05 million in 1993, while the number of employees declined from 1.18 million in 1989 to 671,000 at the end of 1993.

It is impossible to estimate precisely the number of retailing outlets operating across Poland without a permanent location during the early 1990s. Based on field research between 1991 and 1994, though, it would appear that they numbered over 200,000, thus comprising about 40% of the total number of retailing outlets in 1991 (Werwicki and Poweska, 1993). Many such outlets operated in urban markets. Since medieval times, markets had been important privileges of the towns, but under socialism they gradually ceased to exist. However, since the late 1980s they have become revitalized, and by 1993 there were some 2181 urban markets situated across Poland, comprising in total more than 87,700 permanent retail outlets. Another interesting phenomenon has been the emergence of large bazaars on Poland's international borders. In the west, the bazaar customers mostly originated from Germany, whereas in the east the customers were mostly of Polish origin, purchasing goods imported by dealers from beyond the border (Werwicki and Poweska, 1993).

continued

Table 14.3 The changing number of shops in Poland, 1988–95.

Territorial unit		1988		1990		1992		1994		1995	
		Total no.	Private	Total no.	Private	Total no.	Private	Total no.	Private	Total no.	Private
Warsaw	total	13 287	3 869	14 386	13 467	25 417	22 580	23 009	22 538	26 218	–
	in towns			13 844		24 880		21 364		23 704	
Gdańsk	total	7 262	1 544	8 948	8 479	13 226	12 846	15 454	15 262	14 298	–
	in towns			8 168		9 557		13 116		12 152	
Katowice	total	21 448	5 051	23 342	22 249	34 258	33 941	48 346	47 717	50 909	–
	in towns			22 372		30 369		44 451		46 470	
Cracow	total	6 270	1 387	6 167	5 639	11 482	11 287	14 022	13 837	15 344	–
	in towns			5 529		7 877		11 742		12 730	
Łódź	total	6 508	1 954	7 322	6 768	11 144	10 611	13 782	13 527	12 713	–
	in towns			7 195		10 284		13 122		12 091	
Poznań	total	8 787	2 580	9 531	9 303	13 718	13 364	15 395	15 152	14 337	–
	in town			9 036		9 779		13 117		12 097	
Szczecin	total	5 643	597	5 350	4 954	8 203	7 892	12 707	12 481	12 965	–
	in towns			4 751		6 881		10 679		10 777	
WrocŁaw	total	6 727	1 113	6 643	6 283	6 614	6 299	11 097	10 870	12 144	–
	in towns			5 877		5 179		9 181		10 034	
Total for voivodships with big urban agglomerations	total	75 932	18 095	81 689	77 142	124 062	118 820	153 812	151 384	158 928	–
	in towns			76 772		104 806		136 772		140 055	
Other voivodships	total	151 061	24 970	155 736	145 971	228 440	224 069	261 637	256 532	266 672	–
	in towns			124 290		154 176		186 336		188 003	
Total for Poland	total	226 993	43 065	237 425	223 113	352 502	342 889	415 449	407 916	425 600	419 313
	in towns			201 062		258 982		320 408		328 138	

Source: Rynek Wewnętrzny (Internal Market) 1992, 1994 and 1995.

continued

Table 14.4 Enterprises by number of shops in Poland, 1991–95.

Specification	1991	1992	1993	1994	1995
Total trade enterprises	255 787	306 964	328 542	365 713	381 392
1–2 shops	252 001	303 177	324 910	362 026	377 109
3–10 shops	1 631	1 954	2 063	2 279	3 037
11–20 shops	1 195	1 192	1 082	996	907
21–50 shops	857	577	442	374	308
51–100 shops	88	51	36	28	25
101–200 shops	11	10	7	7	4
Above 200 shops	4	3	3	3	2

Sources: Rynek Wewnętrzny w 1995 r. (Internal Market in 1995), Warsaw 1996.

By 1993 the general economic situation in Poland had improved, but this had relatively little influence on the restructuring of retailing, which continued to be characterized by a slackening in the rate of increase of retail outlets (Table 14.3), bankruptcies of many small enterprises, changes in their lines of trade, and the appearance of foreign retailers in the Polish market. The privatization boom in retailing finished around the end of 1992, but subsequently there have continued to be changes in the organizational structure of Polish retailing. First, although cooperative retailing is listed as being within the private sector, it has not in effect changed its economic practices. The number of cooperative shops dramatically declined as a result of private competition, from 42,448 in 1992 to 29,372 by the end of 1995 (Central Statistical Office, Poland, 1992, 1993, 1994, 1995). In rural areas, the previous retailing dominance of the *Samopomoc Chlopska* (Peasant Mutual-Aid) cooperative, once the biggest retailing organization in Poland, has been considerably curtailed as it has had to become a true cooperative. In urban areas, most of the former cooperatives have been privatized, but there do remain some urban cooperatives concentrating mainly on food and hardware.

The first foreign distributor, CLC Maxa, Karsten, appeared in the Polish market in 1991. Four more foreign distributors arrived in 1992, and another four in 1993. By the end of 1995 the 20 foreign enterprises operating in Poland had some 155 shops, with the most numerous being Tesco-Savia (40 shops), McDonalds (22), Rema 1000 (18), and Globi (12) (Central Statistical Office, Poland, 1992, 1993, 1994, 1995). Although for a country of the size of Poland this is not a large number, the total is likely to increase, and their existence may also influence the foundation of new Polish companies similar in style and size to the foreign enterprises.

This brief outline indicates that two stages in the restructuring of Polish retailing have already been completed, the first creating the conditions for the emergence of a free market, and the second representing a period of consolidation. These have brought Poland to the gates of the international market, but with a ballast of outdated fixed assets and an old-fashioned retailing organization structure. The challenge for the third stage is to equalize Poland's economy with that of the European Union, which as far as retailing is concerned will mean consolidation into larger economic units, technical and aesthetic redevelopment, and adaptation to modern marketing principles.

Regional differences

While the above generalizations hold true for Poland as a whole, there are important regional differences in the level of retail restructuring, reflecting three main factors: levels of economic development and population density; the dominant type of economic activity in an area, particularly as it influences unemployment levels and thus demand; and the character of the location, whether urban or rural.

Within Poland it is possible to identify three main population density classes. In urban areas there are generally populations of more than 1200 people per sq km, in wider urban agglomerations densities are around 290 per sq km, and in rural areas densities are in the region of 94 per sq km. Levels of retail privatization do not vary much with overall population density, but as Table 14.5 indicates there have been some differences with respect to the types of trade. Thus, during the first stage of privatization nationwide phenomena included a vast growth in the number of general food stores, butchers' shops, chemists, furniture stores and those dealing with electronic goods.

continued

Table 14.5 Shops by type of trade in Poland, 1991–93.

Territorial unit		Total number of shops	General food stores	Other food stores	Chemists	Textiles clothes and footwear	Furniture and lightning	Electronic goods	Books and stationery	Other shops
Warsaw	1991	18 725	2 386	2 778	848	3 705	465	1 566	1 899	5 078
	1992	25 417	2 269	3 148	1 177	4 441	1 633	3 576	1 038	8 135
	1993	28 171	6 710	2 282	774	2 356	281	569	951	14 248
Gdańsk	1991	11 192	2 392	1 474	441	2 004	320	761	591	3 209
	1992	13 226	3 363	1 241	234	1 080	218	329	466	6 295
	1993	14 170	4 447	1 258	242	1 043	214	300	465	6 201
Katowice	1991	29 791	9 761	5 788	370	2 830	1 098	1 569	318	8 057
	1992	34 258	11 412	6 424	3 222	1 038	1 328	1 847	375	8 612
	1993	35 967	8 620	7 770	1 264	6 178	1 202	2 513	1 837	6 583
Cracow	1991	8 625	1 627	1 342	175	727	215	346	173	4 020
	1992	11 482	2 367	1 189	310	1 114	206	287	332	5 677
	1993	13 741	2 679	2 398	610	2 786	608	255	509	3 196
Łódź	1991	16 177	2 193	1 534	437	2 258	440	545	412	8 358
	1992	11 144	2 136	1 684	408	2 036	307	561	284	3 728
	1993	12 023	1 443	3 307	522	1 740	195	526	793	3 497
Poznań	1991	12 813	2 963	1 199	284	1 178	262	338	306	6 283
	1992	13 718	3 409	1 396	337	1 338	322	456	314	6 146
	1993	14 851	2 964	2 376	757	2 971	399	685	754	3 945
Szczecin	1991	8 160	1 146	1 380	348	1 257	386	503	443	2 688
	1992	8 203	880	685	112	672	135	168	279	5 272
	1993	8 928	1 748	1 842	425	1 802	464	542	293	1 812
WrocŁaw	1991	8 272	1 986	964	186	1 030	189	291	180	3 446
	1992	6 614	2 304	697	125	703	140	194	133	2 318
	1993	10 074	3 290	1 076	191	1 014	179	248	462	3 614
Voivodships with big urban agglomerations total	1991	113 755	24 454	16 468	3 089	14 989	3 375	5 919	4 322	41 139
	1992	124 062	28 140	16 464	5 925	12 422	4 289	7 418	3 221	46 183
	1993	137 925	31 901	22 309	4 785	19 890	3 542	6 338	6 064	43 096
Other voivodships	1991	197 211	49 314	24 003	5 137	26 512	5 070	9 282	1 854	76 039
	1992	228 440	57 274	25 705	5 590	25 449	5 191	9 414	3 719	96 098
	1993	242 657	69 221	35 809	8 220	40 748	7 202	13 301	6 812	61 344
Poland total	1991	310 966	73 768	40 471	8 226	41 501	8 445	15 201	6 176	117 178
	1992	352 502	85 414	42 169	11 515	37 871	9 480	16 832	6 940	142 281
	1993	380 582	101 122	58 118	13 005	60 638	10 744	19 639	12 876	104 440

Sources: 1991 – Statystyczny Województw 1992 (Statistical Yearbook of Voivodeships).
1992 and 1993 – Rynek Wewnętrzny (Internal Market 1992 and 1993).

continued

Table 14.6 Shops by size (square metres) of selling space in Poland, 1992–95.

Territorial unit		Total shops	< 50 m²	50–100 m²	101–200 m²	201–300 m²	301–400 m²	>400 m²	Total sale surface of shops ('000 m²)
Warsaw	1992	25 417	22 636	1 483	740	240	92	226	–
	1994	24 888	22 641	1 211	605	167	65	199	1 359.0
	1995	26 198	23 951	1 124	651	169	66	237	1 212.2
Gdańsk	1992	13 226	12 096	663	291	63	35	78	–
	1994	15 600	14 513	618	272	84	30	83	787.7
	1995	14 298	13 075	684	302	99	44	94	691.2
Katowice	1992	34 256	31 821	1 316	694	167	66	192	–
	1994	48 052	45 080	1 608	825	206	102	231	2 215.6
	1995	50 906	47 584	1 799	916	261	105	241	2 514.1
Cracow	1992	11 482	10 652	453	235	75	19	48	–
	1994	13 996	13 119	475	231	74	35	62	615.1
	1995	15 344	14 444	489	221	76	38	76	662.1
Łódź	1992	11 144	10 261	521	222	52	16	72	–
	1994	13 916	13 006	517	220	69	25	79	729.8
	1995	12 713	11 691	616	230	72	28	76	512.2
Poznań	1992	13 718	12 454	772	303	70	34	85	–
	1994	15 422	14 220	720	287	65	31	99	690.9
	1995	14 337	13 016	762	324	81	38	116	793.3
Szczecin	1992	8 203	7 322	444	231	102	32	72	–
	1994	12 683	11 949	337	206	76	28	87	897.5
	1995	12 965	12 126	384	212	99	36	108	546.6
WrocŁaw	1992	6 614	5 720	434	280	76	42	62	–
	1994	11 121	10 296	395	239	74	44	73	599.5
	1995	12 144	11 170	461	279	86	53	95	656.0
Voivodships with big urban agglomerations total	1992	124 060	112 962	6 086	2 996	845	336	835	–
	1994	155 678	144 824	5 881	2 885	815	360	913	7 595.1
	1995	158 905	147 057	6 319	3 135	943	408	1 043	7 587.7
Other voivodships	1992	228 364	204 462	15 157	5 812	1 366	507	1 060	–
	1994	259 771	238 240	13 540	5 187	1 220	503	1 081	11 582.8
	1995	266 695	244 240	13 949	5 428	1 295	595	1 188	12 204.9
Poland total	1992	352 424	317 424	21 343	8 808	2 211	843	1 973	–
	1994	415 449	383 064	19 421	8 072	2 035	863	1 994	19 177.9
	1995	425 600	391 297	20 268	8 563	2 238	1 003	2 231	19 792.6

Source: Rynek Wewnętrzny za lata 1992, 1994 i 1995 (Internal Market for the years 1992, 1994 and 1995).

continued

However, there was also a recession in the number of textile shops in rural areas and in some urban agglomerations. In the case of Lódź this can be linked with the general decline in its textile industry, but in other areas it was closely related to the general process of privatization and the mass liquidation of socialized textile and clothes stores. Since 1993, though, this process has been reversed, with a rapid growth being apparent in a wide range of retailing activity. Warsaw represents a specific case, where there was a boom in the opening of retail business in 1992, with numbers later declining in balance with levels of demand.

Table 14.6 illustrates that there are also regional differences in the size structure of shops as indicated by their retail surface area. The increase in total numbers of shops was largely accounted for by the expansion in the number of small shops with less than 50 sq m of retail area. However, it is worth noting that shops in the two largest size categories grew in number throughout the period in question, indicating that while shop fragmentation continued in rural areas in the second stage of restructuring, this was much less evident in urban areas. The areas with the most successful retailing development, as indicated in terms of the growth of numbers of shops and their size structure, were the agglomerations of Gdańsk, Katowice, Szczecin and Wroclaw. Warsaw has a very unstable retailing structure in terms of the number of different-sized units, and there would also appear to have been a decline in the number of medium-sized outlets in Łódź, Gdańsk, Cracow and Poznań, as well as a marked decline in rural areas.

Conclusions

Polish retailing has changed out of all recognition since 1989. From a situation where there were small numbers of often deserted retail outlets, there is now a surplus of shops, marts and mobile retailing enterprises. There are also numerous consumer goods of different types, although the quality is not always of the highest level. This is the result of numerous retail changes. Above all, new legislative instruments were introduced, which permitted the private importation of goods and the opening of private retail outlets. The competition thus entailed led to the collapse of many former state and cooperative shops, and the disappearance of state retailing enterprises specializing in specific products such as cars, medicines and jewellery. However, the process of retail restructuring in Poland is far from complete, and the implications of adapting to the needs of the European Union will present a considerable challenge to the Polish people.

Further reading

Werwicki, A. (1992) Retailing in Poland, *International Journal of Retail and Distribution Management*, **20**(6), 34–8

Werwicki, A. (1995) Retailing in Eastern Europe: Poland, *European Retail Digest*, Summer Feature, 9–12

between the shopper and the city. Established in the major capital cities, notably Paris, London and Berlin, as well as the rapidly growing urban industrial cities, such as Milan and Manchester, the department store targeted the *nouveaux riches*, particularly the wives of the emergent industrial and commercial middle classes. Through his interpretation of Emile Zola's novel *Au Bonheur des Dames* (itself based on the Parisian department store *Bon Marché*), Blomley (1996) shows how the department store in the Paris of the Second Empire functioned as a major new locus of retail accumulation besides catering for the 'dreamworlds of the consumer' within a gendered environment.

The department store represented the first of a number of retail innovations which have fashioned and refashioned urban consumer landscapes in European cities. Together with early indoor shopping malls – the Parisian *galeries* – the department store celebrated both conspicuous consumption and the commercial supremacy of the city centre. Mass consumption was to follow, catered for in part by the development of the next round of retail innovation, the emergence of multiple chains. These were to reconfirm the position of city centre retailing, though increasingly retail spaces became socially segregated as commercial enterprises targeted different segments of the consumer market.

While these historical innovations were to (re)affirm the city centre, later innovations have resulted in a more diversified set of consumer landscapes within European cities. Reflecting the accelerating process of suburbanization, supermarkets, and subsequently hypermarkets, out-of-town shopping centres, retail warehouses and discount stores, and the development of leisure-cum-retail parks have emphasized the polycentric nature of the larger cities, particularly of northern Europe. Some developments, such as the planned shopping mall, characterize both types of space, the suburban edge as well as the city centre.

Whether these more recent innovations which characterize the affluent consumer landscapes of northern Europe mark a *real* change in cultural terms from historical practices is debatable. Domosh (1996) has suggested that contemporary retailing is rooted in developments of activity and design beginning two centuries ago. In this sense, department stores of the last century as 'cathedrals of consumption' have been replicated in principle in the contemporary period by the shopping mall – both functioned as a site *for* consumption as well as a sight *of* consumption (Ewen and Ewen, 1982). Similarly, the celebrated *flâneur* of the novel shopping thoroughfares of Paris in the 19th century have their counterpart in the casual and pleasurable forms of shopping written of by commentators in the postmodern city.

Though the development of such institutions is uneven within western Europe, innovations such as supermarket retailing, the hypermarket, and the in-centre and out-of-town shopping mall have become commonplace. Organizational differences in the structure of the retail market continue to distinguish individual countries, and the newer forms of retail institution continue alongside more traditional ones, such as street markets and other forms of small-scale outlet. Yet, where the retail industry has generally been characterized by increasing concentration and capitalization, both forces represented by new consumption sites, what is striking is the extent to which such processes, at various stages of evolution, are apparent throughout western Europe.

The ubiquity of the shopping mall and of decentralized discount retailing in much of affluent Europe has become a part of the taken-for-granted world providing a stark contrast to the consumer landscapes of much of eastern Europe. In comparing the urban impacts of retailing as it developed under the two systems – market-led and state-controlled – Dannhaueser (1994) has shown how in structural terms there were certain similarities in terms of the commercial vitality of the market and in the relative importance of the town centre and edge of town as consumption sites. He compared how retailing had evolved in the post-war period in two small towns on either side of the German divide, Hassfurt in northern Bavaria and Hildburghausen in southern Thuringia in the former German Democratic Republic. The main aim was to assess the extent to which retail trade concentration had occurred, and the effects of this on the commercial viability of the town centres. In Hassfurt, as elsewhere in the Federal Republic of Germany, retail trading was progressively affected by the processes of capitalization and concentration after 1950 – possibly because of the town's size some of these developments were relatively slow to appear, though by the early 1970s several discount stores, previously in the town centre, had relocated to the suburbs with detrimental impacts on the traditional, often family-owned, stores in the *Alstadt*.

In Hildburghausen the structure of retail trading in 1950 was similar to that of Hassfurt, dominated by small to medium-scale family-owned outlets. This structure was soon to be radically reshaped with the establishment in the early 1950s of the first state-owned *Handelsorganization* (HO, the Trade Organization) in the town. As a state-controlled organization, ironically the rapid spread of HO in the German Democratic Republic meant that vertical channel integration became more widespread there earlier than in the Federal Republic of Germany. The establishment of state-owned facilities – HO was later joined by other state-controlled outlets – had similar effects on private trading as in Hassfurt. Yet, with the chronic shortage of goods the retail sector stagnated, as did the town centre generally. Paradoxically, then, the effects of such changes in the two towns – retail innovation in the one and what was increasingly market stagnation in the other – had similar outcomes on the commercial viability of the town centres, albeit for very different reasons. Clearly, the consumer landscapes of the two towns differed radically at the time of reunification, though since 1990 retail trading in Hildburghausen has undergone rapid change. Yet, revival has focused largely on the town centre, resulting in a further twist to the situation in Hassfurt.

Fashion between local and global

Paolo Giaccaria

> Under the sign of commodity, all types of labour become interchanged and lose their singularity – under the sign of fashion, it is both labour and leisure time itself which exchange their signs. Under the sign of commodity, culture is bought and sold – under the sign of fashion, all cultures interplay as simulacra in total promiscuity. Under the sign of commodity, time accumulates like money – under the sign of fashion, it is broken and interrupted in tangled cycles
>
> (Jean Baudrillard, 1976: 132)

Today, the fashion sector tends to identify itself completely with the clothing sector, but, if we look closely, this is a relatively recent phenomenon (Ryan, 1966; Poiani, 1994). In fact, prior to the end of the 1960s, the clothing sector essentially corresponded to mass production. Because of this, *haute couture* and boutique collections were directed respectively towards an international élite and a national bourgeoisie. Thus, the policies of the sector were intended to provide clothing as a basic need and to move the value in exchange as closely as possible towards the value in use. In other words, the dynamics of the sector were determined by what we could define as the quality/price relationship.

Since that period, there has been a constant growth in the role of so-called creativity in the clothing industry. The creation of a recognizable style even in mass production has led to the inclusion of the designer in the production process, to the extent of causing a substantial reversal of roles. The label (*griffe*) of the designer becomes the trademark of the company, under which producers of single goods frequently lose their own identifiability.

As Baudrillard (1976: 17) observes, the hegemony of 'fashion' in the area of the clothing sector is explainable by adding the concept of the *value-sign* to those of *value in use* and *value in exchange*: 'The value in use acts as limit and finality for the value in exchange. The first qualifies the concrete operation of the commodity in consumption (parallel moment to that of the designation of the sign), the second returns to the interchangeability of all commodities under the law of equivalence (a moment parallel to that of the structural organization of the sign)'. In the economics of fashion, value and sign become simulation, lose any reference to reality, fluctuate rootlessly in total indetermination and indifference, and follow a rule of uncontrolled equivalence. In this sense, compared to the anonymity of clothing, fashion brings a more intense circulation of signs and information which are transformed into value and as such exchanged, through the mechanism of the 'designer label', with the recognition of a style and, therefore, its standardization.

The object of this case study is to show how the major or minor affirmation of the value-sign, and therefore of fashion on clothing styles, can be (re)read on a double level within a territory. First of all, the fashion–clothing relationship gives an opportunity to report briefly on the different positions of European countries in the fashion industry. Second, the analysis of the transformation processes of signs into value allows for a reconsideration of the complexity of the relationship between local identities and the dynamics of globalization.

Towards a geography of the fashion industry

It is possible to categorize countries in which the process of identification between fashion and clothing has occurred more rapidly and radically (France and Italy) from others (Germany, the United Kingdom and Spain) in which the *value in use* and *value in exchange*, both properties of the clothing industry, predominate with respect to the *value-sign*. In order to understand, through specific examples, how this theoretical distinction translates itself into a territorial differentiation, it is necessary to refer to the production and distribution structures in the fashion and clothing industries of the two groups of countries (Werner International Inc., 1991).

Production

In terms of production structure, the characterizing aspect in France and Italy is the particular

continued

dynamic in the relationship between the designer and the manufacturing system. This relationship is crucial to the extent that the creation of the value-sign depends strictly on marketing and design, the two aspects most closely tied to the work and image of the designer. This orientation has taken on different forms in the two countries in question. In France, the designer has turned himself or herself into an entrepreneur, setting in motion an unstoppable process of vertical integration and centralization. In Italy, on the other hand, production has remained much more widespread locally, vertically disintegrated and concentrated in traditional industrial districts rather than single companies. Proof of this is the fact that the five largest Italian companies in this sector have barely 7% of the total number of employees. What is more, the Italian textile industry is traditionally spread out into a myriad of small companies. The continually growing competitiveness of the Italian fashion industry has occurred precisely because of its peculiar structure, which has always given a central role in its production process to continuity with the artisan tradition and the employment of local skills and knowledge, built up in layers over the decades. This is a translation of value in use (the capability of producing objects of quality, with innovative techniques, within the context of a consolidated tradition) into a value-sign (*Made in Italy* as a synonym for refined elegance).

This interpretation is reinforced when the situation in Spain is compared with that in Italy (CITYC, 1996; Moda Industria, 1996). In both the Italian and Spanish productive structures, 80% of companies have less than 20 employees. However, such an observation hides the very different positions of the two national industries. What explains this difference is the predominance in the Spanish industry of value in exchange over the value-sign. The Spanish fashion industry, as opposed to that in Italy, has lost continuity with its own great tradition of artisanship, in order to follow a path of development based on low labour costs.

As far as the other countries in the second group are concerned (Germany and the United Kingdom), the pre-eminence of value in use and value in exchange translates itself into productive structures still tied to mass production, illustrated in the presence of very large factories or decentralization into countries outside the European Union. This is the case in the United Kingdom, where 56% of workers in the sector are concentrated in the five largest companies, and also in Germany, which is the European leader in terms of the amount of trade in outsourcing of final assembly, at 3246 million ECU as against 369 million ECU in Italy and 145 million ECU in Great Britain.

Distribution

Europe's different distribution structures reflect those of production. In France and Italy, the predominant distribution channels reveal both a reproposition of their production structure and a reinforcement of the image of national fashion. In France, for example, there is still a prestigious tradition of large department stores, which today cover 15% of the total clothing distribution. In contrast, in Italy the image of fashion is entrusted to a vast network of retailers (equivalent to 70% of clothing distribution), who are frequently specialized, and whose direct relationship with the customer is typical of the entire Italian fashion industry as well as an important factor in competitiveness.

Regarding the countries in the second group, the main task of distribution seems to be to distort as little as possible the relationship between value in use and value in exchange. In fact, the dominant distribution channels are those which have the least effect on cost structures and the greatest reinforcement of mass production. Thus, for example, in the United Kingdom, specialized chains and popular department stores today cover 60% of distribution, while in Germany 14% of retail sales are controlled by purchasing groups of retail buyers formed in order to have greater bargaining power and a similar figure for mail-order business.

The ambiguity of fashion between local and global

The simultaneous affirmation of *Made in Italy* in the world fashion market and the paradigm of

continued

Italian development might appear to suggest a positive causal relationship between the globalization of the circulation of signs and the reinforcement of local identity. For example, the Italian fashion industry seems largely to resist decentralization of production by virtue of its capacity for transforming its own identity into easily circulated signs. In reality, the situation is more ambiguous than it seems. Thus specialized chains are assuming an increasing role in fashion distribution. The creation of these worldwide networks, through which product lines or labels are distributed, is a clear symptom of the affirmation of a worldwide market of signs.

This improbable and frenetic circulation of locally distinctive signs is not, as opposed to how it might seem, a symptom of vitality and autonomy of individual localities with regard to the world system. The immediate recognizability and sharing of a sign presupposes a minimum common knowledge between those who sell and those who buy. In this process of sharing, the identity of each part necessarily undergoes a process of simplification. The circulation of signs requires the reduction and simplification of meanings which have been formed in long processes and rooted in single localities. For example, the affirmation in the 1960s of so-called 'ethnic fashion' and, in more recent times, of young Italian and Oriental designers who cite in their creations out-of-date styles from their own places of origin, indicates how much of its own identity the 'local' has had to sacrifice to mass communication, reducing itself to a game of citations.

Once again, as Baudrillard (1976: 132) writes, fashion 'always presupposes a dead time of forms, a type of abstraction which causes these to become, as if sheltered from time, effective signs capable, as if because of a distortion of time, of returning to assimilate the present with their "old fashionedness", with all the allure of their return in opposition to the ongoing development of the structure'. Taking this reasoning to its extreme consequences, the comprehension of local fashion signs requires neither attention nor intuition, since it bases itself on the minimum common denominator, on a *tabula rasa*, the uprooting of both those who produce fashion and those who consume it (see Weil, 1949).

So here we arrive at the geographical paradox of fashion. On the one hand, the passage from the clothing industry to that of fashion allows for the rooting of production activities in *localities*, as has happened in local Italian systems, and this therefore blocks the process of alienation of production. On the other hand, fashion compels these localities to market their own identities, pulling them into a whirlpool of competition in the production not of commodities but of degraded signs, made banal to the extent of being recognized and shared by the whole world. This paradox leaves the clearest and most painful traces of uprootedness on the territory. We need only consider the environmental degradation which has accompanied the growth of Italian industrial districts (Dansero, 1996), or the effects which lifestyle models of late capitalistic societies have on the populations of the Southern Hemisphere.

In conclusion, the fashion system sharply limits the freedom of action of geographers. It is not enough to promote local economic development in order to save and protect the identity of particular localities. Local production organization does not immediately identify itself with local identity. The need to mediate between the growing complexity of capitalistic structures and the growing simplification of the market of signs requires that the identity of the localities be assumed in its entirety and that greater attention be paid to the codes which regulate the communication between local and global. Every popularization/debasement of local identity into easily consumed signs weakens the autonomy and identity of the local, compromising its very existence.

Further reading

Barthes, R. (1975) *Il Sistema della Moda*, Einaudi, Torino

Conti, S. (1993) The network perspective in industrial geography: towards a model, *Geografiska Annaler*, **7B**(3), 115–30

Ryan, M. (1966) *Clothing: a Study of Human Behavior*, Holt, Rinehart and Winston, New York

Werner International Inc. (1991) *Situation and Perspective of Technical Textiles in the European Community*, Commission of the European Communities, Brussels

Consumption convergence and difference

In a widely quoted article Theodore Levitt (1983: 94), the American marketing scientist, claimed that the 'globalization of markets is at hand, consumer needs and wants had once and for all been homogenized'. Even accepting that Levitt's argument was directed only at the industrial nations, that is to where the bulk of retail consumer spending is concentrated, his vision is at least questionable at the global level; branded goods which have achieved a global reach, in which the same product can be marketed to a similar market segment all over the globe, account for only a fraction of all consumer goods, in spite of their symbolic importance. In the same vein, what evidence there is that the creation of the Single Market in Europe will have similar effects in homogenizing consumer preferences in western Europe is at best contradictory. This reflects the wider uncertainties of the longer-term effects of the European Union, whether it will be a homogenizing force, or whether it will reaffirm the national and local diversities of member states (Zetterholm, 1995).

Evidence supporting the Europeanization of consumption, the common acquisition of branded goods across the continent, is at most patchy. More 'Europroducts' are being marketed, and transnational advertising agencies have sought to standardize how products are promoted across national boundaries, a process which had begun in the 1960s. Yet, national consumer preferences for the 'messages' as to how goods should be promoted are persistent – claims of (for example) French consumers preferring style to the Germans' preference for being shown the quality of a good – so that advertising techniques are by no means easily portable across national boundaries. Similarly, national images remain more powerful for consumer perceptions, so that while goods continue to show their origins through designations such as 'Made in France' or 'Made in Germany', relatively few have adopted the label 'Made in Europe'. Marketing scientists have shown the importance of such labelling where the origins of a product carry buyer perceptions of quality, durability or sophistication, Europe in this sense lacking interpretative fixity.

With the exception of high-quality items, notably prestige cars and fashion goods (see case study), significantly those relatively few goods which have achieved pan-European status tend to be American rather than European in origin. Classic examples include the fast food and soft drinks sectors, goods which in a number of cases have attained global status and which are the representatives *par excellence* of Levitt's vision. In eastern Europe participation in western consumerism was a question of consuming American goods, and particularly those which had otherwise achieved a global market position. The opening of the first McDonald's in Moscow became not just a symbol of the extent to which economic liberalization had taken place, but an icon of Americanized consumerism. As *the* consumerist society, globalized forms of consumption are predominantly American rather than European – for Muscovites being able to patronise a McDonalds, something only available to the rich, is an opportunity to 'consume' America.

Prior to the demise of state socialism most east Europeans were aware of the abundance of western (especially American) consumer capitalism. The growth of international tourism, and the spread of mass media, quite apart from the often conspicuous consumption of the Party élite, had made such awareness inevitable (Belk and Gur, 1994). Television, in particular, was responsible for exposing the gap in the consumer worlds. In much of eastern Europe home ownership of television sets had become the norm, even in rural areas. Ironically, the transmission of American television series – aimed at demonstrating the decadence of western capitalism – had the opposite effect, where subjects with which the east European viewer could identify, an industrial worker or a teacher, were shown to have taken for granted consumer lifestyles which self-evidently were unattainable under state socialism.

These contradictions between the consumer expectations that the two political systems of capitalism and socialism were capable of delivering were at their most extreme in the readings of such American series as *Dallas*, which was transmitted throughout much of eastern Europe. For western audiences the series was a celebration of conspicuous consumption, though even here the series was criticized for its overt Americanization, with the French Minister of Culture describing the programme as *the* symbol of American cultural imperialism (Ang, 1985). Even if the consumer worlds of *Dallas* were more dream than they were ever likely to be reality for most Americans, let alone Poles or Romanians, this is to overlook the point of the consumerist revolution, that it is about desire as well as acquisition.

In the post-socialist order the realities of consumption have borne little resemblance to the desires, further countering any tendencies there may have been towards consumer convergence for all but the new rich between east and west Europe. Inflation and economic insecurity have limited the abilities of most east European consumers to participate in the marketplace, ironically at the very moment that, for the first time, western goods had become widely available. Lofman (1994) describes the frustrations that unfulfilled expectations quickly gave rise to amongst Polish consumers. Even if some of their consumer expectations had been met, that they were continuing to rise (one of the defining characteristics of consumerism) contributed to an increasing 'aspiration gap'. Paradoxically, then, where under socialism consumer expectations and the actual patterns of consumption of the majority were closely linked with one another, under liberalization they have become increasingly dysfunctional.

The inability of many of the new consumers of eastern Europe to afford western goods has brought a measure of disillusionment with the political change – reflected in the return of the Communist Party through the ballot box in several states and even, for some, the expression of a desire to return to the old order – besides accentuating local differences in consumer practices from western modes. Barber (1995) shows how in the former German Democratic Republic disillusionment with reunification and its failure to live up to citizen/consumer expectations has resulted in a partial rejection of westernization. This has become expressed nostalgically, through the demand for local goods – branded soft drinks, beers and cigarettes – manufactured during the socialist period. Ironically, as Barber points out, the production of the former German Democratic Republic branded cigarettes has subsequently come under the ownership of the American corporation, Philip Morris, which in being anxious to expand into the new markets of eastern Europe, given the contraction of the US market, was aware of the cultural significance of the German Democratic Republic brand on local identity. As much as the persistence or return of such local consumer practices may be the product of economic necessity, they demonstrate how cultural factors act as a brake to the emergence of global goods in the sense championed by Levitt (1983).

Regional and local differences of consumer practice in Europe are compounded by the growing *mélange* of consumerisms associated with the immigrant populations, particularly characteristic of most west European countries. The relationships between host and immigrant populations have been mutually interactive – both the consumption modes of the immigrant populations have been hybridized by the effects of the cultural practices of the host, whilst immigrant consumption patterns have become grafted on to those of the receiving societies. In some cases, ethnically defined restaurant and fast food services, the transplantation of novel, often exotic, cuisine, have been of long standing and have become commonplace. In Britain, for example, Chinese and Indian fast food retailing, together with the more recent introduction of Turkish food, have become 'indigenized' sufficiently into British eating habits that they are able to compete with the traditional fast food industry.

For immigrant groups themselves the retention of ethnically distinct consumption practices is a means of maintaining group identity – goods, and their consumption, become the markers of ethnic identity and difference. Yet, while immigrant groups may initially have been sojourners, with time, and particularly once a second generation becomes established, born and raised in the host country, the more likely are cultural practices to become hybridized. As Bouchet (1995: 91) has argued, 'Postmodern ethnicity is a *bricolage* making use of a diversity of cultural motives from different origins and strongly influenced by the mirror image provided by the dominant Other and the overwhelming media'. Asian households, especially those including children, celebrating Christmas, and Moslems consuming alcohol while otherwise maintaining ethnically distinct consumption patterns, are symptomatic of the hybridized consumerism among immigrant consumers, matched by the increasing inclusion of ethnic cuisine among indigenous households.

Caglar (1995) draws out the hybridization processes affecting the marketing of *doner kebab* in Germany and their possible meaning to the wider question of host–immigrant relations. Introduced to Germany by Turkish migrants, *doner kebab* quickly became a staple fast food for the host population, rivalling the traditional forms of fast foods of hamburger and bockwurst. In Berlin, which has the largest concentration of Turkish residents of any German city, there are no fewer than 6000 Turkish fast food outlets, dominated, as elsewhere in Germany, by the street stalls (*doner Imbiss* stands). From the outset *doner kebab* was marketed to meet the preferences of the German consumer, resulting in subtle changes in the ways in which it had been prepared traditionally. More recently, in an attempt to both maintain and indeed extend market share, the marketing of *doner kebab* has sought to reposition its image, *Imbiss* stands calling themselves 'Mac's Doner', 'Mister Kebap' and in the case of one *Imbiss* chain 'McKebap'. The intention behind the introduction of the new logos is in one sense obvious; more fundamentally, though, as Caglar (1995) argues, its purpose is to enhance the external status of the Turkish population, otherwise the victim of stereotyping, or worse, by the host society.

Conclusions

If market processes are forces for homogenization, cultural processes are more forces of diversification. Yet, as has been noted, the way in which even retail markets have developed in Europe reflects the imprint of national regulatory regimes. As much as the capitalization of retailing has become a common theme, how this has developed and how it has become expressed in Europe's cities and towns has been influenced by national legislation aimed at regulating change. Similar problems – of the decline of the small-scale retailer and the increasing dominance of mass retailing, both of which have distinctive urban and rural impacts – have resulted in different types of policy resolution by national and local governments, in turn producing nationally and locally distinct types of retail landscape. Convergent trends in the types of retail outlet and of consumer landscapes across Europe, out-of-town retailing, discount retailing and shopping malls, mask the differences in how these are developing.

Cultural processes are contributory to the diversities which characterize retail markets and consumption in contemporary Europe. Indeed, post-modern accounts suggest that marketing practices in turn may be contributory to cultural difference; that because the meanings attached to the act of consuming are themselves used as the signifiers of difference, marketing increases diversity (Bouchet, 1995). In the increasingly polyglot countries and cities of northern Europe in particular, ethnically distinct patterns of consumption are themselves adding to, and altering, pre-existing national practices.

These complex diversities are masked by the rhetoric of homogenization that has accompanied the European project. 'Barriers fade for European shoppers' claimed the *New York Times* in early January 1992 shortly after the implementation of the Single Market. Admittedly, its advent has been to foster the development of transfrontier shopping across the boundaries of member states, perhaps most transparently across the English Channel by consumers from south-east England taking advantage of cheap French and Belgian goods. Yet, transfrontier shopping, whatever its benefits to the consumer (and disbenefits to the economy forfeiting the trade), hardly constitutes 'Euro-shopping', even assuming it was possible to define such a designation.

Further reading

Costa, J.A. and Bamossy, G.J. (eds) (1995) *Marketing in a Multicultural World; Ethnicity, Nationalism and Cultural Identity*, Sage, London

Davies, R.L. (ed.) (1995) *Retail Planning Policies in Western Europe*, Routledge, London

Eurostat (1993) *Retailing in the European Single Market*, Eurostat, Brussels

Wrigley, N. and Lowe, M. (eds) (1996) *Retailing, Consumption and Capital: Towards the New Retail Geography*, Longman, Harlow

PART IV Social agendas

CHAPTER 15 Demography

TONY CHAMPION

Demography is one of the main driving forces for change in Europe. Over the last three decades the continent appears to have embarked on a 'second demographic transition', with the overall level of fertility moving well below replacement rate and with major alterations taking place in sexual and household-forming behaviour. All European countries have larger proportions of the elderly than ever before, a development which will be strongly reinforced in many countries as their baby-boom generations reach retirement age. The emphasis is also on immigration from other continents rather than emigration overseas, creating much greater racial and ethnic diversity and challenging the strongly held feelings of identity which have resulted from centuries of nation and empire building (see Chapter 6). Furthermore, the continent's long-established tradition of urbanization and city living has been shaken by the effects of deindustrialization and by the emergence of various forms of 'counterurbanization'.

Changing demographic processes

Whether or not one subscribes to the notion of a 'post-industrial society' coming into being in Europe, it cannot be denied that the contemporary demographic behaviour in Europe is very different from that of 30 years ago, let alone from that prevailing during the main period of the demographic transition. Table 15.1 presents a snapshot of the main demographic features recorded in the mid-1990s, including the generally very low rates of fertility and mortality and the widespread occurrence of natural decrease and net immigration.

Mortality and longevity

Mortality is the least trumpeted component of the so-called 'second demographic transition' (van de Kaa, 1987), probably because the theme of 'longer lives' has been a more continuous and regular feature of the European scene than the 'fewer babies' component (Ermisch, 1990). Nevertheless, because of this regularity, the overall increase in life expectancy over the past half-century has been very substantial and there remains considerable scope for further improvement. In eastern Europe mortality rates are significantly higher than elsewhere, while generally across the continent male/female differences in life expectancy are now wider than they have ever been. Meanwhile, the spectre of AIDS (acquired immunodeficiency syndrome) provides an element of uncertainty and reinforces the argument that the future depends more on personal behaviour and lifestyle patterns than on economic conditions and advances in medicine and healthcare.

A remarkable decline in mortality rates has taken place over the past 40 years. Since the early 1950s infant mortality in the 12 pre-1995 European Union (EU12) countries has been falling by around 5% a year on average, down from 49 per thousand live births then to only eight at the end of the 1980s. Overall life expectancy at birth has been lengthened by nine years over the same period, representing an average gain in longevity of one year every four years (Noin, 1995). This progress has not been completely regular, being most rapid in the early post-war period, slowing in the late 1960s but then resuming a fairly steady increase since the mid-1970s.

In geographical terms, across much of the continent, these improvements in life expectancy have been associated with a clear convergence. Among the

Table 15.1 Indicators of the demographic situation in Europe in the mid-1990s.

	Population (m)	Population change (%)	Net migration (%)	Natural increase (%)	Birth rate per 1000	Death rate per 1000	IMR rate	Life expectancy (years)			TFR rate
								Total	Male	Female	
Europe	**727.5**	**0.03**	**0.14**	**–0.11**	**10**	**11**	**11**	**73**	**68**	**77**	**1.5**
Northern Europe	*93.6*	*0.27*	*0.14*	*0.13*	*13*	*11*	*6*	*73*	*68*	*77*	*1.7*
Denmark	5.3	0.67	0.54	0.13	13	12	5	75	73	78	1.8
Estonia	1.5	–1.04	–0.55	–0.49	9	14	15	69	62	74	1.4
Finland	5.1	0.35	0.08	0.27	12	10	4	77	73	80	1.8
Ireland	3.6	0.31	–0.16	0.48	13	9	6	76	74	79	1.9
Latvia	2.5	–1.11	–0.42	–0.69	9	16	19	67	61	73	1.3
Lithuania	3.7	–0.16	–0.05	–0.11	11	12	13	69	64	75	1.5
Norway	4.4	0.49	0.14	0.34	14	10	5	78	75	81	1.9
Sweden	8.8	0.24	0.14	0.11	12	11	4	78	76	81	1.7
UK	58.7	0.33	0.18	0.15	13	11	6	77	74	79	1.7
Western Europe	*180.6*	*0.33*	*0.23*	*0.10*	*11*	*10*	*6*	*77*	*73*	*80*	*1.4*
Austria	8.1	0.19	0.09	0.09	11	10	5	77	73	80	1.4
Belgium	10.1	0.22	0.13	0.10	11	10	6	76	73	80	1.6
France	58.3	0.42	0.08	0.34	13	9	6	78	74	82	1.7
Germany	81.5	0.25	0.39	–0.14	9	11	6	76	73	79	1.2
Netherlands	15.5	0.45	0.10	0.35	12	9	5	77	74	80	1.5
Switzerland	7.1	0.59	0.31	0.28	12	9	5	78	75	82	1.5
Southern Europe	*143.9*	*0.15*	*0.06*	*0.09*	*10*	*9*	*10*	*76*	*73*	*79*	*1.3*
Albania	3.2	–3.19	–5.03	1.84	23	6	33	72	70	76	2.8
Bosnia–Herz.	4.6	0.73	0.00	0.73	14	7	14	72	70	75	1.7
Croatia	4.8	–0.01	0.04	–0.05	11	11	10	71	66	75	1.5
Greece	10.5	0.30	0.25	0.05	10	9	8	77	75	80	1.3
Italy	57.3	0.11	0.16	–0.05	9	10	6	77	74	80	1.2
Macedonia	1.9	4.62	3.63	0.98	16	8	23	72	70	74	2.2
Portugal	9.9	0.09	0.05	0.04	11	10	7	76	72	79	1.4
Slovenia	2.0	0.04	0.04	0.00	10	10	6	74	70	78	1.3
Spain	39.2	0.16	0.13	0.03	9	9	6	77	73	81	1.2
Yugoslavia	10.5	0.30	0.00	0.30	13	10	17	72	69	74	1.9
Eastern Europe	*309.5*	*–0.28*	*0.12*	*–0.40*	*10*	*14*	*16*	*68*	*62*	*73*	*1.4*
Belarus	10.3	–0.21	–0.03	–0.18	11	13	13	69	64	74	1.4
Bulgaria	8.4	n/a	n/a	–0.51	9	14	15	71	67	75	1.2
Czech Rep.	10.3	–0.11	0.10	–0.21	9	11	8	73	70	77	1.3
Hungary	10.2	–0.33	0.00	–0.33	11	14	11	70	65	74	1.6
Moldova	4.3	–0.31	–0.39	0.08	13	12	22	67	63	70	2.0
Poland	38.6	0.07	–0.05	0.12	11	10	14	72	68	76	1.6
Romania	22.7	–0.25	–0.09	–0.16	10	12	21	70	66	73	1.3
Russian Fed.	148.0	–0.22	0.34	–0.57	9	15	18	65	57	71	1.4
Slovakia	5.4	0.22	0.05	0.16	12	10	11	72	68	77	1.5
Ukraine	51.3	–0.76	–0.18	–0.58	10	15	15	68	63	73	1.4

Notes: Data for the latest available year – normally 1 January 1996 for population (first data column) and 1995 for rates (all other columns). Rates may not sum because of rounding and because data for any one country may relate to more than one year. IMR = Infant Mortality Rate per 1000 live births; TFR = Total Fertility Rate. 'Yugoslavia' comprises Serbia and Montenegro.

Sources: Council of Europe (1996), Population Reference Bureau (1996).

EU12 countries, for instance, the smallest gains have been recorded by those countries where the level was significantly above average in the early 1950s, like Denmark and the Netherlands, while the largest increases have been achieved by those where it was originally lowest, most impressively Portugal with a rise of 14 years. As a result, by the end of the 1980s, the range in life expectancy across the EU12 was down to 3.4 years, less than one-third of the 12.8 years range in the early 1950s (Noin, 1995).

The situation in eastern Europe is very different. Though most countries there paralleled the trends of the rest of the continent in the 1950s and 1960s, mortality rates were still significantly below average by the early 1970s and since then the gap has progressively widened, with reductions in infant mortality being more than outweighed by sharp increases in mortality for all ages over 30, particularly working-age men. Explanations include lack of investment in the more advanced medicines needed to fight degenerative diseases, the emphasis on pork and animal fats rather than fruit and vegetables in traditional diets, increases in alcohol consumption and cigarette smoking, and – in the 1990s – the impacts of the radical political, economic and social changes, including higher unemployment generally and the fighting in parts of the former USSR and Yugoslavia (Meslé, 1996) (see case study on the Yugoslav conflict).

The trends are reflected in the latest statistics on infant mortality and life expectancy (Table 15.1). Infant mortality is lowest in northern and western Europe, averaging barely six deaths per thousand live births. At the other extreme, it is more than twice this level in eastern Europe on average and four times this level or more in Albania, Macedonia, Moldova and Romania. Overall life expectancy averages 76–77 years in northern, western and southern Europe, but is eight years less than this in eastern Europe, with the especially low level of 65 years for Russia.

The arrival of AIDS in the 1980s has raised particular uncertainties about future improvements in life expectancy (see Chapter 18). Thus far, its impact has not been as great as once feared, with the highest incidence occurring among drug users and homosexuals. According to Löytönen (1995), however, there remains cause for concern, particularly with the upheavals in eastern Europe, where the incidence has hitherto been much lower. Similar, but as yet less publicized, uncertainties arise from a recent increase in the old infectious diseases, resulting from greater exposure owing to increased worldwide travel and immigration and from higher risk of infection owing to less systematic immunization and the development of new strains resistant to current antibiotics.

Nevertheless, current population forecasts indicate that, on balance, mortality rate reductions will continue to prevail. According to the UN (1991), by 2000–05 infant mortality in the EU12 will have fallen by a quarter from its level at the end of the 1980s, and overall life expectancy will have extended from 75.8 to 77.5 years. In conjunction with recent trends in fertility, this means a continued long-term ageing of Europe's populations and brings a number of important implications.

Fertility and partnership behaviour

Changes in fertility and partnership behaviour lie at the heart of the population debate in Europe. Their central importance is reflected in the impressive phrases used by many authorities. Sporton (1993: 49) observes that Europe has 'the lowest level (of fertility) in the world', Brass (1989: 12) discusses whether Britain is facing 'The twilight of parenthood', Coleman (1996: 48) marvels that people 'choose to have any children at all', and Faus-Pujol (1995: 21) argues that 'Europe is at the beginning of a new demographic transition'. According to Kiernan (1996: 89), 'Men and women are cohabiting more, marrying later, becoming parents at older ages, and having fewer children, as well as terminating their marriages more frequently than was common in the recent past'. In view of these commentaries, it is perhaps surprising to find that, according to Coleman (1996), the pattern of fertility in Europe is characterized by great and increasing diversity, and equally surprising to discover how much uncertainty is being expressed about future trends.

On the face of it, the situation in mid-1990s Europe seems very uniform and straightforward. As shown in Table 15.1, in all four quarters of Europe and in most individual countries, fertility at this time was well below replacement rate (taken to be around 2.1 births per woman at the prevailing mortality rates). The overall European figure was only 1.5, with northern Europe the highest at 1.7, followed by western and eastern Europe both on 1.4 and southern Europe lowest at 1.3 – an impressive reversal of the relative positions at mid-century. Even more impressively, only one country (Albania) lies above replacement rate in the mid-1990s, while even Ireland

Terror in the Yugoslav conflict

James Gow

The Yugoslav conflict of the 1990s was a clash of state projects amid the dissolution of the Socialist Federative Republic of Yugoslavia (SFRY). The SFRY was a communist creation comprising six formerly sovereign states – Serbia, Croatia, Slovenia, Bosnia and Herzegovina (hereafter, Bosnia), Macedonia and Montenegro – with a complex mixture of populations; one state, Serbia, had two autonomous regions within it, Kosovo and Vojvodina. At the beginning of the 1990s, polarized views had emerged of what the future role of the federation should be. With little genuine support for the federation in any of the states, it effectively ceased to function. A major armed conflict ensued of which coercive terror was a principal feature. The essence of that conflict was the Serbian attempt to forge the borders of a new entity, partly based around Serb populations, through force against the aim of other states to maintain their territorial integrity.

In the decade preceding the dissolution of the SFRY in 1991, there were growing signs of inter-communal tension and the emergence of repressive and coercive terror. From 1981 onwards, the province of Kosovo in Serbia, with an overwhelming majority of ethnic Albanians in the population, was a locus of tension. In 1981, elements of the Yugoslav People's Army (JNA) were used to suppress demonstrations by Albanian students. Through the rest of the decade, the JNA and, to a greater extent, Serbian and federal internal security forces imposed a regime of martial law on the province. Ethnic Albanians were liable to be arrested at random, beaten and imprisoned for short terms. In addition, there were occasions where the security forces openly used force, such as the shooting of protesters in the Kosovo capital Pristina in March 1989, when official figures reported 24 killed but unofficial estimates ran to 180.

Serbian action regarding Kosovo sparked tension in other parts of the federation, as Serbian President Slobodan Milosevic orchestrated the removal of Kosovo's autonomy. Governments committed in differing ways to protecting the sovereignty of their states were elected in all the republics and the majority of the non-Serb states. This meant opposing apparent Serbian ambitions to dominate the other states. In particular, this question was acutely felt in Croatia. There, from the spring of 1990, when nationalist Franjo Tudjman was elected President, there were problems with the Serb population in Croatia. There was a series of terrorist detonations, almost entirely carried out by Serb insurgents, attacking variously railway lines, homes and public utilities. In this period there were also terrorist detonations against Serbian properties carried out by Croats. On both sides these attacks were intended both to intimidate those on the receiving end and others in their community and to be acts of provocation.

Violence and unrest in Kosovo and Croatia were precursors to the Yugoslav War of Dissolution, which began in June 1991. Coercion was the central characteristic of that conflict, which had three theatres: Slovenia, Croatia and Bosnia. The conflict in Slovenia was short-lived, lasting ten days, and was generally free from terrorization. The same was not true of Croatia and Bosnia, both of which were subject to the Serbian attempt to establish new borders. That project entailed the terrorization and forcible expulsion of non-Serbs in order to create a set of contiguous territories with new boundaries ethnically purified for the Serbs. The label given to this process was 'ethnic cleansing'.

There were two rationales for 'ethnic cleansing'. One was ethno-nationalist ideology. The other was the strategic imperative of ensuring that the territories of the new entity would be free from political disruption, guerrilla warfare or terrorism: by removing a population, the basis for resistance was also removed. Ethnic cleansing did not rely on direct combat with opponents, but on the demonstrative capacity of the violence which could be brought to bear and the example which was set by partial elimination of a population to induce mass migration. The intended result of this was a contiguous set of ethnically pure territories.

The process of ethnic cleansing was present in the war in Croatia but became far more prominent during the war in Bosnia. One of the focal points during the war in Croatia was Vukovar, where for three months the town was under siege and bom-

continued

bardment. The Croat population was expelled following the surrender of the town. However, some men were not expelled, including those in the town hospital, who faced a mass execution on the site of the Ovcara farm outside the town, where a mass grave of around 300 men was found.

The main thrust of ethnic cleansing came in Bosnia, where the first months of the conflict saw around 130,000 people killed and up to three million forced to leave their homes, or under pressure to leave and receiving assistance from the United Nations. Ethnic cleansing was a feature of the standard Serb pattern of attack, which generally had the following characteristics: (1) preparation, including collecting arms, dismissal of non-Serbs from jobs, propaganda; (2) provocation; (3) use of force; (4) identification of non-Serbs; (5) collection of non-Serbs, expulsion, terror; and (6) camps, forced labour and prisoner exchanges.

In each location, police, broadcast and civil administration were taken over by local Crisis Headquarters, composed entirely of Serbs, that had been organized pursuant to instructions issued in December 1991 by the Serbian political party in Bosnia, the SDS (Srpska Demokratska Stranka – Serbian Democratic Party). The secret preparation was carried out through a Serb security service network based on five security service centres (headquarters), one in each of the 'Serbian Autonomous Regions' to be declared. Acting consistently with a central instruction that required close cooperation and support from JNA units and adopting the appearance of governmental authority, local officials worked with paramilitary units to identify and interrogate non-Serb residents and evict them from their homes.

The separation of Serbs and non-Serbs had two aspects. In some cases, non-Serbs were instructed to mark their homes by flying a white sheet from a window, ostensibly for protection. In reality, the homes then became targets for intimidation, including grenade attacks, or for outright destruction with tanks or armoured vehicles placing direct fire from close range on the properties. The main mode of separation in other places also divided women and children from males over the age of 15. Each group would then be taken to a point of concentration, usually a camp outside the town. Primarily Muslims, these people were often given the opportunity to pay for the right to sign away their property and belongings to the newly established Serb authorities, before paying for a one-exit visa and a one-way ticket out of Serb-controlled areas. In these locations, non-Serb men of military age and many non-Serb women were imprisoned in camps and detention centres. In the camps, there were often conditions of severe hardship and deprivation, with beatings and rape being habitual features. The Serbs established a network of over 40 camps as they seized control throughout eastern and northern Bosnia in the spring and summer of 1992. A small number of the camps, such as Omarska and Luka, were also death camps, where there was evidence of regular and organized killing. According to witness reports, non-Serbs were systematically killed, beaten, raped and mutilated in these camps. By September 1992 most of the survivors were eventually deported to areas beyond Serb control, often in negotiated prisoner exchanges.

Aspects of ethnic cleansing, albeit on a significantly lesser scale, featured in the Croat and Bosnian government (or Muslim) paramilitary operations in the war in Bosnia. Croat units in central Bosnia were responsible for atrocities against Muslims during 1993 at locations such as Stupni Do and the village of Ahmici, where children were apparently burned alive. Muslim units, later to be incorporated into the Bosnian Army, apparently murdered, raped, pillaged and drove Serbs out of villages in eastern Bosnia near the town of Bratunac. Croatian forces and the Bosnian government also established small numbers of camps. Although these appear to have been primarily for prisoners of war, deprivation, brutality and torture were all present.

The final phase of the Yugoslav conflict contained some of the largest-scale manifestations of ethnic cleansing. The final operations by the Croatian Army to end Serb control of nearly one-fifth of the country prompted the exodus, clearly encouraged by the Serb authorities, of up to 200,000 Serbs. Although most fled before Croatian troops arrived, events after they arrived indicated that to have remained might have been unwise.

continued

However, the most blatant act of ethnic cleansing and atrocity occurred near Srebrenica in eastern Bosnia towards the end of the war, under the gaze of Dutch UN troops manning the 'safe area' declared there. In July 1995, the Bosnian Serb army overcame the Srebrenica 'safe area', expelling over 40,000 Muslims – most of them already refugees from other places where they had been ethnically cleansed – and murdering thousands of men of military age. The Bosnian Serb attack was relatively easily accomplished. Although there was some resistance from the small Dutch UN force, the bulk of the Bosnian Government troops in the area had already withdrawn. The Bosnian Serb takeover was terrible. As Serb troops advanced, several thousand Muslims fled in two ways. Around 15,000, mostly men, left through the woods from Susnjari towards the town of Tuzla. This column of flight was interrupted near the village of Buljim and attacked by the Bosnian Serbs. Around one-third of the group escaped to Tuzla, the remainder were trapped on Serb-controlled territory.

A second group of men, women and children sought security in the Dutch UN compound at Potocari, where they remained from 11 until 13 July, and after which they were transported in around 60 buses waiting with Bosnian Serb military drivers to remove them to Bosnian government-controlled territory. As the Bosnian Serb army burned and looted Muslim houses, General Ratko Mladic, the Serb military commander, personally assured the Muslims that they would not be harmed, but would be safely transported out of Srebrenica. However, as the Muslims boarded the buses on 13 July, Bosnian Serb military personnel separated the men from the women and children. The latter were escorted out of Bosnian Serb-controlled territory, while the men were taken to various other locations in the area. Although some of the Muslim men were told they would be exchanged for Bosnian Serbs being held by government forces in Tuzla, the majority were taken to Bratunac and then Karakaj where they were killed at two sites between noon and midnight on 14 July. By 23 July, one way or another, the Muslim population of the Srebrenica area had been virtually eliminated. The massacre at Karakaj was perhaps the most callous act in a litany of atrocities during the war in Bosnia.

In response to the campaign of politically and strategically motivated terror in Bosnia, the international community took a series of limited measures. These included the imposition of a comprehensive regime of sanctions on Serbia and Montenegro for involvement in Bosnia, intended to curtail the Serb campaign there. A second measure was the decision to deploy the largest ever UN peace force to assist in the delivery of humanitarian aid and in creating the conditions for settlement. This had a limited impact, but did allow communities to remain in place. As a consequence, the Bosnian government was often able to reorganize militarily, albeit on a limited scale. The final response offered by the international community was in the field of international humanitarian law. The first ever international criminal tribunal was established to investigate and prosecute war crimes committed on the territories of the former Yugoslavia after January 1991. Based in The Hague, the International Criminal Tribunal for the former Yugoslavia (ICTY) was active in carrying out investigations and issuing indictments against alleged war criminals from all parties to the conflict. By August 1997, ICTY had published indictments against 77 individuals. Of the publicly known indictees, 13 were Croats, four were Muslims and the reminder were Serbs. Of these, four Muslims and one Croat had been detained by ICTY and their trials begun, while five indictees from the Serbian side had been detained and another killed while resisting arrest.

Further reading

Gow, J. (1997) *Triumph of the Lack of Will: International Diplomacy and the Yugoslav War*, C. Hurst and Columbia University Press, London and New York

Honig, J.W. and Both, N. (1996) *Srebrenica: Record of a War Crime*, Penguin, Harmondsworth and New York

Silber, L. and Little, A. (1996) *The Death of Yugoslavia*, Penguin, Harmondsworth and New York

UN (1994) *Final Report of the Commission of Experts Established Pursuant to Security Council Resolution 780 (1992)*, United Nations, New York (UN Doc. S/1994/674, 27 May 1994 and Annexes)

– the highest fertility country in the EU in recent years – has now moved to a sub-replacement level.

On the other hand, this relative uniformity has been reached by a variety of routes, as reflected in the fertility trends of the countries shown in Figure 15.1. In northern and western Europe the present situation comes after a major decline from the 'baby booms' of the 1950s and early 1960s and a subsequent partial recovery in rates, though the latter has been smaller in Germany and most of its neighbours than for northern Europe (especially Sweden) and France. In southern Europe, the higher rates persisted into the 1970s but then experienced an even more spectacular decline, suggesting that 'Catholic fertility' is a thing of the past there. By contrast, most countries in eastern Europe did not experience a baby boom and maintained a fairly constant level of around 1.8–2.4 over most of the post-war period, but have seen a dive in rates since the collapse of their communist regimes at the end of the 1980s.

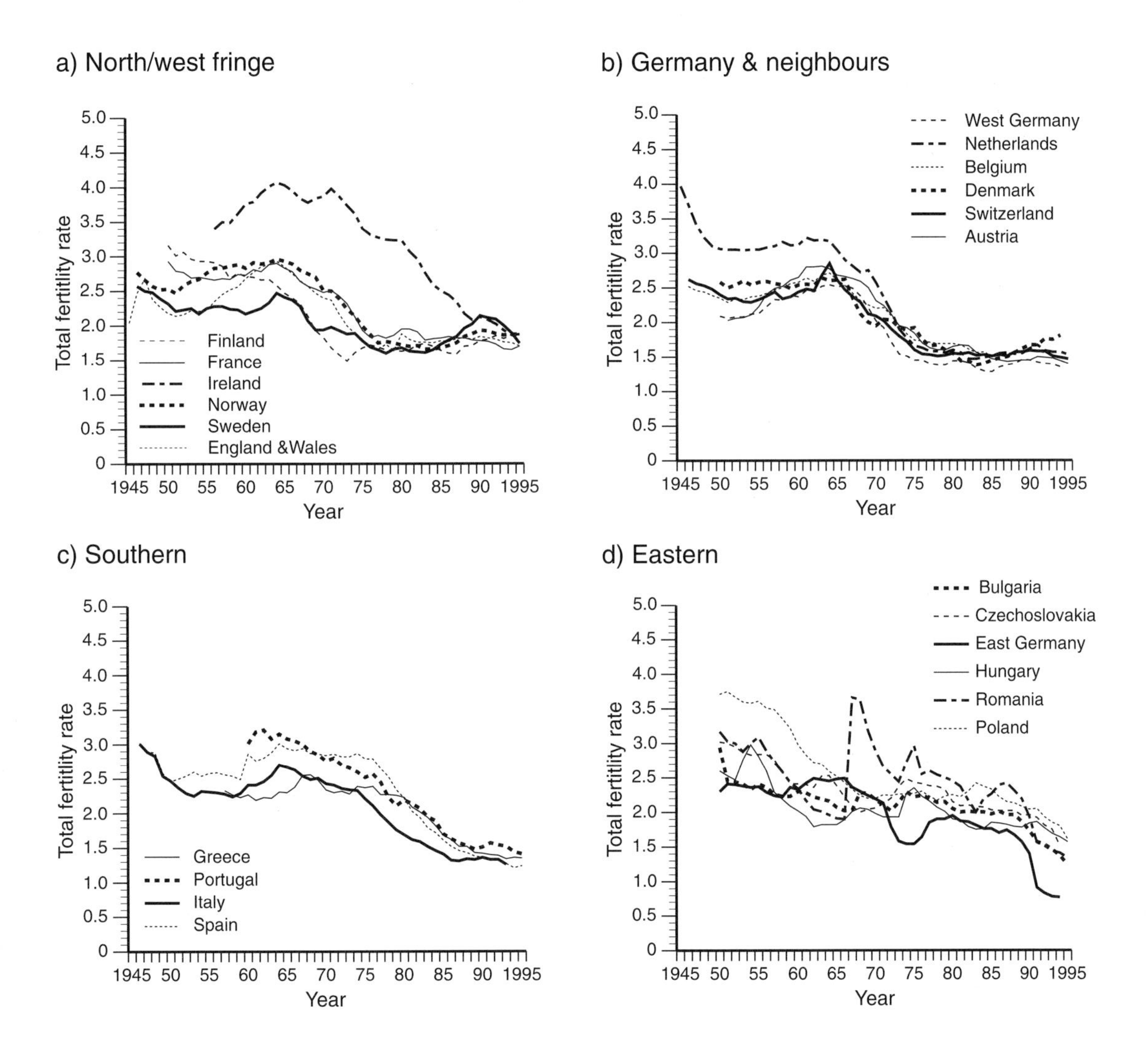

Figure 15.1 Total fertility rates in selected countries, 1945–95 (adapted from D. Coleman (ed.), 1996, *Europe's Population in the 1990s*, pp. 12–15, copyright 1996, with permission from Oxford University Press, Great Clarendon Street, Oxford OX2 6DP, UK; updated from Council of Europe, 1996).

A variety of reasons have been put forward for the reductions in fertility to these low levels. Much attention has been given to economic factors, particularly the general trend towards viewing children as consumer durables and also the effects of recessionary conditions in leading to delayed family building. Alongside this is the idea that fertility will fluctuate according to the job prospects of young workers, which are influenced not only by the current state of the economy but also by past fluctuations in birth rate and can be expected to follow a cyclic pattern. Other cited factors are more social and cultural in nature and might be expected to be more long-term in their impact, including the phenomenal rise in the proportion of working women, the greater emphasis on individual fulfilment and personal gratification and the shift in partnering arrangements away from lifetime marriage towards serial monogamy – all being greatly assisted by more effective contraceptive methods, easier access to abortion, and the breaking of the link between sex and marriage.

Over much of Europe, recent years have seen as dramatic a change in partnership behaviour as in fertility itself (R. Hall, 1995). Since the late 1960s marriage rates have declined in most European countries, beginning in Sweden and Denmark, spreading through most of western Europe and then into southern Europe by the late 1970s and now evident in some countries of eastern Europe (Table 15.2). The reduced stability of marriages is evident in the marked rise in the divorce rate, notably in northern and western Europe (Table 15.2).

These developments have been accompanied by a massive increase in cohabitation in some countries. According to Kiernan (1996), the proportions of 20–24-year-old women in a union that were cohabiting had reached 78% in Sweden by 1985, 49% in France by 1988 and 32% in Britain by 1989. This helps to explain the high level of child-bearing outside marriage recorded by these countries, compared with countries like the Netherlands, western Germany and Switzerland (Table 15.2). In Ireland and most of southern and eastern Europe the level of cohabitation, whilst increasing, is still much lower than elsewhere, though the proportion of births outside marriage varies considerably between countries.

Convergence would thus seem to be more apparent than real. While most countries fell within a historically narrow band of fertility rates in the mid-1990s,

Table 15.2 Trends in partnership behaviour and births outside marriage in selected countries, 1970–95.

	Mean age of women at first marriage		Total divorces per 100 mariages		Extramaritial births per 100 births	
	1970	1995	1970	1995	1970	1995
Northern Europe						
Ireland	24.8	26.8[c]	n/a	n/a	2.7	22.7
Norway	22.7	26.9[b]	13	40[d]	6.9	47.6
Sweden	24.0	28.7[a]	23	50	18.2	52.9
United Kingdom	22.4	26.2[b]	16	42[a]	8.0	33.6
Western Europe						
France	22.4	26.7	12	35[b]	6.8	34.9
Germany (West)	23.0	27.2[a]	16	36[a]	5.5	12.4[a]
Netherlands	22.8	27.4	10	34[a]	2.1	14.3
Switzerland	24.1	27.4	15	38	3.8	6.8
Eastern Europe						
Hungary	21.1	22.2	25	34	5.4	20.7
Poland	21.9	22.0[a]	14	12[a]	5.0	9.5
Romania	22.1	22.8	5	20	n/a	19.8
Russia	23.1[d]	22.4[a]	n/a	n/a	10.8	21.1
Southern Europe						
Greece	22.9	25.4	5	11	1.1	2.9
Italy	24.1	26.5[a]	5	8[a]	2.2	7.7
Portugal	24.3	24.9	n/a	n/a	7.3	18.7
Spain	24.8	26.5[b]	n/a	n/a	1.4	10.8

Note: a=1994, b=1993, c=1992, d=1980.
Source: Council of Europe (1996).

their experiences have differed considerably both in terms of their recent fertility trajectories and in relation to patterns of partnership behaviour.

International and internal migration

Numerically, migration forms a much more important process than mortality and fertility. According to Cruijsen (1996), in the current 15-country EU some 28 million people change address in an average year, including people moving in and out of the EU – much larger than the 4 million births and 3.7 million deaths recorded in 1994. On the other hand, the majority of these moves are over short distances, and the net changes in the size of places' populations are far smaller than the gross movements might suggest, because migrants often take the place of people who have left. The focus here is on the longer-distance movements that produce population redistribution between regions and localities, but even at this level the potential implications are enormous. The numbers involved are substantial, the process can produce significant changes in population profiles where in-migrants to an area differ in their characteristics from out-migrants and – most fundamentally – the types and geographical patterns of migration have altered greatly in recent years, both at the international scale and in terms of migration within countries.

In the context of international migration, White (1993a) refers to the emergence of a 'post-industrial' pattern over the past two decades, representing the latest of three waves affecting post-war Europe following the 'guestworker' phenomenon of the 1950s and 1960s and the subsequent phase of family reunification. This third wave comprises three relatively distinct elements: high-skill labour migration, clandestine movement and asylum seeking. The first of these constitutes an essentially 'invisible' process in that it involves mainly professional, managerial and technically skilled workers who are relatively cosmopolitan in culture, play an important role in the economy and are likely to be on some form of temporary contract or posting from their home companies (Findlay, 1995). Clandestine migration, either through illegal entry or through staying on after the expiry of a short-term tourist, student or work-related visa, is also largely labour-market related, with migrants' motives being mainly to improve their economic situation and with employers often willing to take advantage of their cheap labour, but this process is much more controversial as it threatens the jobs of the domestic population and often introduces very different racial and ethnic influences. Most attention in recent years, however, has been accorded to refugees and asylum seekers, partly because these are closely monitored under international agreements, and also because the numbers applying for entry into European countries exploded during the 1980s and early 1990s (Robinson, 1996; Salt, 1996). In their longer-term impact, these are often little different from the clandestines, in that more are now being drawn from non-European stock and many are believed to be basically economic refugees rather than people fleeing from persecution.

Along with the changing character of international migration, its geographical patterning has also altered greatly since the early 1980s. In particular, the proportion of immigrants coming from other developed countries has fallen markedly. Of the 13 million foreigners living in the EU12 in 1990, 8 million were from outside the EU area, with half of these from North Africa, Yugoslavia and Turkey. In the late 1980s and early 1990s there was a marked escalation of flows into western Europe from two sources: the East and the 'South'. The removal of the Berlin Wall and Iron Curtain led to a sudden jump in outflows from the Warsaw Pact states to the West from an annual average of under 100,000 in the early 1980s to more than 1.3 million in 1989 alone. This flow has eased considerably since then, but the 1990s has seen further large-scale movements within eastern Europe, notably as a result of the break-up of the Soviet Union and the fighting in former Yugoslavia (White and Sporton, 1995; Kupiszewski, 1996). Meanwhile, reflecting the huge economic and demographic gulfs between Europe and most of the Third World, there has been a steady increase in migration pressures from sub-Saharan Africa and Asia (Misiti *et al.*, 1995; Findlay, 1996).

As regards destination areas, Germany has borne the brunt of the surge from eastern Europe in terms of sheer numbers, but relative to their population sizes significant contributions to accommodating asylum seekers have also been made by a number of other countries, notably Sweden and Austria. Perhaps the most significant change since the mid-1970s has been the switch of the larger countries in southern Europe from being net exporters of people to north-west Europe and overseas to being net immigration countries, principally as a result of the large inflows of migrants from Africa and Asia

Moroccan immigration to Spain

Pablo Pumares

Immigration to Spain: a new phenomenon

Spain has undergone a dramatic transformation in recent years, from a country of emigration to one of immigration. Whilst there remain 1,620,000 Spanish citizens living abroad, there has been hardly any new emigration worthy of note over the last 20 years. In 1985, with its sights set on Europe, the Spanish parliament devised the Rights and Freedoms of Foreigners (in Spain) Act (*Ley sobre Derechos y Libertades de los Extranjeros en España*), a specific law to determine a question which until then had not aroused the legislators' interest. As soon as the new Act came into force, a process of regularization was carried out directed towards those foreigners already residing in Spain whose documents were not in order. This process was strongly criticized (rightly so) for its lack of clarity (there were a host of different permits), for the poor publicity it was given, and for the use of police stations to apply for permits (similar to putting one's head into the lion's mouth). Eventually only 35,000 foreigners from the less developed countries (including 7400 Moroccans) applied. This indicated that immigrants accounted for only a small proportion of the total Spanish population at that time.

Over subsequent years Spanish immigration policy did its utmost to prevent Spain from becoming a country of immigration: frontier controls were tightened, unorthodox campaigns to detect and expel immigrants were carried out, and the acquisition and renewal of permits was hindered. In three years, two-thirds of the permits that had been obtained during the process of regularization were lost (Izquierdo, 1991). Nevertheless, the process had been set in motion, and between 1988 and 1991, a period of considerable economic growth linked to Spain's entry into the EC, a substantial number of new immigrants were recorded. These immigrants could not obtain a residence permit according to the Act, and therefore the great majority of them were outside the law and thus unable to benefit from assistance other than that afforded by the so-called non-governmental organizations (NGOs).

In 1990 a switch in the government's immigration policy began to occur on the appearance of a document endorsed by the Parliament which recognized Spain as a country of immigration and pressed for the need to adopt measures contributing to the integration of immigrants residing in its territory, although at the same time it affirmed a determination to intensify frontier controls. The new policy led to a new, more transparent process of regularization (1991) conducted in collaboration with NGOs. It also led in 1996 to the passing of a new regulation of the Rights and Freedoms of Foreigners (in Spain) Act, which puts greater emphasis on the rights of immigrants. Under the 1991 process of regularization statute, some 110,000 work permits were granted, 49,000 of which were obtained by Moroccan immigrants. Other immigrants, although far fewer in number, came from Latin America (Dominican Republic and Peru), Asia (China and the Philippines) and other African countries (Algeria, Senegal, Gambia).

However, following the watershed year of 1992, when the Olympic Games were held in Barcelona, the Universal Exposition took place in Seville, and Madrid was the European Capital of Culture, the full weight of the world economic crisis was felt in Spain. After an initial regrouping of family members of immigrants who had already obtained permits, and coinciding with the way the economic situation was perceived, there was a notable slowing down in the influx of new arrivals. Further contributing to this was an agreement reached with Morocco by which the latter agreed to control departures from its coasts. During the summer of 1996, with growing signs of economic recovery, news of countless small boats (so-called '*pateras*') crossing the Strait of Gibraltar, together with shipwrecks, arrests and expulsions, was back in the headlines. Needless to say, there are also a fortunate number who manage to reach their destination. The controls make the process more difficult and more expensive, but in any event trying to make the frontier foolproof against economic immigrants is practically impossible. A migratory flow from Morocco to Spain has now been created and appears to be self-sustaining,

continued

slowing down at times of economic crisis and speeding up at the slightest upturn, with a constant trickle of people crossing to join family members. Such are the prospects for the medium term, while economic conditions remain substantially the same in the countries of origin and/or destination. This will certainly give rise to the establishment of an ethnic minority group of Moroccans in Spain.

The demand for foreign labour is concentrated in jobs requiring large amounts of low-skilled labour, frequently linked with the informal economy, and in the service and agricultural sectors. Unlike elsewhere in Europe in the period immediately after 1945, the role of industry in attracting labour from abroad is insignificant. Despite a tradition of rural exodus associated with modernization and the introduction of new techniques, modern horticulture requires large amounts of labour that cannot be provided by local workforces. This is necessitating the presence of many foreign workers in rural areas (Figure 15.2). However, due to the hard working conditions, it has generally only been African people, mainly from Morocco, who have dared to work in Spanish agriculture. The demand for services, particularly domestic services in the big cities, has also favoured a growing presence of immigrant women and has increased their economic role, since they can usually find employment more easily than men.

This situation creates a new challenge for Spain in terms of acceptance and coexistence, precisely at a time when its attention is focused on Europe and it has forgotten, or does not wish to recall, its recent past as a country of emigration. If this is mentioned it is often accompanied by the afterthought: 'but we were different, we went to find work ...' European polls place Spain amongst the least racist nations in Europe, although this does not mean that there is an absence of conflict or tension, or a conspicuous attempt to prevent these from occurring. The high unemployment rate (everywhere over 20%) is perceived by Spanish society to be the most pressing problem facing the country, and this makes people very sensitive to the arrival of foreign workers, who are seen as potential competitors for jobs. This may well become a major political argument and certainly facilitates the establishment of discriminatory filters that stress the marginalization of foreign workers. Moroccan immigrants, in particular, are

Figure 15.2 Greenhouse horticulture west of Almería, Andalucía, Spain – the workplace of many Moroccan immigrants (*source:* Tim Unwin, 16 January 1995).

continued

amongst the most rejected by Spanish society due to their different customs and the long history of relations and rivalries between the two countries, which has helped to conjure up the negative stereotype of the 'Moor' in the Spanish imagination (see Chapter 4).

Moroccan immigrants

Morocco, situated on the western edge of the Arab world, the subject of constant contact with and influence from Europe (particularly France and Spain) and having markedly different ethnic groups, is a heterogeneous country containing many diverse ingredients. Bearing in mind the limited nature of its economy, which in the last decade has undergone a critical adjustment, the long tradition of emigration has become the only means by which a large part of Morocco's population are able to carve out a future for themselves. This notion has filtered down through society and helps to explain why Moroccans desperately cling to the idea of leaving the country as the only opportunity open to them.

The Moroccan community in Spain partly reflects the heterogeneity of their country of origin. Although there is a predominance of northerners (from the former Spanish protectorate), especially from the Rif (60% of whom come from rural areas), there has been a gradual diversification of origins, stretching to Oujda and the urban centres of Casablanca, Kenitra, Fez and Marrakech (Lopez Garcia, 1996). This implies a considerable variety of languages (the Berber language spoken in the Rif is very different from Arabic) and of customs (rural/urban, north/south, Berber/Arab), which limits social cohesion and restricts the immigrants to a network of local acquaintances and family members, together with the odd friend they may make as a result of emigration. Consequently it is difficult for Moroccan immigrants' associations to work well or to be fully representative.

Both the educational level and the professional qualifications of Moroccan immigrants are generally low, although even university graduates find themselves destined for precarious jobs (and often at a disadvantage), at times working illegally in domestic service, in small construction companies, in the kitchens of bars and restaurants, as travelling salesmen or in intensive farming. All of this implies not only unstable working conditions, but also a lack of legal stability closely allied to the need for a formal contract of work, which is difficult to obtain (due to the very nature of the sectors in which they are employed) and almost impossible to maintain for any length of time.

The distribution of Moroccan immigrants in Spain is closely related to their quest for employment: they are concentrated in the two largest cities, Madrid and Barcelona (in common with immigrants from other countries), along the Mediterranean basin and in a few inland enclaves where labourers are needed for intensive farming. Given that immigrants of African origin are the only ones to be employed to a significant extent in the agricultural sector (probably the hardest one), it may be said that this factor makes their dispersion, while not excessive, greater than that of other immigrant groups from the less developed countries.

At a local level their scarcity of resources and the resolute aim eventually to return home means that they look for the cheapest accommodation, which leads them to settle in run-down inner-city areas, in neighbourhoods of former emigrants from the rural exodus, or in shanty towns on the outskirts of cities where work may be found. Only those who have settled longest acquire accommodation in dormitory towns around the great cities. Farmhouses or half-abandoned cottages are often rented to them in rural areas, which sharpens their physical segregation and impedes relations with the native population.

As this immigration is a relatively recent phenomenon there is still a notable predominance of young workers, with a majority of males, although a marked presence of women (generally from an urban background) is worth pointing out. The latter have their own migratory routes to cities like Madrid or Malaga, where they find employment in domestic service. This is a clear reflection of social change in Morocco with respect to the relevance of women's salaries, which is forced, to a great extent, by the economic climate. The advantage is that, as

continued

regards integration into Spanish society, women have more opportunity to get to know people and to learn about Spanish customs. However, this has brought about a perverse effect: the fact that a woman does not work outside the home has come to be regarded as a symbol of luxury and social well-being (Ramirez and Gregorio, 1994), with all the connotations that this implies and the desire to emulate that it produces.

The regrouping of families takes place quite rapidly once a residence permit has been obtained. Consequently, since 1992, schools in areas with an immigrant population have witnessed a yearly increase of Moroccan pupils in a range of different age groups. This requires a considerable effort of adjustment at a time when there is a scarcity of specific teaching materials and when the number of support teachers is on the decline due to spending cuts.

Conclusion

As Spain becomes increasingly integrated within Europe, the country has had to face new challenges that are characteristic of more developed countries. Typical of these is immigration from poorer countries. Three key aspects of this question require attention. First, from the political viewpoint, immigration is a particualrly inflammable subject, since it can be used in the medium term by small, extreme right-wing parties as one of the only ways in which they can be seen to be of relevance in Spanish policy making. At the present, there is a wide consensus between the main political parties in Spain not to use immigration as a political missile at the national level, but this agreement sometimes breaks down at the local level. Second, segmentation of labour markets permits the creation of new occupations, mainly in the service sector, and actually becomes essential to maintain the competitiveness of the horticultural sector. This reinforces the third point, which is that Spanish society needs to determine the sort of model within which these minorities can be integrated. At the present there is a pressing need for people to learn how to live together with those from other cultures, and to respect them. Mechanisms to favour inter-ethnic relationships and to promote the settlement of migrants need to be encouraged as a way for better understanding each other. Some programmes are being developed in order to encourage the creation of a less divided society and to prevent the association of some ethnic groups with an underclass. While some NGOs are beginning to play an active role in this field, it is the whole of Spanish society that will have the last word.

Further reading

Gimenez, C. (1993) *La Inmigración Extranjera en Madrid*, Consejería de Integración Social de la Comunidad de Madrid, Madrid

Gozvalez Perez, V. (ed.) (1995) *Inmigrantes Marroquíes y Senegaleses en la España Mediterránea, Generalitat Valenciana*, Conselleria de Treball i Afers Socials, Valencia

Lopez Garcia, B. (ed.) (1996) *Atlas de la Inmigración Magrebí en España*, Universidad Autónoma de Madrid y Ministerio de Asuntos Sociales, Madrid

Misiti, L., Muscarà, C., Pumares, P., Rodriguez, V. and White, P. (1995) Future migration into southern Europe, in Hall, R. and White, P. (eds) *Europe's Population: Towards the Next Century*, UCL Press, London, 161–89

Pumares, P. (1996) *La Integración de los Inmigrantes Marroquíes: Familias Marroquíes en la Comunidad de Madrid*, Fundación La Caixa, Barcelona

Rodriguez, V., Aguilera, M.J. and Gonzalez Yanci, P. (1993) Foreign minorities from developing countries in Madrid, *GeoJournal*, July, 293–300

either seeking permanent settlement there or intending to treat this region as a stepping stone to onward movement to EU member states further north. Italy and Spain have formed the primary destinations for migrants from Africa (see case study) while Greece has been coming under growing pressure from the Middle East and other parts of Asia (Champion, 1994a).

Within-country migration patterns have also undergone radical changes over the past three decades (Champion *et al.*, 1996a; Rees *et al.*, 1996). The early post-war period saw most of Europe still dominated by rural–urban migration, but this pattern had begun to fade in north-west Europe and Italy by the 1960s and in other parts of southern Europe by the 1970s. In its place, other types of migration have grown in volume, as society has become more mobile and people have increasingly been adjusting their housing and broader location in order to suit the stage reached in their life courses and to match the overall greater variety of their lifestyles.

Four sets of changes in people's internal migration behaviour should be highlighted. Retirement migration has become more common, with the introduction of portable pensions and with more people owning their home and being able to release equity by exchanging it for a smaller one in a cheaper area away from the main cities and their commuting zones. Many more young people are leaving home in their teens and early twenties, often well before contemplating marriage and in many cases moving over some distance in order to benefit from higher education and training or to take advantage of better job opportunities. The rapid growth of cohabitation, divorce and remarriage also constitutes a major new force for residential mobility. Finally, families with children, which have traditionally formed the backbone of the suburbanization process, have been moving further away from city cores, as longer-distance commuting has become easier, as services have improved in smaller towns and more rural regions and as work has decentralized.

As with international migration and fertility, there has been a strong temptation to equate these developments in internal migration with the arrival of the 'post-industrial society', both in economic and social terms. Economic restructuring has led older mining and heavy industry areas to suffer an increasing haemorrhaging of population in favour of new commercial and service-oriented places, particularly those situated in western Europe's 'golden triangle' and in national 'sunbelt' zones. Quality-of-life considerations have risen to the fore as more people have been 'voting with their feet', leaving behind the costly, congested and deteriorating conditions of the larger and older cities and thereby contributing to the so-called 'counterurbanization' phenomenon.

Despite this 'post-industrial' tag, however, there exists great uncertainty as to whether these recent developments in migration will continue. With respect to international migration, there appears to be much political support for a 'Fortress Europe' approach designed to minimize the scale of new arrivals from less developed parts of the world. As regards internal migration, the revolutionary interpretation of events has been somewhat dented by the well-publicized 'big city rebound', with the slowdown or even reversal of net migration from city to countryside in the 1980s. Nevertheless, these migration trends have already led to major changes in Europe's population geography.

Changing population geographies

The past quarter of a century of developments in mortality, fertility and migration patterns has had a profound impact on the populations of European countries and regions. Moreover, irrespective of any future trends in these three basic demographic components, these developments have put in train a variety of consequences which will have repercussions lasting well into the 21st century and raise a whole series of important policy issues.

Population size

The single most impressive feature of the 'new' Europe in population terms is its low level of growth and what this means in terms of the continent's place in the world. The key feature of Europe (including all of Russia) in the mid-1990s is that, overall, there were slightly more deaths than births, giving a natural decrease of 1 per thousand in 1995. The long-term rise in longevity in most of Europe has increasingly been offset by population ageing at the same time as the number of births has plummeted. Despite net immigration from the rest of the world, the total population was virtually stationary at this time – truly a situation of 'zero population growth'. By contrast, the rest of the developed world was gaining some 0.7% extra people a year, and the less developed world (including China) an extra 1.5%.

The latest projections reveal the impressive long-term implications of this situation (Table 15.3). Mainly because of age structure changes, Europe's population is forecast to rise by some 17 million people between 1996 and 2010, an overall increase of 2.3% over a period when the rest of the more developed world will have seen its population grow by almost 10% and that of the less developed world by 25%.

Table 15.3 Europe's projected population, 1996–2025, in global context.

Area	1996		2010		2025		% change	
	m	%	m	%	m	%	1996–2010	2010–25
Europe	728	12.6	745	10.7	743	9.1	2.3	–0.3
Other more developed	443	7.7	486	7.0	525	6.4	9.7	8.0
Total more developed	1 171	20.3	1 231	17.7	1 268	15.5	5.1	3.0
Less developed	4 600	79.7	5 742	82.3	6 925	84.5	24.9	20.6
World	5 771	100.0	6 974	100.0	8 193	100.0	20.8	17.5

Source: calculated from Population Reference Bureau (1996).

The outcome is a continuation of the precipitous decline in Europe's share of global population. In 1950 Europe (*excluding* the former USSR) accounted for almost one in six of all people in the world, but this had fallen to below one in 11 by the mid-1990s, and by 2025 it will be down to around one in 16. Even with all eastern Europe and the whole of Russia included, the 'new Europe' is likely to comprise just 9.1% of the world's population in 2025, down by more than one percentage point per decade from its 12.6% level in 1996.

The various European countries fare differently in this process, mainly according to their current levels of fertility. Albania forms an extreme case, with around a 40% increase in population forecast for the 30-year period to 2025. Besides this, the clearest contrast is between most countries of northern and north-west Europe and those of southern Europe. While the Netherlands, Norway, Sweden and France should grow by 9% or more over the period and the UK by 6%, Spain is likely to shrink by 12% and Greece and Italy by around 5%. Eastern Europe presents a much more diverse picture, with overall growth of 4–6% forecast for the Ukraine, Poland and Russia and a decline of 6–9% for Bulgaria, Hungary and Latvia (Joshi, 1996).

Demographic structures

Though there is some uncertainty about the future size of European populations, the direct impacts of the likely changes in numbers of people are modest compared with those arising from changes in the structure and composition of the population. The three aspects of population change described in the first half of this chapter, while in combination producing what is virtually zero population growth, have brought about a fundamental restructuring of the population over the past three decades and have set in motion a pattern of demographic development which appears to contain a high degree of inevitability. The three most important aspects relate to age structure, household composition and ethnic complexion.

The 'greying' of the population is considered by some to be one of the most important changes to be affecting European nations as they move into the 21st century (Kennedy, 1993). Certainly, the statistics on

Table 15.4 Proportion of the population aged 65 years or over, and elderly dependency ratio, 1950–2025.

Area	1950	1970	1990	2010	2025
% aged 65+					
Europe total	8.7	11.4	13.4	16.1	20.1
Northern	10.3	12.7	15.5	16.1	19.8
Western	10.1	11.4	13.4	16.1	22.3
Southern	7.4	9.9	12.7	16.3	20.0
Eastern	7.0	10.4	11.3	13.5	17.6
Dependency ratio					
Europe	13.2	17.9	20.0	24.3	31.7

Notes: Areas of Europe are arranged in order of percentage of persons 65+ in 1990. Dependency ratio refers to the number aged 65+ per 100 persons aged 15–64.

Sources: UN (1992), Warnes (1993).

the growth of the aged population are impressive (Grundy, 1996). As shown in Table 15.4, the proportion of Europeans aged 65 years and over has been rising steadily over the past four decades, being half as much again in 1990 as its 8.7% level in 1950. Moreover, this growth is projected to accelerate as the 'baby boomers' of the 1950s and early 1960s move into retirement age. It can also be seen that the phenomenon is widespread across Europe, though not entirely uniform, the biggest increases being for western and southern Europe (reflecting the very low birth rates there) and the smallest for northern Europe.

This ageing of Europe's population has at least three major consequences. First, it increases the costs of health treatment and personal care (see Chapter 18). It is particularly significant that the numbers of the very old, who impose the greatest costs per capita, are set to double or, in the case of those aged 90 and over, almost triple between 1995 and 2025. Second, there is the problem of raising money for people's pensions, given that an increased pensionable-age population will have to be supported by a shrinking number of people of working age. The dependency ratio (defined as the number of people aged 65 and over per 100 people aged 15–64 years old) had already risen from 13.2 to 20.0 in the 40 years to 1990, but is projected to reach 31.7 by 2025. This represents a massive redistribution of economic, social and political power which governments are currently grappling with. It also raises questions about the extent to which grown-up children will have to choose between a job and looking after an elderly relative, and what impact this will have on the number of children they themselves are prepared to rear. Third, and particularly important for employers, is the ageing of the labour force. Much of western Europe has already experienced a large downturn in number of labour market entrants, but the next two decades will see a reversal in the balance of younger and older workers from around 1.2 20–39-year-olds for every 40–59-year-old in the early 1990s to 1.2 40–59-year-olds for every 20–39-year-old around 2015 (Green and Owen, 1995).

Household composition is now radically different from the past in many parts of Europe and is expected to alter further. Three major changes have taken place since 1960 and can be illustrated by reference to EU averages (CEC, 1996). By 1995 one-person households, including both the elderly and younger people, accounted for one in four households, at 26% being double the level in 1960. Over the same period lone-parent families increased as a proportion of all families (households with dependent children) from 7% to 16%. As a result, by 1995 the traditional household type of a couple with children made up only just over one-third (37.5%) of all households in the 15-country EU.

These household changes have already raised many policy issues. The large increase in the proportion of elderly living alone accentuates the problems of caring for the elderly in terms of social services and income support. Lone-parent families often face a difficult choice between finding gainful employment and devoting more time to raising their children and, given that the majority of lone parents are women (either unmarried mothers or divorcees with custody of the children), will be bringing home well-below-average wages. Most impressive of all is the overall effect on the average size of households, which is currently falling at around 7% per decade: from 2.8 to 2.6 persons per household between 1980/81 and 1990/91 for the EU15 (CEC, 1996). Among other things, this trend has enormous implications for housing requirements and therefore on the scale of urban land take.

All the signs point to a continuation of household fission. In England, for example, it is currently projected that the number of households will have risen by 4.4 million between 1991 and 2016, an increase of 23% compared with a rate of population growth of only 8% over the same period (Department of the Environment, 1995). Across Europe, there are many countries which have seen greater changes in household composition than the UK and others that have not reached that stage yet, indicating the potential for further movement. As shown in Table 15.5, Sweden appears to represent one extreme, with one-person households accounting for two out of every five households and couples with children only 22%, while Greece, Portugal and Spain – despite their very low fertility rates – seem to retain much more of an 'extended family' system at present, with around half of all households containing a couple with one or more child(ren).

Third, the ethnic composition of Europe is becoming more diverse, notably through immigration but also as a result of subsequent family building. The European 'guestworker' system, labour recruitment from former empires and the 'post-industrial wave' of international migration have all contributed to a significant growth in numbers of foreign residents. Particular issues are raised by the arrival of less

Table 15.5 Houshold composition (%) in selected countries, 1990/91.

	One person	Couple with child(ren)	Lone parent with child(ren)	All with children
Sweden	39.6	21.8	3.9	25.7
Denmark	34.4	26.3	5.8	32.1
Norway	34.3	30.8	8.2	39.0
Germany	33.6	30.5	6.3	36.8
Switzerland	32.4	32.1	5.1	37.2
Finland	31.7	26.2	4.1	30.3
Netherlands	30.0	33.5	6.3	39.8
Austria	29.7	35.3	8.1	43.4
Belgium	28.4	35.7	9.2	44.9
France	27.1	38.1	7.2	45.3
United Kingdom	26.7	32.9	9.0	41.9
Luxembourg	25.5	38.4	7.9	46.3
Italy	20.6	46.7	8.5	55.2
Ireland	20.2	47.9	10.6	58.5
Greece	16.2	49.1	6.0	55.1
Portugal	13.8	49.9	6.8	56.7
Spain	13.4	55.8	8.2	64.0

Notes: Countries ranked in order of proportion of one-person households.

Sources: Eurostat (1996).

skilled immigrants and of those of non-white racial origin and non-European culture. These problems are often exacerbated by the uneven spatial patterns of immigrants' destinations, traditionally in areas needing low-wage labour in declining industries and more recently in the largest cities, where cheaper housing is more plentiful and job openings are expected to be more numerous (White, 1993b; Champion, 1993; Champion *et al.*, 1996b).

Geographical distribution

The geographical patterning of population growth and decline across Europe has undergone some substantial changes during the post-war period, but these lack consistency and provide a rather weak basis for anticipating future trends. One important development has been the convergence of birth and death rates across most of Europe, not only at national level but also intra-regionally within countries. In particular, the higher fertility of rural areas is largely a thing of the past, as family size in traditional rural families has fallen and urban influences have been spreading into the countryside, while the birth rate of many urban areas has been boosted by the above-average fertility of recent immigrants. As a result, the inter-regional range of natural change rates is much smaller now than in the past, and migration has become relatively more important in accounting for regional differences in overall population change.

The most remarkable, yet also most controversial, change in population distribution trends of the post-war period has been the switch from urbanization to 'counterurbanization'. Strong urbanization tendencies prevailed between 1950 and 1970, with the proportion of people residing in 'urban places' in Europe as a whole growing from 56.2% to 66.6%. Since then, while the proportion has continued to rise, it has done so at a considerably slower rate, reaching 73.4% in 1990 (Champion, 1995). More importantly, in contrast to the earlier trend for the urban population to become more concentrated in the larger cities, people have been redistributing themselves down the urban hierarchy. By the early 1970s for north-west Europe and by the 1980s in southern Europe, the traditional positive relationship between net migratory growth and city size had become weaker or had been replaced by a negative one (Fielding, 1982; Champion, 1989).

More recently, however, this phenomenon appears itself to have waned, if not gone into reverse (Champion, 1995). The evidence for Great Britain shows that net migration losses from London and the other major conurbations actually peaked in the early 1970s, shrank steadily for a decade and then experienced a marked, but rather short-lived, recovery in the mid-1980s (Champion, 1994b). France saw a strengthening of its deconcentration relationship between the two decades, but Norway and Sweden had resumed their traditionally strong urbanization pattern by the 1980s. Several other countries, notably Austria, Belgium and the Netherlands, experienced the switch back to urban concentration between the first and second halves of that decade (Champion, 1995). Indeed, the overall result of these changes was that, for Europe as a whole, the period 1980–89 was characterized by a markedly lower level of net migration between cities and regions than in previous decades and by the lack of any dominant spatial dimension of population redistribution (Champion *et al.*, 1996a).

Various ideas have been put forward to make sense of these patterns and trends (Champion, 1995). The 'anomaly' explanation gives prominence to period-specific effects, though there is no universal agreement about whether it was the special events of the 1970s that temporarily arrested the traditional urbanization process or those of the 1980s that interrupted the movement to a new 'post-industrial' pattern of more dispersed settlement. The 'cycle' approaches include those who believe that the recent ups and downs in urban concentration and deconcentration tendencies can be related to fluctuations in economic activity, birth rate and other factors, as well as those who conceive of urban regions and wider urban systems as having their own 'life cycles' involving a 'reurbanization' stage. In a broader theoretical perspective, it is suggested that there are two major sets of processes at work, one relating to the restructuring of economic activity and influencing migration principally between macro-regions, the other involving deconcentration from larger urban centres to less congested areas mainly within these broad regions and being promoted by the search for more pleasant and cheaper conditions by both residents and employers.

Given the lack of agreement on the causes of past trends, it is not surprising that there is little consensus about the future population map of Europe. Indeed, the main attempts at looking forward have been in the form of alternative scenarios. For instance, Kunzmann and Wegener (1991) have discussed whether the 'banana' or the 'grape' analogy is the more appropriate, comparing the likelihood of Europe being dominated by growth in a single 'core' zone extending from south-east England to north Italy as opposed to a patchy map of growth and decline resulting from competition between individual city-regions across the continent. Rees (1996) has produced a series of regional population projections based on alternative assumptions about future migration patterns, notably a 'growth regions' scenario in which net migration is related to current economic strength measured in terms of production per capita, and a scenario of (counter)urbanization based on the relationship between net migration change and regional population density. Masser *et al.* (1992) have adopted a purely qualitative approach which attempts to visualize the impact of three fundamentally different political stances stressing respectively economic growth, regional equity and environmental sustainability.

Whatever transpires, it is sure to have important implications for society and governments. A lack of economic potential and population dynamism continues to plague the more remote rural areas, particularly in rapidly restructuring eastern Europe. By contrast, the more accessible rural areas, together with small and medium-sized towns and cities in Europe's varied growth zones, have for some time been experiencing strong net in-migration and have been subject to patterns of new development which may not prove sustainable in the longer term. Probably the greatest challenge is posed by the larger cities and older mining and heavy industry conurbations, where the large-scale programmes for urban regeneration and associated regional development operated up to now appear in most cases to have had rather limited success in staunching the outflow of better-qualified residents (Cheshire and Hay, 1989). While continuing to wrestle with the serious economic problems caused by deindustrialization, many of these cities have not benefited greatly from the 'gentrification' phenomenon and have seen the recent waves of immigrants swell their already sizeable low-income populations.

Conclusion

Demographic factors are fundamental in influencing the changing map of Europe. Their role has been particularly evident over the last two decades because of the major alterations taking place in Europe's demography, notably in terms of aggregate effects of changes in individual people's behaviour in relation to living together, having children and moving house. Moreover, compared with the main industrial era, people now appear to exercise more power. They have far greater economic resources at their disposal, they possess a much wider range of labour-market skills, and they constitute a major political force that can, if it chooses, persuade governments to adopt policies other than those dictated by big business. In general, people in Europe now have greater freedom to choose where they live, expecting that private-sector services, high-skill labour employers and what remains of the welfare state will follow them. For the future, the main geographical questions concern whether the steps being taken to produce a more integrated Europe will raise migrants' sights further and how effective will be any reaction to the recent widening of the economic and demographic gulf between Europe and its African and Asian neighbours.

Further reading

Coleman, D. (ed.) (1996) *Europe's Population in the 1990s*, Oxford University Press, Oxford

Council of Europe (1996) *Recent Demographic Developments in Europe 1996*, Council of Europe Publishing, Strasbourg

Hall, R. and White, P. (eds) (1995) *Europe's Population: Towards the Next Century*, UCL Press, London

King, R. (ed.) (1993) *The New Geography of European Migrations*, Belhaven, London

King, R. (ed.) (1993) *Mass Migrations in Europe: The Legacy and the Future*, Belhaven, London

Noin, D. and Woods, R. (eds) (1993) *The Changing Population of Europe*, Blackwell, Oxford

Rees, P., Stillwell, J., Convey, A. and Kupiszewski, M. (eds) (1996) *Population Migration in the European Union*, Wiley, Chichester

CHAPTER 16 Education and welfare

MICHAEL BRADFORD

This chapter focuses on young people, whose development and welfare are of major importance to the futures of all European countries. It begins by comparing Europe's varying education systems and traditions, which are seen as reflecting the complex intertwining of social, economic and political processes over many decades. They also influence access to different forms of work and the social mobility of young people from different social backgrounds. The chapter then explores the variability in social backgrounds and welfare within countries, through an examination of the differences in the proportion of their children living in poverty. The differences between countries partly reflect the response of their welfare systems to recent economic and social changes. The varying gaps between the affluent and the poor also pose differential problems for the education systems in the extent to which the influence of widely varying social backgrounds will have to be offset within the schools.

One of the recent economic and social changes which has consequences for the relationships between school, home and work is then considered: the growth in both parents being in paid employment. Their needs and the needs of an increasing number of lone-parent households have led to a growing demand for childcare for school-age children. Recent changes in this welfare area are examined in depth for three case study countries showing the varying responsibilities that are taken by schools, child-care institutions and parents. This is a welfare area which is highly geographical as care is relocated from home to schools or to care institutions, and another set of non-family adults have an influence on the emerging identities of the children. The extent and way this service is provided in different countries present revealing insights into their educational and welfare systems and more widely their attitudes to children and the role of the local and national states.

Different education systems and traditions

There are major differences in the educational structures of European countries which reflect the playing out of tensions between a number of social, cultural and political processes (see, for example, Brock and Tulasiewicz, 1994). There are, for example, tensions over the extent to which an education system should be oriented towards universal rather than élite education. Where the focus is on élites, which usually emerge even in a universal system, there are differences in how the élite is viewed. Some education systems place more emphasis on the promotion of individual social mobility, while others advance the collective improvement of particular social groups. There are tensions over the extent to which education on morals and social values should be provided publicly in schools or privately in the home by the family or through other institutions. There are tensions between the extent to which education should be organized and decided upon locally as against centrally. The varied education systems of European countries reflect different histories and philosophies towards education and to society. Recent approaches and priorities represent some general trends, such as decentralization to involve parents, while again revealing important differences. These approaches and priorities have become interwoven with past structures to produce the present mosaic which characterizes European education systems.

Education in England, for example, has been much more oriented to the production of an élite, historically through its private schools, which, in contrast to nearly all other industrialized countries, still set the tone for the rest of the education system. The selection of the élite and the norm of social mobility was not seen to be the open contest characteristic of North American education but, as Turner (1971) called it, a 'sponsored norm' whereby élite recruits were chosen by the established élite and its agents. In this case, a few pupils were identified by teachers as inherently worthy of joining and reproducing the established élite. The tripartite system of schools (grammar, technical, secondary modern) introduced after the Second World War changed the composition of the élite in a more meritocratic way to include very able children from all social classes in its grammar schools through selection at 11. In the 1960s and 1970s the introduction of comprehensive schools, and later a single set of exams at 16, ostensibly changed the orientation towards universal education. The same sets of examinations, however, still operated at 18. Recently, tiering has been reintroduced at 16 for many subjects, which divides young people and classifies them into two streams before they enter the exams. Outside observers suggest that there is still a tendency in schools to dismiss large numbers of pupils as though they did not have any capacity for higher education. The superimposition of neighbourhood comprehensive schools onto socially segregated housing areas in cities also meant that the anticipated collectivist upgrading of whole social groups did not occur, and access to educational qualifications in general, and specifically to higher education, which expanded rapidly in the early 1990s, showed for 1988 to 1996 a growing gap between the richer and poorer areas of the country (Walker and Walker, 1997).

In France, universal secondary education was not accepted until 1947 and not implemented until after 1959 (McLean, 1995). France has a much more collectivist, state-oriented approach to education than England. The mass and élitist elements of its system began to merge when in 1989 it was decided that 80% of 18-year-olds should enter university. Its universal component is exemplified by the minimum standards required at various levels. Its élitist component is reflected in the high failure rate at all levels. The selection of its élite follows more the collectivist tradition whereby the most suitable people needed to lead society are chosen, rather than individuals being rewarded for their success.

The egalitarian orientation to Swedish society is illustrated in its early adoption of common secondary schools in 1947 (see Swedish case study). Elsewhere in western Europe selection between the ages of 10 and 12 prevailed into the 1960s (McLean, 1995). By 1975, however, only Germany, the Netherlands, Austria and Switzerland retained largely segregated schools. Educational segregation did not superimpose itself onto income groups as closely as in England, and the reputations of types of school were not in any way so stigmatized as in England. In the Netherlands the more vertical (pillared) rather than horizontal structure of society is reflected in the structure of schooling with Protestant, Catholic and non-confessional public schools. The separation of funding and delivery may also be important here. In the Netherlands all schools are state-funded with teachers' salaries the same in each type of school, but the religious schools are privately organized and run. Even here, though, the recent association of some immigrant groups, particularly the low-achieving Surinamese, with the secular public schools in some cities has contributed to a more horizontal stratification of schools in these areas.

The degree to which public education takes over the role of the family is partly demonstrated by the length of the school day. Germany and Scandinavia have much shorter school days, with schools open, until recently, only in the morning. Familial education is left for the afternoons. In France and England there has always been a tradition of fulfilling some of the familial roles in schools. In England, school starts early at five years of age or younger, and the 5–7 infant schools have been dominated much more than in Germany and Scandinavia by teachers acting as substitute parents. The educational culture has been one of replacing what have sometimes been viewed as incompetent parents, so that moral and social education can be provided. Even the English private schools for the wealthy were historically also places where children boarded and learned moral and social codes, which in other countries were learned in the family home.

There are major organizational differences between countries which are prominent in any study of comparative education (McLean, 1995; Broadfoot, 1996). In general, France, southern and eastern European countries have been much more centrally controlled (Tomiak, 1986a; Brock and Tulasiewicz, 1994). The central state has been identified with progress as the leader of economic and social change. Education is seen as one of the ways in which society is transformed. This view of education and the state is deeply embedded and it still affects relationships between teachers, students, parents and employers. Even more recent local organizations such as School Councils in France, where parents have a greater say over decisions affecting the schools, are fully integrated into the state framework and have not become the focus

Swedish welfareism – a model in question?

Roger Andersson

In the 1930s, just after the Social Democrats came to power in Sweden, bringing to the political agenda issues concerning economic and social justice, the American journalist Marquis Childs (1936) described Sweden as '*The Middle Way*'. Since then a great number of politicians, journalists and researchers with leftist sympathies have both devoted great attention to Sweden and admired its achievements. Childs's book was the starting point for a long-lasting international interest in the 'Swedish model'.

However, this model has been given rather different meanings by different people. In its broadest sense it signifies a political culture characterized by consensus and cooperation connected to an ideology of rationalistic social engineering. Those who would like to narrow the use of the term often want to restrict it to some basic features of Swedish labour-market institutions. In the latter case it was two circumstances that used to draw attention to Sweden: first, the existence of an active labour-market policy, where the state effectively prioritized attempts to find new jobs in new sectors for redundant workers, instead of having to pay cash for the unemployed; and second, the overall consensus between the organized employers and the strong and well-organized labour movement. For several decades, the state played a passive role in the wage-bargaining system, for example. Wages were settled at the national level and without many strikes or other labour conflicts.

Although the 'Swedish model' stands for something more than a notion of a specific type of welfare state, it is useful to approach the Swedish welfare state via the notion of a specific Swedish model, not least because the ongoing changes of the role of the state relate, directly or indirectly, to a crisis for the entire Swedish model, in both its broad and its narrow meaning.

The central wage-bargaining system in Sweden has more or less collapsed over the last ten years. This collapse was initiated by the Swedish Employers' Organization (Svenska Arbetsgivareföreningen, SAF), which wanted wages and other benefits and responsibilities to be settled closer to the level of the individual firm. Although many trade unions did not agree with this idea, they soon found out that there was no partner to discuss and compete with at the central level.

These changes brought with them a stronger element of individualism in the former collectively organized Swedish working life. One reason for employers wanting to have a less centralized bargaining system was the increased possibility for firms to introduce more flexible forms of production and to differentiate wages between different groups of employees. In this context, representatives of private business put forward the argument that small nominal wage differences due to some 30 years of struggle for a 'solidaristic wage policy', and high income taxes, reduced incentives for employees to work hard and to improve their levels of performance.

As a result of the combined effects of a series of badly timed and perhaps ill-conceived reforms in the late 1980s – notably the deregulation of financial markets and an underfinanced tax reform – Sweden experienced a series of economic disasters in a very short period of time: the housing and real estate market collapsed, as subsequently did the financial system; unemployment quadrupled from 2% to 8%; the state's budget deficit exploded; and the Swedish krona lost 30% of its value compared with the major European currencies.

Although somewhat questioned before, predominantly from the political right, the Swedish welfare state came under severe pressure, and social benefits had to be cut in order to bring the state's budget into better balance. Not one single part of the welfare system has since been left unchanged, and the complexity of these changes together with the high speed of reform make it difficult to evaluate the effects for the ordinary Swedish citizen. But in order to shed light on some of the more important aspects of these changes, it is necessary to take a closer look at the structure of the Swedish welfare state.

Therborn (1991) identifies several types of welfare state by asking three questions:

1 How is the system of social rights and welfare provision organized in terms of institutions (i.e. the role of the state itself)?

continued

2 How are the social rights allocated with respect to the citizens (general or selective allocation)?
3 What is the extension of these rights (i.e. their social and economic importance)?

According to Therborn (1991), describing the situation in Europe at the end of the 1980s, Sweden and Denmark were the only two welfare states characterized by the following combination: the state (or more correctly, public bodies at different administrative levels) had almost exclusive control of the finance and provision of social benefits and social services; social rights were primarily of a general character and they were therefore not as in many other European countries designed for special groups of citizens; and the politically controlled welfare sector was very extensive, with social expenditures reaching about one-third of the GNP. Although Belgium, France and the Netherlands had similar levels of social expenditures, these countries differed quite substantially from Sweden and Denmark with respect to the other two dimensions. During the 1990s, the measures taken by the Swedish government and the Parliament have primarily affected the first and third dimension (the role of the state and the level of social expenditures).

While healthcare for children and the elderly, as well as education, used to be almost exclusively within the public-sector domain, several measures have recently been taken to open up these social services to competition. It is today more common – at least in the major cities – to find private alternatives within all these subsectors, although political control remains in terms of finance (i.e. the local state now often buys these services on contract from publicly or privately owned service providers). But money allocated for the provision of these social services has also been cut back, making many predominantly female employees redundant within day nurseries, hospitals and public service houses for the elderly. Productivity has thereby increased rapidly, but few would say that the service quality has remained as high as it used to be. And the public sector still has to support these newly unemployed, but now without receiving anything back from them. There is therefore a vivid political debate in Sweden about how to find new solutions, whereby people's demand for (public) social services could be met without increasing budget deficits at the central and local level and without having to increase taxes even further.

The downsizing of state expenditure in the 1990s has also come to affect the level of cash transfers to the households. Pensions for the elderly have thus been reduced. As a result of this, and also because of a proposed plan for a radical reform of the pension system, many people have started to save in private pension funds. To become old, sick or unemployed did not mean very much to the family economy during the 1980s, but now it means a substantial reduction in economic standard of living (20–40%). The general cash transfers to families having children aged 0–16 or to parents on parental leave (12 months) have also decreased, as have almost all other social benefits. Although tax levels have been and are still high in Sweden – only Denmark is comparable in this respect (both countries having more than 50% of their GNP passing through the public sector, according to 1995 OECD statistics) – the majority of the population seems to accept this as long as social security payments are kept at a high level. However, many people's confidence in the public sector and in the political sphere has been affected by these changes. As a consequence, those who can afford to save more money increase their private savings. Domestic demand for consumption is still very low, with signs of a 'dual' or 'two-speed' economy becoming increasingly visible since 1992. Firms within the export-oriented sectors of the economy have made huge profits, while sectors producing for final consumption in Sweden have faced great difficulties due to the majority of the population's decreased purchasing power. Unemployment remains at a European, not a 'Swedish' level. For many people, and in particular women, those living in the north of the country, and others who have been extremely dependent upon the welfare state both as a service provider and as an employer, Sweden's membership of the European Union in 1995 meant both a confirmation of a process where Swedish politicians seemed to have given up the Swedish model, and a big disappointment in terms of the possibilities of re-establishing the 'old and good' welfare state. As pointed out by Ekstedt *et al.* (1995), the Swedish welfare state had become a strong element of the country's national identity

continued

for a whole generation of many ordinary Swedish people. What we seem to have witnessed in contemporary Sweden is therefore an identity crisis, and this is of a much more profound and long-lasting nature than the question of a few percentage points up or down in the cost of the welfare schemes. It remains to be seen if the present Social Democratic government (1994–98) succeeds in its rather radical efforts to re-establish the balance of the state budget by raising taxes and cutting back benefits, while at the same time keeping up standards in the production of services. At present, it seems that this could be easier than for many Swedes to recover from their loss of faith in the country's own middle way between socialism and capitalism.

Further reading

Dahlström, M. (1993) *Service Production. Uneven Development and Local Solutions in Swedish Child Care, Geografiska Regionstudier*, **26**, Department of Social and Economic Geography, University of Uppsala, Uppsala

Esping-Andersen, G. (ed.) (1996) *Welfare States in Transition: National Adaptations in Global Economies*, Sage, London

Korpi, W. (1991) *The Development of the Swedish Welfare State in a Comparative Perspective*, Swedish Institute for Social Research, The University of Stockholm, Stockholm

Marklund, S. (1988) *Paradise Lost? The Nordic Welfare States and the Recession 1975–1985*, Arkiv, Lund

for local opposition to the state system (McLean, 1995). In the 1970s, there was some regional decentralization in France with the 25 regions now controlling about 20% of total expenditure, but it is still within the overall centralized framework. In Sweden there is also a strong centralist collectivist tradition, but here too there has been some recent decentralization with local education management units, which control three or four schools, having considerable autonomy (McLean, 1995).

In England, in contrast, there has been a history of much greater decentralization with many powers being vested at the local government level. Only in the late 1980s was there a major move towards centralized state powers through the introduction of a national curriculum, the removal of some powers from local education authorities and the central direct funding of newly created grant-maintained schools (Bradford, 1995). Even this was accompanied by greater powers for the governing bodies of schools, again at the expense of those of the local education authorities. Indeed, some authors have argued that there has been a 'hollowing out' of the local state with its powers going upwards to central government and downwards towards schools, particularly to head teachers and governing bodies (Bradford, 1995).

Elsewhere there are midway positions. In the German federal system, the *Länder* have many powers (Phillips, 1995). In the Netherlands state and church schools coexist, while in Norway and Denmark elementary schools are locally organized and secondary schools are organized by the central state.

While there have been some important decentralizing tendencies, even in countries with centralist systems, there have also been some marked increases in central power, in part through the increased importance of assessment. This has emerged alongside the more utilitarian approach to education which emphasizes the competitive position of countries in the global economy. Performance indicators are sought that allow comparisons of educational outcomes or processes across countries, so that simplistic associations between educational indicators and economic performance can be made into causal explanations of relative economic failure. There has also been a move within countries to compare pupils and schools on national examination results against norms so that individual progress can be discussed with pupils and their parents, as for example in France and England (Broadfoot, 1996). In some countries such as England this has also been introduced so that schools can be compared. France has so far avoided the national publication of tests, which has been used in England and Wales to promote competition between schools for pupils and, through pupil numbers, resources. France has not yet accepted the assumption that such competition leads to improved standards. National assessment, whether published or not, does lead to increased central powers relative to those of the professionals. The greater control is argued as being necessary for comparability and efficiency as well as being in the interest of the economic success of the country. These powers have been taken on partly to legitimize the public expenditure associated with education and partly to legitimate the state in its role of facilitating economic and social development. Whether these powers are needed and whether they have any effect at all on economic and social progress remain open questions.

The different powers of the central state and centralist traditions influence other important parts of educational and welfare provision. For example, they partly account for the variation in pre-school provision. In France, Spain and Italy pre-school provision is very high. In France pre-school participation was universal by 1993. In contrast in England it is still relatively low, which was the reason for the introduction of a pilot nursery voucher scheme by the Conservative government in 1996 and the priority given by the new Labour government to pre-school provision in 1997. The provision of nursery education is seen as a key component in providing equal opportunities in education because it prepares and socializes children, especially those from poor backgrounds, for schooling. Given the political consensus in England in the 1960s for equal opportunities, it is surprising that this key element was neglected by successive governments, both Labour and Conservative. The location of such decisions at the local rather than the central level partly accounts for this. It also accounts for the great variety in amount and mix (public, private, voluntary) of pre-school provision in England (Pinch, 1984; Ross, 1993). Nowadays it is also regarded as important in providing care while parents, particularly mothers, find paid employment outside the home.

The varying welfare of children

The different educational systems and welfare structures across Europe have responded in varying ways to economic, social and political changes which directly or indirectly affect children (Brock and Tulasiewicz, 1994). There have been some major social changes in the last decades which reflect fundamental economic changes. The next section examines some of these social changes with particular attention to their effects on the position of children.

During the last two decades a major social change has occurred in many countries: a widening of the gap between the most and the least affluent households. This phenomenon has also been associated with a relative increase in the number of children living in poverty (Kennedy *et al.*, 1996). In the UK, for example, the proportion of children living in households with incomes below 50% of the average (after housing costs) increased from 10% to 32% between 1979 and 1991/92. At the same time average living standards increased by 36% (Department of Social Security, 1994). This widening gap is arguably a joint effect of economic change associated with globalization and governments' responses to that change through their social and welfare policies.

The increased number of children living in poverty can partly be explained again by economic changes through the relative increase in the number of households with no or low-income earners, a modern indicator of deprivation (Bradford *et al.*, 1995). It is also influenced by the relative increase in lone-parent households, a significant social change through increased separation, divorce and single-parent families. Such households have more difficult access to the job market and in some countries are receiving reduced welfare benefits. Finally, it is argued that there may be an unequal intergenerational allocation of welfare benefits, such that the increasing number of the dependent elderly are receiving a disproportionate share of welfare benefits relative to the dependent young. There is some evidence that children are at greater risk of poverty than older people (men over 65 and women over 60) in the UK, France and the Netherlands (Kennedy *et al.*, 1996), but the reverse is the case in Luxembourg. In Germany, Italy and Sweden, there is very little difference between the two groups. There would therefore seem to be possible evidence of unequal intergenerational allocation of welfare only in some countries.

There are other important differences between children's welfare in European countries. A comparison between countries using data from the mid- to late 1980s (Luxembourg Income Study) showed that there are major contrasts from Sweden at one end of the spectrum to Italy at the other. In Sweden there are very few children in poverty (4.4% in families with below 50% of average income). Luxembourg (6.1%), Germany (9.3%) and the Netherlands (9.7%) also have relatively few children in poverty compared with France (14.6%), the United Kingdom (16.7%) and Italy (17.6%). Even in these worst western European countries, however, child poverty seems low in relation to the level in the USA (29.6%) (Kennedy *et al.*, 1996).

There are some similarities between countries in that usually the proportion of children in poverty is related to the number of children in the family, the number of parents present, the number of earners and the presence of young children. Sometimes these differences between types of household are very marked. For example, even though Germany has a relatively low overall level of children in poverty, children in lone-parent households are four times as likely to be in poverty than those in two-parent households. While this difference is greatest for Germany, again it is least for Sweden. If the lone parents are not earning, their average income can drop dramatically relative to overall average income, as in France (from 77% to 45%) and Italy (from 87% to 46%). In contrast, in the Netherlands it reduces relatively little (from 75% to 70%), reflecting the country's more generous treatment of this group in its welfare system at that time.

As indicated earlier, the proportion of children in poverty has grown recently in some countries as inequality in general has increased, and it is possible that, if these cross-country comparisons were repeated for the 1990s, they would show even greater differences between countries. These increasing proportions of children in poverty are of considerable concern and they pose difficult questions for education and welfare systems as well as for the future social mobility and social structure of the countries.

The development of school-age childcare

Some of the children in poverty come from households where it is difficult for the parents, usually the mother, to take up employment opportunities because of the lack of care facilities for their children. This is particularly important for parents with preschool age children and, of course, lone parents, but it is also significant for those with school-age children. Such are societies' concerns with the risks that children may be exposed to outside the home that many parents are loath to be absentees in the period between the end of school and the end of their work day. Even though these risks may be exaggerated, there has been a growth in most European countries in the provision of school-age childcare.

The relationships between school, home and work have received relatively little attention from geographers. This section on school-age childcare, therefore, explores one aspect of such relationships. The growing interests in women, and more recently children and identity, make this a particularly rich aspect for geographers to consider. The different and changing relationships between schools and childcare provision among countries adds further geographical interest. The general situation is reviewed first, and then three case studies of Sweden, Germany and France (Petrie, 1996) are discussed in order to show the varying responses to a similar issue which are superimposed upon different existing structures and relationships between schools, childcare institutions and homes.

In the European Union the proportion of employed mothers increased rapidly between 1985 and 1993, rising from 42% to 51% for mothers with children under the age of ten. The level, however, varies greatly between countries, with the proportion of employed mothers (with children under 15) being as high as 76% in Denmark and 70% in Portugal, while only 34% in Ireland and Spain. The prevalence of part-time work, which may be able to fit more closely to school times, varies markedly among countries too, with the Netherlands having a particularly high rate of 88%, the UK 64% and Germany 60%; while the southern European countries have very low rates (less than 20% in Italy, Greece, Portugal and Spain). There is, therefore, a growing demand for school-age childcare in Europe, but the level of the demand varies between countries. The growth of mothers in full-time employment, in particular, has led to a restructuring of the child's day. Children used to leave school and go home to play inside or play with friends in their neighbourhood. Now, for many children there is a substantial period of time between the end of school and the arrival home of a working parent. As a consequence of the provision of childcare facilities to bridge this gap, some of them now spend a significant time with non-family adults, which impacts on the development of their identities.

The length of the school day varies across Europe. It is short in Sweden and Germany. It is relatively long in France. The duration of the period between school and the return home of a working parent thus varies between countries, with consequences for childcare. Regardless of the differences between countries in the length of the school day or the prevalence of part-time working, the increasing numbers of women in the workforce has meant that childcare is also needed during the school holidays.

In some countries, notably Denmark, France, Sweden and some of the Austrian and German states, there has been a strong proactive policy and a high level of provision of childcare for these school-aged children. In other countries, such as Luxembourg, Portugal and Finland, provision and policy are developing. In contrast, in Greece, Italy, Spain and Ireland there is little to note as yet in either provision or policy (Petrie, 1996). Where the development of school-age childcare provision has occurred, it has been mainly in response to the growth of employed mothers, but it also reflects the growing concern about the risks to unaccompanied children in public space. In Sweden school-age care is well-established, as part of a well-developed welfare state (see Swedish case study). The service was developed during the 1970s, mostly for under-tens. In response to the short and irregular school day, the *fritidshem* (free-time home) opened both before and after school and in school holidays. By 1993, 40% of 7–9-year-olds attended a *fritidshem* and a further 12% were in family day care, organized by local authorities (Petrie, 1996). The provision continues to grow. From 1995, local authorities have had to provide facilities for children aged 6–12.

There has been an interesting division between the type of education provided within schools and *fritidshem* (Petrie, 1996). In schools the curriculum is more formal and subject-centred. In the *fritidshem* it is much more child-centred with the name 'free time' giving some clue to the arrangement. They learn in many different ways, with group play being encouraged. With the late starting age of seven and the short day, schools have much work to pack in. The *fritidshem* have 'facilitated expressive activities, play and informal learning, and emphasized the interpersonal and collective dimensions of children's activities in the comparatively long hours which surround the short school day' (Petrie, 1996: 227).

There has been an interesting recent change in the relations between school and childcare. Rather than being separate institutions, the school and *fritidshem* are now being integrated, with the child's time at school being extended and the trained childcarer in the *fritidshem* being involved in lessons in the school. As a result the schools are becoming more child-centred, in contrast to the situation in England and Wales, where, as a consequence of recent reforms, schools are becoming less so. There are still separate roles for the teacher and the childcarer, but the child's day is now more oriented to the school and the difference between the two institutions has become blurred.

In Germany, since unification, there have been a number of changes in school-age childcare (Petrie, 1996). In eastern Germany the number of mothers in full-time employment has declined from a very high level, with an accompanying decrease in the provision of childcare. In most of the western states, however, the characteristic European increase in mothers in employment has been matched by increased childcare provision. This has been mainly through *Hort*, day shelters, over half of which are provided by local authorities. The rest are organized by voluntary groups. Schoolchildren's houses (*Schulekinderhauser*) are a new development on school premises but with their own separate accommodation.

There are many similarities between the Swedish and German systems, but the interaction between school and childcare provision has been rather different. In Sweden there has been a two-way flow between the school and *fritidshem*, with child-centred education going into the schools from the *fritidshem* and the more formal teaching of schools being taken up and developed through activities in the *fritidshem*. In Germany the flow has only been one way, with the activities of the school penetrating those of the *Hort* through homework.

Homework is central to the German education system with every child having some homework as soon as they start school at the age of six. Again a short and irregular day has necessitated education to be provided outside the school. German mothers have been expected to fulfil that role of supervising homework and looking after their children outside the irregular school hours. This has tied them to the house and to childcare, reinforcing their domestic role and their inequality with men (Enders-Dragasser, 1991). This unpaid teaching role for mothers has also extended social inequalities among children within Germany. Those mothers whose first language is German and who have been through the German school system, and those with higher educational achievement, are more able to perform this extra-school educational role. When children attend a *Hort* the mother's responsibilities for homework are transferred to the *Hort*, where it becomes a central task. A similar transfer of responsibilities occurs for the schoolchildren's houses, where there is often greater cooperation with the school.

In contrast to Sweden and Germany, in France the school day is long. Here, recent policy has been developed for children and young people which takes into consideration childcare in the absence of parents, but it is set within a wider context than child protection. The context is broadly educational with provision for cultural, sporting and recreational activities. The child is seen as a member and citizen of the local community. Funding is provided by central government, but implementation is by contracts between central government and local partnerships consisting mainly of voluntary organizations and local authorities (Petrie, 1996). The policy is called 'looking after the life rhythms of the child and the young person'. The underlying belief is that children are rather lethargic in the early afternoon and that it is better to allow them to choose other activities then and return later in the day to more demanding activities. Non-teaching staff may be brought in for these other early afternoon activities or children may be taken out on visits to places of local interest. After the later, more demanding activities, children play and are cared for until a parent arrives to pick them up. In France, then, much more is centred on the school. The policy coordinates the child's educational and leisure activities and provides care services without any other institution being involved. There is no equivalent to the Swedish *fritidshem*. The school is also more oriented to the community, with which it takes responsibility for introducing the child to civil life and its related opportunities.

These three examples show the effects of different social contexts in influencing childcare policies which involve the school. In other countries, such as Austria, Denmark and Norway, there is also a developing relationship between the school and school-age childcare. This is unlike Flanders, where a short-term government initiative used unemployed women, after limited training, to staff childcare centres but discouraged any inclusion of homework in their activities. The link with schools was not developed, with the emphasis being on free play and creative activities. In Flanders two separate institutions deliver two separate curricula, in contrast to Sweden, where different people, now within the same institution, deliver them.

In all three examples discussed here, the domain of the school has been enlarged recently, but much more so in France than in Sweden or Germany. The school has become more important in structuring the time and activities of children. There has also been a transfer of responsibilities away from the parent to the school or to childcare institutions. In Germany this has concerned the overseeing of homework. Children are spending less time in a domestic situation, but in these countries at least, they are not spending proportionately more time in unattended public space where their parents feel they are at risk. The structure of children's lives and the influences on their identity are therefore changing across Europe as countries respond in different ways to global economic forces and social change.

While these examples have mainly been drawn from western Europe, fundamental changes have also been taking place in the relationships between school and home in eastern Europe and the former Soviet Union. As Tomiak (1986b: 1) has stressed with reference to education in the former Soviet Union, 'it is necessary to emphasise that in communist theory and practice the educational system is seen as being at one and the same time an integral part of Soviet society, the Soviet economy and the Soviet political system'. Soviet and east European educational laws and statutes were thus specifically designed 'to produce not only harmoniously developed persons but also determined communists unquestionably loyal to the political leadership in the country, efficient workers and individuals with life-styles sufficiently similar to make social divisiveness impossible' (Tomiak, 1986b: 1). With the replacement of communist ideology by the twin pillars of the free market and liberal democracy, such aims have had to be dramatically overhauled (Kitaev, 1994; Daniels *et al.*, 1995). As Führ (1995: 260) has emphasized in the context of the former German Democratic Republic, 'Replacing the socialist unified school by a flexibly structured school system and restructuring higher education created great problems'. The same is true throughout eastern Europe. As the case study of Petržalka New Town in the Slovak Republic illustrates, it was frequently the case that insufficient attention was paid to the development of schools and welfare services in urban developments that occurred in eastern Europe during the 1960s and 1970s. This has meant that considerable investment is now necessary not only in housing and transport renewal, but also in improvements to the physical fabric of schools. This must take place against a backdrop of restricted funding and demand for dramatic overhaul of the curriculum structures in these countries.

Changes in education within eastern Europe and the republics of the former Soviet Union have been severely hampered by the lack of financial resources. However, as Kitaev (1994) has emphasized with respect to Russia, changes in the education system have not so much been introduced by educators

The development and transformation of Petržalka New Town, Bratislava, Slovak Republic

Jozef Mládek and Dušan Šimko

The breakdown of totalitarian rule in the eastern part of central Europe in 1989 paved the way for fundamental changes in the political, economic and cultural conditions of the post-communist countries. After the peaceful 'divorce' of the Czechoslovak Federation, two new states were created in 1993: the Slovak and Czech Republics.

Over many centuries, the town of Petržalka has developed as an independent unit with close and strong connections to the city of Bratislava. Historically, Bratislava has developed asymmetrically, being built predominantly on the left bank of the Danube, leaving the right bank relatively undeveloped. The right bank has traditionally been less favourable for the establishment of settlements, because it consists of a low young terrace resulting from the depositional activities of the Danube. This is cut through by numerous active and dead river branches, and there are also many marshy areas with difficult access. Most of the area is flooded several times a year by the high waters of the Danube. Another barrier to the development of the city on the right bank has been the proximity of the frontier. Bratislava, although eccentrically situated in Slovakia, could nevertheless develop because of its extensive hinterland.

The settlement of the right bank of the Danube has a long history in spite of these difficult natural conditions. The oldest settlement in the place where Petržalka now exists was founded in AD 13. The first settlers were Germans and Croatians, and by AD 18 Petržalka consisted of two independent parts – Flezyndorf and Engerau – which combined to make a single German village. This was connected with Bratislava by a so-called 'flying bridge'. The population consisted of peasants, who grew vegetables and fruit and sold their products in Bratislava's markets. Some of them also worked as carters and craftsmen. Later, Petržalka became a recreational place for the inhabitants of Bratislava, and by 1866 Petržalka had some 594 inhabitants (103 houses). New impulses for the development of Petržalka were the building of the iron bridge, which was opened in 1892, and the railway line. These led first to the construction of new quarters for the workers in Bratislava's factories, and then later came the industrialization of Petržalka itself. The number of Petržalka's inhabitants thus increased from 904 in 1890 to 2947 in 1910. In 1914 an electric railroad was built connecting Bratislava with Petržalka and Vienna, and then in 1919 Petržalka was allotted to the Czechoslovak Republic.

During the interwar period, Petržalka's population increased significantly and by 1930 its inhabitants numbered 14,164. Most of them worked in the local industries or those in Bratislava, although some continued to make a living from the growing of vegetables and fruit. Petržalka's population increased mainly through in-migration, which also led to a rapid change in its nationality structure. Thus, in the 1930 census, 55.4% declared themselves as Czechoslovak, 22.4% as German and 14.3% as Hungarian.

After the Vienna arbitration of 1938, Petržalka was annexed by Nazi Germany, being given back to Czechoslovakia only after the Second World War. In 1946 Petržalka became a part of Greater Bratislava, and in 1950 it was joined administratively to Bratislava. From 1950 to 1970 its population fell from 15,966 to only 13,899.

Designing and constructing new Petržalka

Before the Second World War, Slovakia was mainly rural; only about a quarter of the whole population lived in towns. The Stalinist regime after the communist takeover in 1948 started a huge industrialization programme that accelerated urban growth. After the 14th Congress of the Communist Party of Czechoslovakia in 1969 a megalomaniac plan of national urbanization in Slovakia was drawn up, involving the construction of several medium- and large-scale new towns. One of these was the New Town of Petržalka.

The growth of Bratislava after 1945 required ever more construction sites because the concept of building big residential satellites generated extremely high demands for large compact territories. Suitable sites for centralized mass housing

continued

construction using traditional technologies on the left bank of the Danube were gradually occupied. So in the mid-1960s attention turned more and more to the right bank and Petržalka.

In 1966 an international competition was announced to produce an 'Ideal Urban Study for the Southern District of the City of Bratislava (Petržalka)'. The competition conditions required the creation of an urban unit which solved all of the problems concerning housing, employment and services for a residential area of 100,000 inhabitants. The international competition drew unusually broad attention. Almost 700 potential applicants from some 35 countries requested the competition details, and 84 competition entries were received from 19 countries. The international jury, however, decided that none of the entries fulfilled the parameters of the project and so did not award a first or second prize, although they did present five third prizes. With hindsight, the problems that were to emerge with the scheme can now be seen to have resulted in large part from this design stage. The fact that none of the competition designs was accepted as simply the best influenced the development of the whole urban structure.

The new urban structure for Petržalka was prepared in a very short period of time (Figure 16.1), with its characteristic features focusing on the spatial order of the basic functional elements, the connections to the rest of Bratislava, the demographic structures and processes, the transportation system, and the quality of life for its inhabitants. From the point of view of spatial and functional organization, Petržalka can be considered as a linear zonal town organized along a basic north–south axis, with a secondary cross-axis. Housing construction was carried out with 4-, 6-, 8- and 12-storey housing units made out of prefabricated elements. Most of the designed basic service facilities were built, together with some of the higher-level public facilities, the basic communication system and technical infrastructure. One of the basic problems of Petržalka as a city district, though, is the lack of a central axis that could have served as a spine for the whole territory. Such a basic compositional and functional axis would have had an indispensable role in the public transport system and would have served as an element integrating the whole structure.

The figures that depict the structure and dynamics of Petržalka's population are very impressive. Over the period 1970–90, the population increased by nearly 115,000, and by 1991 Petržalka was the largest district of Bratislava. Most of the population growth is due to migration, and the share of the inhabitants in the active age group is high at 59.3%, with children accounting for 34.9% of the population. Economically, Petržalka has become strongly dependent on the activities located in the rest of Bratislava. In 1991 there were about 19,000 jobs in the primary, secondary and tertiary sectors, which meant only about 14.9 jobs for every 100 inhabitants.

The construction of Petržalka New Town created serious environmental pressures, and although some attempt was made to landscape the city, the absence of greenery is very evident. Throughout the whole urban space the artistic and aesthetic needs of the population were underestimated.

The need for rehabilitation and humanization of the living environment

The development of Petržalka outlined above had serious implication for the quality of life of its inhabitants. The preference for residential construction provided accommodation for many inhabitants, but led at the same time to neglect of other key urban functions, particularly public service facilities and schools. The creation of new jobs has now slowed down, public transport services can only function with heavy subsidies, and the living environment has been devastated. The inhabitants find it difficult to put up with the indifference, anonymity and inhumanity of the living environment. Questions about the renewal and humanization of Petržalka's living environment are now at the centre of attention for many scientific, research, planning and design institutions. The complexity of such a task, however, requires the participation of many specialists from the spheres of urban planning, architecture, geography, transportation, social sciences and ecology.

Five key areas require attention. First, it is necessary to increase the number of jobs within the

continued

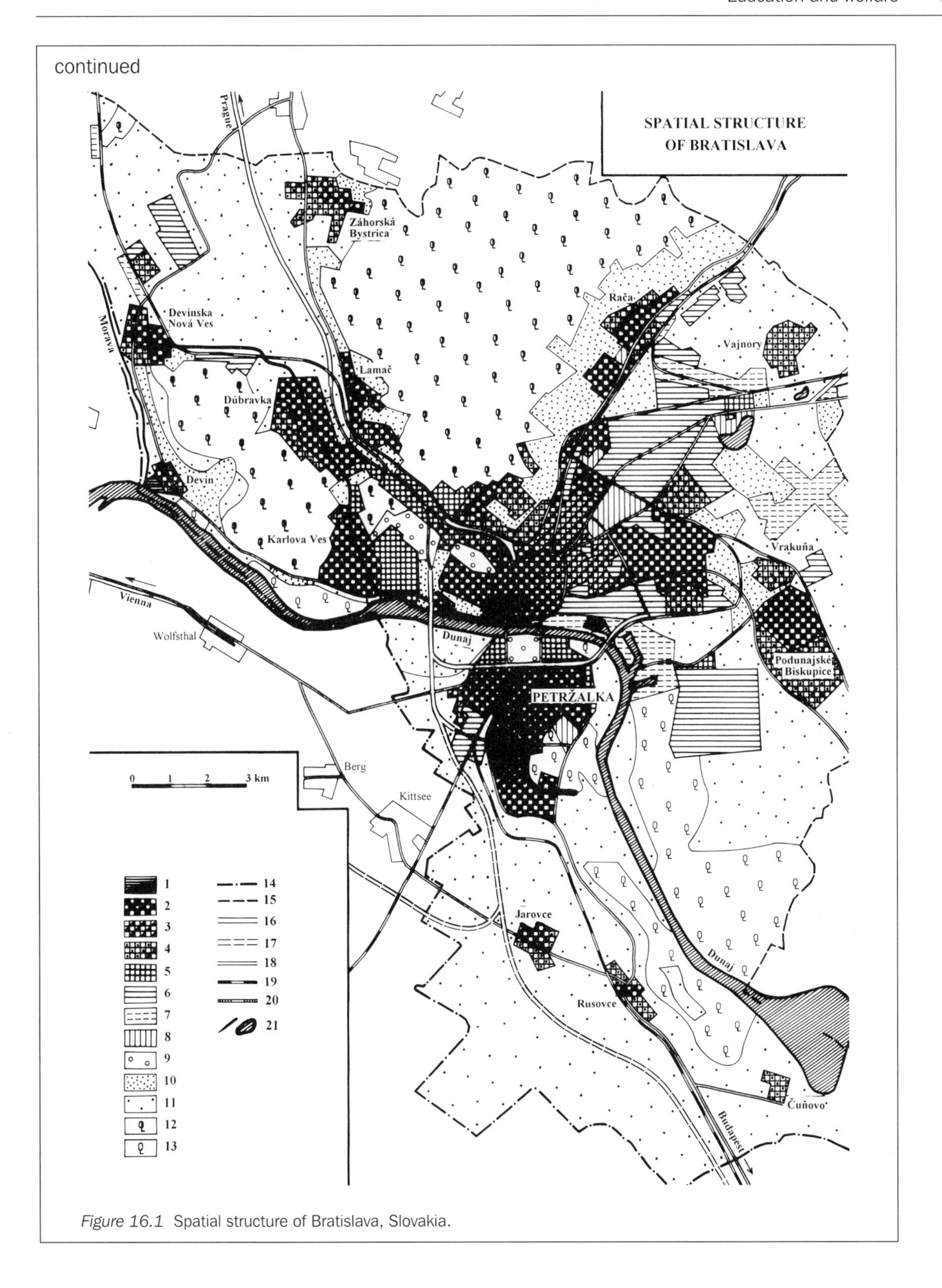

Figure 16.1 Spatial structure of Bratislava, Slovakia.

continued

Figure 16.1 Key

1. Historical centre of the city
2. New high-rise flats complexes
3. High-rise flats areas with services
4. Low-rise flats areas and villages
5. Services
6. Industrial and building areas
7. Transport areas
8. Sport and tourism areas
9. Green space
10. Vineyards and gardens
11. Agricultural areas
12. Forests
13. Floodplain forests of Danube
14. State boundary
15. Administrative boundaries of Bratislava
16. Highway
17. Highway under construction
18. Roads
19. Railway
20. Railway under construction
21. River, lake

urban area of Petržalka itself, particularly in the tertiary sector. This should lead to a reduction in the demands for public transportation to other parts of Bratislava. At the same time the function of Petržalka will change from that of a passive residential district to an active one with an economic base of its own. Second, it is necessary to complete a third transport radial paying particular attention to the development of a mode of transport with low levels of ecological damage. Third, substantial attention needs to be paid to the landscaping of the city, and the introduction of greenery to the urban environment. It is essential to develop a regime to use and protect the remains of the water forests along the Danube. Fourth, the quality of housing needs to be enhanced, focusing particularly on improvement in the quality of the façades and roofs, and improving the design and aesthetics of the buildings. Finally, considerable attention needs to be paid to the improvement of schools, cultural facilities, sports facilities, service facilities and health facilities.

Further reading

Šimko, D. and Mládek, J. (1993) Petržalka – environment and its humanization, *Slovenský Národopis*, **41**, 201–15
Steis, R. (1988) Urbanistische Entwicklung der Hauptstadt der SSR, *Architektur, Urbanismus*, **22**(1), 3–19

themselves but have rather resulted from the wider crisis of the communist system and the centrally planned economy. Paradoxically, despite greater democracy and transition towards a market economy, the Russian education system thus remains strongly centralized and dependent on national-level decision making. While there have been changes in curricula, both private and parastatal institutions are above all profit-oriented and many seem to be copying western European and North American systems. Another example of the chaotic situation facing educational reform in eastern Europe is reported by Halasz (1993), who notes that the generally negative effects of the educational decentralization policies introduced in Hungary as a result of the 1985 Education Act have to some extent prejudiced the case of those arguing for further decentralization in the 1990s.

Conclusions

Educational systems in Europe display many differences, which reflect much wider political, social and cultural differences between the countries. The need for these systems to offset the variations in social backgrounds of children varies from country to country. In some, such as England, there is a relatively wide and growing gap. In others, such as Sweden, the gap is narrow and social backgrounds are of less importance. However, even in Sweden, strike action has been taken against the way that the children of Finnish immigrants have been treated in the school system (Honkala *et al.*, 1988). Improved access to employment can obviously help narrow the socio-economic gaps, and the provision of school-age and pre-school childcare can facilitate such access.

Yet such provision is very variable across Europe and tends to be less where the gap is wider.

In this chapter school-age childcare has been discussed in detail because it links schools and work through parents' employment rather than the school-to-work transition of young people, about which much has been written. It also shows some of the varied relationships between the school, home and welfare institutions, in this case those providing childcare. The examples chosen reveal the changing domain of the school and the transfer of responsibilities from the home to either the school or the childcare institution. These subtle changes in responsibilities and roles and their differences between countries, and sometimes areas within countries, lead to new contexts in which children grow up and their identities develop. Such contexts contribute to new geographies, and the children who have been fostered within them will themselves help create further new geographies for their children to inherit.

Further reading

Bernstein, B. and Brannen, J. (eds) (1996) *Children, Research and Policy*, Taylor and Francis, London

Brannen, J. and O'Brien, M. (eds) (1996) *Children in Families*, Falmer Press, London

Brock, C. and Tulasiewicz, W. (eds) (1994) *Education in a Single Europe*, Routledge, London

McLean, M. (1995) *Educational Traditions Compared: Content, Teaching and Learning in Industrialized Countries*, David Fulton Publishers, London

Walker, A. and Walker, C. (1997) *Britain Divided: the Growth of Social Exclusion in the 1980s and 1990s*, Child Poverty Action Group, London

CHAPTER 17 Gender, geography and Europe

CLAIRE DWYER AND FIONA M. SMITH

This chapter addresses the importance of gender in the structuring and negotiation of social relations in Europe, and it illustrates some of the many ways in which *gendered geographies* of Europe can be explored. While the chapter examines the extent to which the experience of gendered differences structures the lives of men and women in Europe, we also emphasize that gender is a constructed category which is produced, and contested, differently across time and space. We want to begin, however, by thinking about why a book on the geography of Europe should have a chapter which focuses specifically on *gender*, and what this might mean.

As a starting point it must be emphasized that gender is not a self-evident or 'natural' division, but is instead a category which is socially constructed. Based upon the recognition of biological difference – a male and female *sex* – human beings are treated differently depending upon whether they are defined as male or female. In other words, individuals are *gendered* by society as girls and boys, men and women. While individuals are usually born either male or female, what it means to be male or female – a gender identity – is acquired over time (WGSG, 1997). If gender can be recognized as a socially constructed category it is also a category which is *mutually constituted* with other social categories like race. What is meant by this is that individuals are both *gendered* and *racialized* by society, and these processes work together rather than independently. In other words, if you are defined as male your gendering as a man will be different depending on whether you are recognized as a black man or a white man. Since these categories are not fixed, but are socially constructed, what it means to be a man or woman will vary in different places and times and so will what it means to be a white man or woman.

Although we have emphasized that gender is a socially constructed rather than self-evident category, it is nevertheless highly important, and in the next section we examine why this is so. However, before we do that we want to reflect on the other category which we are using in this chapter – Europe. As other authors in this volume have suggested, Europe is not a self-evident category but like other geographical categories is constructed or imagined. In this chapter, we want to emphasize that Europe is *itself* gendered and racialized – it is constructed through ideas about gender and race. This is well illustrated by Blake's image of 'Europe supported by Africa and America' produced in 1796 to illustrate John Stedman's *Narrative* (Figure 17.1) (Vine, 1994). Continents are represented as female figures, albeit with racial differences. The idea of 'Europe', which we often take simply as a geographical place, is gendered as female and racialized as white. Thus while gender is a social construction, geographical categories, like Europe, are also implicated in the construction of gender (see also Radcliffe, 1996; Nash, 1994; Sharp, 1996). We will return to the question of how geographical categories are constructed and how focusing on gender may cause us to challenge the construction of other categories in the conclusion.

Theorizing gender

In this section we want to develop further the definition of gender offered in the introduction and outline some of the ways in which gender is important in understanding European geographies. Individuals are gendered by society, which attributes particular gender characteristics to men and women. What is important about these characteristics is that they are defined in opposition to each other but are not equally valued. It is a particular feature of European cultures that characteristics defined as masculine are

Figure 17.1 Blake, 'Europe supported by Africa and America' (originally published in 1796 in Steadman's *Narrative*).

generally valued more highly than characteristics defined as feminine. Thus gender is a relational concept – women are defined *in opposition to* men. The social constructions of the two genders relate in a way that works to the general advantage of men and the general disadvantage of women, although this can be disrupted at different geographical scales (region, household, workplace).

There are a number of different ways in which the concept of gender can be used to explore the social geographies of Europe. One way is simply to argue that gender is an important category of differentiation, and to look at the differences between the lives of men and women in different parts of Europe. While this might be interesting in itself, it risks simply remaining at the level of description rather than analysis. An alternative approach is to ask how are *gender relations* between men and women constructed, contested and negotiated in different European countries? In other words, what are the processes underlying the power relations between men and women? Thus one focus of analysis has been to think through the processes underlying power relations between men and women, which usually work to ensure that male dominance, or *patriarchy*, is maintained. A number of theorists (Walby, 1994; Duncan, 1994a, 1996) have set out to compare the geographies of patriarchy in Europe. One of the difficulties for such an analysis, however, is providing a satisfactory theorization of patriarchy. Walby (1990: 20) defines patriarchy as 'a system of social structures and practices in which men dominate, oppress and exploit women'. It is recognized that patriarchy is highly differentiated and varied in how it emerges and works. Yet it can be difficult to understand both *how* patriarchy works, and how useful it is as an overarching concept which can be applied in different times and places. Walby (1994) argues that patriarchy can be recognized as having variable forms – depending upon the variable interaction of six structures of patriarchy: paid work, the household, the state, male violence, sexuality and culture. She argues that patriarchy varies in form and degree and seeks comparisons between European states through data sources relating to these different forms of patriarchy. While this comparison is useful, it concentrates primarily on the economic sphere because of difficulties in obtaining suitable data for some of the other forms of patriarchy she outlines. This comparison suggests that there is a diversity of gender relations in Europe, but it does not necessarily explain this pattern. Duncan (1994a) expands upon Walby's theorization of patriarchy to incorporate the idea of a *gender contract*. This describes the particular constellation of gender relations negotiated between men and women, as for example in the household unit, and mediated by the state, through taxation structures, childcare provision or implementation of equal rights legislation. Duncan (1994a) argues that this enables comparisons of gender relations between different European states to be better understood. In the third section we therefore look at how gender relations intersect with different geographies by focusing on the varied patterns of women's employment across Europe.

One of the problems of comparing gender relations is a tension between seeking commonalities and recognizing diversity. While these theorists argue that it is possible to develop an analysis which can offer a general structural model of gender relations by focusing upon the concept of patriarchy, it is clear that social differences *between* men and women may be more significant in some contexts than their shared gender positions. Thus not only may the social construction of masculinity and femininity vary considerably historically, culturally and geographically, but within different contexts gender differences may not be the *most* significant social category. As argued in the introduction, gender is mutually constituted with other social categories.

Thus another way in which we might analyse the category of gender is to consider the ways in which gender is being 'disrupted' in different ways and in different contexts. What it means to be a man or a woman, to be masculine or feminine, is being negotiated differently in and through different places. Individuals may resist both the roles required of them or the geographies associated with them. Gendered identities may be 'performed' differently within different places and places may become gendered in different ways. Thus in the fourth section of this chapter we look at the ways in which the category of gender is 'disrupted' by focusing upon how gender is negotiated differently in and through different places and the ways in which places become gendered.

Mapping gender: employment inequalities in Europe

At the outset, it is important not to lose sight of the common experiences which women share across Europe. One of these is their consistently poorer status in paid employment in comparison with men: 51.9% of the EU's population are women (Glasner, 1992) but consistently fewer women than men are recorded in the labour force, with female activity rates ranging from 87% down to 40% of male levels (Figure 17.2). As access to paid employment generally provides an indication of equal opportunities, this suggests levels of inequality for women across the EU. Women's involvement in the paid labour force has increased considerably in the last few decades, reducing differences between women and men, but the overall numbers also point to 'persistent and significant' inequalities between men and women in the nature of their jobs, terms and conditions of employment, levels of remuneration, and levels of unemployment (Perrons, 1994).

Nature of work

Even where participation rates are high, there is considerable 'gender segregation' of employment (Glasner, 1992). Women tend to be concentrated in certain jobs in certain sectors ('horizontal segregation') and to occupy the lower grades of employment in each sector ('vertical segregation'). This is particularly evident in relation to the service sector, where over 70% of women's employment (compared with around 50% of men's) is concentrated, where women are disproportionately concentrated in lower levels and in certain sectors (catering, 'caring', sales). In industry, women also tend to be employed in a limited range of jobs, while women's work in agriculture is often in less mechanized sectors or on family farms, where their work is not counted officially (Oxfam, 1996). Desegregation comes largely from men moving into female-dominated sectors, rather than vice versa.

Terms and conditions of employment

Women are more likely to be involved in part-time work than men, although the actual proportion varies from high levels in the UK and the Netherlands to low levels in France, where women tend to work full-time, and in western Germany, where women tend to be less involved in paid employment (Pfau-Effinger, 1994). Levels of part-time and full-time employment depend on the forms of work available. Generally 'atypical' employment (temporary contracts, flexible hours, casual employment) has been increasing in the EU with processes of globalization, or with the flexible specialization of post-Fordism (Perrons, 1994). Overall, though, it is women who constitute a large proportion of this 'atypical' workforce.

Cultural expectations of women's role in household and family are often reinforced by state policies, such as the provision of childcare or the arrangement of the school day (Garcia-Ramon and Monk, 1996). UK childcare provision is low and many women adopt a part-time work strategy to cope with family responsibility. In France provision is high and most women remain in full-time employment, whereas in Sweden part-time work entitles workers to equal rights and conditions and many families adopt a twin-income strategy (Duncan, 1994b). Why, though, is there such variation even between countries with similar 'welfare regimes' (Walby, 1994; Kofman and Sales, 1996)?

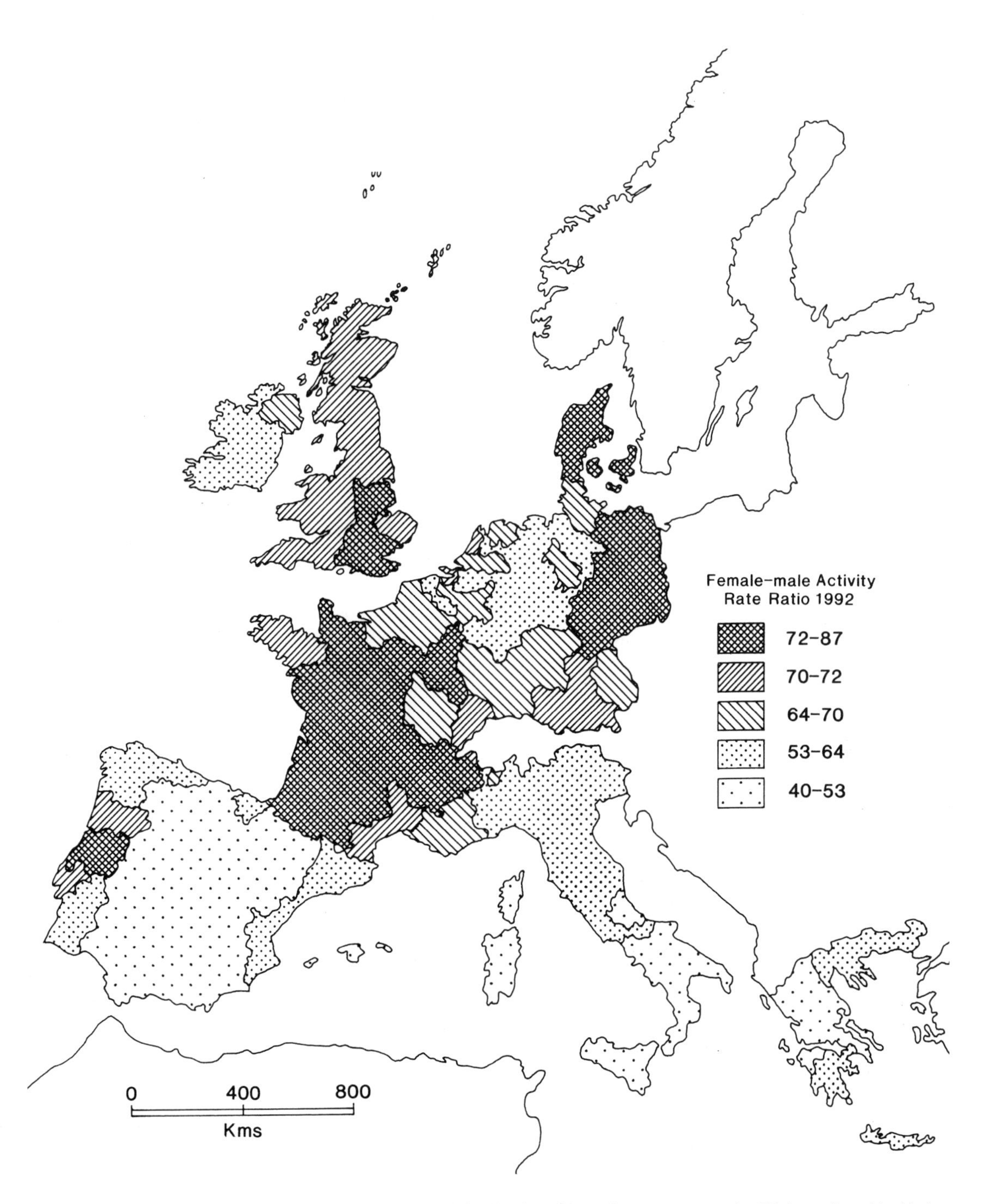

Figure 17.2 National and regional levels of unequal participation in paid employment across the EU (reproduced by kind permission of the author from D. Perrons, Gender as a form of social exclusion: gender inequalities in the regions of Europe, *Research Papers in Environmental and Spacial Analysis*, **37**, Department of Geography, London School of Economics and Political Science; data from Eurostat, 1994).

Such diversity can be seen as reflecting different patriarchal structures, or 'gender contracts' (Pfau-Effinger, 1994). The term 'gender contract' originated in Sweden as a counterpart to the 'social contract' of the social democratic welfare state and describes the ways that state, economy and individuals divide up responsibility for household and family. Perrons (1994, 1996) found that, generally, women experience more equality in more regulated employment systems with stronger welfare provision than in more liberal market economies. Differences between such gender contracts can be seen in the age-specific participation rates of men and women (Figure 17.3). In some countries most women combine family and work, and their profile is similar to men's. Others exhibit marked drop-off rates, particularly among 25–34-year-old women, although in some cases women return to work when children are older (Perrons, 1996).

Variations in existing 'gender contracts' also help to explain differing responses by women to European integration. For women in the UK it could mean access to better employment and social benefit rights. On the other hand, women in Sweden were ambivalent towards the referendum on joining the EU, as they expected to lose their welfare provision to a European standard lower than their own (Duncan, 1994b).

Remuneration

The Treaty of Rome included Article 119 on equal reward for equal work. Ever since, EU policies have promoted equal pay and conditions legislation among member states (Glasner, 1992). Despite this, lower levels of employment for women, higher levels of part-time work and lower cultural values attached to sectors in which women's employment is concentrated mean women's overall income levels are lower than men's across Europe.

Unemployment

Women not only experience poorer working conditions than men, but across the EU, with the exception of Britain, women make up the majority of the unemployed. This is often ignored in British accounts, such as that by Wise and Chalkley (1990), but lower female unemployment rates in the UK may be because the benefits system makes it economically less attractive for women to register as unemployed. For example, partnered women or older women who have paid insufficient national insurance contributions may not be eligible for unemployment benefit. Levels of unemployment are particularly high for young women in southern Europe (42% in Spain and 34% in Greece in 1990). Despite increased labour force participation, levels of female unemployment and differentials between women's and men's unemployment rates have not generally decreased. Women may be more often unemployed because cultural assumptions about the domestic and familial roles of women and the breadwinner role of men mean it is more acceptable to consign women to a non-employment role. In this view women constitute a pool of 'reserve labour' which can be mobilised when required by the economy, but for cultural reasons can also be removed from the labour force in times of surplus labour supply (Glasner, 1992).

Women's employment

Many women share common experiences of poorer employment conditions, indicating patriarchal structures across the EU. The idea of the 'gender contract' helps to account for national-level differences in the ways these are played out. However, Figure 17.2 also shows clear regional variations within countries. It is not so much that national legislation varies regionally (although there may be differential access to services). Rather, the types of women's employment and the relation of such employment to particular places mean that women are less well placed to deal with changes in labour demand, and with the uneven effects of globalization and economic change associated with EU integration (CEC, 1994; Perrons, 1994).

Work in Spain, Portugal, Greece and Italy illustrates this issue well. A series of studies (André, 1996; Sabaté-Martinez, 1996; Stratigaki and Vaiou, 1994; Vaiou, 1995, 1996) have shown that here women are involved much more in 'atypical' employment, including family labour on farms, homeworking for small manufacturing enterprises, self-employment in small commercial firms, or seasonal work in tourism. These forms of work are both culturally conditioned (combining family and work) and economically necessary (where female unemployment is high, any form of work is better than none). At the same time it is less regulated and is not offically recognized as 'employment'. This makes much women's labour

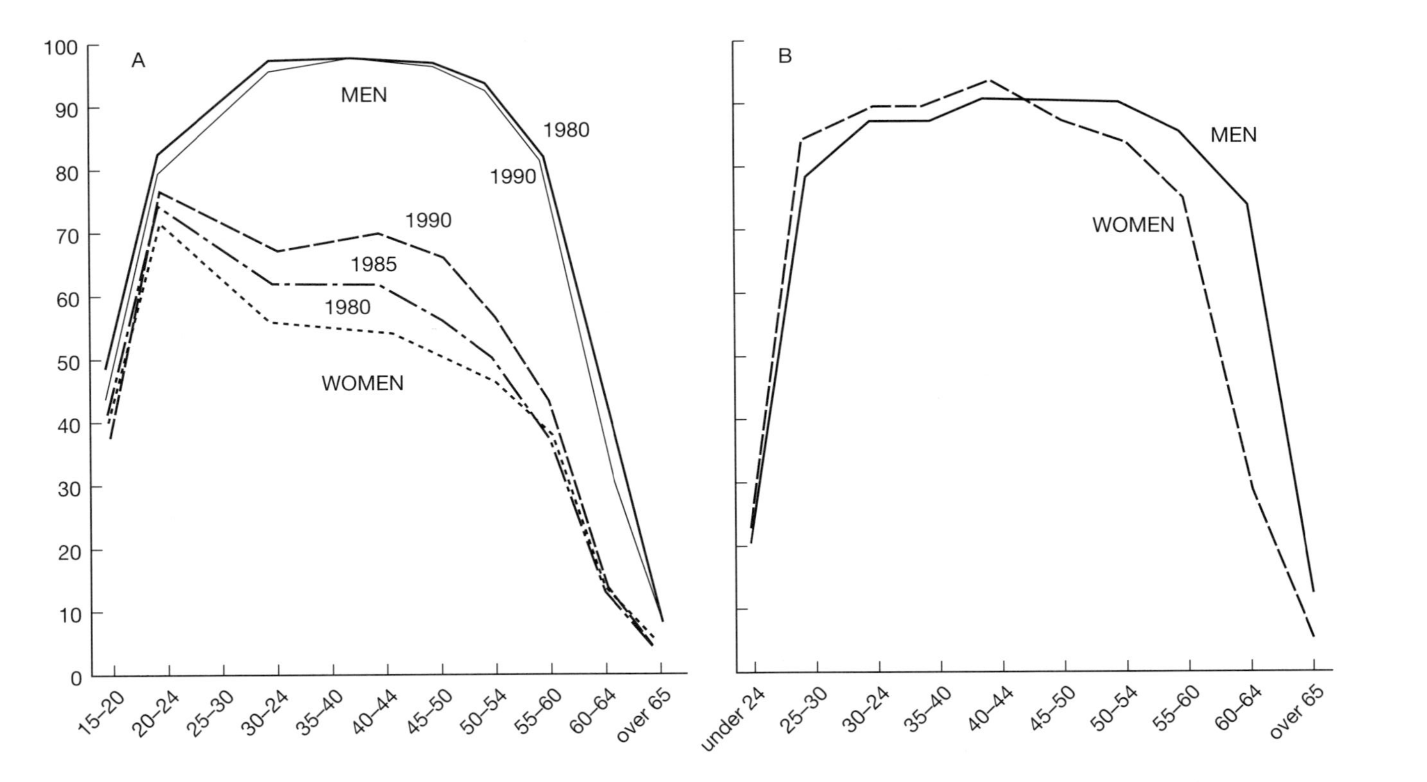

Figure 17.3 Activity rates (participation in paid employment) by age groups in: (a) German Federal Republic, 1980–89; (b) German Democratic Republic, 1989 (reproduced from S. Quack and F. Maier (1994) From state socialism to market economy – women's employment in East Germany, *Environment and Planning A*, **26**, 1257–76, Fig. 1, © 1994, with permission from Pion Ltd, 207 Brondesbury Park, London NW2 5JN).

invisible in official documents. Vaiou (1995: 40) thus notes that 'Many women registered as housewives ... may be involved in farming for part of the year, in tourism during the season, in a family shop for some hours every day, and in homeworking, all without ever gaining the status of a "working person"'.

In addition, women's labour is often highly segregated and locally based. Particular local constellations of employment (declining industrial employment and the related increase in service-based employment and in self-employment in Catalunya, Spain; or industrial restructuring in Marche in Italy, which involves many women combining homeworking and agriculture) tie women into particular local economic structures. Regional economic change therefore disproportionately affects women, whose work is concentrated in a small range of industries, such as textiles and clothing. Local informal networks which give access to work and to childcare cannot simply be transported elsewhere (Stratigaki and Vaiou, 1994). This makes women less able to adopt the EU ideal of the mobility of labour in response to increasingly mobile employment (Cecchini, 1988).

Vaiou (1995) argues that policies on EU integration do not specifically ignore women. Rather the principles of the Maastricht Treaty assume a citizenship based on the worker, who can then be part of a 'social partnership'. This worker is assumed to be in formal, 'typical' employment, with full social benefits and rights, and is assumed to be able to move in order to take advantage of the Single European Market, something which applies to (perhaps a minority of) men in the 'core' economies of western Europe. The view of 'woman-as-worker' positions much women's employment as 'atypical', reflecting a narrow view of 'social partnership' in which many women's experiences of work are simply ignored. It also results in policies where the response to lack of equality in women's access to employment is tackled in an individualized way, as with training initiatives under the EU Action Programmes on Women and Men, rather than any change in the policies of European integration (see also Close, 1995). Economic change and policies of European integration, including those aiming for the 'convergence' of European economies before the introduction of the single currency, are having and will have strong impacts on many women. Increased competition and restructuring tend to affect the sectors and regions in which women's employment is concentrated, as do cuts in expenditure on public services, which reduce job availability and access to social services. Even the Common Agricultural Policy concentrates on modernizing the more traditional branches of agriculture where women's labour predominates. Changing patterns of work for women in rural areas of Spain are explored further in the case study.

Women in the transformations of eastern Europe

In the previous section we suggested that important links exist between women's work, their position in relation to family and household, and the state. This is not unique to capitalist economies, but experiences in western Europe should not simply be applied to the transitional economies of eastern Europe. Recently there have been a series of publications on the experiences of women in the transforming countries of post-communist eastern Europe and the post-Soviet states (Aslanbegui *et al.*, 1994; Einhorn, 1993; Funk and Mueller, 1993). Generally under state socialist systems women were faced with a 'double burden' of paid employment and family responsibility (Corrin, 1992), despite official declarations that women achieved emancipation through their equality in participating in the labour force and through measures socializing domestic tasks (Einhorn, 1993). However, this constituted a state-supported patriarchy as it assumed that women were still responsible for family and household, for example by giving only women days off for housework.

High levels of female labour force participation meant that women were deemed to have gained equality (nearly 90% of all women worked in the German Democratic Republic), but here too gender-segregation continued. Just as in western Europe, women were over-represented in lower-ranking jobs and their employment, although there had been inroads in many professions and sectors, remained concentrated in services, particularly the 'caring' sectors, in less technical and more labour-intensive sectors of industry, and in agriculture, where women's work was (and is) typically in lower-skilled jobs and therefore is less valued (Figure 17.4). As a result, women have been particularly hit by economic restructuring since the late 1980s. Cutbacks by companies unwilling to keep non-productive functions, reductions in public expenditure and rationalization

Work alternatives for Spanish women in rural areas

Mireia Baylina

The Spanish rural environment underwent very important social and economic changes during the 1960s and 1970s, and the pace of these changes has increased since Spain's entry into the European Community in 1986. The guidelines of the new Community agricultural policy indicate a rigorous restructuring process in farming and in rural areas, creating marked regional differences. Some of the most significant results have been the dismantling of non-competitive types of farming, a very considerable drop in agricultural employment and the appearance of a subsidized agricultural sector. To this panorama, we must add the effects of economic restructuring at the international level and the consequent transformation of rural areas – due to their attractive location – into places for the consumption of new demands, such as leisure and housing, which proceed for the most part from the urban population. Thus the clearly functional, food-producing specialization of the rural environment has disappeared, and the area has emerged as a multi-functional space where farming coexists with other productive functions.

Such changes have had a considerable impact on the rural population, and particularly on women, who now face a new stage in their search for additional income for supporting the family unit. In short, if the survival of family-run farming in Spain is closely tied to the participation of women, as various studies have shown, women are also indispensable in the rural world's economic restructuring process.

One of the new activities being carried out in the rural environment that is most closely connected to farming and in which women play a leading role, is *rural tourism*. As an alternative to Spain's traditional mass tourism, the rural variant began in the country's mountainous areas in the 1980s. Accommodation is offered in rural homes where guests may carry out activities which acquaint them with the natural and cultural aspects of country life. Directed towards consumers in search of peace and quiet and a traditionally rural atmosphere, this phenomenon requires first and foremost a spacious location, and usually involves national heritage areas with outstanding landscapes. In numerous cases, rural tourism is being launched with the purpose of diversifying economic activity with a view to supporting the population, increasing agricultural income and conserving the national heritage. What is more, it is compatible with farm work, since guests are accommodated either in the same house as the rural workers or in adjacent apartments, eat the same meals as the family, and buy such things as crafts and fresh produce from them.

Rural women, traditionally educated to care for the family, are fundamental to this enterprise, since they both run the household and attend to the tourists. The housewife is responsible for cleaning the house, looking after the tourists' linen and clothing, and doing the cooking; the husband, on the other hand, sees to obtaining subsidies and coordinating the restoration and/or maintenance of the house. The wife, who usually devotes an average of 8.5 hours a day to this work, may have another female family member to help her (daughter, mother-in-law, mother) and on occasion – at weekends, for example – paid female help. Even though work in rural tourism has restricted her participation in the farming process, she continues to tend the market garden, small livestock, the flower garden, in short, the immediate surroundings, which is perhaps what guests value most. In this way, women are becoming active agents in the conservation of nature, agriculture and the landscape.

Typical of the activities of women involved in rural tourism is the following extract from an interview given by a woman in Catalonia:

> The person who sees to the children's schoolwork is me; I'm also the one who looks after their clothing. If the school calls someone, it's me; if somebody has to help the children, it's me; if the accountant has to be seen, I'm the one who does so and the same goes for the bank. It's not just the pension . . . it's running the household but on a larger scale, and this requires a great deal of time and effort. And the accounting . . . until now I've done it, but this year I said, 'Enough is enough'. So I told my husband, 'Either do it yourself or take it to the accountant'. The shopping is my affair and so is paying the bills. The same goes for the paperwork, which I can't keep up with. Even though I take it to the accountant, first it has to be classified . . .

continued

one of these days soon, I'll just have to sit myself down in the office and spend a week straightening it all out. You know what happens. For a whole week you go accumulating little pieces of paper and then they pile up. And there comes a time when someone has to do something about it. So who does it? Me, of course. And so it goes. I'm not making it up, you know. It's really true. Either I do it myself, or it doesn't get done. Do you understand what I mean? It's not that I do it because I want to. I do it because it's a job no one else can be bothered with. I look for volunteers but so far I haven't found any. Neither adults nor children. Now, I like to be active, I really do, but you get to the point where it's all just too much. Clothing and linen for seven people, food for seven, work for seven, the bathrooms, the dining-room and guest rooms, dealing with the guests, preparing jams, marmalades and desserts . . . It's a great deal of work and if you don't do a bit of planning, then it's impossible to see to the guests properly. On the other hand, if you manage to organize things, you can pull it off.

(Caballé, 1996: 62)

A second occupational alternative for rural women is that of *homeworking*, which has made a comeback in some areas and has emerged in others as one of the 'new' and 'atypical' forms of employment. Homeworking is a productive, salaried job which is performed in the home of the worker, but under conditions which are very precarious if not illegal. Homeworking constitutes an increasingly widespread practice which affects different economic sectors and geographical contexts. Its growth goes hand in hand with the decentralization of production and the option of subcontracting. Within this context, homeworking supposes the final link in a chain of contractual relations which make up an interconnected network of enterprises descending hierarchically from formal to informal economic levels. In general, it is the clear expression of the tendency to utilize occasional employment quickly to adapt the number of workers to market demands while at the same time avoiding the cost of maintaining permanent staff.

Within the rural job market, women provide an ideal type of labour force that fulfils all the requirements of industrialization in terms of low salaries, temporary work, 'obedience' as a socially acquired value and ignorance of workers' rights. Indeed, homeworking is paradigmatic in the level of feminization. This panorama is made even bleaker by decreased agricultural income and the lack of more interesting alternatives for women. It is within such a context that industrial homeworking appears as one of the 'flexible' forms of work carried out to a large degree on an informal basis and almost entirely by women.

In Spain, industrial work within the home is now being carried out both in rural areas with an industrial tradition and in other areas where industrial activity is practically non-existent. Although very closely linked to the more traditional sectors which depend heavily on labour, such as clothing and footwear, it is also carried out in the electronics and services sectors. At-home female workers can work directly for an intermediary, a workshop or a small company, which in turn depends on larger companies, and they are paid according to the amount of work done. While the amount earned is small, it can make up roughly one-third of the total income entering family units. Not only do these female workers lack a job contract or a minimum wage; they also have to do without social security, pension, holiday or unemployment benefits.

In the opinion of employers offering this type of work, homeworking is an ideal type of work for certain women with family responsibilities, or for those who want to work part-time. According to this way of thinking, working in one's own home provides flexibility in the use of time, thus enabling women to organize their working day freely and to alternate remunerative work with domestic work. Yet homeworking in its present conditions is neither a flexible type of employment nor can it be generalized as part-time work. In the first place, what really determines the allotment of time for women is the irregularity of the offer and delivery deadlines. More often than not this results in intense work, even at antisocial times, in order either to reach a pre-established quota or to step up the rhythm of work for early delivery. Moreover, a job in which no instructions are given as to the quantity to be delivered or the period of time in which it is to be done cannot be considered *a priori* to be part-time employment. While it is true that a female worker can, in theory, decide upon her own volume of production, if her decision does not coincide with the needs of the potential employer, no negotiation is possible and she will lose the job.

continued

These features are exemplified in the following interview given by a female homeworker in the clothing sector in Andalusia:

> When there's work to do, I get up at about 6 a.m. I sew until 8 o'clock which is when I wake up the children, give them their breakfast and take them to school. Then I come home, do some housecleaning, put the lunch in the oven and sew until lunchtime, which is when they come home from school. I give them their lunch, finish tidying the house and go back to my sewing until it's time for the children's afternoon snack. After that, I sew until 8:30. Then I stop to see that the children get their shower and I make supper. This is a normal day, always the same routine. That's my sewing timetable. From 6 o'clock in the morning to 8 o'clock at night. Ten hours of sewing. Then there's getting the main meal and breakfast, doing the washing and hanging it out . . . I manage to squeeze this in with my work. Somehow I manage to get it all done.
>
> (Baylina 1996: 206)

Work in rural tourism and homeworking are two sources of employment for rural women which are at present expanding in Spain. In spite of the main difference – work in tourism is an autonomous business initiative, whereas homeworking is dependent work – both affect women favourably in that they are remunerative occupations. While being paid for work does not give women economic independence, it does help them to make decisions and raises their self-esteem. Both forms of employment, though, have a great deal to do with activities that women have traditionally carried out, or represent values 'naturally' associated with the same. This, together with the home workplace, contribute to a social underestimation of these activities, more marked in homeworking since it is an informal job performed in precarious conditions and therefore more invisible at the social and institutional levels. Seen from a business perspective, both occupations perfectly fulfil the requirements which the new strategy demands: a flexible labour force with a great capacity for adaptation. Thus, women's work is functional to the system and crucial to the development of both activities. From the perspective of gender, the traditional sexual division of labour is reinforced in both cases, not only in the domestic sphere, for which they continue to be responsible, but also in the productive sphere, in that their professionalism is limited.

Further reading

Baylina, M. and Garcia-Ramon, M.D. (1995) Informal economy, gender and flexibility of the labour market: industrial homeworking in rural Spain, *Working Papers of the International Geographical Union*, Commission on Gender and Geography, **32**, 1–16

Garcia-Ramon, M.D. and Cruz, J. (1996) Regional welfare policies and women's agriculture labour in Southern Spain, in Garcia-Ramon, M.D. and Monk, J. (eds) *The Politics of Women's Work and Daily Life in the European Union*, Routledge, London, 247–62

Garcia-Ramon, M.D., Canoves, G. and Valdovinos, N. (1995) Farm tourism, gender and the environment in Spain, *Annals of Tourism Research*, **22**(2), 267–82

Garcia-Ramon, M.D., Villarino, M., Baylina, M. and Canoves, G. (1993) Farm women, gender relations and household strategies on the coast of Galicia, Spain, *Geoforum*, **24**(1), 5–17

within the privatized service industries have all particularly affected female employment. Industries in which women's employment was higher, such as textiles and optics, have suffered greatly from international competition and have received less structural support from governments than male-dominated sectors, such as mining, car production and shipbuilding (Quack and Maier, 1994).

Both women and men have had to change their employment, but the changes have differed. One study in Moscow found that while many women opt to work for the state or for international companies, men prefer to engage in entrepreneurial activities (Bruno, 1995). In line with western Europe, women are not only more likely to become unemployed but are also more likely to remain unemployed. Women thus constitute two-thirds of those unemployed in eastern Germany (Quack and Maier, 1994). There is some disagreement, though, about the responses to this. Some, such as Bruno (1995), argue that for many women the return 'home' has different significance than in western countries, since the home was formerly one of the few places where one could retreat from the politicization of everyday life which the state/Party aspired to (although it is also possible to overestimate the extent of 'totalitarian control'). An

Figure 17.4 Tobacco worker, Bulgaria. 'In Bulgaria 117 women for every 100 men work in farming and forestry. Many of these are migrant workers from Turkey such as this woman harvesting tobacco leaves' (Oxfam, 1996) (*source*: Oxfam, 1996; photograph by Melanie Friend/Format).

opportunity to return to the home, albeit occasioned by unemployment, might offer at least a temporary respite from the double burden. Others, such as Bütow *et al.* (1992) and Quack and Maier (1994) argue that women in eastern Germany actually exhibit an unbroken attachment to the importance of employment, by registering as unemployed rather than simply removing themselves from the register, by investing in training, by refusing to have children and by being involved in local initiatives (see Chapter 15). They are engaged in 'collective, but silent' protest against the expectation by the united German state that eastern German women will conform to the 'conservative housewife' model of western German patriarchy (Quack and Maier, 1994: 1273).

That there is less protest than might be expected about women's unemployment reflects various factors: that protest tends to be about all job losses, because the issue is seen as more an issue of restructuring as a whole rather than a women's issue; that there is a strong anti-feminist trend in most eastern European countries, as 'equality' was seen as a function guaranteed by the old state regime (Funk and Mueller, 1993). The situation in eastern Europe illustrates that gender does not necessarily always constitute a shared identity and that assumptions in western geographical analysis that categories such as 'home' or 'work' are uniformly understood are inadequate (WGSG, 1997). The following extract from an interview with an eastern German woman, a local councillor for the civic movement with grown-up children, who works full-time, illustrates the impact of economic and social restructuring on women. At the same time it highlights the difficulty of assuming that gender creates common agendas between women in eastern and western Germany, or between younger and older women, and suggests that western assumptions about the meaning of geographical categories, such as home and work, need not necessarily apply:

> Many, many young people migrated who are urgently needed here now ... So I am really glad that my children are still here. But many people's children leave and they don't come back. When they have left once then they don't come back. That's why, for me, for example, it is immensely important that the availability of Kindergarten places remains. You know, some people in the council say, 'in Nord-Rhein-Westfalen there's only this or that availability'. Well I have to say that doesn't interest me. I want the Kindergarten availability to be good here. What other people do is their concern. [...] For me Kindergarten has two important functions: children learn a lot in Kindergarten, to share, to move about, to be considerate, to prepare for school, and at the same time they allow the mothers unrestricted opportunity to work. And for a long time here, I hope, many more women will work than [in western Germany]. They keep telling us it will end up that 'with you too only 40% of women will work'. And for us, 90% of all women worked and really despite all the stresses felt well doing it. Well, it's just rubbish when someone tries to say now that we had to work. Of course we had to but we also wanted to work. And that applies to me now too. I think it's terribly bad when they act as if there was no place for the fifty year olds, and the older ones, and that they aren't given a chance any more. That is just so depressing for them. Supposedly only young, beautiful, who-knows-what, people with unlimited time are wanted. And ideally they shouldn't have any more children. But at some point a woman must be able to have her children. And I think they should have them early and not when they are already forty and they have to convalesce so that they can even have one. Because the parents are too old, too worried. When I sometimes look at the child problem in the West, I mean of course it wasn't ideal with us, but the children over there, they all seem like .. they are all single children, all dressed-up. And our system had very many disadvantages, but the employment of women was not a disadvantage.
> [...]

> Supposedly, in the future the Kindergartens will only be open for seven hours a day. But that is an idiotic demand. [...] They should be open as long as we need them. And if the hours are cut, I don't know when women are supposed to work full-time. We can't just take the West's model where the women only work five hours a day – then a seven hour opening time is sufficient. Here they work on average eight hours, because we still have the forty hour working week, you know? And then they still have to get their children there and fetch them. Really nine hours are still too few. Up to now several were open ten or eleven hours a day. [...] If we reduce that then the women will have no chance to work full time any more.
>
> (interview conducted by Fiona M. Smith, 30 April 1992)

Engendered geographies

The intersections of geography and patriarchy discriminate against women in different ways across Europe. However, women may also resist their prescribed gender roles and challenge gender assumptions including their allocation to 'appropriate geographies', such as the home. This resistance to the ways in which particular spaces are gendered suggests that places and spaces, such as work, home and nation-state, are not unproblematic and that space is in turn socially constructed: 'since social relations are created and re-created by human interaction, this fundamental statement also means that space is gendered, constantly changing, and imbued with power' (Dahlström, 1996: 1). Spaces are experienced differently by those who hold 'different positions as part of ... the social relations of space' (Massey, 1994: 3). Such positioning includes gender (as well as age, class, sexuality, ethnicity and other identities).

Places can be viewed as intersections of networks of social relations stretching across space and time which in turn help to create identities and social relations in and of places and spaces. Places therefore become gendered in particular ways, by the activities taking place there, by the employment structures of the area, by media images of the nature of the place, by local traditions of gender relations, by the linkages the area has to other areas (Dahlström, 1996). Not only that, but gender identities are performed in and through space and place in ways that may reinforce or disrupt particular places, definitions of spaces and/or the nature of gendered identities.

Gendering place/performing gender

Several studies illustrate how particular spaces and environments are gendered. Research of high-tech work environments and of the international finance sector (Massey, 1995; McDowell, 1995) suggests that such work environments are strongly gendered, and that both men and women who work there tend to adopt gendered identities which conform to a dominant heterosexual masculinity. Rural areas too are gendered. A study by Dahlström (1996: 7–8) of remote areas in northern Norway argues that 'the environment and society in [this] periphery are gendered male'. There is a range of reasons for this: there are more men than women (indeed female out-migration is seen as a key 'problem'); male activities (farming, fishing, hunting, 'cruising' for miles in cars) are 'visible and highly valued' whereas female activities such as employment in the service sector or handcrafts and riding are 'less visible and appreciated'; and 'male' activities are seen from the outside as representative of these areas, 'something typically rural'.

For young women growing up here, life involves contradictions between modern views of gender presented in the mass media and promoted through the Norwegian education system and the lack of role models and opportunities for being a 'modern, independent woman' in the periphery, whereas young men are socialized into male adult life through local hunting teams and car-based activities. They also see their fathers as role models. Young women face choices about their occupation and the possibility of combining it with parenthood which seem more difficult to combine in this male periphery. Many decide to leave. The choice to leave may be less a 'problem' with young women and be more about the dominant gendering of the periphery as 'male' and the actions and opinions of young men in the periphery which are out of step with those of young women in the periphery and of young men and women elsewhere. For many young women, refusing to perform the gendered identities allocated to them in a male place involves a choice to move.

In the following case study, this question of the construction and contestation of gender identities in different places is discussed further in the context of the ways in which masculinities might be being negotiated and renegotiated within the deindustrializing city of Sheffield, in the UK.

The (re)negotiation of masculinity in Sheffield, UK

Matthew Shepherd

The term 'masculinity' is commonly used to refer to those characteristics of behaviour and attitude typically associated with men that are subjectively constructed in opposition to women's identities of 'femininity'. Such traits include being unemotional, independent, competitive, logical and aggressive (Morgan, 1992). In recognition that there is no universal identity of men, it is more accurate to speak of 'masculinities', by which is meant that age, social class, race, sexuality and geographical location influence the construction of masculine identities, and that some individuals are actively trying to redefine their own identities away from traditional stereotypes (Connell, 1995).

Given the increasing attention being paid by geographers to issues of masculinity (McDowell and Massey, 1984; Jackson, 1991, 1993; Sommers, 1996), this case study examines the negotiation of maculinity in the context of the English city of Sheffield, and questions whether men's attempts to 'restructure' their masculine identities have significantly affected gender relations. Since the last century, Sheffield has been globally renowned as the 'City of Steel', with its physical, economic, social and cultural landscapes dominated by the steel and cutlery industries. These provided a work arena within which masculine identity could be constructed. The steel industry and steel-producing towns have characteristically maintained a dominating 'macho' image (Zukin, 1991) and a male dominance of production jobs. Likewise, male workers have been typified as physically strong, emotionally tough and the epitome of 'real men' – 'men of steel' (Taylor and Jamieson, 1996). However, since the early 1980s changes in production techniques, government funding and global demand for steel have prompted radical economic restructuring. Employment has declined rapidly – 41,000 steel manufacturing jobs were lost between 1981 and 1983 alone (Taylor *et al.*, 1996) – leaving the unemployment rate uncharacteristically higher than the national average. Not surprisingly, economic restructuring has impacted upon gender relations as the traditional sphere of male identity formation – the steelworks – has been lost. The consequence has been not only a transformation of Sheffield's visual geography and its economic prospects, but also of its social relations.

Goodwin *et al.* (1993) note that the economic decline has altered the complexion of local party politics, as women have achieved a greater parity in the local labour market and the labour movement. However, this new outlook has affected not only party politics, but also the sexual politics of individuals; it has even taken a distinctly urban form. Employment and 'breadwinning' have been identified as singularly important in the formation of masculinity (Bertaux-Wiame, 1981; Bowl, 1985; Brittan, 1989; Miles, 1989), since they have afforded men the opportunity to maintain positions as the 'head of the family', to be 'productive', and to be absent from the 'feminine' sphere of the home during the day. This has led Campbell (1993) to relate violent disorders in urban Britain in the early 1990s to male unemployment and the effects of joblessness on masculine identities. Similarly, Tolson (1977) and Wheelock (1991) discuss how men have felt emasculated by unemployment and unable to validate their masculinity given their absence from the workplace. Although such studies may be locating new spaces of masculinity and highlighting how identity formation is in flux, their focus has remained almost exclusively upon men, leaving the problem of unemployment and poverty for women unexplored.

Like Newburn and Stanko (1994), who locate crime as a means through which to prove masculinity, Taylor and Jamieson (1996) have drawn links between the once flourishing male culture of steel production in Sheffield and the identity formation of unemployed young offenders in the city. They parallel the 'hard work' ethic of the 'Little Mesters' (steel craftsmen) with the new criminal form of 'grafting' employed by disaffected teenagers. They argue that these youths are emulating the 'men of steel' masculinity through criminality to perpetuate the city's masculine heritage and thereby uphold male dominance in the city. However, Sheffield has also been a site of pro-feminist sexual political activity by men trying to create new masculinities which break from this heritage. *Achilles Heel* – an anti-sexist radical

continued

men's magazine – was originally coproduced by male collectives in London and Sheffield, with the latter city hosting, for example, the 1990 *Changing Men, Changing Politics* conference, and being home to a number of men's groups, similar to the 'women's consciousness-raising' groups popularized in the 1970s. Such groups have aimed to provide forums for men to discuss gender issues, emotions and sexual oppression (Tolson, 1977). The underlying ethos is that masculinity can be transformed in ways which may contribute to more progressive sexual politics (Segal, 1990; Christian, 1994). Recently, one of the groups in Sheffield has been organized around an educational function (sponsored by the local education authority) for men to discuss what is 'good' and 'tough' about being male, and to question 'how we can both realize pride in ourselves and our identity *and* respond positively to stop oppressive behaviour'. The group encourages men to change their masculinities to more positive forms.

Interviews conducted with some of the men who attended this course found that these men's consciousness of 'masculinity' – most commonly expressed in terms of their 'manhood' – far outweighed any awareness amongst men and women who had not attended such a course. These men were able to connect aspects of their own lives with concepts discussed in the masculinity literature. For example, Terry expressed that his job did not relate to his manhood, Bob felt that the group had been 'emotionally challenging' and had contributed to his 'personal growth and development' as a man, whilst Tony commented that his period of unemployment and subsequent voluntary work had allowed him to escape 'male-dominated' industry. All three men noted that the group had helped them better understand sexism and their relationships with women in their personal lives. In such ways, these men appeared to have been encouraged, and helped, to shape for themselves a masculinity which they were both comfortable with and which contributed more positively to gender relations.

The extent to which such apparently 'changed' masculinities have any substance to them can, however, be questioned. Bob had been sent on the course by his female line manager in the social services. He felt it had not been of any 'major use' and talking about their 'hearts' had tried his patience. Tony still has trouble accepting financial support from his female partner despite his long-term unemployment. Terry was able to conceptualize behaviours associated with 'new manhood' – decorating, washing pots, housework, wallpapering, child care – but said that he will not do them himself – 'but I don't expect my woman to do it' – as doing housework is not in his personality. His decision to end his domestic violence was not out of a desire to stop abuse of women or to reflect his 'new' masculinity, but to retain a 'peaceful life'. Such accounts conflict sharply with the course's aims and reflect the entrenchment of particular discourses of, and attitudes towards, women.

Further interviews in a local metal products manufacturing company revealed that, amongst both the men and women workers, any consciousness of 'masculinity' or 'manhood' was much less easily expressed or recognized. However, 'masculinity' was still apparent within this workplace. The company's deliberate sexual division of labour was based upon discourses of men and women as distinctively different and as possessing complementary skills. The foundation for this was the discursive construction of bodies and bodily capabilities of women and men as different: men were constructed as physically stronger than women and as therefore superior, whilst women were assumed to be obsessed with having children and consequently unable to undertake shop floor work as adequately as men (even though most of this work only involved button pushing). The sexes were described as physical, psychological and emotional opposites. These constructions were used to justify men's occupation of highly paid production work and managerial posts within the firm, and women's location in vastly lower paid administrative functions. Discourses of appropriate behaviours for men and for women were held and conformed to by both the women and men interviewed. Identity construction remained tied to Sheffield's industrial legacy.

This case study has illustrated that Sheffield functions as a space for the negotiation of masculinity and for men's sexual politics. It has also suggested that the key stimulus to the changing

continued

social geography of the city has been the massive changes to its economy and employment structures. Although 'masculinity', men's identities and men's roles in the creation, maintenance and reproduction of sexual inequalities have become increasingly questioned and problematized, it is not possible to claim that masculinities have changed to more positive forms or that gender relations have undergone any significant transformation. The organized form of men's politics may have been more pronounced in Sheffield than in many other towns and cities, but it is difficult to assess whether it has had any noticeable consequences. The actual number of men involved in such movements is very low. Those men who question themselves and their identities as men are exceptional; those who subsequently change their practices seem more exceptional still.

Further reading

Connell, R.W. (1995) *Masculinities*, Polity Press, Cambridge
Segal, L. (1990) *Slow Motion. Changing Masculinities, Changing Men*, Virago Press, London
Taylor, I., Evans, K. and Fraser, P. (1996) *A Tale of Two Cities*, Routledge, London

Bodies and boundaries

The previous section considered some of the ways in which the meanings of gender are being negotiated in different parts of Europe. This section uses the example of the so-called *L'affaire du foulard* (the Headscarf Affair), which unfolded in France in 1989, to explore the complex intertwining of discourses of gender, race and nation and to suggest the ways in which dominant understandings of gender might be destabilized or resisted (Moruzzi, 1994).

In October 1989 three young women of North African heritage were suspended from their public secondary school in Creil (50 km north of Paris) for wearing headscarves. Their headteacher, Ernest Cheniere, argued that such attire was a contravention of French laws of secularism (*laïcité*), which denote a formal separation between state and religious institution and prevent proselytizing (or winning converts) in schools. While, as Moruzzi (1994) argues, the events could be interpreted within a specific local context within which Jewish pupils were also threatened with suspension for refusing to attend Saturday classes, the events quickly became the focus of a national debate. Lionel Jospin, the Minister of Education, responding to an appeal by a French anti-racism organization, SOS-Racisme, intervened to insist that the young women should be allowed to return to school. He argued that the *laïcité* laws do not specifically forbid the wearing of headscarves or other religious symbols such as crucifixes to school, and that the wearing of headscarves should not necessarily be interpreted as an act of proselytizing.

Jospin's intervention provoked a critical response from groups on both the Right and the Left in France. While criticism from Le Pen's anti-immigrant *Front National* was unsurprising, it was a letter in *Le Nouvel Observateur* from five prominent leftish intellectuals (Elisabeth Badinter, Regis Debray, Alain Finkelkraut, Elisabeth de Fontenay and Catherine Kintzler) which accused Jospin of betraying the French national tradition, likening his 'capitulation' to that by France in Munich in 1938, which ensured that the events in Creil provoked a national debate about the meanings of Frenchness. This debate focused particularly on the threat which the girls wearing headscarves represented to the Republican secular values of France.

While there is insufficient space here to develop the complexities of this debate, we want to suggest that this specifically French example of contested cultural politics illustrates the ways in which intersecting ideas about gender, ethnicity, race and religious identity are used in the construction and contestation of the nation-state. What was particularly important in the discussion of the actions of the three young women was that their multiple subjectivities – as young women, as French residents and as North African Arab immigrants – were both ignored and simplified. Their headscarves were interpreted as straightforward symbols of radical Islamic fundamentalism, backwardness and oppression. Their decisions to wear headscarves were seen as the result of paternal pressure, or as false consciousness. Such interpretations relied upon a singular normative construction of gender and did not allow the possibility for other gender identities through which the headscarf might have different meanings. As Moruzzi (1994: 663) points out, this is particularly surprising in the French post-colonial context given the symbolic role of the veil in the Algerian Revolution (see Fanon, 1965; Moruzzi, 1993). Thus French feminists interpreted the headscarf as a sign of Islamic women's patriarchal domination, ignoring the possibility that it might represent a means through which young women constructed their gendered identities in relation to competing constructions of femininity. As the interviews recorded by Gaspard and Khosrokhavar (1995) suggest, individuals may choose to wear a headscarf for many different reasons.

By focusing state control upon the bodies of young women, reinforced through the Bayrou circular in 1994, a particular gendered and racialized version of French national identity was being forged (Hargreaves, 1995). The uneasy alliance between the Left and the Right emphasized the inadequacy of dominant constructions of the French state, founded upon the ideals of Republicanism, to grapple with complex and multiple belongings. It is perhaps significant that these same ideals were being challenged and reworked by the young women themselves – protesters at a school in Mantes-la-Jolie where similar expulsions were planned marched under a banner with the motto '*Liberté, Égalité, Fraternité*' (Hargreaves, 1995: 131).

There are two important issues that we want to draw from this example. The first is to emphasize that European states have used particular ideas about gender in the construction and consolidation of the nation-state. As *L'affaire du foulard* illustrates, European ideals of citizenship are forged through gendered and racialized discourses. Young Muslim French women may challenge the working of these discourses as they seek to assert their own gender identities in resistance to racism and anti-Islamic prejudice. Second, this example highlights the *differences* between women. To what extent are the gendered lives of left-wing French European women similar to those of French women of North African origin? How can these differences be understood and worked with to forge alliances? Both this example and the one quoted earlier from eastern Germany suggest the ways in which the category of gender can be destabilized and fragmented. However, they also offer examples of how 'other' experiences, of communism or of a non-Christian heritage, highlight often unspoken assumptions about the definition of 'Europe' or 'Europeans'.

Conclusion: gender in/and Europe – scales and sites

The arguments and examples presented in this chapter have emphasized the significance of the initial questions which we formulated – namely how gender is an important aspect of life in Europe, how gender as a category is produced through space and place, and how geographical categories are themselves constituted through gender. The chapter predominantly uses women's experiences to highlight the issues raised when gender is taken seriously as a category through which to think about the geographies of Europe. This is not because 'gender' is the same as 'women', but because it is often the absence of consideration of women's experiences which leaves many geographical analyses gender blind – that is they assume they are including everyone, but are actually ignoring experiences other than predominantly male ones (Garcia-Ramon and Monk, 1996).

Engaging with gender as a socially constructed category shows how geographies help to constitute gender relations and gender identities in different ways across Europe. It allows us to show how the geographical categories which we use are not simply neutral labels for types of space or place, but are also constructed through gender, amongst other things. 'Europe', 'non-Europe', the 'nation', the 'state', the 'region', 'place', 'work', 'home' and the 'body' are not just 'containers' for different actions but are constructed in meaningful ways. These constructions may work against the interests of women, and it is possible to identify commonalities and differences in the forms of patriarchy which affect women in Europe. However, we have also seen that women and men can resist and disrupt these geographies, gendered identities and gender relations.

One of the questions raised in our discussions is whether gender is a meaningful category at all. We have stressed both the ways in which gender is a socially constructed category, and that differences *between* women (and men) may challenge the usefulness of a category like gender and the use of all-encompassing theoretical concepts like patriarchy. In conclusion we want to point out that, notwithstanding these difficulties, gender is a social category which can be invested with meaning which may be powerful in creating a sense of identity or in forging resistance. For example, the Women's Forum in Northern Ireland is a political alliance which seeks to unite women through a shared gender identity and to cross-cut other identifications such as religion, ethnicity and class. It is a resistance movement created specifically through an evocation of a shared gender identity as women, and often as mothers (see also Radcliffe, 1993). Another example, which works gender identities in a slightly different way, is the organization Women Against Fundamentalism (WAF). WAF emerged in the UK as an alliance amongst women from many different ethnic backgrounds to challenge both racism and religious fundamentalism in response to public debate around the so-called 'Rushdie Affair', following the publication of Salman Rushdie's *The Satanic Verses* (Yuval-Davis, 1992). It has subsequently developed links with other women's movements across Europe (and beyond) to campaign on issues of oppression against women. The organization combines a recognition of the ways in which women share experiences of oppression, particularly in relation to religious fundamentalisms, with an emphasis on the differences between women. Thus they suggest the ways in which the category of gender might be used to create or evoke a sense of shared identity but acknowledge that gender may be only one aspect of an individual's identity in particular places and scales.

Further reading

Aslanbegui, N., Pressman, S. and Summerfield, G. (1994) *Women in the Age of Economic Transformation: Gender Impact of Reforms in Post-Socialist and Developing Countries*, Routledge, London

Einhorn, B. (1993) *Cinderella Goes to Market: Citizenship, Gender and Women's Movements in East Central Europe*, Verso, London

Garcia-Ramon, M.D. and Monk, J. (eds) (1996) *Women of the European Union*, Routledge, London

Hargreaves, A. (1995) *Immigration, 'Race' and Ethnicity in Contemporary France*, Routledge, London

WGSG (Women and Geography Study Group) (ed.) (1997) *Feminist Geographies: Explorations in Diversity and Difference*, Longman, Harlow

CHAPTER 18

Health and health policy in Europe

GRAHAM MOON AND SARAH CURTIS

This chapter considers both the health of Europe's population and health policy within the European context. Because the case studies accompanying this chapter focus on specific health problems, our emphasis here is on health policy. We begin with a brief examination of the evidence concerning disparities in the patterning of health trends at the scale of the nation-state, focusing particularly on the clear divisions in levels of health status between the countries of western Europe and those of the former Soviet Union and its one-time satellites. These disparities are considered in the context of the epidemiological transition (Omran, 1971). In the second part of the chapter, we discuss the health policy response. We consider differing national strategies for the promotion of 'health gain' and the scope for pan-European approaches through agencies such as the World Health Organization and the European Union.

Health in Europe: patterns and trends

On a global scale, the countries of the European continent exhibit some of the highest levels of 'health' in the world. The World Bank, in its division of the world into demographic regions, allocates European countries to one grouping comprising the Former Socialist Countries of Europe (FSE) and to another called the Established Market Economies (EME), which includes all other European countries as well as the USA, Canada, Japan, Australia and New Zealand. Table 18.1 compares three indicators across all demographic regions: life expectancy at birth, the child mortality rate (CMR – the probability of dying by the exact age of five), and the global burden of disease as measured by lost disability-adjusted life years (DALYs) per 1000 population. The first indicator is a straightforward record of the typical longevity of a population while the CMR was chosen in recognition of the view that 'The health of children is an important indicator of the nation's overall health and childhood is a time when opportunities for maximizing health potential are likely to be great' (Smith and Jacobson, 1988: 165). The calculations required to estimate DALYs are summarized in World Bank (1993) and provide an estimate of the years of healthy life lost due to death or disabling illness.

The disparity between the EME together with the FSE and the rest of the world is striking, though it must be noted that such global comparisons are problematic owing to inadequacies in the published

Table 18.1 Global health disparities, 1990.

Region	Life expectancy at birth	Child mortality rate	Lost disability-adjusted life years
Sub-Saharan Africa	52	175	575
India	58	127	344
China	69	43	178
Other Asia	62	97	260
Latin America and Caribbean	70	60	233
Middle East	61	111	286
FSE	72	22	168
EME	76	11	117

statistics. Notwithstanding such problems, it is clear that European countries enjoy substantially superior levels of child health and a generally better life expectancy than most other countries. The only comparable countries, outside those non-European members of the EME group, are Israel, Cuba, those of the Asia–Pacific rim and the southern cone of Latin America. Even the worst-performing country within Europe, the Republic of Moldova with a life expectancy of 69 and a child mortality rate of 32, returns figures substantially above the majority of other countries.

Although comparisons continue to be problematic, particularly within a Europe where boundaries and political regimes have changed extensively over the past decade, it is equally clear that there is also a substantial division within Europe with regard to health status. This division is captured most starkly in Figure 18.1, which shows a general west–east reduction in life expectancy to which Greece and Finland, owing to their more easterly locations, are only minor anomalies; life expectancy falls off starkly through eastern central Europe and into Russia. This pattern is replicated with other indicators. Figure 18.2 selects two for illustrative purposes: the age-standardized death rate for all causes of death per 100,000 population, and the infant mortality rate per 1000 live births. In both cases, there is a rise from left to right across the graph and an acceptance of the contention that eastern central Europe and the former USSR experience significantly worse health status (McKee, 1991; Boys *et al.*, 1991; Moon, 1994) is unavoidable.

Explaining these patterns is a problematic exercise at this geographical scale. Nevertheless some obvious regularities and associations can be discerned, not least the general equation between better health status and higher levels of economic development. Popularly cited explanations would include the positive impacts of the 'Mediterranean diet' with its low saturated fat content and the Nordic commitment to a high level of state intervention in healthcare. On the negative side, attention might be drawn to the impact of economic transition, relative deprivation, war (although health data are not currently available for much of the former Yugoslavia), poor diet and environmental pollution. These explanations are, however, at best superficial and generalized at such a crude geographical scale. The link between material deprivation and health is more convincingly demonstrated by the even stronger associations between health status and deprivation, economic development and perceptions of health-related behaviour which are found at the finer geographical scale and in data for individual people.

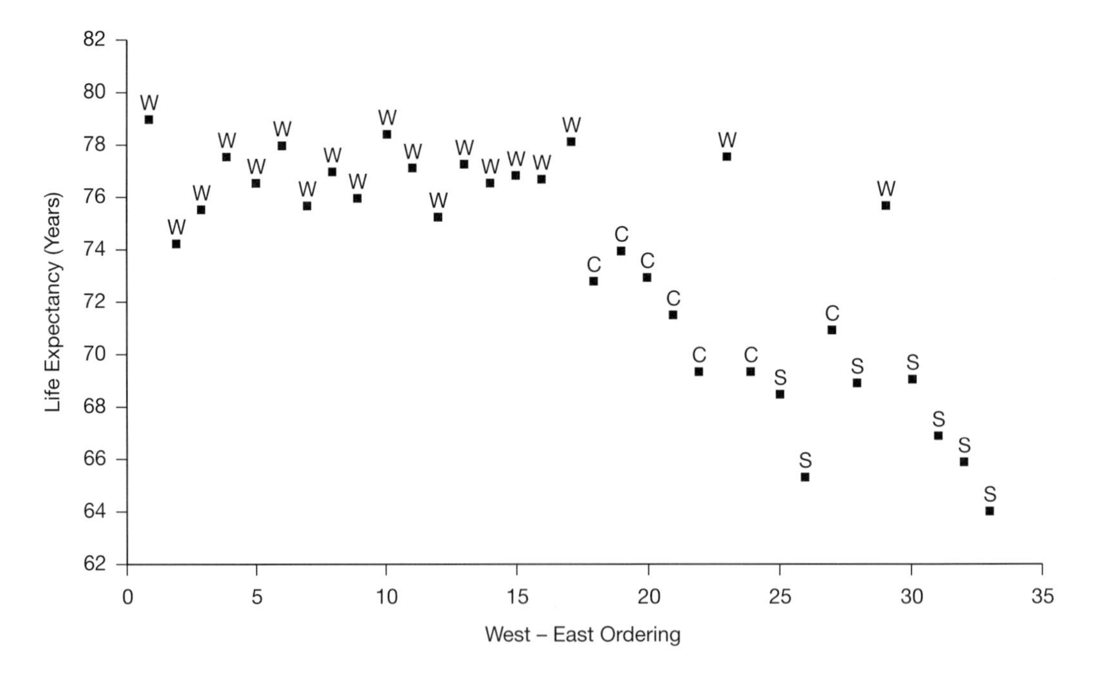

Figure 18.1 Life expectancy in Europe. Countries are sorted from west to east: W = western Europe; C = central Europe, former Warsaw Pact countries excluding former Soviet Union; S = former Soviet Union (data from WHO Health-For-All Database).

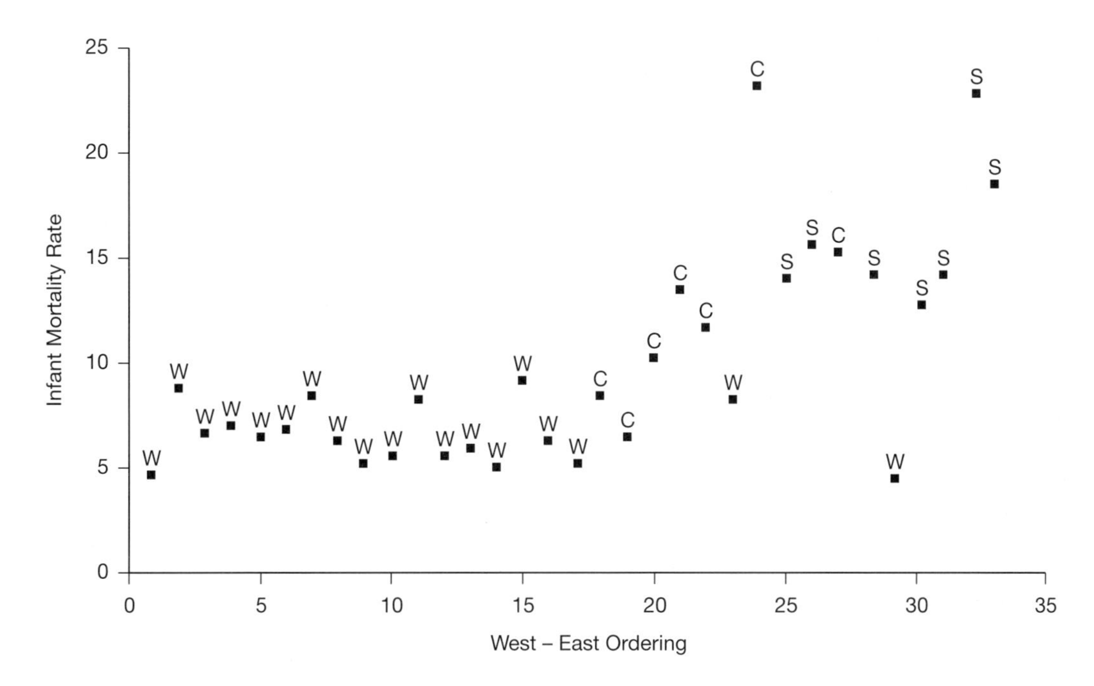

Figure 18.2 (a) Infant mortality in Europe (deaths under one year per 1000 live births).

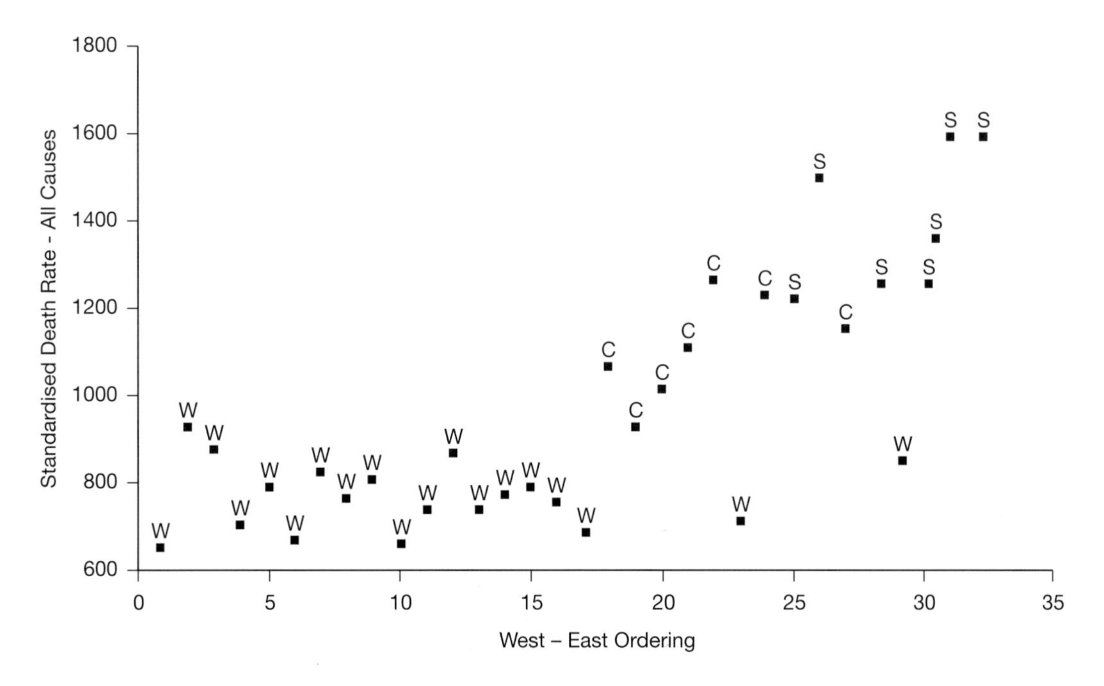

Figure 18.2 (b) Standardised death rates (all causes, all ages) in Europe (per 100,000 people). W, C and S are defined in Figure 18.1 (data from WHO Health-For-All Database).

European Union (EU) recognition of a 'Europe of Regions' and increasing involvement in health matters through the incorporation of public health within the EU social chapter have brought much improved statistical information on health and disease (Statistical Office of the European Communities, 1996). Even before these developments, however, there was a clear interest in regional variations in mortality (Fox, 1989; *Social Science and Medicine*, 1990) and there is quite a long history of studies of health inequality which have not been confined only to the EU. Thus Tonnellier (1990) has noted a broad north–south regional divide in France with the south being favoured, while Aase (1989) identifies a similar pattern to variations with regard to ischaemic heart disease in Norway (see also case study of cancer in Norway). In Hungary, Orosz (1990) identifies a more complex pattern involving both urban–rural differences and a relationship to regional levels of development. In England and Wales, the substantial evidence of the Black Report and its successor documents (Townsend *et al.*, 1992), as well as analysis of geographical health variations (OPCS, 1981; Britton, 1990; Duncan *et al.*, 1993; Jones and Duncan, 1995; Shouls *et al.*, 1996; Bentham *et al.*, 1995), have underlined both social and spatial inequality demonstrating linkage of disadvantage and poverty to levels of mortality. This complex relationship of health to deprivation, behaviour and other factors continues to be sustained at the intra-regional and local level. Work in north-east England provides a good example of such research. Phillimore and colleagues (Phillimore and Morris, 1991; Phillimore and Reading, 1992) have studied the different mortality experiences of Newcastle-upon-Tyne and Middlesbrough, noting the effect of industrial pollution levels upon health and their synergistic relationship with deprivation.

There is clear evidence that this pattern of inequality, replicated at different geographical scales, has existed for many years and has, in some senses, been becoming more pronounced (Benzeval *et al.*, 1995). Confining our attention to the national scale, Figure 18.3 takes six arbitrary marker countries and considers the changes in the age–sex standardized death rates for diseases of the circulatory system, infectious and parasitic diseases, and external causes of injury and poisoning. All rates are per 100,000 people and the time period under consideration stretches from 1970 to the latest available return. Two general conclusions are evident. First, Hungary and, to a markedly greater extent, the Russian Federation have usually returned higher standardized death rates on each of the indicators; this is typical of the other formerly socialist economies of Europe. Second, this 'health gap' is not always narrowing. For all three indicators, the trends in the Russian Federation stand out; the transition has brought worsening mortality from chronic, infectious and external causes. More specifically, in the case of circulatory diseases, there has been a steady reduction in the death rate in all other marker countries except Hungary, where the reduction has been marginal and from a much higher start point. With infectious and parasitic diseases, a steady low rate in the UK, the Netherlands and Sweden can be compared with marked reductions in Hungary and Spain, which now seem to be stabilizing towards a European norm of around eight deaths per 100,000 people. A similar process of convergence is evident with deaths from external causes and poisoning, but in this case the rates for Hungary are contrary to the trend.

These conclusions can best be summarized by reference to the concept of the 'epidemiologic transition' (Omran, 1971; Jones and Moon, 1992) – the gradual replacement of infectious disease by 'chronic' disease as the major cause of death and the relationship of this process to that of economic development. In Europe, this process is clearly generally well advanced. However, the reduction in infectious disease is evidently less well established in the marker countries of Spain and Hungary, where economic transition has been more recent, and the situation in the Russian Federation illustrates clearly that it is possible for the epidemiological transition to be reversed. 'Progress' to a stage of reducing degenerative disease consequent upon health service and socio-economic development has clearly yet to be achieved outside western Europe and, even when achieved, it has to be sustained. Generalizing from the experience of the chosen marker countries, the picture for European health is one where, in eastern countries, continued attention to the control of infectious diseases is needed alongside substantial action on the problem of chronic disease; in the west, attention needs to focus on sustaining programmes addressing chronic, degenerative health problems.

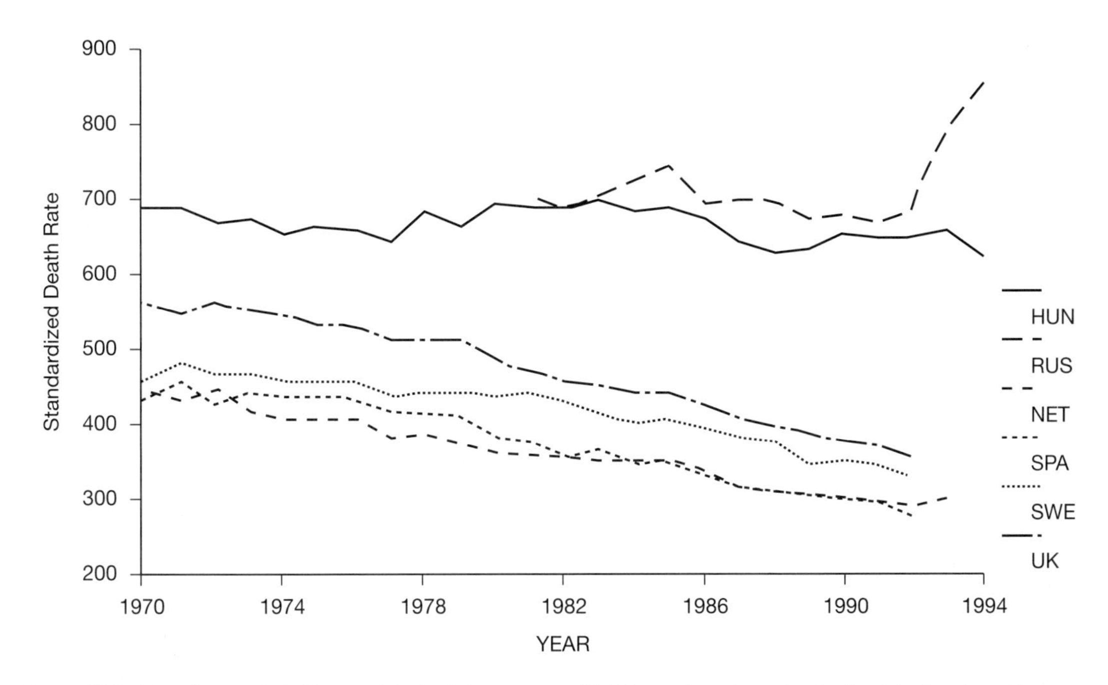

Figure 18.3 Mortality trends in Europe: (a) circulatory causes; All data are based on standardized death rates, all ages, per 100,000 people. HUN = Hungary; RUS = Russian Federation; NET = the Netherlands; SPA = Spain; SWE = Sweden; UK = United Kingdom (data from WHO Health-For-All Database).

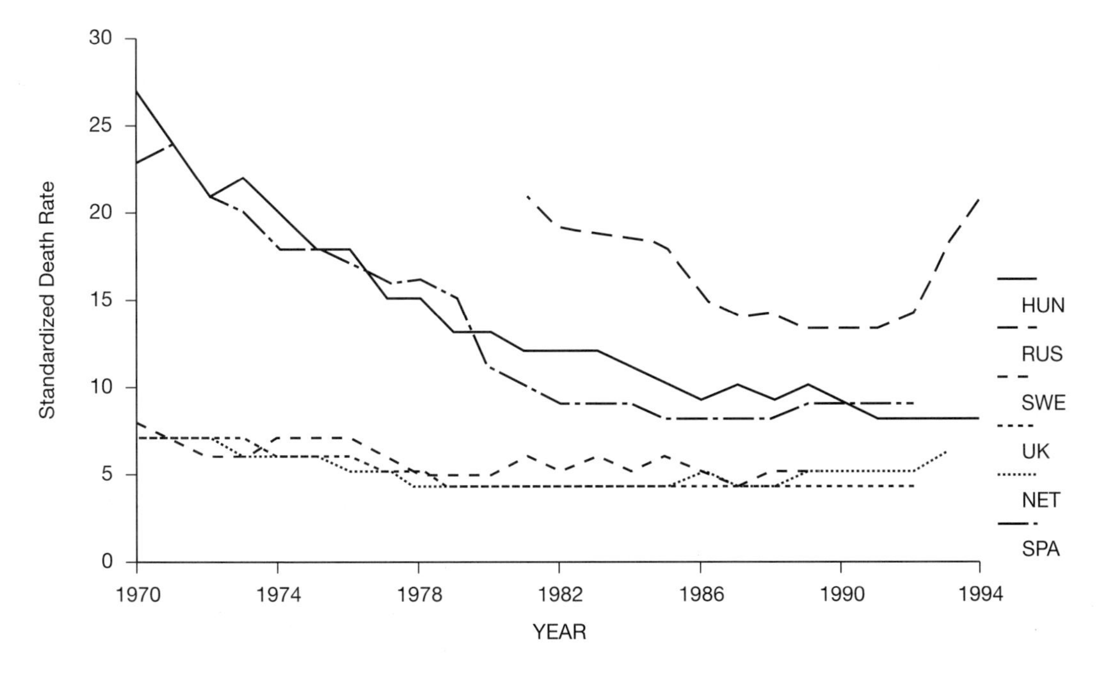

Figure 18.3 Mortality trends in Europe: (b) infectious and parasitic causes.

Cancer in Norway

Asbjørn Aase

Norway is a country of remarkable contrasts in terms of its health geography. Using life expectancy at birth as a measure, there is a difference of 5.0 years for males and 3.1 years for females (1991–95) between the best county (Sogn og Fjordane in the west) and the worst (Finnmark in the far north). At lower geographical scales even greater contrasts exist between inner fjord districts in the west and some fishing communities in the north, and between 'west' and 'east' in the capital city of Oslo.

This pattern of inequality may seem surprising. Norway has a fairly small population of 4.4 million inhabitants which is genetically homogeneous, apart from a Saami minority in the north and recent immigrant groups. It is also a welfare state with a strong commitment towards social and geographical equity. But it is also easy to point to factors working towards differentiation: the country covers a latitudinal range of 13°; it contains some of the wettest as well as the driest climates of Europe; and it spans environments from large cities to very remote fishing and farming settlements. These contrasts have led to the emergence of socio-economic and cultural differences which are reflected in attitudes and lifestyles which may have an impact on health.

Cancer accounts for about 22% of all deaths in Norway. This is fairly normal for a country having reached a stage in the epidemiological transition where degenerative diseases are dominating the mortality pattern. Cancer is not, however, *one* disease. There are many types, manifesting themselves in different body sites, with different aetiologies, geographical patterns and secular trends.

The present case study discusses selected cancer types from a geographical viewpoint, based on incidence data from the Norwegian Cancer Registry. The maps use the 19 counties as reference units, subdivided into urban (settlements with more than 10,000 inhabitants) and rural areas. Urban populations are shown as squares, rural ones through choropleth mapping.

Malignant melanoma is an example of a cancer which is causally linked to the natural environment. The risk factor is ultraviolet radiation from the sun, indicating that the risk from exposure increases from the poles towards the tropics. Due to their skin pigmentation, black people are less at risk than white, with the fair-haired and fair-skinned genotypes being especially risk-exposed. Intermittent recreational sun exposure seems to be more dangerous than regular exposure. The incidence map of Europe appears as an anomaly with regard to the expected latitudinal pattern, with higher rates for the Nordic countries than in southern Europe, and with Norway at the top of the list. This could be due to pigmentation characteristics, but also to some extreme sun-worshipping habits.

Figure 18.4 shows the incidence of malignant melanoma for women in Norway for the period 1982–91. The geographical pattern is closely related to differences in natural ultraviolet (UV) radiation, with a fivefold difference from the south-eastern coast to the far north, and with some differences also between the cooler and wetter west, and the warmer and sunnier south-east. Urban areas show higher incidence than rural. Melanoma is the fastest-increasing cancer in Norway. The rise started at the end of the 19th century with economic progress and more time for outdoor leisure activities. There are now signs of a levelling off in the younger age groups, but increased travel to sunny destinations abroad and the impact of ozone depletion on UV radiation may pose new risks.

Figure 18.5 shows the incidence of cancer of the tongue, mouth, throat and oesophagus. Rates are higher in urban areas – especially Oslo – than in rural, and there is a regional dimension with Finnmark showing high incidence, and the west and some other areas having favourable rates. The main risk factors for this group of cancers are high tobacco and alcohol consumption, with possible synergistic interactions between them. The geographical pattern is correlated with maps of other cancers with a similar aetiology, such as cancers of the lung and pancreas. Cancer of the *cervix uteri* in women, which may be related to sexual behaviour, also has a similar geographical distribution. There are some cultural characteristics underlying these regional patterns. Thus in

continued

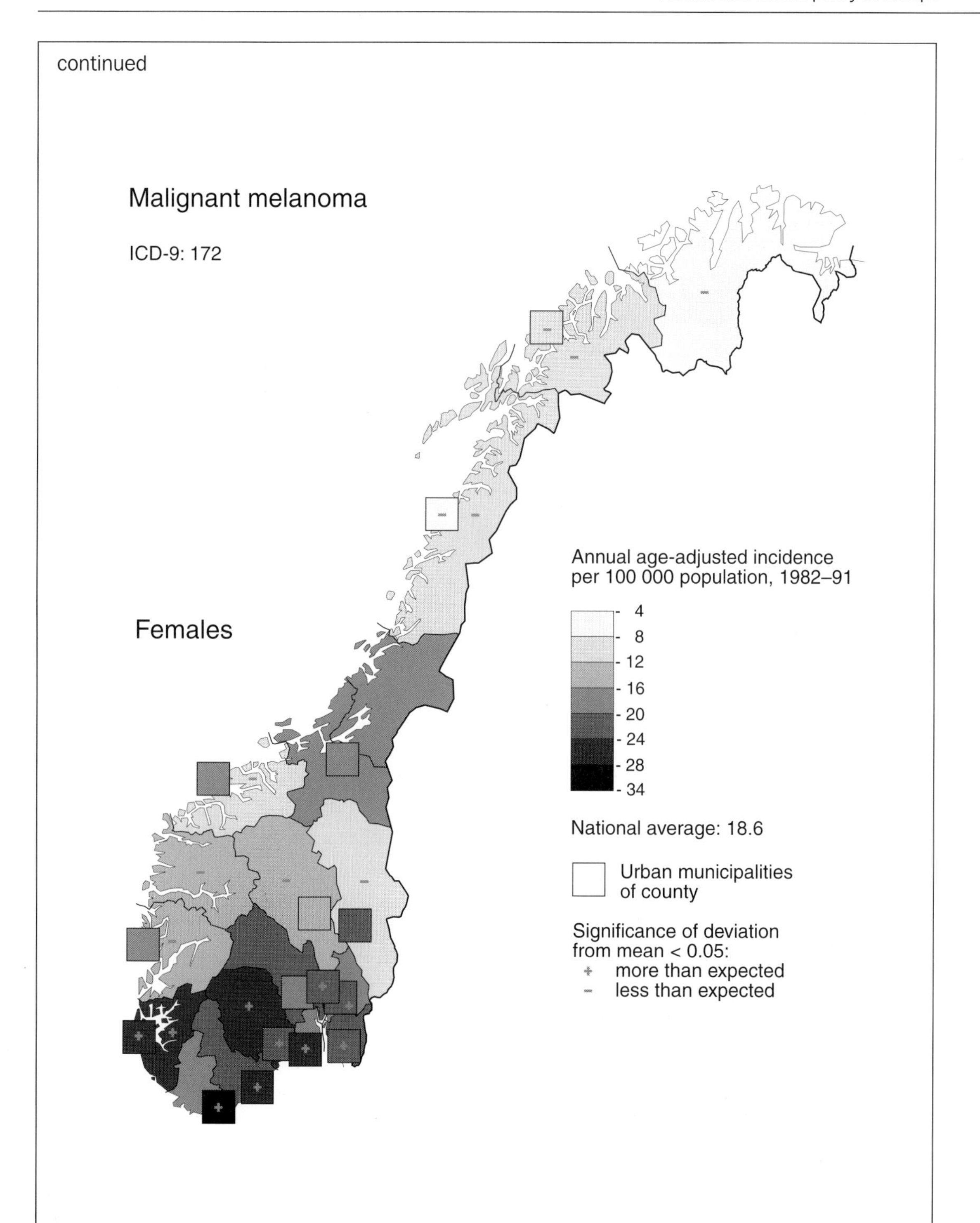

Figure 18.4 Incidence of malignant melanoma in Norway (*source*: National Atlas of Norway, copyright 1996 Norwegian Mapping Authority, reproduced by permission).

continued

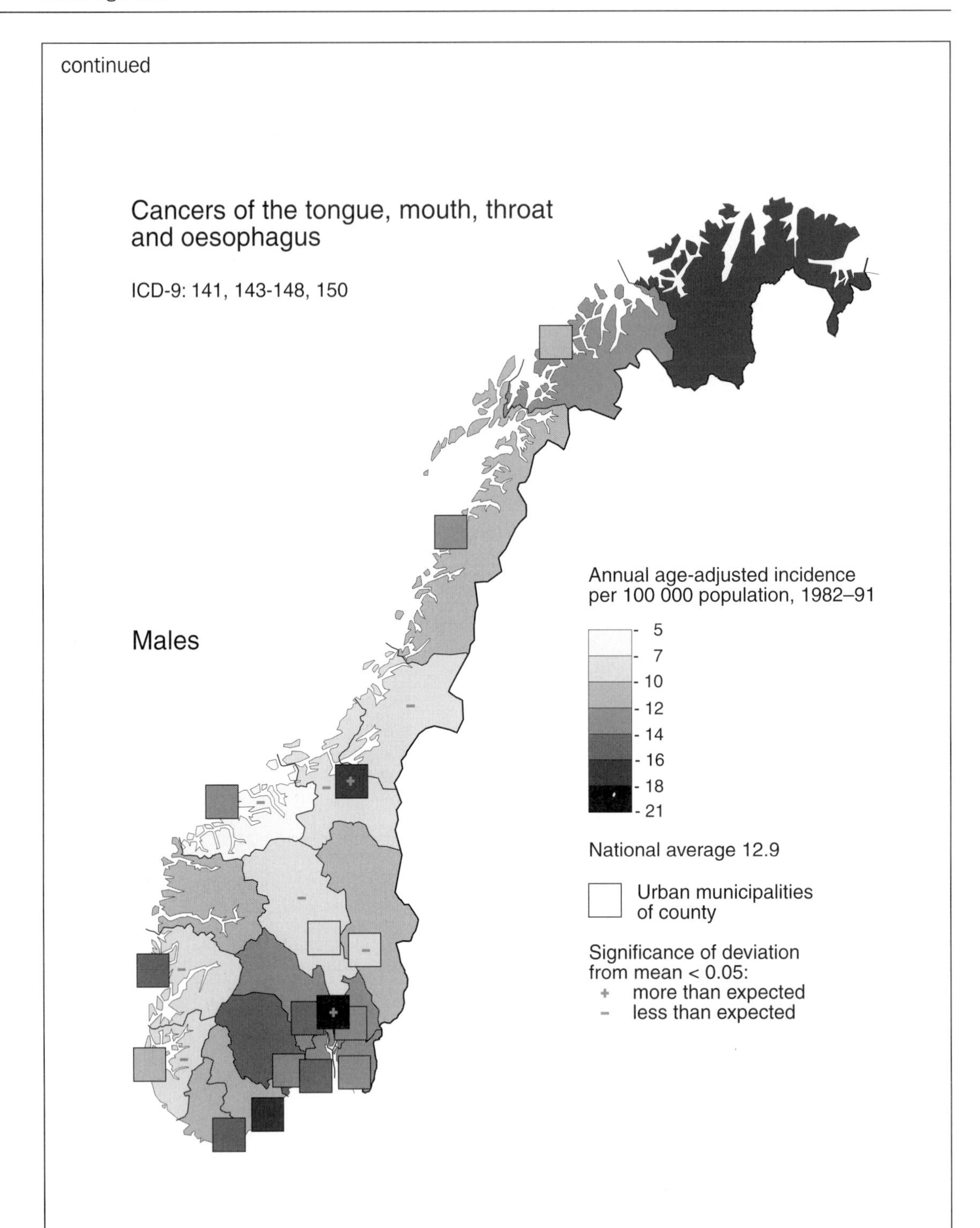

Figure 18.5 Incidence of cancers of the tongue, mouth, throat and oesophagus in Norway (*source*: National Atlas of Norway, copyright 1996 Norwegian Mapping Authority, reproduced by permission).

continued

western Norway, religious and family values are held in high regard and these have given rise to self-imposed restrictions concerning alcohol and tobacco consumption. At the other end of the scale is Finnmark and, to some extent, other areas in the north, where hard and risky physical conditions may be the cause behind a more mobile and turbulent lifestyle. These differences are also reflected in death rates from coronary heart disease, alcohol-related diseases, suicide and traffic accidents.

Many of these disease patterns are undergoing change. The spatial diffusion model is a useful approach for studying such time and space transformations. Cancers of the digestive organs may serve as an interesting case. Among these, stomach cancer shows a high incidence in the periphery, particularly in the far north, and low incidence in Oslo and most other urban areas. It has shown a marked decline over recent decades. The main cause behind this decline is assumed to be diet change following the introduction of the fridge and the freezer, and better internal transportation with the result that smoked and salted foods have been replaced by fresh or frozen meat, fish, fruit and vegetables. When such innovations occur, they are first introduced in the central and most urbanized areas and are gradually diffused towards the periphery. The present map of stomach cancer is therefore reflecting – with a time lag – a certain stage in a process of structural and behavioural change.

Colon and rectum cancers show the opposite trends and patterns. They are on the increase and incidence is higher in urban than in rural areas, and lowest in the far north. The increase is assumed to be linked to some unfavourable dietary changes, notably the introduction of 'fast food', and other foods with a low fibre content. It seems that these diet habits have again first been taken up in the most central parts of the country and have then spread towards the periphery. There is one exception to these inverse patterns of stomach versus colon and rectum cancers. Western Norway has a high incidence of all three types. It may be that some peculiarities of dietary habit in this part of the country remain to be clarified.

Lung cancer, which accounts for the highest number of cancer deaths among men and the third highest among women, has passed the peak in incidence for males, as a result of lower numbers of men smoking. For women, however, no similar reduction in smoking has been observed, and incidence continues to rise. The trend towards reduced incidence among males started with a turnaround in Oslo. There are thus diffusional trends, both in the spread of smoking from men to women, and in the geographical pattern in the onset of falling incidence among males.

Breast cancer is the most important cancer among women, in both incidence and number of deaths. Incidence shows a moderate increase. One element in the aetiology of breast cancer is reproductive behaviour; it is assumed that early pregnancies and a high number of births act as protective factors. The map of breast cancer seems to reflect this. The incidence is lowest in Finnmark, which had, until recently, the highest fertility, and it is generally higher in urban than in rural areas, which may reflect a preference for a smaller family size in the cities.

Prostate cancer, which is first in incidence and second in number of cancer deaths among men, has an aetiology which is to a large extent unknown. The lowest incidence in the country is found in northern Norway and particularly in Finnmark. Since genetic factors are known to play a role, this may explain the extremely low rates in Saami-dominated communities. However, it is also low in other parts of northern Norway. If the causes behind this could be identified, it would contribute to a better general understanding of factors behind prostate cancer.

Different types of cancer have markedly different geographical patterns. There is a UV-determined pattern with high melanoma incidence in the south-east, a smoking- and drinking-related pattern with high incidence in the cities and the north, a traditional diet pattern with high rates in the periphery, and a modern diet pattern which largely takes the opposite form. There are also different patterns related to various aspects of sexual and reproductive behaviour. Finally, there are some cancers where the geographical patterns are hard to explain. In maps of total mortality or incidence from cancer, many of these patterns tend to balance each other out. The only dimension which

continued

clearly remains is a pattern of higher rates in urban than in rural areas. It seems that differences in lifestyles may be a more determining factor than the characteristics of the external environment.

Further reading

Aase, A. (ed.) (1996) Helsetilstand/Health Conditions, in *Nasjonalatlas for Norge/National Atlas of Norway: Helse/Health*, Statens kartverk, Hønefoss

Bentham, G. and Aase, A. (1996) Incidence of malignant melanoma of the skin in Norway, 1955–1989: associations with solar ultraviolet radiation, income and holidays abroad, *International Journal of Epidemiology*, **25**, 1132–8

Engeland, A., Haldorsen, T. and Tretli, S. *et al.* (1993) *Predictions of Cancer Incidence in the Nordic Countries up to the Years 2000 and 2010*, Munksgaard, Copenhagen

Glattre, E., Finne, T.E., Olesen, O. and Langmark, F. (1985) *Atlas over Kreftinsidens i Norge 1970–79/Atlas of Cancer Incidence in Norway 1970–79*, Kreftregisteret/Landsforeningen mot kreft, Oslo

Ross, A., Aase, A. and Nymoen, E.H. (1990) Exploring cancer patterns in Norway through cluster analysis, *Norwegian Journal of Geography*, **44**(4), 189–99

Health gain

How might these twin tasks be best tackled? Almost 20 years ago, McKeown (1979) concluded that health-related behaviour, social and economic conditions and the state of the built and industrial environments were the key determinants of health. Thus it might be argued that improvements in health flowed largely from state or private action to improve these factors. Medical interventions and organized biomedical healthcare were not irrelevant but were certainly less important than might be assumed given the reliance which was placed on them. McKeown's thesis was challenged by demographers, and recent evidence suggests that the role of organized healthcare may have been underestimated (Schofield *et al.*, 1991). Bunker *et al.* (1994), writing from a US per-

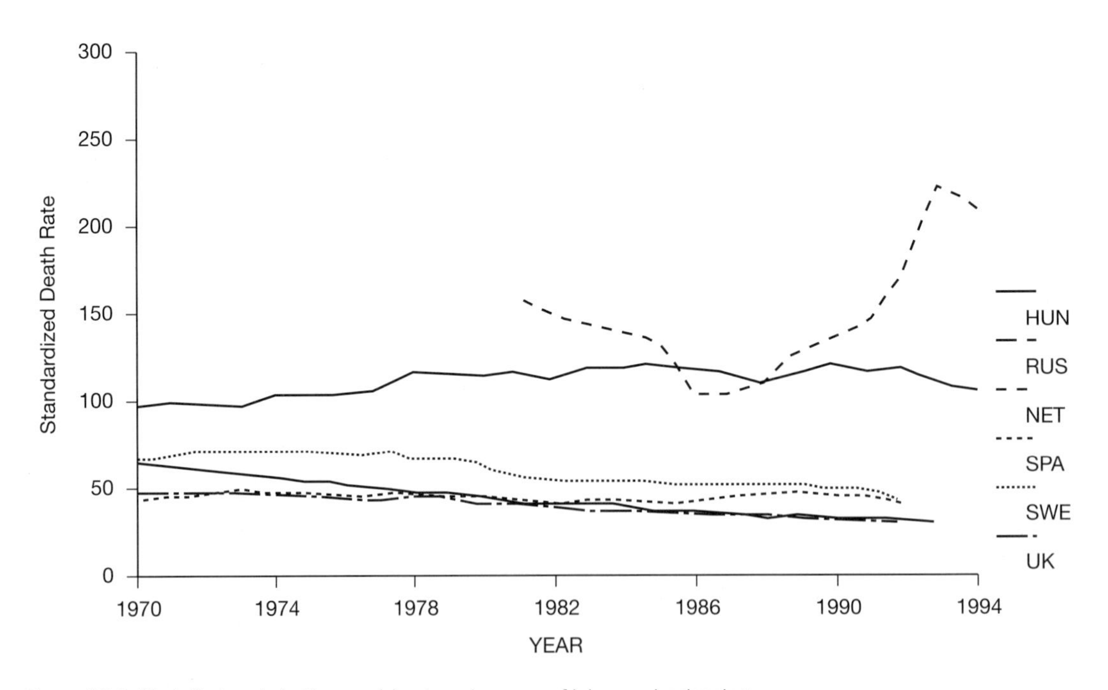

Figure 18.3 Mortality trends in Europe: (c) external causes of injury and poisoning.

spective, argue that preventive services may add some 1.5 years to average life expectancy, with the parallel figure for curative services extending perhaps to four years. Furthermore, a substantial body of work has drawn attention to the scope for reducing mortality by focusing on conditions amenable to medical intervention (Charlton and Velez, 1986; Humblett *et al.*, 1986; Poikolainen and Eskola, 1986; Gill and Rathwell, 1989; Boys *et al.*, 1991).

Notwithstanding the McKeown thesis, the strategic approach to modifying health variations chosen in most European countries has emphasized the development of biomedical healthcare services. It has been assumed that, through the more effective delivery of healthcare, higher levels of health will result. The strategy has therefore been one of equality through healthcare policy rather than through health policy. This situation is, however, changing and the structural actions proposed by a strategy of health through public policy are increasingly advocated as necessary complements of biomedical healthcare. The implication is not just that curative medicine needs to be allied with illness prevention and health promotion by screening and surveillance through health and related services. In addition, such strategies need to be complemented by access to healthy living conditions and positive health-related behaviour. Health policy must aim to influence not only individual choice but also the spatially differentiated structural constraints and opportunities of cultural and politico-economic circumstance. Change is needed at the level of society as well as at the individual level. The following discussion focuses firmly on this structural level and its relevance for health policy and health service organization. We confine our attention mainly to the articulation of policies for health gain in Europe at the international and national scales.

Health policy: the pan-European level

Supra-national health policies reflect those factors which are thought to be important in determining health difference and the actions which are thought likely to reduce health variation across a region. Such health policy is especially important in those domains which benefit from action and coordination at an international scale (Stein, 1993). This may apply, for example, where there is a significant health problem which affects more than one state and which any one state cannot satisfactorily tackle on its own. At the grandest scale, this can even involve the articulation of health policy and international relations. One of the basic prerequisites for health identified by the World Health Organization is thus 'freedom from war'. More straightforward examples might include the advantages of coordinated measures against the spread of infections. In a region such as Europe, populations frequently move across international boundaries, leading to communication and potential for cross-infection between populations. Similarly, national policies for protection of air and water quality may not be effective unless neighbouring countries also implement complementary policies; pollution effects can often spread through natural systems across national boundaries as in the case of the (East) German/Polish/Czech 'Black Triangle' (Herzman, 1994). International health policy also often concerns those actions which will produce more benefits for all countries if they are undertaken collaboratively rather than by separate countries, or where the action will be more cost-effective if it is undertaken at the international scale, as in the case of international collaboration in health research.

The international agency which is most directly concerned with health policy formulation at the supra-national scale in Europe is the World Health Organization (WHO), especially its European Bureau, based in Copenhagen. The WHO (part of the United Nations) can exercise influence on governments only indirectly, through international diplomatic activity, but it has been influential in promoting a structural focus on the health of populations. Its European Bureau covers both east and west Europe. In western Europe, the European Union (EU) also has a significant part to play, since it has the potential to influence the policies of member nations fairly directly through European Parliamentary debate and European legislation. Unlike the WHO, the EU was not established primarily to develop international collaboration over health policy. It has, however, added public health and health promotion to its terms of reference. The state socialist principles formerly evident in the regimes of central and eastern Europe provide a third pan-European element to health policy. The philosophy and mode of operation of the socialist influence in eastern Europe was very obviously different from the role of the EU in the west. However, we suggest there was also a parallel: in both the east and the west, health gain for the populations was not a primary aim of the organizations which had the greatest power over international policy making. Both in the EU and under eastern European state socialism (before its disintegration), priorities for economic growth tended to dominate over priorities for health. We consider here in more detail the different 'strands' of health policy for Europe, as articulated by the WHO, EU and socialist regimes, and how they may be more effectively woven together in future.

To consider first the contribution of the WHO, Kickbusch (1993) has suggested that the European health strategy of the WHO involves the organization adopting a number of roles. As an innovator, the WHO can help to articulate new frameworks for international collaboration in Europe. As advocate, the WHO may help to ensure that some of the less politically powerful populations and countries in Europe are fairly represented in international debate, which is especially important since lack of political and economic power often goes together with poorer health status. As consultant, the WHO can ensure information exchange between countries, notably between east and west Europe, recognizing that the west may learn from the east as well as vice versa. The WHO may also perform a facilitating role, for example, through its healthy cities programme, which has created a global banner under which local actions in cities throughout Europe have been promoted (Ashton, 1992a).

One of the best-known innovations by the WHO in Europe was the action to bring about international consensus in 1984 on a set of targets towards which all European countries might move in order to achieve 'Health for all by the year 2000'. The perspective on public health on which the targets were based was not exclusively from the point of view of medical services, but was strongly directed towards the social, political and environmental factors which are important for health. Perhaps the main aspect of the innovation was the introduction of shared, measurable objectives for action to improve population health, against which progress in each country could be measured. These targets have evolved since the first list was agreed in 1984 to reflect agreed changes in priorities over time (WHO, 1993; Curtis and Taket, 1996).

The current list, agreed in 1991, is briefly summarized in Table 18.2. This illustrates the wide perspective which is taken by the targets. Some are concerned with monitoring health, since the fundamental aim of the strategy is to improve health for the whole population of Europe. Others relate to the conditions which affect health, such as the environment, health-related behaviour, national health policy, health services and health-related research (WHO, 1993). An important aspect of this strategy is that it focuses particularly on health gain rather than on improvement to health service delivery as an end in itself. Nevertheless, changes to health services are called for in the health strategy. The WHO emphasizes especially changes which will promote a 'front line first' approach to healthcare delivery: primary healthcare services outside hospitals, which are the first point of health service contact for most people and which have the greatest role to play in helping to keep people well, rather than treating them once they become sick.

Some of the Health-For-All targets set important political and moral agendas for the WHO European health strategy. For example, Whitehead (1994) has commented on Target 1 (promotion of greater equity in health) and Target 38 concerning ethics. She points to the significance of these targets for the ways that health is debated in European countries and the way that policies are decided by government. Earlier in this chapter we drew attention to patterns of health inequality and their socio-economic dimensions. For the WHO, one implication of this inequality is the need for 'health impact assessments' of all public policies, especially in the social and economic spheres. Their 'Investment in Health' agenda focuses specifically on how such policies may influence health. Policies which can help to promote better and more equitable health in national populations include family support payments to increase resources in low-income families, education programmes for young people, and pricing and taxation policies which restrict access to addictive substances such as alcohol and especially tobacco and provide resources for health promotion (Kickbusch, 1993). Whitehead's (1994) analysis also points to the significance of human rights issues and the democratic empowerment of disadvantaged populations for the improvement in health in Europe.

Turning to the role of the EU in health policy, most commentators note that 'economic and commercial policies have been and will be the driving force' (Kokkonen and Kekomaki, 1993: 35); social policy and health have been regarded as subordinate and complementary. Ashton (1992b) and Verwers (1992) observe that EU legislation relating to public health arises from economic issues and the need to

Table 18.2 Elements of the World Health Organization Health-For-All (HFA) targets: the health policy for Europe.

Type of target	Area of action	Brief summary of specific targets
Goals	Equity, quality of life	Reduce differences in health between and within countries; opportunity for people to develop and use their health potential
	Health of vulnerable groups	Opportunities for people with disabilities; healthy ageing; health of women; health of children and young people
	Better health status	Prevention/control of chronic disease, communicable disease, cancer, accidental death and injury, mental disorders and suicide
Strategies	Lifestyles conductive to health	Promotion and support of healthy living; reduction in health-damaging consumption of tobacco, alcohol, psychoactive drugs
		Settings of social life and activity in settlements should provide opportunities for health promotion; accessible and effective education and training to improve competence in promoting health
		States shoud have developed and be implementing intersectoral policies for health promotion including public participation
	Healthy environment	Physical and social environments supportive to health; improve health of people at work
		Access for all to safe drinking water; water or air pollution should not pose a threat to health; control risks to health from solid and hazardous waste and soil pollution; improved food quality and safety
		States should have developed and be implementing policies on environment and health for sustainable development and prevention/control of health risks; also effective management systems to implement policy
	Appropriate care	Universal access to quality healthcare based on primary care, supported by secondary and tertiary (hospital) services
		Wide range of primary care to meet basic health needs; cost-effective hospital care to improve health and patient satisfaction; access to appropriate high-quality services for people needing long-term care and support
		States should have developed and be implementing policies ensuring universal access to services; cost-effective management of health services with resources distributed according to need
Support	Public health development and mobilization	Management structures to inspire, guide and coordinate development in line with HFA principles; a wide range of organizations in public, private and voluntary sectors actively contributing to HFA; mechanisms to strengthen ethical considerations
		Health research and development to support HFA; health information systems to support HFA; education and training of health and other personnel
		States should have developed and be implementing policies in line with HFA, balancing lifestyle, environment and health service concerns

Source: adapted from WHO (1993).

support and regulate health-related industries. Examples of such issues include the requirements for quality, safety and product labelling in industries producing pharmaceuticals, medical equipment and foods, and the need for mutual recognition of medical qualifications as part of a system of free movement of workers throughout the EU. Policies concerning the impact of industries on the environment, and consumer protection, also have useful effects for population health.

Article 129 of the Maastricht Treaty placed more emphasis on health protection as a focus for EU policy. This included a somewhat utopian vision of incorporating a health component into all other EU policies, as well as promoting a comprehensive EU health policy with strong similarities in scope to the WHO health strategy for Europe. The EU was to identify priorities for action programmes in relation to public health and involve the Council of Ministers in decision making on health issues. However, it was agreed that health services should remain outside the scope of the Maastricht Treaty in accordance with the principle of 'subsidiarity' by which some issues are identified as being best dealt with at the national level (see Chapter 9). There are some advantages to this principle with regard to healthcare. Each country is free to develop policies which its government considers best suited to the needs of the national population and which are sustainable with the available national resources. As we have already seen, there are substantial health variations between the different countries of Europe; it is appropriate that EU health policy should allow for the need to pursue different objectives in different parts of the Union.

There are, however, also disadvantages in subsidiarity. The willingness and capacity of each country to develop appropriate health policy rests on the political, economic and social conditions in the country. There are likely to be continuing national differences of opinion across the EU regarding the relative effectiveness of different systems of healthcare and consequent variations in health outcome, which may not necessarily be desirable from the point of view of international health equity. These differences may also present barriers to economic goals such as international exchange of labour and goods. Furthermore, without general agreement on health priorities in the EU, and especially on the need to reduce health differences, it may be difficult for individual countries to take action at the national level. Thus, countries giving priority to equalizing levels of health status through social investment might, in terms of narrow economic criteria, appear less competitive on grounds of efficiency and productivity.

Our third 'strand' of pan-European health policy concerns the model of health and welfare policy which was until recently prevalent in central and eastern Europe. This 'socialist' health policy has now largely been swept away as one symptom of the transition from socialism. Its key characteristic was a very strong dominance by central government administration in line with the command economy approach to all dimensions of society. There were some advantages for health in this approach. For example, access to basic health services was, in theory, assured for the whole population, and the connections between primary care (provided in polyclinics) and hospital care were ostensibly well-coordinated. In many countries the dictatorial nature of health administration enabled screening and immunization programmes to achieve very high levels of coverage of the population and, consequently, considerable success in controlling infectious disease.

On the other hand, the health of the population seemed to be given a lower priority in terms of national spending than in western European countries. Ensor (1993) comments on the proportion of gross national product in 1991 which was estimated to be spent on health in former socialist countries to be on average 3–4%, compared with an average of 7–8% in the established market economies, although it was also noted that individual countries vary and that, for example, Hungary, Czechoslovakia and the United Kingdom all spent close to 6% of GNP on health in 1991. The command economy model for provision of health and welfare services tended to adopt a welfare productivity approach, so that there was more emphasis on the level of activity of health services than on whether or not they were beneficial for health (Manning, 1992). Furthermore, the use of rather rigid planning norms for health service provision was not sensitive to local conditions and did not ensure that at the local level, healthcare was appropriate or effective for the population (Virganskaya and Dimitriev, 1992). In practice, there was considerable inequity in access to healthcare and in standards of living, for example between urban and rural areas and between ordinary people and individuals in privileged positions in the political hierarchy (Mezentseva and Rimachevskaya, 1990; Telyukov, 1991; Ryan, 1978). The emphasis on growth and expansion in economic policy gave rise to major concentrations of noxious industrial activity and serious environmental degradation. Ostrowska (1993), commenting on the Polish case, also describes how the socialist state undermined the capacity of lay people to take initiatives with respect to their own health and healthcare. Medical professions were relatively weak, compared with their position in western Europe. Taken together, these processes probably contributed substantially to the relatively poor health which has been experienced by the populations of eastern Europe over the last two decades.

Since the collapse of the socialist regimes in central and eastern Europe a chaotic situation has developed. Each state has generated its own plans for health policy and health service reform (see case study of the AIDS epidemic in Russia). Ideas from western Europe and the United States have been very influential in the ensuing debates over how to tackle the health problems faced in eastern Europe, and the

WHO European Office has played a continuing role. Some models and targets have, however, been adopted rather uncritically without proper consideration of their appropriateness for countries undergoing radical economic and social transition. In many countries it has been difficult to develop coherent health policies in the prevailing conditions of rapid change, fundamental transition and economic difficulty. Commentators note the slow progress towards comprehensive integrated programmes which go beyond basic biomedical intervention; any constructive changes will be difficult to achieve in current conditions (Baranov, 1991; Telyukov, 1991; Ostrowska, 1993; McKee, 1991).

Overall, there are tensions and inconsistencies in international health policy which need to be resolved. Ashton (1992b) suggests that there is potential to address this issue through a parallel partnership in health policy development on the part of the WHO and the EU. Stein (1993) points out that other international organizations such as the Council of Europe may also have a role to play, but the forum for debate needs to be broadened to recognize that health policy requires professional and 'consumer' input as well as the often utopian and élitist ideas of international organizations. A further factor limiting the scope for supra-national heath policy, however, is the principle of subsidiarity and the scope for national self-determination of health policy.

Health policy: the national level

Most individual countries, particularly in western Europe, have now developed national policies to improve the health of their populations. In many cases these national policies have responded to the emphasis placed by WHO on 'health gain', and WHO Health-For-All targets provide a consistent theme. However, there is enormous diversity in the approaches adopted by each country. Some countries, especially Nordic countries such as Finland and Sweden, have adopted WHO strategy fairly comprehensively. They now have national 'health for all' policies which reflect the same breadth of agenda for action as indicated in Figure 18.2. In Finland, for example, municipalities formulate health policies which integrate health and social care and demand high-level intersectoral cooperation with regard to health promotion (Anon., 1991). Other countries have taken up the WHO recommendations more selectively. For example, Greece, Spain and Portugal have chosen to place most emphasis on health service change and on the development of primary healthcare. In Spain, an unpopular attempt to introduce managerialism and impose an entrepreneurial culture was superseded in 1991 by a largely successful primary care-led reform.

In the United Kingdom, the policy document *The Health of the Nation* (Department of Health, 1992) sets out targets for improvement in five selected causes of ill-health and death: cardiovascular disease, cancer, sexually transmitted diseases and AIDS, mental illness; and accidents. These have become priority areas for action on health across the country. This strategy has merits in that it is clear that its most important objective is to improve health and that health services should be assessed in terms of their effectiveness in improving health. Rathwell (1991), however, is among those who have criticized it for being over-selective, concentrated only on a limited number of diseases. Other criticisms include its tendency to focus on individual responsibility for health rather than the responsibilities of society at large (Farrant, 1991) and the fact that it was slow to address the large body of evidence on inequalities in health (Radical Statistics Health Group, 1991).

Reforms to healthcare services

The pan-European organizations, particularly the WHO, stress health through public policy, and national health policies have tended, to a greater or lesser extent, to follow this position. However, there remains an abiding belief at the national level that, notwithstanding health through public policy, improved health can also be achieved through reforming the healthcare system. Thus, in parallel with health policy developments, there have also been important changes in the organization of health services in many European countries in recent years. In both western and eastern Europe debates are taking place concerning the effectiveness, efficiency and equity of health services and reforms are proposed which are claimed to be aimed ultimately at improving the health of the population.

The health service systems of European countries have grown up based on three quite different models. Most commentators describe western European health systems as positioned somewhere along a continuum between, on the one hand, the Beveridgean

The AIDS epidemic in Russia

Markku Löytönen

Europe divided

For 45 years, Europe was divided into two competing political and economic blocs. To the west were market economies and parliamentary democracies; to the Soviet-led east was central planning and a communist system. The fall of the Soviet system launched the largest rearrangement of political, economic and ethnic boundaries in Europe since 1945. The former socialist countries are now trying to distance themselves from the structures of the communist era, both in practical terms and mentally, by emphasizing their independence and sovereignty. Economically, eastern Europe is seeking ways to develop a working market economy and become an integrated part of the industrial world.

Despite violent episodes within some parts of the former Soviet region and many alarming signs of political restlessness, Russia itself has managed to take the first steps on the way to a political system based on parliamentary democracy and a market economy. The latest news of economic growth, a slowing down of inflation, and a working yet fragile democracy signal that the worst years of restructuring may be over.

For a majority of people living in the west, economic growth over the past five decades has meant a steady increase in their quality of life, whether measured by socio-economic variables or by health indicators. The collapse of the Soviet system and subsequent rapid liberalization revealed that the eastern bloc was in a worse than expected state of social and economic depression and backwardness. Damage to the environment proved to be far more serious than anticipated, and in many cases difficult to rectify. A great part of the industrial capacity was outdated and uncompetitive in an emerging market economy. Perhaps the most striking differences, however, were found in healthcare provision, accessibility and quality. For example, the current average life expectancy in some eastern European countries is over 10 years less than in comparable countries in the west. Although Europe is no longer politically divided, economically and epidemiologically the division into two still exists.

Crisis and diseases

Any crisis – whether due to war, economic depression, political restlessness, or an earthquake – is known to cause an increase in the incidence of contagious diseases. It is, thus, not surprising that the restructuring of the economic and political systems in Russia has been followed by a marked increase in the incidence and prevalence of contagious diseases. This has been attributed to malnutrition, poor living conditions, poor healthcare provision and inadequate supplies of drugs (Burns *et al.*, 1994). Although it is difficult to obtain exact and reliable figures for Russia, let alone any statistics depicting regional differences, there is strong evidence of several growing epidemics (Romanus *et al.*, 1995). Some reliable data are available for limited regions through collaborative monitoring programmes. For example, within the St Petersburg district the overall mortality caused by infectious diseases has increased from 10.9 per 100,000 in 1991 to 28.4 per 100,000 in 1993 (Collaborative Research Plan ..., 1996).

The AIDS epidemic

Until 1989, the AIDS epidemic had not created serious problems in the Soviet Union (Medvedev, 1990a, b). The number of HIV carriers among the population of 285 million people was reported to be 112, of whom only two had progressed into AIDS. The prevalence was slightly higher among foreign students, who often came from developing countries in Africa. If found HIV positive, foreign students were routinely deported to their home countries. All known Soviet citizens with HIV were isolated in certain hospitals in Moscow. The AIDS epidemic was seen as a minor problem to be addressed through administrative or judicial means.

Contrary to most other countries in the industrial world, no serious public awareness campaigns were launched in the Soviet Union, and even

continued

healthcare workers remained only poorly informed of proper preventive measures. Despite two less-than-systematic screenings in 1986 and 1987, an outbreak of nosocomial paediatric cases involving 31 children and seven of their mothers in one city in 1989 caught both the authorities and the public by surprise. The early fumbling was then replaced by more systematic screening and monitoring. Information on AIDS and its modes of transmission was made available to the public, but in a more conservative way than in most western countries.

In plain figures, the number of reported AIDS cases in Russia has been increasing slowly but steadily. It is currently only 224, even with adjustments for reporting delays (HIV/AIDS Surveillance in Europe, 1996). The figure for the former Soviet region is 376 with Ukraine having the largest number of reported AIDS cases (93) after Russia. The estimated percentage of under-reporting is currently 5% (HIV/AIDS Surveillance in Europe, 1996).

According to the Europe-wide monitoring programme, the great majority of all reported adult and adolescent AIDS cases in Russia have been contracted through either homosexual (48%) or heterosexual (23%) behaviour. The proportion of cases arising from an undetermined mode of transmission (23%) is surprisingly large, and the proportions of cases from transfusion (2%) and from injecting drug users with AIDS (0%) are both surprisingly small. Since the proportion of nosocomial paediatric cases is very large (39%) and since the reporting system may be prone to errors due to the judicial consequences of diagnosis, especially regarding drug users, it may be justifiable to question whether these figures reliably depict the true proportions of the modes of transmission of the virus in Russia. Even if questionable, the characteristics of the AIDS epidemic in Russia differ to a certain extent from those found in western Europe, a feature also found in most other former socialist countries in eastern Europe (HIV/AIDS Surveillance in Europe, 1996; Priimägi, 1991).

Despite the rapid increase of many infectious diseases, including most sexually transmitted infections, it is surprising that the prevalence of AIDS remains at a relatively low level not only in Russia but in all eastern European countries (Figure 18.6). The same pattern is repeated when looking at incidence data (Figure 18.7). The most recent statistics, however, indicate an accelerating growth. With incidence figures levelling off in most countries in western Europe, it is reasonable to forecast that the prevailing differences are diminishing between east and west Europe.

Several factors have contributed to the existing differences in the modes of transmission and the slower emergence of the epidemic in Russia. Perhaps the most important factor has been that during the Soviet era travelling was strictly regulated. Travelling abroad was limited to very few people with a majority of trips being business-related only. Travelling as well as migration within the Soviet Union was also regulated in many ways. The systematic and instant deporting of all foreign HIV carriers and isolation of Russian carriers in the earliest stages of the epidemic also contributed to the slow emergence of the epidemic.

The opening of borders and the abolishing of regulations regarding internal migration and travelling have now led to a substantial increase in internal migration. Reliable figures depicting international travel are difficult to obtain for the whole country. Finnish customs statistics, though, show a tenfold increase in travel between Finland, the St Petersburg district of Russia and Estonia in less than five years, with well over 3 million travellers in 1995 (Löytönen and Maasilta, 1996).

These changes, as well as economic and political crises, are now reflected in most health indicators – increasing prevalence and incidence in AIDS statistics being only one among many alarming signals (Conradi, 1992a, b). Once the epidemic has really taken off, then the epidemiological situation may rapidly become much worse than it is today. In such a situation, the Russian healthcare system may not necessarily be able to provide proper treatment or (often very expensive) medication to HIV-infected people. Efforts to prevent a sexually transmitted disease with an exceptionally long incubation period also provide a major challenge

continued

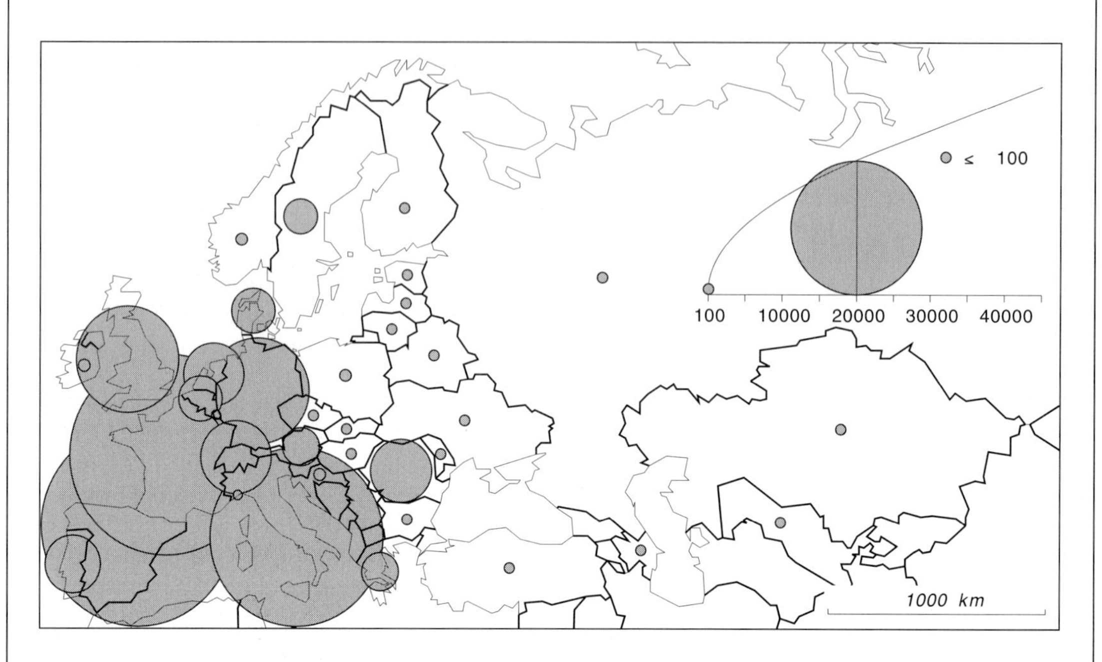

Figure 18.6 Cumulative total number of reported AIDS cases in European countries by March 1996, with adjustment for reporting delays (*source*: HIV/AIDS Surveillance in Europe, 1996).

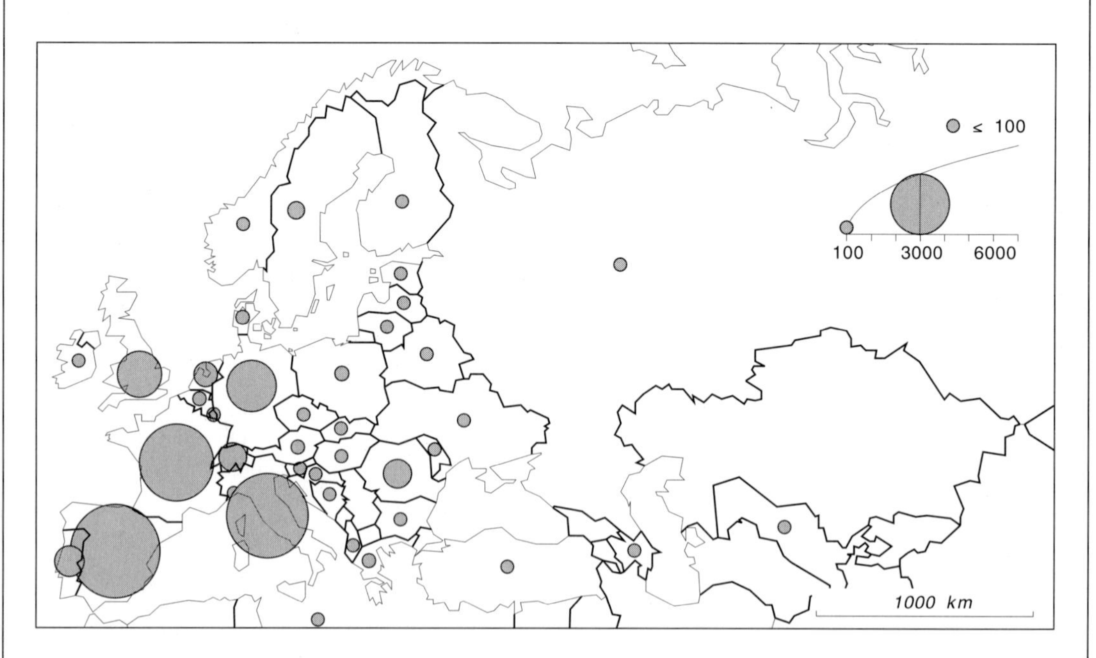

Figure 18.7 AIDS incidence rate in 1995 in European countries, with adjustment for reporting delays (*source*: HIV/AIDS Surveillance in Europe, 1996).

continued

for public health authorities. And an AIDS epidemic is not the only threat on the horizon – the first decade of the new nation may well be shadowed by a major increase in morbidity and mortality due to several epidemics caused by other communicable diseases.

Further reading

Conradi, P. (1992) Crisis in Russian health service, *British Medical Journal*, **304**, 937

Medvedev, Z.A. (1990) Evolution of AIDS policy in the Soviet Union I. Serological Screening 1986–7, *British Medical Joumal*, **300**, 860–1

Medvedev, Z.A. (1990) Evolution of AIDS policy in the Soviet Union II. The AIDS epidemic and emergency measures, *British Medical Journal*, **300**, 932–4

(National Health Service) model of integrated state provision funded by general taxation (Denmark, the UK and Ireland are more extreme examples) and, on the other hand, the Bismarckian model run on the basis of social health insurance and a system of providers in the public and private sectors which may be quite independent from government (as in France, Germany and the Netherlands). The third model, to some extent an extension of the Beveridgean but involving greater central planning and coordination, was developed in central and eastern Europe under the former socialist regimes.

While the socialist model was abandoned as a consequence of political transition, and was arguably failing in any case to deliver in terms of improved health, the Beveridgean and Bismarckian models are also being questioned as ideal types of healthcare system. Changing social expectations, an ageing population, the growth of chronic disease needing continuing care, the expense of technical developments in healthcare, and the lack of incentives for cost containment (Taylor-Gooby, 1996; Abel-Smith and Mossialos, 1994) are pressures incapable of easy resolution. Yet health reform is endemic in Europe, with many countries introducing changes designed to reduce state responsibilities for the cost and organization of health services and some convergence between the different models on which health systems are based in European countries. The precise strategies being adopted reflect the different starting position of each country in terms of the historical background to its health service system. In eastern Europe, health service planners are looking for models to replace the socialist healthcare system (Curtis *et al.*, 1995; Scheiber, 1993; Ostrowska, 1993; Ensor, 1993). Meanwhile western European countries are striving to integrate the best features of the Bismarck and Beveridge models (Elola, 1996; Creese, 1994).

Although access to health services is an important consideration for the health of the population in Europe, we have already underlined the view that other factors, especially living conditions, are equally important. Thus health service reform alone is unlikely to have a major impact on improving health. In addition, since most of the impetus for reform seems to arise from the needs for governments to constrain healthcare costs in the face of rising demands, there can be little confidence in the view that narrow reforms of healthcare systems will generate much more than marginal reductions in the health inequalities noted at the start of the chapter.

Conclusion: Challenges to health in Europe

In 1994, the member states in the European region of the WHO met and agreed the 'Copenhagen Declaration', which recognized that, in the countries of Europe, 'our societies stand at a historic crossroads' (WHO, 1996: 13) and expressed concern about persistent health inequalities, the resurgence of communicable disease, the health effects of increasing migration, the environmental impact of events such as the accident at the Chernobyl nuclear power plant, and a situation in which 'health in some countries is facing a crisis due to a deterioration in the prerequisites for health and the economic constraints that affect healthcare services' (WHO, 1996: 14). The declaration called for preferential support to protect and improve the health of vulnerable and high-risk groups in European countries, for priority to be given to the support of countries in eastern and central Europe, and for support to victims of armed conflict and natural and human-generated disasters.

The declaration also underlined the need for better-managed healthcare for health gain and more dialogue and coordination.

The Copenhagen Declaration recognizes the need to consider health inequality as well as trends in average health of the population in Europe. Geographical analysis is often a powerful tool in the analysis of inequalities, so that geographers have a contribution to make to this debate. The need for more recognition in Europe of the susceptibility of health to the effects of social and economic policy is also clear. Until the health impacts of the economic organization of European society are recognized in political debate it seems unlikely that we will see great reductions in health variation.

Further reading

Curtis, S. and Taket, A. (1996) *Health and Societies: Changing Perspectives*, Arnold, London

Ensor, T. (1993) Health system reform in former socialist countries of Europe, *International Journal of Health Planning Management*, **8**, 169

Herzman, C. (1994) *Environment and Health in Central and Eastern Europe*, World Bank, Washington DC

Normand, C. and Vaughan, P. (eds) (1993) *Europe without Frontiers: the Implications for Health*, Wiley, Chichester

Ryan, M. (1978) *The Organization of Soviet Medical Care*, Blackwell, Oxford

CHAPTER 19 Tourism and travel

DEREK R. HALL

The aim of this chapter is to examine and exemplify the contribution of international tourism and travel development to the contemporary social geography of Europe. It does this through an evaluation of the nature, extent and impacts of international tourism and travel, an examination of contemporary debates surrounding European tourism, and an identification of the key literature sources informing those debates. As the chapter concentrates on pan-European dimensions and trends, it focuses largely on international rather than domestic tourism and travel, although this by no means suggests that the domestic dimension is not significant.

European tourism development and current trends

It has been suggested by Pompl and Lavery (1993) that the relationship between Europe and tourism development has three important dimensions: it was in Europe, particularly in Britain, that the tourism industry was invented, refined and developed; Europe has dominated world tourism for the past 40 years; and tourism development and integration has been encouraged by the EU and its predecessors, with the European tourist industry employing over 35 million people, representing over 8% of the workforce (although estimates vary between 6% and 16%).

As an essential component of European production and consumption, the tourism and travel industry plays a significant role in employment, the economy and trade, and acts as an instrument of social and cultural change and contact, assisting the interaction and transfer of values and lifestyles (Williams and Shaw, 1994). In addition to being one of the fastest growing and largest elements in the global economy, representing a key ingredient in contemporary globalization and restructuring processes, tourism and travel has also become a major growth area for academic research.

The deficiencies of tourism statistics in assisting our understanding of the spatial, structural and social patterns of tourism in Europe have been well rehearsed (Hall, 1991; Williams and Shaw, 1991; Burton, 1994; WTO, 1994). But it would appear that the continent now receives less than two-thirds of all international tourist arrivals (Table 19.1), declining to an estimated 59.4% for 1995 from 70.8% in 1970 (WTO, 1996c: 8). It also receives only just over half of all tourist receipts (Table 19.2). As Europe's most visited countries (Table 19.3) (WTO, 1996b), France, Spain, Italy, Hungary and the UK accounted for more than half of all Europe's international arrivals (52.9% in 1994), and the four west European countries alone accrue half of the continent's tourism receipts (49.46% in 1994). Significantly for the European economy, however, in terms of levels of receipts per tourist, only three European countries ranked within the world's highest 25 in 1994, and all three were Scandinavian (WTO, 1996c: 15; Nyberg, 1995).

Although the Single European Act, which came into effect in January 1993, embraced the tourism industry by virtue of policies designed to provide freedom of movement and the protection of tourists, measures introduced in a number of areas over several decades have had some significance for the tourism industry, including those to reduce restrictive trade practices, harmonize fiscal policies, provide consumer information and protection, facilitate labour mobility, provide work-based protection to employees, support environmental protection and standards, and facilitate investment and trade (Baum, 1995).

The Commission of the European Communities has been preparing community policies on tourism since 1982. Yet European policy for tourism has been conspicuously slower to evolve than have policies in many other economic and social sectors, and, partly because of the diffuse range of activities which comprise the industry, has developed in a somewhat piecemeal and *ad hoc* way (Robinson, 1993; Baum, 1995). In 1986 the main objectives for the EC's policy towards tourism were stated as facilitation and promotion, improving the seasonal geographical distribution, providing better information and protection, improving working conditions in the industry, and increasing awareness of practical problems including those relating to the collection, collation and nature of tourism statistics and to the use of the Community's financial instruments (CEC, 1988; Barnes and Barnes, 1993).

The EU view of tourism is that of a dynamic sector with great growth potential and a crucial dimension of European integration, as reflected in the designation of 1990 as European Year of Tourism (CEC, 1991); it is one of the bright spots of the European economy, representing around 6% of both EU employment and GDP. It is viewed positively as assisting economic development of rural and peripheral regions, helping small and medium-sized enterprises (SMEs), and assisting protection of the environment through such mechanisms as the Mediterranean Partnership (Pompl and Lavery, 1993; Montanari, 1995). As Williams and Shaw (1994) have suggested, tourism's contribution to uneven regional development and its potential to both threaten and assist environmental conservation have raised the profile of tourism as these concerns have received a higher priority in Brussels.

Some of the contradictions of EU environmental and structural fund policies, in terms of encouraging more passenger travel and goods movement, and the provision for it, while exhorting the adoption of environmental sustainability, are highlighted in the ambivalent role of tourism. They are also reflected in contrasting national-level attitudes towards tourism, ranging from the views of Alpine Convention members France, Italy and Slovenia on encouraging tourism growth, compared with the conservationist Austria, Germany and Switzerland. Further, policy is some way behind addressing the implications for tourism of a number of structural, technological and spatial changes taking place in Europe, such as the continued growth of business tourism despite videoconferencing; the critical importance of accessible hub centres, increasingly focusing activity on centres with major international airports such as London, Paris and Amsterdam (WTO, 1994); the growth of mega-events and attractions, such as EuroDisney; the impact of the Channel Tunnel (Page and Sinclair, 1992; Gibb, 1994); and the emergence of central and eastern Europe as both tourism destination and tourist generator (D.R. Hall, 1995).

European approaches have been most effective in three particular areas of policy, which have supported:

- requirements to extend the tourism season for a continent whose climate is generally markedly seasonal;
- the development of rural tourism as an alternative source of income for farming communities, as a means of improving rural services, and, particularly in tandem with heritage tourism, as a means of encouraging higher-spending niche tourists, albeit indirectly through regional policies and structural funds rather than necessarily through strictly tourism-related policies; and
- the means of attracting wider tourism markets, notably those of Japan and the former Soviet bloc countries (Pearce, 1988).

From mass market to niche segmentation?

The emergence of a post-war mass tourism market from earlier forms of tourism was stimulated by socio-economic factors in part influenced by transatlantic trends: increasing leisure time, growing disposable incomes, paid holiday entitlement, earlier retirement age (coupled with improved health and longevity), technological change including mass consumption of motor vehicles, computerization of reservation systems, larger and faster forms of intercontinental transport, a structural consolidation of the holiday tour industry and changes in occupational structure (Edwards, 1981; Steinecke, 1993; Gilbert, 1994).

Tourism development articulates a number of the characteristics central to debates on European restructuring (Britton, 1991; Montanari and Williams, 1995). Mass tourism of the 1960s represented mass consumption (Urry, 1990; Shaw and Williams, 1993) of a standardized product, with producer-dominated markets located in such purpose-built environments as Torremolinos and Benidorm. The latter, for example, grew from a fishing village of just 2726 inhabitants in 1950 to a high-rise urban centre with more than 60,000 permanent residents in the 1980s. This process thus generated new urban structures dedicated to tourism consumption which represented a spatial if

Table 19.1 Europe: international tourist arrivals, 1980–94.

Region (as defined by the WTO)	Total arrivals ('000)			Av. annual rate of increase (%)	Share of world total (%)		
	1980	1990	1994	1980–94	1980	1990	1994
Western Europe	71,410	113,838	118,428	3.7	25.0	24.8	21.7
Southern Europe	60,624	92,076	95,681	3.3	21.2	20.1	17.5
Central/East Europe	36,001	46,723	74,313	5.3	12.6	10.2	13.6
Northern Europe	17,921	26,648	31,443	4.1	6.3	5.8	5.8
East Mediterranean	2,390	7,423	9,942	10.7	0.8	1.6	1.8
Europe total	188,346	286,708	329,807	4.1	65.8	62.4	60.4

Sources: WTO (1996a: 205; 1996c: 5).

Table 19.2 Europe: international tourism receipts, 1980–94.

Region	Total receipts (US$) millions)			Av. annual rate of increase (%)	Share of world total (%)		
	1980	1990	1994	1980–94	1980	1990	1994
Western Europe	27,870	59,502	67,078	6.5	26.5	22.5	19.4
Southern Europe	19,528	48,254	56,667	7.9	18.6	18.2	16.4
Central/East Europe	3,499	4,849	14,404	10.6	3.3	1.8	4.2
Northern Europe	11,115	25,510	26,712	6.5	10.6	9.6	7.7
East Mediterranean	1,433	5,865	8,321	13.4	1.4	2.2	2.4
Europe total	63,445	143,980	173,182	7.4	60.3	54.4	50.1

Sources: WTO (1996a: 230; 1996c: 6).

Table 19.3 Europe: major recipients of international tourist arrivals, 1994.

	Arrivals ('000)	% change from 1993	% share of European market	Tourism receipts as % GNP (1990[a])	Tourism receipts per tourist(US$)
France	61.312	1.23	18.59	0.02	403
Spain	43.232	7.85	13.11	4.36	506
Italy	27.480	4.17	8.33	1.39	871
Hungary	21.425	–6.05	6.50	–	67
United Kingdom	21.034	8.43	6.38	1.34	722
Poland	18.800	10.59	5.70	–	327
Austria	17.894	–1.99	5.43	7.44	735
Czech Republic	17.000	47.83	5.15	–	116
Germany	14.494	1.02	4.39	0.72	730
Switzerland	12.200	–1.61	3.70	–	621
Greece	10.713	13.81	3.25	3.73	365
Portugal	9.132	8.28	2.77	6.94	419
Netherlands	6.178	7.31	1.87	1.36	908
Turkey	6.034	2.20	1.83	–	716
Belgium	5.309	3.69	1.61	1.98	976

[a]CEC DG-XXIII Tourism Unit (1994: 148) includes only EU members.
Sources: WTO (1996c: 96–8, 102).

not structural shift from the earlier generation of mass consumption resorts located closer to their markets. Those resorts, such as Blackpool, Ostend, Boulogne and Scheveningen had been created to serve spatially, temporally and financially restricted domestic industrial workforce markets. Indeed, the spatial shift to the new mass resorts of southern Europe was often more real than apparent: package tourists often travelled by night, particularly by charter flight, to arrive in hotels, and perhaps whole resorts, dominated by their own national groups (Pearce, 1987a, b). Further, 'Package holidays to popular resorts tend to be in cultural terms more akin to what Lenin saw of Germany during his famous return journey to Moscow in a sealed train' (Williams and Shaw, 1994: 309).

The new centres of mass international tourism were dependent upon private-sector investments, but state intervention was required to regulate production and apply minimum standards for accommodation, catering and other service provision, as well as providing infrastructural investment. The model which emerged was one of host state support for, in effect, market dominance by a decreasing number of tourism source companies consolidating their hold on the market through vertical integration of the agencing, tour, air travel and accommodation sectors. However, the functioning of these resorts also relied on the existence of large numbers of indigenous small enterprises, many of which were locked into subcontractual relationships with the major tour companies, contributing to a form of dependency relationship which exposed a vulnerability to both fashion and further structural change in the tourism industry (Montanari and Williams, 1995).

With a relatively standardized product emphasizing the general attributes of sun, sand and surf, or snow and ski-slopes, rather than the particularities of places, competition in mass tourism was based on price sensitivity, which demanded low costs and thus helped to sustain low-wage labour markets. The convergence of a bad press, low-status image, increasing environmental awareness, growing numbers of critiques of tourism impacts, and the self-reappraisal of host governments, complemented a growing structural shift in the tourism industry (Nash, 1996).

Just as many consumers in more-developed societies became dissatisfied with standard, mass-produced goods and sought more varied and customized products to meet their individual needs (Featherstone, 1991), as an escape from the pressure and pollution of over-regimented urban living (Krippendorf, 1987), many tourists rejected standard mass-package tours in favour of holidays which responded to their desire for learning, nostalgia, heritage, make-believe, action, and a closer look at the 'Other' (Boissevain, 1996: 3). A relative shift in the tourism product from Fordism to post-Fordism (Urry, 1990) was the response to such changes in consumption aspirations. The greater emphasis on more individualistic or specialized niche-market holidays required smaller-scale and more flexible provision, thereby creating high value-added market segmentation – opportunities for new and opportunistic small companies just at a time when the large conglomerates were extending their dominance of the industry through vertical integration. Many of these smaller firms benefited from lower overhead costs and more flexible labour, while improvements in information technology allowed them to compete relatively successfully in sophisticated and internationalized markets.

Urban and heritage tourism

Metropolitan areas became Europe's major tourism magnets (Table 19.4) (van den Berg *et al.*, 1995). With a decline in the role of state intervention, many urban and regional economic regeneration strategies embraced tourism as a major policy instrument, as exemplified in Glasgow, Liverpool and Barcelona (Law, 1994; Wynne, 1992). This often involved major infrastructural investments such as waterfront redevelopment (Jansen-Verbeke and van de Wiel, 1995; Hoyle, 1996) and the promotion of 'sense of place' through major events and festivals (Gratton and Taylor, 1995).

Table 19.4 Tourism in Europe's major metropolitan areas, 1991.

	Tourist numbers		Average length of stay
	Arrivals	Nights spent	(days)
London	14,700	82,600	5.62
Paris	12,602	28,269	2.24
Munich	3,243	6,608	2.04
Rome	2,684	12,019	4.48
Vienna	2,638	6,718	2.55
Berlin	2,542	6,405	2.52
Milan	2,135	5,579	2.61
Brussels	2,046	3,035	1.48
Frankfurt	1,863	3,443	1.85
Barcelona	1,819	4,090	2.25

Sources: van den Berg *et al.* (1995: 187) Shacher (1995: 153).

While such developments may be rooted in the present and look to the future, an increasingly dominant form of recreation has come almost to symbolize postmodern tourism – the appreciation and promotion of 'heritage' and 'cultural' tourism (Richards, 1996). For Ashworth (1993: 14), heritage comprises 'historic resources, whether the conserved built environment of historic architecture and urban morphology, associations with historical events and personalities and the accumulations of past cultural artefacts, artistic achievements and individuals', which are seen to represent the single most important primary attraction for tourists, rendering heritage sites and the cities they cluster within Europe's most important contemporary tourism resorts (Ashworth and Tunbridge, 1990). As a consequence, heritage has taken on a leading role in the promotional tourism images of most European countries. Ashworth (1993) points to the volume sales of 'Michelins' and 'Baedeckers', and the substantial growth in subscriptions to heritage trusts and museum associations as evidence of a mass European interest in the recreational consumption of heritage. It is often the case, however, that heritage resources attract predominantly local residents on repeat visits.

Although these trends were largely initiated within the developed market countries of the north, much of the Mediterranean littoral has responded to them by, for example, upgrading the promotion of non-coastal historical and natural attractions. Promotion of city breaks and the construction of heritage parks and golf courses have become notably competitive, spatially diffusing and complementing tourism employment opportunities by generating facilities for all-year leisure pursuits (Priestly, 1995b). The promotion of such varied destinations and activities is targeted at 'quality tourists', perceived, accurately or otherwise, as more affluent and cultured, and who may be 'widely viewed as the key to liberating destination communities from enervating dependence on hordes of low-spending package tourists' (Boissevain, 1996: 3). Boissevain illustrates this point by noting Spain's adoption, in 1992, of a new national tourism slogan, 'Spain: passion for life' (together with a motif designed by Catalan artist Joan Miró) (Albert-Piñole, 1993), to replace the long-established 'Spain: everything under the sun'.

The Spanish case is emblematic of pan-European trends in that tourism development has taken place within the context of substantial political change. A post-Franco process of liberalization and decentralization transferred tourism administration to autonomous regional governments from the late 1970s (Pearce, 1997). As each authority has defined its own objectives and policies, improvements in public investment and planning, legislative regulation and training provision have taken place; at the local level, municipal plans have attempted to fit regional and national guidelines (Barke and Newton, 1995; Priestly, 1995a).

In Majorca, for example, policies were introduced in 1985 which limited building permits to four-star or higher quality properties, with a maximum height of three storeys, and a minimum of 30 square metres of space per guest. Accompanying environmental enhancements included tree planting, pedestrianization schemes and the development of a public access seafront esplanade in the resort of Magaluf. Also in Majorca, wider customer mix was sought from new markets, large single-sex groups were discouraged, and tour company representatives' roles were redefined in order to improve tourist behaviour and the tourism image (Laws, 1995). These developments were subsequently consolidated and incorporated into the Spanish government's four-year tourism improvement plan (1991–95), which focused on the country's heritage and other cultural and natural resources with the overall objective of improving product quality rather than expanding capacity (Jenner and Smith, 1993).

Rural tourism

In a number of European countries leisure-related activities now dominate rural employment growth. This has usually arisen from one of two contrasting situations: where the attractions of the 'countryside' have acted as a basic resource for tourism organized and sustained through small enterprises that can be locally owned; and where a rural area has been transformed by major inflows of capital and people, whether Alpine ski centres (Pearce, 1995) or self-contained coastal complexes (Schweizer, 1988), EuroDisney or a Center Parc, which have profoundly transformed the spatial and economic structure of the area (Figure 19.1). Such processes not only transform the immediate rural landscape, but their impact on other economic and social sectors, both within and beyond the tourist centre, may be considerable (Cavaco, 1995).

Employment may be generated not only in the accommodation, food and leisure sectors, but also in local craft and other manufacturing industries. Growth of rural employment in tourism can encourage population retention or even repopulation in areas that might otherwise experience depopulation (Ciaccio, 1990), although the source areas of migrants to tourism employment may themselves suffer structural imbalances in the labour force as a consequence.

Reflecting their lower population densities and relatively narrow range of economic activities, rural areas tend to render significantly lower multiplier effects from tourist expenditure. Seasonality may further limit income and employment gains (Hannigan, 1994). However, the degree to which local communities regard such development as positive or negative can depend heavily upon local circumstances.

Generally, tourism is viewed in much of Europe as a major development option, even for remoter rural areas (Ciaccio, 1990). For example, 69% of Spanish project proposals under the EC LEADER programme for rural regeneration have been for tourism. In Greece, areas of second-home and tourist demand have shown some of the most important economic growth (Getimis and Kafkalas, 1992). In Ireland, the relatively poor performance of manufacturing-led policies aimed at rural and national development induced the government to emphasize policies favouring tourism, and this sector provided 75% of new employment in the 1989–92 period.

The potential for, and relatively low cost involved in establishing SMEs in rural tourism and recreation provides former urban dwellers with an opening to rural life, particularly where proceeds from the sale of a home in a more costly urbanized region can be deployed (Buller and Hoggart, 1994a). There is, however, limited evidence to indicate the long-term viability of such enterprises and their local impact. Although the ownership of tourist enterprises by those who had migrated into the Scottish Cairngorms was resented by locals (Getz, 1981), Buller and Hoggart (1994b) found no evidence of this in the attitudes towards British migrants in rural France. Earlier evidence from Ireland suggested that one of the more significant social impacts tourism brought to isolated rural communities was an enhancement of local self-esteem (Messenger, 1969).

While mass tourism may dominate particular locations (Pearce, 1987a), in rural areas relatively small numbers of tourists can have a disproportionate impact, although local hosts' responses will vary considerably (Boissevain, 1996), and the seasonality of tourist movement can be used to advantage if it complements other economic activity such as aspects of farming or fishing, so that tourist employment can be generated during otherwise slack periods (Puijk, 1996).

Figure 19.1 EuroDisney, Paris (*source*: Tim Unwin, 1 March 1997).

Ecotourism in Austria

Erlet Cater

Product or principle?

Successful ecotourism relates not only to its prospects for sustainable tourism development but also to its potential for contributing towards sustainable development in general (Cater, 1994). Whether or not it succeeds on both these counts depends on its interpretation. As a product, it is identified as a niche or market segment, generally equated with nature or ecologically based tourism. It may also include the cultural attractions of the destination; 'natural' attractions are often a product of many centuries of indigenous land management. This is particularly apposite in Austria, where over the past 1000 years the Alpine farmer has created many new biological interfaces with a variety of valuable natural features (Broggi, 1988).

Ecotourism interpreted merely as a product, however, may be ecologically and culturally based, but not necessarily ecologically sound, responsible or sustainable. To incorporate these vital characteristics it must adhere to three essential principles: environmental and cultural integrity, the enhancement of local livelihoods, and economic viability. If ecotourism embodies these essential principles it should produce symbiotic relationships, with environmental protection resulting both *from* and *in* improved local standards of living, sustained visitor attraction, continued profits for the industry and conservation revenue.

The rationale for ecotourism in Austria

Whilst ecotourism has been eagerly seized upon by new tourist destinations, anxious to avoid the mistakes of antecedents elsewhere, it has also captured the interest of long-standing destinations such as Austria, which are seeking both to diversify their product and to ensure greater sustainability.

The economic significance of tourism

Tourism in Austria contributes significantly to economic growth, employment and balance of payments, with tourism receipts constituting around 20% of export earnings (Embacher, 1994). Including domestic tourism, the total turnover of tourism and tourism-related activities accounts for an estimated 15% of GDP, with some 400,000 people employed.

Variation and change

The economy is therefore vulnerable to overall tourism market variations. Changes in market characteristics also condition these trends. Table 19.5 shows that both tourist arrivals and tourism receipts declined in the 1990s, reflecting expanding international competition and the failure of

Table 19.5 Austria: trends of tourist arrivals and tourism receipts, 1990–95.

Year	Tourist arrivals (thousands)	% change over previous year	Tourism receipts (m US$)	% change over previous year
1990	19 011	4.44	13 410	25.13
1991	19 092	0.43	13 800	2.91
1992	19 098	0.03	14 526	5.26
1993	18 257	–4.40	13 566	–6.61
1994	17 894	–1.99	13 160	–2.99
1995	17 173[a]	–2.40[a]	12 896[b]	–2.00[a]

[a] Department of Tourism, Austrian Federal Ministry for Economic Affairs, 1996.

[b] Estimate obtained from applying Austrian Department of Tourism figure of percentage change to WTO (1994) data.

Sources: WTO (1996c), Austrian Department of Tourism (1996).

continued

Figure 19.2 The Alpine ski resort of Obergurgl in the summer (*source*: Tim Unwin, 10 August 1980).

Austria, until recently, to adapt to changing demands. Changing market segmentation is also evident, with the so-called 'post-tourists' (Urry, 1990) or 'new tourists' (Poon, 1993) railing against the characterisics of mass consumption implicit in mass tourism. The changing fortunes of Austrian tourism must, therefore, be viewed in the light of these overall trends, but it is necessary to examine the dynamics of the situation to understand how Austria is attempting to try to reverse them. Austrian tourism displays distinctive sectoral, temporal and spatial features (Figure 19.2).

Zimmermann (1991) notes several features, associated with a post-industrial society, which have had a lasting effect on the structure of tourism in Austria. First, changing conditions of employment (unemployment, low wages and reduced overtime) have mainly affected those social classes which are predominant in Austrian summer tourism. Second, the visitors are likely to include fewer younger people, more single households and also more, and more active, senior citizens. These groups are more independent and flexible, inclined to travel further afield than Austria. Third, the demands of 'new' tourists for more diversified holidays will impact negatively on the traditional stucture of Austrian tourism, and fourth, increasing ecological consciousness implies increasing attention to environmental management.

Whilst Austria enjoys a favourable double-peaking of international tourist arrivals, with 41.1% of foreign bednights in winter and 58.6% in summer (EIU, 1993), this aggregate picture masks two areas of concern. First, Austria has lost its competitive edge more in summer than in winter. In 1995 the number of bednights declined for the fourth consecutive summer; international demand dropped by 7.4% compared with the winter 1994/95 shortfall of 2.7% (Austrian Department of Tourism, 1996). Second, capacity is heavily under-utilized in the shoulder months of spring and autumn.

Tourism in Austria is also spatially uneven. The western provinces of Tyrol, Salzburg and Vorarlberg are most popular, their share of overnight stays increasing from 42% to 61% between 1952 and 1995. Correspondingly, the three eastern provinces' share decreased to 13% by 1995 (Table 19.6). Consequently, saturation point has been reached in some Alpine areas, with a negative effect on growth, whilst simultaneously the non-mountainous areas of Lower Austria, Vienna and Burgenland are not achieving their tourism earnings potential.

continued

Table 19.6 Regional distribution of tourism in Austria, 1995.

Province	Bednights in all accommodation		
	% domestic	% foreign	% total
The High Alps	*30.9*	*70.8*	*60.5*
Salzburg	18.4	19.4	19.1
Tyrol	9.7	43.2	34.6
Vorarlberg	2.8	8.2	6.8
The Eastern Provinces	*20.7*	*10.0*	*12.7*
Vienna	3.1	6.9	5.9
Lower Austria	12.8	2.3	5.0
Burgenland	4.8	0.8	1.8
The rest of Austria	*48.4*	*19.2*	*26.8*
Carinthia	16.4	11.3	12.7
Styria	19.2	3.9	7.8
Upper Austria	12.8	4.0	6.3
Total bednights (m)	30.12	86.99	117.11

Source: Austrian Department of Tourism (1996).

Ecotourism initiatives in Austria

An examination of several ecotourism initiatives in Austria highlights their relevance to improved standards of living for local communities, cultural and environmental integrity, economic viability and visitor satisfaction.

Enhancing rural livelihoods

Whilst tourism has played a vital role in certain rural areas of Austria for over a century, ecotourism is seen as key strategy for economic renewal in declining rural areas. The OEAR (Austrian Association for Regional Self-Reliant Development), a non-governmental agency promoting and supporting endogenous local development, advocates tourism development in peripheral areas. Environmental soundness, social acceptance and maximization of local added value are central principles (Hummelbrunner and Miglbauer, 1994).

Farm tourism has existed in Austria for over a century. Indeed Alpine tourism began with farmers letting rooms to tourists and combining tourism with agriculture (Burton, 1995). Letting rooms on farms is the most important touristic income source for Austrian farmers: 30,000 farms let approximately 250,000 beds (Pevetz, 1991). Farm tourism in Austria has gained a new impetus, however, with the Austrian Farm Holidays Association, created by the eight provincial associations as a coordinating and integrating body in 1991 (Embacher, 1994).

Ensuring cultural and environmental integrity

As well as increased emphasis on farm tourism, rural tourism has recently taken a new turn in Austria with an initiative aimed at ensuring sustainability.

The self-regulatory Holiday Villages in Austria Association was established in 1991 to encourage high environmental standards compatible with the growing demands of tourism. Membership criteria concern village character, minimum ecological standards and minimum social and touristic standards coupled with maximum load thresholds. Strict membership vetting is undertaken by a national committee. By July 1995 there were 32 member villages, five awaiting assessment and 15 rejected applications (Parker, 1996).

The Hohe Tauern National Park, receiving 4 million visitors annually, includes 29 communities. In an attempt to conserve nature, preserve traditional living space and promote tourism, the national park area has been subdivided into a largely unspoiled high-alpine core zone and an outer zone including landscapes modified by human impact. Local participation in the planning process has reversed initial widespread rejection of the park concept. The national park thus preserves a beautiful authentic mountain environment; provides sustainable living space, maintaining a traditional but evolving cultural landscape; and establishes a recreation area based on a concept of 'gentle tourism' (Stadel *et al.*, 1996).

A clear commitment to the principle of a symbiotic relationship between tourism and the environment is evidenced by the success of the Weisensee scheme in south-west Carinthia. This strategic concept, drafted with full local participation, involved a comprehensive analysis of the region's economic, cultural, social and environ-

continued

mental problems. Amongst some of the far-reaching policies are the transportation of visitors into the valley from car parks at the valley entrance, and a moratorium on the construction of a lakeside thoroughfare.

Economic viability

Tourism is an economic activity, which must be capable of making a profit in order to survive and benefit the community (Murphy, 1994). Successful promotion of the ecotourism product in Austria includes two initiatives: eco-marketing and eco-labelling. With reference to eco-marketing, the Austrian National Tourism Office aims to promote and present high-quality tourism which maintains environmental quality and the quality of local livelihoods. As well as launching 'Green Villages', it is concerned with promoting alternative tourism in the shoulder months, such as environmentally oriented and environmentally sound cycling holidays during spring and autumn.

Environmental seals of quality fulfil several important aims. As well as informing tourists on environmental practice (facilitating holiday selection), eco-labelling creates and maintains a competitive advantage, providing a basis for environmental management and thus greater sustainability. The Tyrolean Environmental Seal of Quality is a major initiative along these lines. Establishments which meet a set of obligatory criteria, monitored on an annual basis, are listed in the Tyrolean Tourist Board publication *Travelling with Nature* (Williams and Shaw, 1996).

Visitor satisfaction

The delivery of a successful product by tourism enterprises is dependent on visitor satisfaction. All of the above initiatives are aimed at enhancing the quality of the tourism product. In particular, the Green Villages scheme, in setting a maximum tourist–resident ratio, recognizes not only social, ecological and physical carrying capacities, but also the psychological carrying capacity; this threshold ensures a degree of exclusivity and an uncrowded holiday experience for visitors.

Future prospects

There is a marked reorientation towards a new tourism identity for Austria in an attempt to attract the 'visitor 2000' (Embacher, 1994). How successful this strategy will be in regaining a competitive edge is highly dependent on a number of exogenous factors. What is evident, however, is that the various initiatives are essentially 'eco-realist' in that they attempt to reconcile the many interests involved and thus move towards a more sustainable tourism industry in the country.

Further reading

Embacher, H. (1994) Marketing for agri-tourism in Austria: strategy and realization in a highly developed tourist destination, *Journal of Sustainable Tourism*, **2**(1&2), 61–76

Hummelbrunner, R. and Miglbauer, E. (1994) Tourism promotion and potential in peripheral areas: the Austrian case, *Journal of Sustainable Tourism*, **2**(1&2), 41–50

Pevetz, W. (1991) Agriculture and tourism in Austria, *Tourism Recreation Research*, **16**(1), 57–60

With the growing emphasis on alternative forms of tourism, ecotourism and other country-related activities are finding an increasing number of adherents (Eadington and Smith, 1992). This development has been grafted onto the ideal in many northern European city dwellers' perceptions of serene rural idylls synonymous with the good life, and which can act as a strong attraction for tourists (Krippendorf, 1987) (see case study of Austria).

The last 15–20 years have also witnessed a growing number of northern Europeans taking up residence in warmer southern Europe (King and Rybaczuk,

1993), with, for example, Spain being the home for 135,000 Germans, 82,000 French, 51,000 Belgians and 30,000 Britons, and Portugal acting as the residential location for almost 430,000 French, 93,000 Germans and even 34,000 Luxembourgeois (OECD, 1994c). Although some of these inflows will result from employment-related moves, a significant number are driven by consumption considerations. Many of the inflows are targeted towards tourist zones, whether for retirement migration or across the broader age range (Buller and Hoggart, 1994a). But the character of tourist destinations does vary: coastal zones retain a high priority, but trends towards more upland, scenic areas and more remote places are significant in 'repopulating' areas that have experienced previous depopulation (King, 1991). With abandoned properties being taken up, one consequence of such inward investment can be a marked improvement in local housing quality, along with increased employment opportunities in construction and home maintenance (Hoggart and Buller, 1995).

It is too simplistic to argue that there has been a linear and universal shift from Fordism to post-Fordism to produce a dominant postmodern tourism in Europe. Any such transition has been uneven and relative, and the coexistence of different modes has persisted. Some sectors, such as rural tourism, have been far less influenced by mass production and consumption, while 'coastal' tourism embraces a sufficiently wide range of environments and activities such that it could never be characterized as a wholly mass-tourism product or experience (Montanari and Williams, 1995). Further, many centres of mass-tourism – such as Benidorm – appeared to experience a revival in the 1990s, just when there was growth in more specialized holiday markets elsewhere in Spain. This reflected the inward restructuring and attempted reimaging of former mass-tourism locations, but it also embraced a rejuvenation of the mass-tourism cycle in Europe due to the growing demand for package holidays from recently 'liberated' central and east Europeans and other emerging markets.

Indeed, one notable characteristic of post-communist restructuring and privatization processes within the central and eastern Europe tourism industry has been the rapid growth of travel agents and holiday companies providing Mediterranean coast packages (Gibson, 1996). In the summer of 1994, for example, the Czech airline CSA was involved in a large number of charter flights out of Prague, including 95 to Split and Dubrovnik to convey some 10,000 Czech tourists to these still inexpensive Dalmatian coastal resorts not long after they had been liberated from the Balkan war. With a long-term consolidation of central and eastern Europe's economic stability and standards of living, increasing numbers of households will improve their mobility and access to international travel, a process which, nonetheless, will be both socially and spatially uneven and dependent on a wide range of factors. But it can be suggested that the process of political, economic and social restructuring in central and eastern Europe and the lands of the former Soviet Union has created a series of demand waves for mass tourism just at a time when product demand in the West was thought to be on the wane (D.R. Hall, 1995), thus complicating any conception of European tourism smoothly transforming itself from a modern to a postmodern consumption experience.

Tourism as a motif of restructuring in central and eastern Europe

This chapter is based on the premise that tourism development is an integral component of social, economic and political restructuring across Europe (Hall, 1991; Montanari and Williams, 1995). Not least in central and eastern Europe, a number of restructuring processes have implications for, and are in their turn components of, tourism development. These include the loosening of constraints on personal spatial and social mobility, deregulation, large- and small-scale privatization, divestment and price liberalization, reorientation of foreign trade, inward investment, globalization (of, for example, accommodation and transport sector elements), internal currency convertibility, upgrading and reorientation of infrastructure, adoption of new technology, and skill training programmes (D.R. Hall, 1995).

A fifth of Europe's international arrivals are recorded for central and eastern Europe (22.5% in 1994), but only a twelfth of the continent's tourism revenues are generated there (8.3% in 1994, up from a mere 3.4% in 1990) (WTO, 1996c: 89). Improved access, changing image projection, penetration of Western inward investment and a growth of service industry employment and training have attracted growing numbers of international arrivals to central Europe (Hungary, Czech Republic, Poland, Slovakia), but these tourists are comparatively low spenders and short stayers. Large numbers of excursionists – cross-border day-trippers and petty traders, both taking advantage of differentials in currency rates – and transit travellers inflate 'visitor' figures considerably. For example, numbers of recorded international arrivals in Poland quadrupled between 1990 and 1994, from 18 to 74 million, yet in the latter year only nine million overnight stays were recorded (WTO, 1996a: 137). Although international tourism is largely stagnating in the Balkans and the former Soviet Union, as continuing political and economic instability deters potential visitors and market investors, with the assistance of the international banks and the EU a number of central and eastern European countries, as in the rest of Europe, are looking to higher-income generating, season-extending niche products with minimum social and environmental impacts, which can exploit comparative advantage, such as business and conference tourism, rural tourism and the promotion of cultural heritage (see Caucasus case study).

Rural and nature tourism have also received substantial promotion in the region in recent years with professional marketing undertaken by local and central government, NGOs and private sectors, extolling, for example, Poland's national parks (Witak and Lewandowska, 1996), Estonia's protected areas (Ruukel, 1996) (Figure 19.3) and even Serbia's 'landscape painted from the heart' (Popesku and Milojević, 1996), the latter with the assistance of Saatchi & Saatchi. Like Spain, Slovenia explicitly reconfigured its tourism emphasis by replacing its 'Sunny side of the Alps' slogan in 1996 with 'The green piece of Europe'. Its rural–cultural emphasis is encapsulated in promotions for 'wine journeys' (Fujs and Krašovec, 1996).

Areas of debate

'Sustainability'

The paradoxes surrounding concepts of 'sustainable tourism' are amply articulated in central and eastern Europe (Hall and Kinnaird, 1994). Any restriction of personal freedom – in terms of spatial proscription in sensitive areas, banning the pursuit of certain activities or discouraging the development of certain types of businesses, and indeed physical and spatial planning generally – might be seen to echo the half-century of post-war communist imposition. Certainly the new 'freedoms' are being pursued by an emergent private entrepreneurial sector, not least in the tourism and travel sectors, sharply focused on short-term profit (Koulov, 1996).

Figure 19.3 Farm tourism accommodation in rural Estonia (*source*: Tim Unwin, 17 June 1995).

The Caucasian spa resorts: ecological context and problems of development

Dmitri Piterski and Isolde Brade

The Caucasian Mineral Waters region is now one of the largest and the most important recreation regions in Russia. Before the disintegration of the USSR 7% of the total health facilities of the country were situated in the region of the Caucasian spa resorts. The official date of birth of these famous Russian resorts is 1803, although their recreation function has been developed since the 18th century.

The Caucasian Mineral Waters region is an area of some 400 sq km, situated in the northern foothills of the Caucasus, not far from Elbrus (Figure 19.4). Four famous resorts are situated here: Kislovodsk, Yessentuki, Pyatigorsk and Zheleznovodsk. The total recreation potential at the beginning of the 1990s was more than 45,000 places in rehabilitation and health facilities, and about 2500 in tourist hotels (Table 19.7). The

Table 19.7 Key characteristics of the Caucasian spa resort towns.

	Places in rehabilitation and health facilities early 1990s	Height above sea level; therapeutic specialization of resorts	Population (thousands) 1 Jan 1996	Industrial specialization of towns
Kislovodsk	17,000	890 m, cardiovascular diseases, nervous diseases	116.8	Light industry, food industry, woodworking industry
Yessentuki	12,500	614 m, gastric diseases	89.6	Light industry, food industry
Pyatigorsk	9,500	510–630 m, gastric diseases, bone diseases, nervous diseases	185.4	Machine-building, production of building materials, light industry, food industry
Zhelesnovodsk	6,500	630 m, gastric diseases, internal diseases	48.1	Light industry, food industry
Mineralnye Vody	—	—	86.0	Machine-building, production of building materials, light industry, food industry
Lermontov	—	—	25.0	Machine-building, production of building materials, chemical industry
Total	45,500[a]		551.6[b]	

[a] At present the recreational potential of the region is less than half used.
[b] Without refugees and involuntary immigrants.

Sources: Goskomstat Rossii, statistical institutions of corresponding towns, authors' calculation based on sources cited in Further reading.

continued

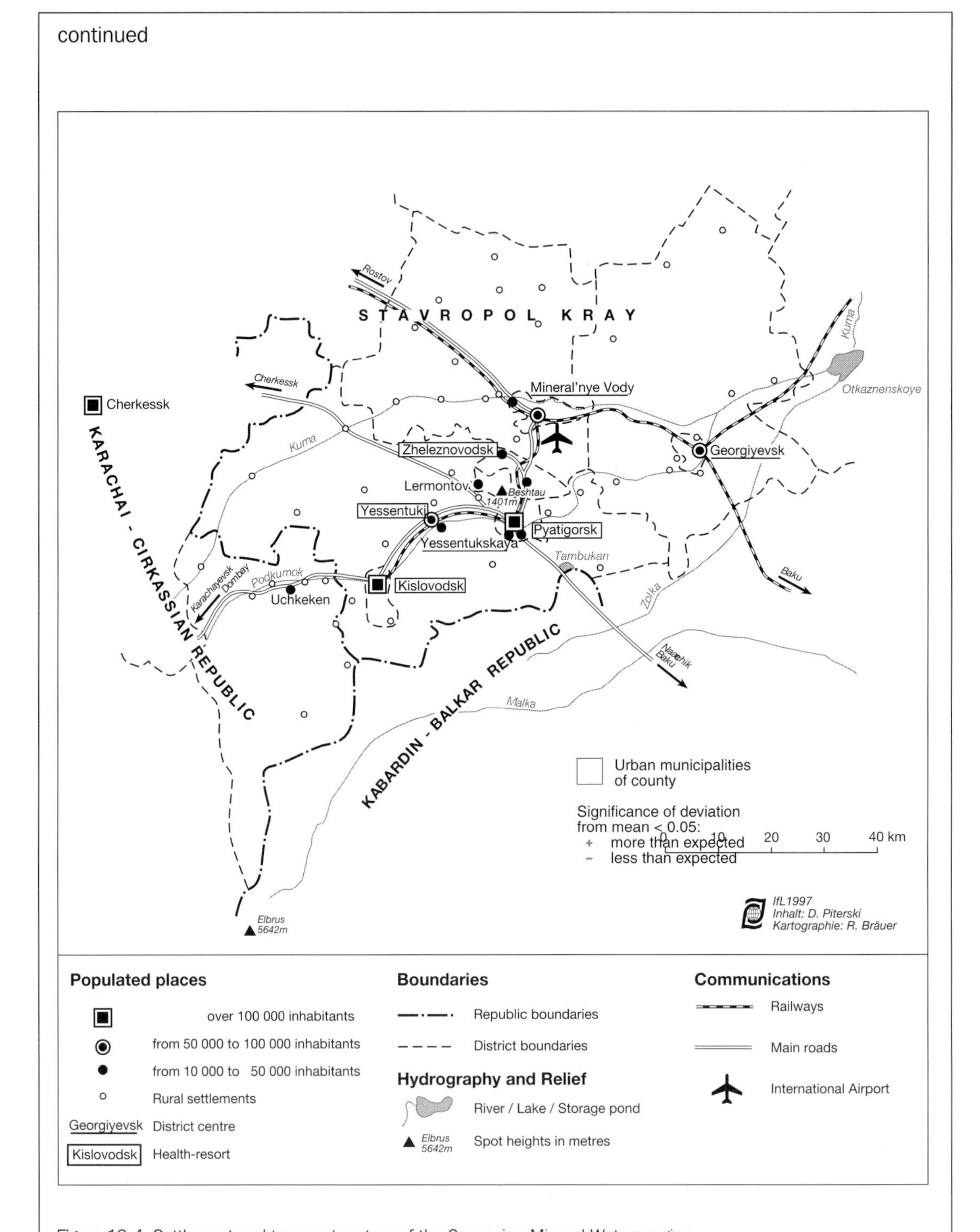

Figure 19.4 Settlement and transport system of the Caucasian Mineral Waters region.

continued

largest resort is Kislovodsk, with 40% of the region's recreation potential.

Kislovodsk differs from the other Caucasian spa resorts in its pure, dry air and lack of mists, plenty of sunlight all the year round, a unique *terrainkur* (the route for therapeutic walking) and the therapeutic mineral water. The resort's main function is the prevention of cardiovascular ailments, the most common human diseases today. The resort of Kislovodsk is also well-known because of its mountain park, which was laid out in 1823. The total area of this park is 1200 hectares, and the total length of the famous Kislovodsk *terrainkur* is over 80 km.

The future development prospects of the Caucasian Mineral Waters region are closely connected with the rich reserves of different mineral waters (10,000 cubic metres per day) and mud (1.7 million cubic metres). The favourable natural recreation conditions, their concentration in a relatively limited territory, the considerable potential of health resorts and the recognized high efficiency of the therapeutic treatment provide good perspectives for the further development of the health-resort complex of the Caucasian spa towns. But this region also faces numerous problems which will adversely affect its development perspective: pollution from industrial establishments, a critical ecological situation, oversized resort hotels below international standards, an uncertain geopolitical context, and recent socio-economic problems. Owing to such problems, the total number of holidaymakers in the Caucasian spa resorts has fallen during the 1990s and by 1996 comprised less than 50% of the 1990 figure.

The ecological context of the region is of special interest from the geographic point of view. The Caucasian Mineral Waters region is a multifunctional city agglomeration with about 550,000 urban dwellers, and the population density along the Mineral'nye Vody–Kislovodsk city axis is about 900 inhabitants per sq km. It is notable that these figures are comparable with those of many city agglomerations of non-resort regions in Russia. Besides four health resorts, six more towns are situated in the region. Among them are the traffic centre of Mineral'nye Vody with its international airport, and the industrial town of Lermontov. The agriculture of the region has mainly concentrated on specialization to support the resorts.

Over the last 25–30 years the total population of the region has increased considerably, due to the non-resort activities of the area. During this period the industrial zones in the spa resorts and other towns have also grown considerably to their present area of more than 1800 hectares. The largest city and the most important industrial centre of the Caucasian spa resorts is Pyatigorsk with more than 185,000 urban dwellers.

The pollution from some of the industrial and agricultural enterprises of the region is very high. Examples of the most significant sources of pollution include the machine-building plants and chemical enterprises in Pyatigorsk and Lermontov, and the poultry farms in Yessentuki. The industrialization of this region has created important conflicts of interest between economic growth, resort development and acceptable living conditions.

It is important to emphasize that the bad ecological situation in the Caucasian Mineral Waters region is partly a result of the peculiarities of the regional planning system of the Soviet Union and Russia, notably the low degree of actual regional and urban planning. During the implementation of the so-called complex social–economic development regional plans, urban development plans were substituted by plans of ministries, departments and enterprises. This planning system enabled territorial bodies of management to have a comparatively small influence on the economic and social development of their own regions, because the most important aspects of regional development were under the strong control of ministries, departments and union managerial bodies.

In the Caucasian Mineral Waters region all stages of the regional and urban planning documents were implemented. However, during the implementation of only the general plans for the health resorts since 1970, more than 200 violations of the plans have been recorded. As a result, the Caucasian Mineral Waters region with its unique natural and climatic conditions, as well as resources, has been transformed into a specific zone with a relatively unplanned settlement structure where industrial establishments, public

continued

enterprises and warehouses, facilities for transport and agriculture extend directly into the residential areas and even into the health-resort zone.

More than 1300 potential sources of environmental pollution are situated in the region. Due to these unsatisfactory ecological conditions, it is not possible to make use of the entire range of mineral waters for therapeutic purposes without the introduction of expensive purification facilities. Recent scientific and planning studies suggest that a complex set of environmental protection measures should be implemented to provide opportunities for the stabilization of the ecological situation and for the restoration of the Caucasian Mineral Waters region.

As marketing of the health-resort potential is needed, the development of general plans and schemes for the environmental protection of the region were conceived which were specifically oriented towards the dislocation of industrial, transportation and agricultural activities, which are the main environmental polluters. The most important planning proposal is the introduction of a special regional regime for using the natural conditions and resources (the territory, mineral waters and mud, and natural climatic conditions), as well as the organization of an appropriate environmental monitoring system.

In this case, the Caucasian Mineral Waters region could be held up as a model for environmental protection among resort and recreation areas elsewhere and not only in Russia. Nevertheless, there is an important 'but' to be taken into account. For the present, the implementation of the new health-resort concept will only be possible with capital investments (domestic and foreign) and modern technology from abroad. The limiting factor in this case is the unstable geopolitical situation in the northern Caucasus and the correspondingly bad investment climate in this part of Russia.

Further reading

Brade, I. and Piterski, D. (1994) Die Kaukasischen Mineralbäder – Möglichkeiten und Grenzen der Entwicklung einer Tourismusregion (The Caucasian spa resorts. Possibilities and limits in the development of a tourism region), *Europa Regional*, **4**, 10–19 (in German, abstract in English)

Goskomstat Rossii (1996) *Size of the Population of the Russian Federation by Cities, Workers' Settlements and Rayons on 1 January 1995*, Goskomstat, Moscow (in Russian)

Perzik, J.N., Piterski, D. and Brade, I. (1997) Regionalplanung und Stadtentwicklungskonzeptionen in der früheren UdSSR und in Rußland (The regional planning and conceptions of the settlement systems of Russia and of the former USSR), *Beiträge zur Regionalen Geographie* (forthcoming) (in German, abstract in English)

Romanov, N.E. (ed.) (1980) *Natural Resources and Productive Forces of the Northern Caucasus. Recreation Resources*, Rostov University, Rostov-on-Don (in Russian)

Vartanyan, G.S. and Plotnikova, R.I. (1993) Estimation of the prospects for the development of the mineral waters resources of the Caucasian Mineral Waters region, *Razvedka i Okhrana Nedr*, **11**, 29–32 (in Russian)

Vedenin, Y.A. and Savel'yeva, V.V. (eds) (1985) *Problems of the Use of Recreation Resources in the Caucasian Mineral Waters Region*, Academy of Sciences of the USSR, Moscow (in Russian)

Paradoxically, however, in peripheral regions of ecological importance, 'traditional' local practices such as hunting, poaching and smuggling were often carried on under previous regimes, often within the 'shadow' economy, because of relative isolation from central authority and/or implicit if illicit compliance of local figures of authority. Such areas as Romania's Danube Delta, one of Europe's most important wildlife habitats, which are now subject to conservation and protection, have required a considerable effort to involve the local population in processes of sustainable development and planning. Employment- and income-generating processes, such as the recruitment of park wardens and the encouragement of hosting bed and breakfast-type accommodation, have been pursued particularly in an attempt to preclude the community feeling that its livelihood is threatened by conservation designations, which are then, by definition, not sustainable (Hall, 1996).

In central Europe, sheer weight of numbers of visitors, even in pedestrianized areas such as central Prague, have forced planners to think seriously about restricting access to such famous structures as the Charles Bridge. To relieve congestion for both residents and tourists in the much visited Old Buda area of Budapest, the city authorities introduced a shuttle minibus service and comprehensive parking restrictions in the 1980s. In a number of cities, hotel developers are now required to provide substantial (often underground) car parking facilities: in Bratislava, for example, the Hotel Danube, completed in 1993, can accommodate 400 vehicles. Such developments and requirements are, of course, also likely to induce further traffic growth.

The literature on tourism impacts is now extensive (Young, 1973; Mathieson and Wall, 1982), and the major arguments need not be repeated here. However, as Ashworth (1993: 16) has noted, with the close coexistence of large numbers of tourists and heritage resources, 'feet, breath, even digestion of visitors are now seen as posing a serious threat to the physical survival of Stonehenge, Lascaux cave paintings and the Sistine Chapel respectively'.

Heritage as promotion

While many explicit examples of both positive and negative direct tourism impacts can be cited (Archer and Cooper, 1994), the more subtle interaction of tourism with other processes of modernization and development has been caught up in the wider challenges of pan-European restructuring. Not least in the tourism and travel industry, 'sustainability' can be interpreted in a number of different ways, whether for altruistic or marketing purposes. Neither is 'heritage' a value-free concept: economic power and politics influence what is preserved and how it is interpreted. Hewison (1987) blamed the 'heritage industry' for encouraging Britain's 'climate of decline' by promoting an obsession with a 'comforting' past rather than facing the challenges of the present and future. Tourism was seen as parasitical and, through its failure to reinvest in culture, ultimately destructive.

Heritage as nationalism

Promotion of heritage and a recognition of the importance of cultural history was a feature of the communist period in all countries of central and eastern Europe, as exemplified in Bucharest's open air village museum (Focsa, 1970) and the large-scale refurbishment of central Cracow (Dawson, 1991). This was often not primarily for international tourism purposes, but rather to inculcate a sense of identity and achievement amongst each country's citizens, most notably seen in the priority given to the faithful reconstruction of Warsaw's Old Town after Second World War destruction, employing old plans, photographs and paintings by Canaletto and Bellotto as guides.

The resurgence of the expression of nationalism and a (re)creation of new state systems, most notably in the Balkans and the former Soviet Union, is witnessing an employment of the heritage industry as a means of reinforcing national or particular ethnic identity (King, 1996). The synergy between nationalism and heritage was forcefully illustrated in the wars of Yugoslav succession through the roles of those two symbols of the tourism industry of a multi-ethnic Yugoslav state: the Turkish bridge over the River Neretva at Mostar, and the medieval walled city of Dubrovnik. Centuries of Turkish domination of much of the Balkans, and continuing Islamic culture and faith, were embodied in the cultural and ethnic symbolism of the Mostar bridge. Although physically linking Croat and Muslim communities across the river like a hand of peace and friendship in the ethnic melting pot that was Bosnia-Herzegovina, the bridge had increasingly come to symbolize for some Christian Slavs a perceived Islamic threat. The bridge's destruction through shelling by Bosnian/Herzegovinan Croats graphically confirmed the end of the multi-ethnic dream. It highlighted the powerful imagery of heritage for cultural antagonisms, as slow-motion replays of the bridge's destruction were televised around the world.

The bombarding of Dubrovnik by Serbs and Montenegrins was aimed primarily at destroying the tourism industry as a major plank of the Croatian economy (Oberreit, 1996). Four-fifths of recorded Yugoslav tourism earnings had been previously generated in Croatia, mostly from the long Dalmatian coast and offshore islands. Serbs had long coveted an access corridor to the sea, a factor which had split major European powers in the 19th century. That the elongated, southward thrusting Croatian territory had become a source of considerable income generation as a result of the rise of the tourism industry from the 1960s had only strengthened Serbian resentment. Dubrovnik, with its Venetian heritage, became viewed, paranoically, as anti-Orthodox and anti-Serb.

That the bombardment of the city raised shrill voices in the West from the great and good concerned with the need to protect the heritage value of Dubrovnik's *built* environment only helped to bring into sharp relief the apparent relative lack of will in the West to stop the *human* suffering wrought by the Yugoslav conflicts (Hall and Danta, 1996). Heritage was portrayed almost as if it was a Western plaything which had to be protected from the 'Other'. Dubrovnik's value as a UNESCO World Heritage Site appeared to be abstracted from its contemporary social setting. Only the assemblages of buildings and their fabric were promoted as worthy of protection for future generations, with the people living in or by such sites being unnoticeable and dispensable unless offering some readily digestible cultural 'performance'. In a conflict characterized by both extreme human brutality and collective enfeeblement, the Western response to the plight of Dubrovnik starkly confounded the received wisdom that if sustainability is to mean anything in urban and heritage tourism, local communities must be actively involved in, and must benefit from, processes of protection and promotion.

The gendered nature of tourism employment

The gendered horizontal segregation of occupations is particularly noticeable in semi-skilled, domestic and service-type occupations, such as those in the tourism industry (Kinnaird and Hall, 1994; Swain, 1995), especially where they mirror functions carried out in the home. Conventional explanations have implied that, if the best jobs and highest rewards are linked to an accumulation of human capital, women are inevitably disadvantaged because their process of accumulation is interrupted by marriage, birth and child-rearing (Coppock *et al.*, 1995). However, the position of women who continue to work without such breaks and yet who remain in low-status, low-paid occupations is not explained by such conventional approaches. Even where there is evidence of men and women starting with equal skills, qualifications or experience, the distribution of higher-status and higher-paid grades remains uneven. The implications and structural consequences of gendered employment in tourism-related activity in Europe are substantial.

Where tourism-related activity is pursued as a stimulus to economic development, it has been shown that employment opportunities for the local population are often typified by a predominance of unskilled, low-paid jobs, such as cleaners, clerks and kitchen staff. As in most forms of employment, these categories of tourism work tend to reinforce and exacerbate gender divisions of labour, resulting in probably inferior prospects for women's potential income attainment, job security, work satisfaction, access to resources, social mobility and socio-economic status. This has been exemplified within the hotel and catering industry in Britain and Ireland, for example (Breathnach *et al.*, 1994), where women are recruited into work which is deemed to represent an extension of their traditional domestic responsibilities, for which they will be inherently skilled. Although men are often employed as porters and stewards, they tend to be over-represented at professional, managerial and supervisory levels (Crompton and Sanderson, 1990). Utilizing tourism as a strategy for development (and the gender division of labour it reinforces) creates a situation in which women, otherwise marginalized in the workforce, are very much part of the prevailing capital and patriarchal social and economic structures (Ireland, 1993).

Early advocates of tourism as a strategy for development viewed tourism employment as a positive way of integrating underprivileged subgroups of society into the mainstream economy, but such notions are in danger of encouraging and reinforcing overt gender and ethnic divisions of labour within the tourism industry (Britton, 1991). Lever's (1987) study of Spanish migrant workers showed that much seasonal, unskilled, low-income and insecure tourism employment is undertaken by rural women who migrate as a result of poor rural employment opportunities. Female employment represents a lower cost than male because, for example, they sweep and tidy at the end of the day and perform other 'women's' tasks which men refuse to do. While tourism migration may bring temporary improvement for individual migrants, it acts to postpone the need to address long-term rural development questions, not least employment provision for women. Further, in debates on the pursuit of small-scale, 'sustainable' tourism, advocacy of local control, sensitivity to indigenous cultural and environmental characteristics and direct involvement of the local population have rarely addressed gender considerations as being central.

Changing gender relations can also be expressed through the way in which tourism impinges upon family structures. For example, family situations and household status will often determine women's access to employment opportunities. The demonstration effects of tourism development for the institution of the family may vary considerably according to geographical and cultural context. In Crete, Kousis (1989) found that, thanks to mass tourism, change in rural family structure reflected a more widespread control of decision making among family members and the possibility of increased autonomy for women. However, determining that economic rather than cultural factors induced change, she suggested that profoundly gendered practices, such as the importance of marital arrangements and the dowry system, have lost little of their significance in such a Mediterranean society. Further, the development of relationships between local men and female tourists required a revision of local moral codes which, applying only to male Cretans, thereby widened the gap of behavioural 'norms' between local men and local women.

Host–guest relations involve at least some exchange of social and economic values. The extent and degree of symmetry depends upon the nature and context of the interaction. Unless tourism is managed in a comprehensively 'prescribed' manner as in some of the (erstwhile) more totalitarian state socialist societies, some form of interaction will take place.

Conclusion

The development from mass to niche/postmodern tourism and travel experiences, while clearly a phenomenon of note, both within Europe and relating to the experience of Europeans travelling outside Europe, is by no means shared by all Europeans. Indeed, although the data are patchy, perhaps a fifth of citizens in west European countries, and often many more further east, enjoy no holiday experiences away from home at all. This 'tourism underclass' has grown substantially in central and eastern Europe with the privatization of tourism and travel facilities and loss of former subsidies (Hall, 1991). Its implications for social relations and identity in Western source countries should be a matter of concern. The resurgence of the mass-tourism product cycle, enmeshing the citizens of central and eastern Europe and helping the reinvigoration of some Mediterranean resorts, further confounds any simplistic notion of a linear transformation in Europe from mass to niche tourism.

Tourism and heritage promotion have become important ingredients in national and regional restructuring strategies and the attraction of inward investment. Yet questions of cultural and ecological sustainability, and the impacts of tourism employment on rural–urban, local–incomer, gender, household and other social relationships have barely begun to be addressed within the framework of such processes.

Small-scale 'sustainable' projects may in the longer term distinguish at least part of Europe's approach and response to tourism's environmental, cultural and other pressures. If the vulnerable natural and cultural environments particularly of the peripheral edges of Europe can be protected and enhanced through the medium of sustainable rural and ecotourism, then important models may have been established for others to adopt and adapt.

For better or worse, tourism will remain an integral part of regional reconstruction in much of Europe. Throughout the continent, and particularly in central and eastern Europe, there is a desperate need for appropriate empirical studies to better inform those involved in tourism strategy development and plan implementation.

Further reading

Boissevain, J. (ed.) (1996) *Coping with Tourists: European Reactions to Mass Tourism*, Berghahn Books, Providence RI and Oxford

Hall, D.R. (ed.) (1991) *Tourism and Economic Development in Eastern Europe and the Soviet Union*, Belhaven Press and Halsted Press, London and New York

Montanari, A. and Williams, A.M. (eds) (1995) *European Tourism: Regions, Spaces and Restructuring*, John Wiley, Chichester

Pompl, W. and Lavery, P. (eds) (1993) *Tourism in Europe: Structures and Developments*, CAB International, Wallingford

Williams, A.M. and Shaw, G. (eds) (1991) *Tourism and Economic Development: Western European Experiences*, 2nd edition, Belhaven Press, London

PART V Conclusions

CHAPTER 20 European futures

TIM UNWIN

It is the evening of 15 May 1997; the place is Barcelona – just off the Ramblas. The bars are full; the streets teem with people, waving the blue and red flags and banners of their football team. Last night Barcelona won the European Cup Winners' Cup. In one of the many bars, young men are smoking and drinking beer – Estrella Damm; lovers talk, heads close together. On the faded, yellowing walls are countless equally fading pictures and photographs. Posters: 'Toros en Madrid', 'Feria de San Isidro 1972'; a bull's head set on a red and green shield, pierced by swords – 'Los de Gallito y Belmonte. Epoca de Oro. Barcelona'. Yet bull fighting is not typical of Catalunya. From the kitchen – smoke and smells of frying fish and potatoes; debris lies on the floor – a circular plastic ring seal from a bottle of water, cigarette butts, beer stains. Behind the bar, a balding man with wisps of finely combed, grey hair leans on the counter trying to make sense of the accounts. Waiters, with rounded faces and jet black hair, rush to and fro: more beers, some tapas. And behind the bar, row upon row of bottles – Tio Pepe and La Ina sherries, Beefeater Gin, Bacardi.

Place and period are crucial to any understanding of Europe. This book, written in 1996 and early 1997, reflects the issues and agendas of these years. Almost a decade after the dramatic upheavals that took place at the end of the 1980s in eastern Europe, its contents thus reflect the transformations and restructuring that have been taking place across the region ever since. However, within a couple of months, and by the time the book is actually published, significant new agreements and new agendas will have been decided. The leaders of the European Union are shortly to sign the Amsterdam Treaty making significant revisions to the Treaty of Rome, and paving the way for a new era of political and economic change in the European Union. The NATO–Russia Act was likewise signed in Paris on 27 May, opening the doors to membership of the North Atlantic Treaty Organization (NATO) for certain eastern European countries. How these events are seen and experienced by Europe's peoples depends very much on their identities and on the places where they live. There is neither a single common European identity nor a common European interest. It is the diversity of Europe and its great complexities that this book has sought to explore.

Such diversity of identity and meaning is well exemplified by the city of Barcelona (for a presentation, see http://www.bcn.es/english/ciutat/iciutat.htm) (Figure 20.1). Someone living in the city may consider themselves not only as an inhabitant of Barcelona, but also as a Catalan, a Spaniard, and a European. People have nested layers of identity with regard to place, and these are used as elements of identity on different occasions and in different contexts. Barcelona characterizes many other aspects of contemporary Europe. The hoarding surrounding the restoration work on the burnt-out Opera notes the financial support that has come not only from Catalunya and Spain, but also from Dutch and German companies, as well as from the European Union. The selection of bottles lining the shelves of the bars reflects the importance of global alcoholic beverage corporations, and the numerous shops along the main streets, with names ranging from Benetton and Naf Naf to McDonalds and Marks and Spencer, likewise reflect the increasingly global character of European retailing. Barcelona's football team is even managed by an Englishman.

Historically, too, Barcelona has experienced numerous different and contrasting waves of European experience. The outlines of the Roman city, with its grid plan and forum, still remain at the heart of the medieval Barri Gòtic. Between the 9th and the 11th centuries Barcelona lay in the Spanish March at the very border between Christian Europe and Islamic Iberia. By the end of the 13th century, following the reconquest, it became necessary to construct

Figure 20.1 Barcelona: Barri Gòtic (*source*: Tim Unwin, 16 May 1997).

a new city wall to enclose the expanded urban area that had resulted from the growing role of the city as a political, commercial and religious centre, and in subsequent centuries its leaders sought to maintain its specific political identity and economic vitality through a delicate balancing act of negotiation between the growing powers of a Castilian Spain centred on Madrid and a France centred on Paris. Following the embellishment of the city with neoclassical façades and buildings during the 18th century, Barcelona experienced the riots and confrontations that were common to many other parts of Europe in the 19th century. These saw the burning of many of the city's monasteries and convents, but they also created the opportunity for substantial urban redesign following the demolition of the walls.

Moreover, the second half of the 19th century witnessed significant economic expansion, with the city becoming known as 'la petita Manchester'. The first railway in Spain was constructed in Barcelona in 1848, and by the end of the century the city had a thriving economy, based largely on its industry and on overseas trade with the Spanish-speaking lands of the Americas and the Pacific. This period also witnessed a remarkable explosion of architectural design and construction, represented above all by the Modernista tradition associated with Antoni Gaudí (1852–1926) and his students. Gaudí's buildings, such as La Pedrera on the Passeig de Gràcia, or the Sagrada Familia, while reflecting all the potential and strength of 'modern' materials and principles, nevertheless have a flamboyance and character distinctly lacking from north European modernist architecture.

Theorizing European change in a global context

Several of the chapters of this book, particularly those exploring Europe's cultural heritage, have stressed the varying meanings attached to the idea of Europe and European identity in the past. Despite half a century of increasing political and economic integration, Europe today remains an area of remarkable diversity and differential change. Although there have been numerous theoretical attempts to understand the relationships between the local, the national and the global, particularly in the European context, there remains little certainty or agreement. As Amin (1997: 1) has emphasized, 'The more we read about globalization from the mounting volumes of literature on the topic, the less clear we seem to be about what it means and what it implies'. In this context he draws attention to the very contrasting conceptualizations of globalization held, for example, by W. Robinson (1996), Hirst and Thompson (1996a, b) and Giddens (1996).

On the one hand, there are those who see capitalist globalization as having become a dominant force, tying the world together in an intricate web of exploitation and profit expropriation (Figure 20.2). As W. Robinson (1996: 13–14) has argued, 'Capitalist globalisation denotes a world war' which 'was incubated with the development of new technologies and the changing face of production and of labour in the capitalist world, and the hatching of transnational capital out of former national capitals in the North'. In contrast, there are others, such as Hirst and Thompson (1996a), who suggest that such ideas are wide of the mark, and that 'the notion of globalisa-

Figure 20.2 McDonald's, Toruń, Poland (*source*: Tim Unwin, 24 September 1994).

tion is just plainly wrong'. A third view is that represented by Giddens (1996), who suggests that globalization is both an 'out there' and an 'in here' phenomenon. As he notes, 'Globalization invades local contexts of action but does not destroy them; on the contrary, new forms of local cultural identity and self expression, are causally bound up with globalizing processes' (Giddens, 1996: 368).

The evidence of this book would go some way to supporting Giddens' approach, or what Amin (1997: 8) has characterized as 'out there – in here connectivity' (see also Held, 1991, 1995). In understanding the development both of the *idea* of a unified Europe, and also of the social, economic, political and cultural changes that have occurred *within* Europe over the last fifty years, it is helpful to consider not only global processes but also how they interact with particular sets of local processes and identities.

Such arguments intersect closely with another set of theoretical debates, namely those concerning the character of the changes that have taken place in eastern Europe and the former Soviet Union over the last decade (Callinicos, 1991; Milliband, 1991). While some, such as Eisenstadt (1992: 21), see these changes as 'one of the most dramatic events in the history of mankind', others are more sanguine about their significance. Tilly (1993: 234), for example, suggests that 'In the perspective of 500 revolutionary years, the collapse of Eastern European regimes loses some of its close-up magnitude', and Markoff (1996) likewise comments that they reflect merely the latest in a series of democratic waves that have swept over the global political system since the late 18th century. It is nevertheless important to link together the changes that have taken place in both western and eastern Europe in recent years, and to situate them within a broader reconceptualization of the linkages between the local, the national and the global. While some liberal and left-wing idealists might have wished for the emergence of an alternative 'third way' between the evils of capitalism and communism, the events that have unfolded in eastern Europe since the end of the 1980s have revealed very clearly the underlying pressures that have drawn the states of eastern Europe into the interconnected global political economy of the contemporary world. Whether this system is termed capitalist, free-market, or liberal democracy, and there are subtle but important contrasts between these terms, makes little real difference to the farm labourer living in Poland, or the entrepreneur in Estonia.

In theoretical terms, the reincorporation of the newly independent states of eastern Europe and the former Soviet Union into the mainstream of European politics and economic activity has been most closely associated with the concept of *transition*, involving the twofold introduction of a so-called liberal democracy and a free-market economy (see, for example, Pickles and Smith, 1997). These two elements owe much to the shock-therapy model proposed by Sachs (1990), which was designed to create states in eastern Europe which were as open as possible to international economic forces, and which had institutional structures specifically designed to facilitate integration with the mainstream global economy (Gowan, 1995). It is no coincidence that such arguments emphasize essentially economic and political agendas, while leaving social and cultural considerations relatively ignored.

Two important sets of interconnected debates have arisen over the character of transition, the first concerning the relationships between modernization and political freedom, and the second pertaining to the relative roles of global and national processes and institutions. Habermas (1994a: 84) has argued cogently that, despite the rhetoric associating the free market with liberal democracies, 'the evidence continues to mount that there is no automatic relationship between capitalist modernization and political freedom'. Indeed, he suggests further that 'the cultural peripheral conditions for European modernity, capitalist developmental poverty, ecological devastation, political repression, and cultural disintegration braid together into an increasingly desperate feedback loop' (Habermas, 1994a: 84). This raises important questions not only as to whether continued capitalist modernization within eastern Europe will remain associated with broadly defined democracy, but also as to whether further integration of the countries of western and northern Europe into the European Union will require new systems of legitimation in order to guard their inherited democratic traditions. In this context, Held's (1995) observation that the links between liberal democracy and national sovereignty have usually been accepted without question is highly pertinent. His analysis of the many different kinds of political regime that are claimed to be democratic highlights the need for close examination of the precise systems of linkage between global interconnectedness, systems of political representation, and issues of national identity. Elsewhere, Held (1991: 148) suggests that despite the development of international and intergovernmental links, the age of the nation-state is by no means over. The extent to which this will remain true, as the political leaders of many European states seem willing to give up increasing amounts of their traditional power to centralized European institutions, is a matter of considerable debate and conjecture.

This raises complex issues concerning the role of European nation-states in the global political–economic system of the future. The great epoch of the nation-state was the 19th century, and it is no coincidence that the dominance of this form of political unit emerged at the same time as the new forms of economic activity associated with industrial and financial capitalism. While the apparently growing strength of regional political and economic institutions, such as the European Union, in the latter part of the 20th century, might appear to suggest a decline in the role of nation-states, to the benefit of both local and supranational political entities, it is by no means clear that such 'hollowing out' is occurring (Ohmae, 1990, 1995). Hirst and Thompson (1996a, b), for example, argue vehemently that economic processes at the national level remain central to the global economy, and that nation-states play a crucial role in providing the necessary governance for this economy to function successfully. As they comment, 'The central function of the nation-state is that of distributing and rendering accountable powers of governance, upwards towards international agencies and trade blocs like the European Union, and downward towards regional and other sub-national agencies of economic co-ordination and regulation' (Hirst and Thompson, 1996a: 408). However, as Amin (1997) has stressed, this argument fails fully to account for the complex web of economic, social, cultural and political linkages that is emerging at the global level, and the restructuring of connectivities that is taking place between specific states and their societies (see also Held, 1995). Moreover, the proposed changes to the political functioning of the EU, including the introduction of majority decision making, are further evidence of the renegotiation that is taking place between the roles of the global, the regional and the national (Williams, 1994). This is particularly reflected in the political and institutional changes that have recently been taking place in the context of European integration.

European integration: institutional contexts

The political and economic restructuring of eastern Europe in the 1990s has been closely associated with a reorganization and expansion of European political alliances and institutions (Allum, 1995). Key among these have been the proposed expansion of NATO and the European Union, as well as the considerable increase in membership of the Council of Europe. A brief analysis of these changes highlights the complex diversity of interests represented by each institutional structure.

NATO expansion

NATO was founded in 1949 as the key military and political bastion of western interests following the Second World War (http://www.nato.int/). Although essentially a military alliance, designed to defend the dominant capitalist economies of North America and western Europe against possible aggression from the Soviet Union and its satellites, it has nevertheless also maintained broader cultural, social and economic interests designed to enhance cooperative security structures for Europe as a whole. Initially an alliance between 12 nations, Greece and Turkey joined NATO in 1952, the Federal Republic of Germany joined in 1955, and Spain became the 16th member in 1982.

Two sets of political events in the early 1990s, however, forced the members of NATO to reconsider their role in European political affairs. First, with the fall of the Berlin Wall in 1989, German reunification in 1990, and the dissolution of the Warsaw Pact organization in 1991, the nature of the perceived military threat from the former Soviet Union changed dramatically. The end of the so-called Cold War between West and East was initially heralded as having the potential to bring an enormous peace dividend to the countries of western Europe and NATO. However, this optimism was soon tempered by the recognition that a period of relative stability and balance of power between NATO and the Warsaw Pact was to be replaced by one of considerable uncertainty, reflected in the outbreak of violent conflict in the former Yugoslavia. Second, the Treaty on European Union signed at Maastricht in 1991 aspired to the development of a common defence policy for the Union, and sought to place European defence responsibilities much more directly within the hands of the Western European Union, consisting of the original six members of the EEC together with the UK. Article J4 of the Maastricht Treaty thus specifically stated that 'The Union requests the Western European Union (WEU), which is an integral part of the development of the Union, to elaborate and implement decisions and actions of the Union which have defence implications'. While the Treaty in Article J4 also stated that 'The policy of the Union in accordance with this Article shall not prejudice the specific character of the security and defence policy of certain Member States and shall respect the obligations of certain Member States under the North Atlantic Treaty and be compatible with the common security and defence policy established within that framework', this refocusing of Europe's defence aspirations has necessitated a reappraisal of the relative roles of NATO and the WEU.

As a result of these changes, NATO has sought to develop new strategies to enhance the potential for long-term peace and stability in Europe. In particular, in 1991 the North Atlantic Cooperation Council (NACC) was established as a forum for cooperation between member states of NATO and those in central and eastern Europe as well as the former Soviet Union. Subsequently, in January 1994, NATO established the Partnership for Peace (PFP) programme within the framework of the NACC, in order to develop a new security relationship across Europe as a whole. With 27 partner countries from eastern Europe and former Soviet republics joining the programme, the Russian government saw the possible expansion of NATO as a considerable threat, and yet another reminder of its rapid fall from global power. However, subsequent detailed negotiations between NATO and Russia led to the signing on 27 May 1997 in Paris of the Founding Act on Mutual Relations, Cooperation and Security between the two entities. While many in Russia saw the signature of this Act by President Yeltsin as indicating the Kremlin's final capitulation to the West, and Mikhail Gorbachev, the former Soviet leader, claimed that 'the West had betrayed Russia, going back on assurances made in 1990 that the alliance would not expand' (*The Times*, 28 May 1997: 17), the Act was welcomed by many of the PFP partners, who saw it as an opening of the doors to their membership of NATO. The Founding Act, while not a legally binding document, establishes the ground rules for a new security partnership that gives Russia theoretically equal status in peacekeeping and conflict prevention. It also provides the context within which NATO will be able to invite new members to join the alliance by 1999. While it seems likely that countries such as Poland, Hungary and the Czech Republic will rapidly join NATO, Russia remains vehemently opposed to the membership aspirations of the former Soviet republics such as Estonia.

Economic and monetary union

For many states in eastern Europe, their aspirations to join NATO are closely linked to their desires to become full members of the European Union. While the restructuring of the European Union following the Maastricht Treaty of 1991 has been examined in some detail in Chapter 9, recent political events have emphasized the potential fragility of this agreement.

There are compelling arguments for greater European economic and political integration, and these are clearly reflected in the European Commission's promotion of the project of economic and monetary union. The Commission thus argues that 'the story of the last 25 years is one of a constant search by the Member States of the Union for deeper monetary cooperation as a means of strengthening the political bonds between them and to protect their central creation, the common market' (European Commission, 1996: 3). The single currency is designed to create a more efficient single market, to stimulate growth and employment, to eliminate the costs connected with the existence of separate currencies, to increase international stability, and to enhance joint monetary stability for the member states. Underlying these arguments is the key assumption that the economies of participant European states should be converging, and that they should be based on the principles of an open market economy and free competition. This therefore seeks to enshrine a single type of economic system within a political and legal framework. Once a single currency is created, the Commission argues that the change will be irrevocable. In its words, 'A single currency means irrevocably fixed interest rates' (European Commission, 1996: 14). However, attempts by national governments to meet the required convergence criteria, as measured by interest rates, inflation and budget deficits (see Chapter 12), have caused serious economic and political problems.

During 1996 and early 1997 it became increasingly clear that even France and Germany, the countries most strongly advocating the creation of a single currency, would have difficulties meeting the convergence criteria necessary for monetary union on 1 January 1999. In December 1996, on the same day as the launch of the official designs for the euro by the European Monetary Institute, the French and German governments eventually agreed to resolve their long-running dispute over the precise rules of monetary union. However, within months, both countries were to encounter serious political and economic crises, which at the time of writing in mid-1997 make any predictions over the eventual fate of the single currency highly uncertain. In France, the austerity measures introduced by the centre-right coalition in order to satisfy the criteria for joining the single currency proved to be increasingly unpopular, and President Chirac thus called for early elections in May 1997 in order to seek a renewed parliamentary mandate. Contrary to most expectations, the Socialist Party under Lionel Jospin swept to victory in the second round of voting, on a platform which ruled out further stringency measures to meet the Maastricht criteria. While Jospin remains committed to the single currency, his pledge to put economic growth and employment in France above the requirement to satisfy the convergence criteria makes it difficult to see how he will be able to steer France into monetary union, unless these criteria are relaxed. The victory of the French Socialists, coming only a month after the Labour Party's rout of the Conservatives in Britain, highlights an important shift to the left in European politics, and suggests that the people of Europe are by no means willing to sacrifice all in the drive to economic and monetary union so beloved by the Commission.

To make matters worse for the single currency, at the same time as France was rejecting President Chirac's policies, a row broke out between Chancellor Kohl of Germany and Hans Tietmayer, President of the Bundesbank. With rising unemployment, falling tax revenues and declining inward investment, it seems likely that Germany's budget deficit will exceed the figure of 3% of gross domestic product required for joining the single currency. The solution proposed by the German government to resolve this difference was to revalue its gold reserves. However, the Bundesbank responded by condemning such a proposal as interference in the bank's traditionally independent status, as being a breach of the Maastricht criteria and as a threat to the credibility of the euro. Germany, having regularly accused the Spanish and Italian governments of trying to fudge the required criteria, now appeared to be doing exactly the same thing. The decisive summit to determine which countries will be eligible for entry into the single currency is not to be held until spring 1998, but it seems likely that growing popular pressure will lead to at least some renegotiation of either the criteria themselves or the proposed start date of 1 January 1999.

The Council of Europe

While the European Union has largely been driven by economic and political interests, designed to ensure that Europe's economy remains competitive on an international stage, it is by no means the oldest or broadest European political institution. This status belongs to the Council of Europe, which was founded by ten countries in 1949 with the main objectives of creating an intergovernmental organization designed to protect human rights and pluralist democracy in Europe (Figure 20.3). By 1980, a further 11 countries had joined the Council, with San Marino and Finland becoming members in the 1980s. Its three main institutions are the Committee of Ministers, the Parliamentary Assembly, and the Congress of Local and Regional Authorities of Europe. In addition to its focus on democracy and human rights, the Council of Europe also places special emphasis on the promotion and development of a European cultural identity, and on seeking to develop solutions to European social problems such as minorities, xenophobia and bioethics. In many ways this provides an important cultural and social counterpart to the economic and political focus of the European Union.

With the break-up of the former Soviet Union, the Council of Europe has played a major role in seeking both to develop a political partnership with the countries of eastern Europe, and to help them with political, legislative and constitutional reform. The key condition for membership of the Council is an irrevocable decision by a state to defend human rights, democracy and the rule of law. Following such decisions, Hungary was accepted into the Council of Europe in 1990, followed by Poland in 1991, Bulgaria in 1992, Estonia, Lithuania, Slovenia, the Czech Republic, Slovakia and Romania in 1993, and Latvia, Albania and Moldova in 1995. Andorra also joined in 1994, and Russia was accepted as the 39th member of the Council in 1996.

The pursuit of democracy and human rights lies at the very heart of the Council of Europe, with its 1950 European Convention on Human Rights establishing a sophisticated mechanism for monitoring and protecting human rights. In 1961, this was matched by a European Social Charter, which sought to guarantee the effective exercise of some 19 social rights, including the rights to work, to strike, and to receive social security, which were deemed to be fundamental to European society. A later protocol added four further rights including the equality of men and women in matters of employment and occupation, and the right of the elderly to social protection. While these provide a fundamental ethical dimension to the idea of Europe, it is in the field of cultural heritage that some of the Council's most interesting work is done. The European Cultural Convention of 1954 provides an overall framework within which intergovernmental cooperation in the fields of education, culture, heritage, sport and youth can be achieved. One of the most interesting aspects of the Council of Europe's cultural policy, though, is that it is premised on an argument that 'Diversity lies at the heart of Europe's

Figure 20.3 The Council of Europe's headquarters in Strasbourg (*source*: Tim Unwin, 9 November 1995).

cultural richness which is our common heritage and the basis of our unity' (Council of Europe, 1995: 37).

The Council of Europe's emphasis on unity in diversity, and on the importance of the multi-faceted character of European cultural identity (Shelley and Winck, 1995), provides an important balance to the European Commission's drive for economic and political unity. Since the Maastricht Treaty in 1991 formal citizenship of the European Union has existed; according to Article 8 of the Treaty, 'Every person holding the nationality of a Member State shall be a citizen of the Union'. While this has largely been designed to permit voting and residential rights for people across Europe, there has been surprisingly little debate about the broader cultural significance of the dual citizenship now possessed by people living within the European Union, and what it actually means. Many people feel that their national culture is now threatened by increasing economic and political integration, and the role of the Council of Europe in retaining a sense of cultural diversity at the centre of the meaning of being European is therefore likely to become increasingly important in the future.

Popular support for the European Union

The May 1997 elections in France emphasized that people living within specific European states are by no means always willing to sacrifice their own interests on behalf of the aspirations of those seeking ever closer European economic and political union. Furthermore, it is also abundantly clear that people living in different parts of Europe have very contrasting views concerning the future of the European Union.

This is well illustrated by the results of a poll conducted in December 1996 on behalf of the French newspaper *Le Figaro*, the German business paper *Handelsblatt*, the British newspaper *The Daily Telegraph*, and the Italian news magazine *L'Espresso* (Table 20.1). Although this poll focused mainly on issues surrounding the planned introduction of a single currency, it indicated both the great uncertainty of people within the Union about its future, and the very differing views of people in the four states towards the Union. The overall highest level of support for the EU was found in France, where 73% of the sample interviewed considered membership to be a good thing, but even in Britain an equal proportion (34%) thought membership to be good as thought it bad; not all Britons are as against the Union as some people elsewhere in Europe might like to caricature them. It is somewhat ironic, though, that only five months after this poll the French electorate turned so dramatically against the stringency policies that their government deemed necessary to enable them to satisfy the Maastricht criteria; in December 1996 some 58% of those polled in France had claimed to have felt that these sacrifices were indeed worth it. While French and Italian views on many issues were broadly similar, these contrasted quite markedly with those of people living in Germany and Britain. This is particularly apparent in the responses to questions concerning the single currency, where British and German views were much less supportive than those of the sample interviewed in Italy and France. While Germans trust the strength and stability of the mark, they are somewhat sceptical about the likely future strength of the euro.

One of the most striking findings of the poll was the antipathy felt towards the idea of Germany becoming the dominant power in the EU. Over 50% of all those interviewed in each country other than Germany felt that Germany was likely to become the dominant European power, and over 70% disliked the idea (King, 1997). However, divisions in attitudes towards the EU are based not only on national cleavages, but also on such divides as gender and class. The same poll, for example, revealed that across Europe women were considerably more likely to be opposed to a single currency than were men; 'in Germany women would oppose EMU by 49 to 37 per cent while men would support it by 50 to 39 per cent' (King, 1997: 13). Older people, particularly in Britain, also tend to be less enthusiastic about Europe than the young.

In explaining these differing attitudes towards the political and economic future of the EU, it is significant to note that the editors and political analysts of the newspapers in which they were published continually referred back to cultural traditions as reference points in their arguments. King (1997: 13) thus asserted that

> The British people's view also needs to be understood against the background of history.
>
> ... Britain has not been invaded – successfully – since 1066. Moreover, unlike France, Germany and Italy, Britain has a continuous history of liberal and democratic evolution going back at least to the 18th century.

Table 20.1 British, German, French and Italian views on Europe.

	Britain	Germany	France	Italy
People hold different views about how they would like to see the European Union develop. Which of these alternatives comes closest to your own view?				
A fully integrated Europe with all major decisions taken by a European government	10%	16%	17%	25%
No European government but a more integrated Europe than now, with a single currency and no frontier controls	17%	37%	39%	42%
The situation more or less as it is now	18%	31%	14%	12%
A less integrated Europe than now, with the European Union amounting to little more than a free trade area	22%	12%	17%	8%
Complete withdrawal from the European Union	23%	4%	9%	5%
Don't know	11%	0%	4%	8%
Generally speaking, do you think your country's membership of the European Union is a good thing, a bad thing or neither good nor bad?				
Good thing	34%	51%	73%	66%
Bad thing	34%	14%	23%	10%
Neither	22%	26%	3%	18%
Don't know	11%	8%	1%	7%
If ... a referendum were to be held during the next few months, how would you vote, in favour or against the creation of a single European currency?				
In favour	26%	43%	61%	71%
Against	56%	44%	33%	12%
Wouldn't vote	5%	2%	2%	3%
Don't know	14%	11%	4%	13%
Most European countries are cutting government spending and borrowing in order to meet the so-called 'Maastricht criteria' for joining a single European currency. In the case of your country, do you think the sacrifices that people are having to make are worth it or not?				
Are worth it	17%	39%	58%	54%
Are not	59%	44%	34%	33%
Don't know	24%	17%	8%	13%
If the European Union draws ever closer together and there is a single currency, which country, if any, do you think will become the dominant power within the Union?				
No dominant power	9%	28%	30%	12%
Britain dominant	4%	4%	2%	3%
France dominant	5%	9%	5%	4%
Germany dominant	56%	28%	50%	51%
Italy dominant	0%	0%	0%	2%
Other/not sure/don't know	25%	31%	12%	27%

Source: derived from *The Daily Telegraph*, 10 January 1997:8.

Figure 20.4 Windmills at Angla, Sasaremaa, Estonia: a rural European landscape (*source*: Tim Unwin, 18 June 1995).

Figure 20.5 Toledo, Spain: Islamic, Christian and Jewish urban landscape (*source*: Tim Unwin, 10 April 1987).

> There have been no violent revolutions, no interludes of fascism, no episodes of enemy occupation. As a result, the British are used to their own independence.

Likewise, the editor-in-chief of *Handelsblatt*, Rainer Nahrendorf, commented that

> The question of possible German dominance of the EU belongs to the old way of thinking and goes against the post-war approach which led to the creation of the European Coal and Steel Community and the EEC.
>
> It is precisely because the Germans have learned the lessons of the Nazi dictatorship that all German governments since the war have been in favour of tying the country ever more closely into a union of European nations.
>
> (King, 1997: 13)

French national independence and identity is emphasized in the following account by *Le Figaro*'s political analyst, Charles Rebois:

> the Europe of French dreams can only be one in which each nation keeps its own national characteristics. It must be neither federalist nor freemarket.
>
> (King, 1997: 13)

And in explaining the Italian findings, Antonangelo Pinna, executive editor of *L'Espresso*, suggests that

> Italians are rather realistic, conservative, concrete and have a sceptical opinion on the ability of their politicians to run public affairs. They have been disappointed too many times by broken promises and unattained goals. European union appears to be seen by them as a chance, maybe the last chance, for a new start in a country with too much imagination and too few rules.
>
> (King, 1997: 13)

These images, reinforced by their promulgation in the media, are reminders of the powerful hold that culture has over popular imaginations, and that political and economic judgements alone will not determine the future character of the European Union. Moreover, they also emphasize the strong role that editorial comment in the press can have over national self-consciousness.

A European geography

This book has stressed throughout the need to integrate an understanding of the cultural, economic, political and social dimensions of Europe, and to identify the ways in which these give rise to particular expressions in different places. In part the character of such places is influenced by the nature of their underlying physical environment, but it is also a result of the complex tapestry of human actions that have been woven upon such a base (Figures 20.4 and 20.5). Moreover, as Massey (1994: 154) has commented, such places can be imagined as 'articulated moments in networks of social relations and understandings...where a large proportion of those relations, experiences and understandings are constructed on a far larger scale than what we happen to define for that moment as the place itself'. How we come to understand any one place thus reflects our wider understanding of the whole of which it is a part; it reflects our appreciation of other places. By exploring and writing about different European places, the authors of this book have thus sought not only to shed light on the range of global and local processes that have influenced them, but also to express their uniqueness and their vitality. Europe is a place made up of a myriad of other places. It is this unity in diversity that *A European Geography* has sought to elucidate.

Bibliography

Aase, A. (1989) Regionalizing mortality data: ischaemic heart disease in Norway, *Social Science and Medicine*, **29**, 907–11

Abel-Smith, B. and Mossialos, E. (1994) Cost containment and health care reform: a study of the European Union, *Health Policy*, **28**, 89–132

Acton, E. (1990) *Rethinking the Russian Revolution*, Edward Arnold, London

Agnew, J.A. (1995) The rhetoric of regionalism: the Northern League in Italian politics, 1983–94, *Transactions of the Institute of British Geographers*, **20**, 156–72

Agnew, J.A. (1996a) Time into space: the myth of 'backward' Italy in modern Europe, *Time and Society*, **5**, 27–45

Agnew, J.A. (1996b) Mapping politics: how context counts in electoral geography, *Political Geography*, **15**(2), 129–46

Agocs, A. and Agocs, S. (1994) Too little too late: the agricultural policy of Hungary's post-Communist government, *Journal of Rural Studies*, **10**, 117–30

Agulhon, M. (1981) *Marianne into Battle: Republican Imagery and Symbolism in France, 1798–1880*, Cambridge University Press, Cambridge

Albert-Piñole, I. (1993) Tourism in Spain, in Pompl, W. and Lavery, P. (eds) *Tourism in Europe: Structures and Developments*, CAB International, Wallingford, 242–61

Aldcroft, D.H. (1980) *The European Economy, 1914–1980*, Croom Helm, London

Alexinsky, G. (1968) Slavonic mythology, in *New Larousse Encyclopedia of Mythology*, 2nd edition, Paul Hamlyn, London, 281–98

Allum, P (1995) *State and Society in Western Europe*, Polity, Cambridge

Althusser, L. (1969) *For Marx*, Penguin, Harmondsworth

Althusser, L. and Balibar, E. (1970) *Reading Capital*, New Left Books, London

Altuna, J. (1983) La race basque, in Haritschelar, J. (ed.) *Etre Basque*, Privat, Toulouse, 89–105

Amin, A. (ed.) (1995) *Post-Fordism: a Reader*, Blackwell, Oxford

Amin, A. (1997) Placing globalisation, Paper presented to the RGS-IBG Annual Conference, Exeter, 7–9 January 1997

Amin, A. and Thrift, N. (1994) *Globalization, Institutions and Regional Development in Europe*, Oxford University Press, Oxford

Anderson, B. (1983) *Imagined Communities: Reflections on the Origin and Spread of Nationalism*, Verso, London

Anderson, B.G. and Borns, Jr, H.W. (1994) *The Ice Age World*, Scandinavian University Press, Stockholm

André, I.M. (1996) At the centre on the periphery? Women in the Portuguese labour market, in Garcia-Ramon, M.D. and Monk, J. (eds) *Women of the European Union*, Routledge, London, 138–55

Ang, I. (1985) *Watching Dallas*, Methuen, London and New York

Anon. (1991) *Health for All Policy in Finland: WHO Health Policy Review*, WHO Regional Office for Europe, Copenhagen

Anon. (1995) France prepares for EMU, *The Economist*, 9 December, 11–12

Anon. (1996) EMU and what Alice found there, *The Economist*, 14 December, 23–5

Ansart, P. (1970) *Naissance de l'Anarchisme*, Paris: PUF

Apalegui, A.C. (1975) *La Foralidad Guipúzcoana*, 2nd edition, Ediciones de la Caja de Ahorros Provincial de Guipúzcoa, Zarauz

Archer, B. and Cooper, C. (1994) The positive and negative impacts of tourism, in Theobald, W. (ed.) *Global Tourism: the Next Decade*, Butterworth-Heinemann, Oxford, 73–91

Ardanza, J.A. (1991a) *Discurso del Lehendakari en la Sesión de Investidadura*, Eusko Jaurlaritza, Vitoria-Gasteiz

Ardanza, J.A. (1991b) *Debate de la cuestion de confianza planteada por el Lehendakari José Antonio Ardanza al Parlamento vasco*, Eusko Jaurlaritza, Vitoria-Gasteiz

Arnaud, E. (1710) *Histoire de la Glorieuse Rentrée des Vaudois dans leur Vallées*, Estienne Garni, Cassel

Arter, D. (1995) The EU referendum in Finland on 16 October 1994: a vote for the West, not for Maastricht, *Journal of Common Market Studies*, **33**, 361–87

Artis, M.J. (1994) European Monetary Union, in Artis, M.J. and Lee, N. (eds) *The Economics of the European Union: Policy and Analysis*, Oxford University Press, Oxford, 346–67

Ashton, J. (1992a) *Healthy Cities*, Open University, Milton Keynes

Ashton, J. (1992b) Setting the agenda for health in Europe: What the United Kingdom could do with its presidency of the European Community, *British Medical Journal*, **304**, 1643–4

Ashworth, G. and Larkin, P.J. (1994) *Building a New Heritage: Tourism, Culture and Identity in the New Europe*, Routledge, London

Ashworth, G.J. (1993) Culture and tourism: conflict or symbiosis in Europe?, in Pompl, W. and Lavery, P. (eds) *Tourism in Europe: Structures and Developments,* CAB International, Wallingford, 13–35

Ashworth, G.J. and Tunbridge, J.E. (1990) *The Tourist-Historic City,* Belhaven Press, London

Aslanbegui, N., Pressman, S. and Summerfield, G. (1994) *Women in the Age of Economic Transformation: Gender Impact of Reforms in Post-Socialist and Developing Countries*, Routledge, London

Aston, T.H. and Philpin, C.H.E. (eds) (1985) *The Brenner Debate: Agrarian Class Structure and Economic Development in Pre-Industrial Europe*, Cambridge University Press, Cambridge

Austrian Department of Tourism (1996) *Tourism in Austria in 1995,* Austrian Federal Ministry for Economic Affairs, Vienna

Baetens Beardsmore, H. (1981) Linguistic accommodation in Belgium, *Brussels Pre-Prints in Linguistics*, Linguistics Circle, Vrije Universiteit Brussel and Université Libre de Bruxelles, March, 1–21

Bagehot, W. (1873) *Lombard Street. A Description of the Money Market*, Henry King, London

Bagnasco, A. (1977) *Le Tre Italie: La Problematica Territoriale dello Sviluppo Italiano*, Il Mulino, Bologna

Bahn, P.G. and Vertut, J. (1988) *Images of the Ice Age*, Windward, Leicester

Bakounine, M. (1973–82) *Oeuvres Complètes*, Champ Libre (8 volumes), Paris

Balassa, B. (1961) *The Theory of Economic Integration,* Irwin Press, Ill.

Balibar, E. (1991) Es gibt keinen Staat in Europa: racism and politics in Europe today*, New Left Review*, **186**, 15–21

Balkir, C. and Williams, A.M. (eds) (1993) *Turkey and Europe*, Pinter, London

Banque Indosuez (1990) *Les Echanges Commerciaux Européens, Index, Revue Trimestrielle Economique*, Banque Indosuez, Paris

Baptista, A.M. (1989) Perspectivas de desenvolvimento económico da Área Metropolitana de Lisboa, *Sociedade e Território*, **10/11**, 43–8

Baranov, A. (1991) Maternal and child health problems in the U.S.S.R., *Archives of Disease in Childhood*, **66**, 542–5

Barber, B.J. (1995) *Jihad vs. McWorld*, Random House, New York

Barber, L. (1996) France in warning to Italy over ERM re-entry, *Financial Times*, 15 October, 20

Bardi, L. (1996) Transnational trends in European parties and the 1994 elections of the European Parliament, *Party Politics*, **2**(1), 99–114

Barke, M. and Newton, M. (1994) A new rural development initiative in Spain: the European Community's Plan LEADER, *Geography*, **79**(4), 366–71

Barke, M. and Newton, M. (1995) Promoting sustainable tourism in an urban context: recent developments in Malaga City, Andalusia, *Journal of Sustainable Tourism*, **3**(3), 115–34

Barnes, I. and Barnes, P. (1993) Tourism policy in the European Community, in Pompl, W. and Lavery, P. (eds) *Tourism in Europe: Structures and Developments,* CAB International, Wallingford, 36–54

Barnett, C.R. (1958) *Poland: its People, its Society, its Culture*, Grove Press, New York

Barrett, D.B. (1982) *World Christian Encyclopedia: a Comparative Study of Churches and Religions in the Modern World AD 1900–2000*, Oxford University Press, Nairobi

Bartolini, S. and Mair, P. (1990) *Identity, Competition, and Electoral Availability. The Stabilisation of European Electorates 1885–1985*, Cambridge University Press, Cambridge

Basque Studies Program Newsletter (Department of Basque Studies, University of Nevada, Reno (1968–present)

Bateman, D. and Ray, C. (1994) Farm pluriactivity and rural policy: some evidence from Wales, *Journal of Rural Studies*, **10**, 1–13

Baudrillard, J. (1976) *L'Echange Symbolique et la Mort*, Editions Gallimard, Paris

Baudrillard, J. (1988) *America*, Verso, London

Baum, T. (1995) *Managing Human Resources in the European Tourism and Hospitality Industry*, Chapman & Hall, London

Baylina, M. (1996) *Trabajo industrial a domicilio, género y contexo regional en la España rural*, Doctoral thesis, Department de Geografia, Universitat Autónoma de Barcelona

Bayly, C.A. (1989) *Imperial Meridian: the British Empire and the World, 1780–1830*, Longman, London

Beckford, J.A. (ed.) (1986) *New Religious Movements and Rapid Social Change*, Sage-UNESCO, London and Paris

Begg, I. (1995) The impact on regions of completion of the European community internal market for financial services, in Hardy, S., Hart, M., Albrechts, L. and Katos, A. (eds) *An Enlarged Europe: Regions in Competition?* Regional Studies Association, Jessica Kingsley Publishers, London, 145–58

Belk, R.W. and Gur, G. (1994) Problems of marketization in Romania and Turkey, in Schultz, C.J. III, Belk, R.W. and Gur, G. (eds) *Research in Consumer Behavior Volume 7*, JAI Press, New York, 123–55

Bell, D. (1980) *Sociological Journeys, Essays 1960–1980*, Heinemann, London

Bell, D.S. and Shaw, E. (1994) *Conflict and Cohesion in Western European Social Democratic Parties*, Pinter, London

Bellanger, S. (1990) Toward an integrated European banking system: 1992 and beyond, in Gillespie, I. (ed.) *Banking 1992*, Eurostudy Special Report, Eurostudy, London, 35–44

Bellona (1996) Miljøstiftelsen Bellona, http://www.grida.no/ngo/bellona/index.htm

Bentham, G., Eimermann, J., Haynes, R., Lovett, A. and Brainard, J. (1995) Limiting long term illness and its associations with mortality and indicators of social deprivation, *Journal of Epidemiology and Community Health*, **49**, Supplement 2, S57-S64

Benzeval, M., Judge, K. and Whitehead, M. (1995) *Tackling Inequalities in Health: an Agenda for Action*, Kings Fund, London

Bercé, Y-M. (1987) *Revolt and Revolution in Early Modern Europe: an Essay on the History of Political Violence*, Manchester University Press, Manchester

Berger, P.L. (1969) *The Social Reality of Religion*, Faber, London

Bergier, J.F. (1984) *Histoire Economique de la Suisse*, Payot, Lausanne

Bergmann, T. (1992) The re-privatisation of farming in eastern Germany, *Sociologia Ruralis*, **32**, 305–16

Bertaux-Wiame, I. (1981) The life history approach to the study of internal migration, in Bertaux, D. (ed.) *Biography and Society,* Sage, Newbury Park, Calif., 252–66

Beynon, H. (1995) The changing experience of work: Britain in the 1990s, Paper to the Conference on Education and Training for the Future Labour Markets of Europe, 21–24 September 1995, University of Durham

Bladen-Hovell, R. (1994) The European monetary system, in Artis, M.J. and Lee, N. (eds) *The Economics of the European Union: Policy and Analysis*, Oxford University Press, Oxford, 328–45

Blaut, J.M. (1975) Imperialism: the Marxist theory and its evolution, *Antipode*, 7(1), 1–19

Bloch, M. (1962) *Feudal Society*, Routledge & Kegan Paul, London

Block, F. (1977) *The Origins of International Economic Disorder: a Study of the United States International Monetary Policy from World War II to the Present*, University of California Press, Berkeley

Blomley, N. (1996) I'd like to dress her all over: masculinity, power and retail space, in Wrigley, N. and Lowe, M. (eds) *Retailing, Consumption and Capital: Towards the New Retail Geography*, Longman, Harlow, 238–56

Bociurkiw, B.R. and Strong, J.W. (eds) (1975) *Religion and Atheism in the U.S.S.R. and Eastern Europe*, Macmillan, London

Bocock, R. (1993) *Consumption*, Routledge, London

Boissevain, J. (1996) Introduction, in Boissevain, J. (ed.) *Coping with Tourists: European Reactions to Mass Tourism,* Berghahn Books, Providence, RI, and Oxford, 1–26

Bouchet, D. (1995) Marketing and the redefinition of ethnicity, in Costa, J.A. and Bamossy, G.J. (eds) *Marketing in a Multicultural World; Ethnicity, Nationalism and Cultural Identity*, Sage, London, 68–104

Bourdon, J. (1991) *Formation et Développement Régional en Europe*, Rapport à la DATAR, Paris

Bowl, R. (1985) *Changing the Nature of Masculinity – A Task for Social Work?,* University of East Anglia, Norwich (Social Work Monograph 30)

Bowler, I.R. (1985) *Agriculture under the Common Agricultural Policy: a Geography*, Manchester University Press, Manchester

Bowler, I.R. (1986) Intensification, concentration and specialisation in agriculture: the case of the European Community, *Geography*, **71**, 14–24

Bowler, I.R. (1992) Sustainable agriculture as an alternative path of farm business development, in Bowler, I., Bryant, C. and Nellis, M. (eds) *Rural Systems in Transition: Agriculture and Environment*, CAB International, Wallingford, 237–53

Bowler, I.R. and Ilbery, B.W. (1997) The regional consequences for agriculture of changes to the Common Agricultural Policy, in Laurent, C. and Bowler, I. (eds) *CAP and the Regions: Building a Multidisciplinary Framework for the Analysis of the EU Agricultural Space*, INRA, Versailles, 105–15

Bowler, I.R., Clark, G., Crockett, A., Ilbery, B.W. and Shaw, A. (1996) The development of alternative farm enterprises: a study of family labour farms in the northern Pennines of England, *Journal of Rural Studies*, **12**, 285–95

Boys, R., Foster, D. and Jozan, P. (1991) Mortality from causes amenable and non-amenable to medical care: the experience of eastern Europe, *British Medical Journal*, **303**, 879–83

Bradford, M.G. (1995) Diversification and division in the restructured English education system: towards a post-Fordist model?, *Environment and Planning A*, **27**, 1595–612

Bradford, M.G., Robson, B.T. and Tye, R. (1995) Constructing an urban deprivation index: a way of meeting the need for flexibility, *Environment and Planning A*, **27**, 319–33

Brass, W. (1989) Is Britain facing the twilight of parenthood? in Joshi, H. (ed.) *The Changing Population of Britain*, Blackwell, Oxford, 12–26

Braudel, F. (1985) *Civilization and Capitalism 15th–18th Century, Volume II: the Wheels of Commerce*, Fontana, London

Braverman, H. (1974) *Labor and Monopoly Capital*, Monthly Review Press, New York

Breathnach, P., Henry, M., Drea, S. and O'Flaherty, M. (1994) Gender in Irish tourism employment, in Kinnaird, V. H. and Hall, D. R. (eds) *Tourism: a Gender Analysis*, John Wiley, Chichester and New York, 52–73

Brett, E.A. (1983) *International Money and Capitalist Crisis: the Anatomy of Global Disintegration*, Heinemann, London

Brettell, C.B (1986) *Men who Migrate, Women who Wait: Population and History in a Portuguese Parish*, Princeton University Press, Princeton, NJ

Briggs, D. and Kerrell, E. (1992) Patterns and implications of policy-induced agricultural adjustments in the European Community, in Gilg, A. (ed.) *Restructuring the Countryside: Environmental Policy in Practice*, Avebury, Aldershot, 85–102

Brittan, A. (1989) *Masculinity and Power*, Blackwell, Oxford

Britton, M. (1990) *Mortality and Geography: a review in the mid-1980s. England and Wales*, HMSO, London

Britton, S. (1991) Tourism, capital and place: towards a critical geography, *Environment and Planning D: Society and Space,* **9**, 451–78

Broadfoot, P.M. (1996) *Education, Assessment and Society*, Open University Press, Buckingham

Brock, C. and Tulasiewicz, W. (eds) (1994) *Education in a Single Europe*, Routledge, London

Broggi, M.F. (1988) The Alpine region in danger, *Naturopa,* **59**: 18–20

Browne, H. (1983) *Spain's Civil War*, Longman, Harlow

Bruno, M. (1995) The second love of worker bees: gender, employment and social change in contemporary Moscow, paper presented at the Institute of British Geographers' Conference, Newcastle-upon-Tyne, UK, January 1995

Budd, A. (1993) *The EC and Foreign and Security Policy*, University of North London, London (European Dossier Series, No. 28)

Bugge, P. (1993) The nation supreme. The idea of Europe 1914–1945, in Wilson, K. and van der Dussen, J. (eds) *The History of the Idea of Europe*, Routledge, London, 83–149

Buller, H. and Hoggart, K. (1994a) *International Counter Urbanization: British Migrants in Rural France*, Avebury, Aldershot

Buller, H. and Hoggart, K. (1994b) Social integration of British home owners into French rural communities, *Journal of Rural Studies*, **10**, 197–210

Bunker, J., Frazier, H. and Mosteller, F. (1994) Improving health: measuring the effects of health care, *Millbank Quarterly*, **72**, 225–58

Burke, E. (1968) *Reflections on the Revolution in France*, Penguin, Harmondsworth

Burkert, W. (1985) *Greek Religion*, Harvard University Press, Cambridge, Mass.

Burnham, R.E. (1938) *Who are the Finns?*, Faber, London

Burns D.N., Gellert G.A. and Crone R.K. (1994) Tuberculosis in Eastern Europe and the former Soviet Union: How concerned should we be?, *The Lancet*, **343**, 1445

Burt, S. (1995) Retail internationalisation; evolution of theory and practice, in McGoldrick, J. and Davies, G. (eds) *International Retailing Trends and Strategies*, Pitman, London, 51–73

Burton, R.C.J. (1995) Geographical patterns of tourism in Europe, in Cooper, C.P. and Lockwood, A. (eds) *Progress in Tourism, Recreation and Hospitality Management, Volume 5*, John Wiley, Chichester, 3–25

Butlin, R.A. and Dodgshon, R.A. (eds) (1998) *The Historical Geography of Europe*, Oxford University Press, Oxford

Bütow, B., Heidreich, H., Lindert, B. and Neuke, E. (1992) *Frauen in Sachsen: zwischen Betroffenheit und Hoffnung*, Rosa-Luxemburg-Verein e.V., Leipzig

Buttimer, A. (1994) Edgar Kant and Balto-Skandinavia: *Heimatkunde* and regional identity, in Hooson, D. (ed.) *Geography and National Identity*, Blackwell, Oxford, 161–83

Button, K. and Maggi, R. (1995) Videoconferencing and its implications for transport – an anglo-Swiss perspective, *Transportation Journal*, **26**(2), 12–20

Bylund, E. (1960) Theoretical considerations regarding the distribution of settlement in Inner North Sweden, *Geografiska Annaler*, **42**, 225–31

Byrne, P. (1988) Religion and the religious, in Sutherland, S., Houlden, L., Clarke, P. and Hardy, F. (eds) *The World's Religions*, G.K. Hall & Co., Boston, Mass., 3–28

Caballé, A. (1996) Dona i reestructuració a les àrees rurals: el cas de l´agroturisme a les comarques del Solsonès, el Bages i el Berguedà, *Treballs de la Societat Catalaña de Geografia*, **11**(42), 37–100

Caliagli, M., de Winter, L., Mintzel, A., Culla, J. and de Brouwer, A. (1995) *Christian Democracy in Europe*, Institut de Ciències Politiques i Socials, Barcelona

Caglar, A.S. (1995) *McDoner: Doner Kebap* and the social positioning struggle of German Turks, in Costa, J.A. and Bamossy, G.J. (eds) *Marketing in a Multicultural World: Ethnicity, Nationalism and Cultural Identity*, Sage, London, 209–29

Cahill, T. (1995) *How the Irish Saved Civilization*, Doubleday, New York and London

Cain, P.J. and Hopkins, A.G. (1993a) *British Imperialism. Innovation and Expansion 1688–1914*, Longman, London

Cain, P.J. and Hopkins, A.G. (1993b) *British Imperialism. Crisis and Deconstruction 1914–1990*, Longman, London

Callinicos, A. (1991) *The Revenge of History: Marxism and the East European Revolutions*, Polity, Cambridge

Cameron, E. (1991) *The European Reformation*, Clarendon Press, Oxford

Campbell, B. (1993) *Goliath*, Methuen, London

Cândia, V. (1987) Iconoclasm, in Eliade, M. (ed.) *The Encyclopedia of Religion, Volume 7*, Macmillan, New York, 1–2

Caprio G. Jr and Levine, R. (1994) Reforming finance in transitional socialist economies, *World Bank Research Observer*, **9**, 1–24

Carr, R. (1982) *Spain 1808–1975*, 2nd edition, Clarendon Press, Oxford

Carter, F.W. and Turnock, D. (eds) (1993) *Environmental Problems in Eastern Europe*, Routledge, London

Carter, F.W., Hall, D.R., Turnock, D. and Williams, A.M. (1995) *Interpreting the Balkans,* Royal Geographical Society, London (Geographical Intelligence Paper No. 2)

Cashmore, W. (1979) *Rastaman: the Rastafarian Movement in England*, George Allen & Unwin, London

Cassirer, E. (1951) *The Philosophy of the Enlightenment*, Princeton University Press, Princeton, NJ

Castrén, M.A. (1852–70) *Nordiska Resor och Forskingar*, Helsingfors

Cater, E. (1994) Introduction, in Cater, E. and Lowman, G. (eds) *Ecotourism: A Sustainable Option?* Royal Geographical Society and John Wiley, Chichester, 3–18

Cavaco, C. (1995) Rural tourism: the creation of new tourist spaces, in Montanari, A. and Williams, A.M. (eds) *European Tourism: Regions, Spaces and Restructuring*, John Wiley, Chichester, 127–49

CEC (Commission of the European Communities) (1985) *Completing the Internal Market, COM(85), 310 Final*, CEC, Brussels

CEC (Commission of the European Communities) (1988) *Package Travel: Protection for the Consumer*, CEC, Brussels

CEC (Commission of the European Communities) (1989) Annex, *Official Journal of the European Communities*, No. **L 386**, 13

CEC (Commission of the European Communities) (1991) *Report by the Commission to the Council and the European Parliament on the European Year of Tourism*, CEC, Brussels

CEC (Commission of the European Communities) (1994) *Competitiveness and Cohesion: Trends in the Regions*, Office for Official Publications of the European Communities, Luxembourg

CEC (Commission of the European Communities) (1996) *The Demographic Situation in the European Union 1995*, Office for Official Publications of the European Commission, Luxembourg

CEC DG-XXIII Tourism Unit (1994) *The Evolution in Holiday Travel Facilities and in the Flow of Tourism inside and outside the European Community*, Office for Official Publications of the European Communities, Luxembourg

Cecchini, P. (1988) *The European Challenge – 1992: the Benefits of a Single Market,* Wildwood House, Aldershot

Central Statistical Office, Poland (1989) *Rocznik Statystyczny GUS 1989* (Statistical Yearbook of the Central Statistical Office), Central Statistical Office, Warsaw

Central Statistical Office, Poland (1992–93) *Roczniki Statystyczne Województw 1992 i 1993* (Statistical Yearbooks for Voivodships, 1992 and 1993), Central Statistical Office, Warsaw

Central Statistical Office, Poland (1993–96) *Rynek wewnetrzny w: 1992, 1993, 1994, 1995, Informacje i Opracowania Statystyczne GUS* (Internal Market in: 1992, 1993, 1994, 1995, Statistical Information and Analysis, Central Statistical Office, Warsaw

Champion, A.G. (ed.) (1989) *Counterurbanization: The Changing Pace and Nature of Population Deconcentration*, Edward Arnold, London

Champion, A.G. (1993) Introduction: key population developments and their local impacts, in Champion A.G. (ed.) *Population Matters: the Local Dimension*, Paul Chapman Publishing, London, 1–21

Champion, A.G. (1994a) International migration and demographic change in the Developed World, *Urban Studies*, **31**, 653–77

Champion, A.G. (1994b) Population change and migration in Britain since 1981: evidence for continuing deconcentration, *Environment and Planning A*, **26**, 1501–20

Champion, A.G. (1995) Internal migration, counterurbanization and changing population distribution, in Hall, R. and White, P. (eds) *Europe's Population: Towards the Next Century*, UCL Press, London, 99–129

Champion, T., Mönnesland, J. and Vandermotten, C. (1996a) The new regional map of Europe, *Progress in Planning*, **46**, 1–89

Champion, T., Wong, C., Rooke, A., Dorling, D., Coombes, M. and Brunsdon, C. (1996b) *The Population of Britain in the 1990s: a Social and Economic Atlas*, Clarendon Press, Oxford

Chaney, D. (1994) *The Cultural Turn*, Routledge, London

Charlton, J. and Velez, R. (1986) Some international comparisons of mortality amenable to medical intervention, *British Medical Journal*, **292**, 295–301

Cheshire, P.C. and Hay, D.G. (1989) *Urban Problems in Western Europe*, Unwin Hyman, London

Chick, V. and Dow, S.C. (1988) A post-Keynesian perspective on the relation between banking and regional development, in Arestis, P. (ed.) *Post-Keynesian Monetary Economics*, Edward Elgar, Aldershot, 219–50

Childs, M.W. (1936) *Sweden: the Middle Way*, Yale University Press, New Haven, Conn.

Christian, H. (1994) *The Making of the Anti-Sexist Men*, Routledge, London

Ciaccio, C. (1990) L'urbanisation touristique des espace ruraux en Sicile, in Korcelli, P. and Galczynska, B. (eds) *The Impact of Urbanization upon Rural Areas,* Polish Academy of Sciences Institute of Geography, Warsaw, 135–43

CITYC (Centro de Información Textil y de la Confección) (1996) *La confección en España*, Centro de Informacion Textil y de la Confección, Barcelona

Clark, G.J.D. (1966) The invasion hypothesis in British prehistory, *Antiquity*, **40**, 172–89

Clark, R.P. (1994) *The Basque Insurgents: ETA, 1952–80*, University of Wisconsin Press, Madison

Clarke, P. (1988) Introduction: new religious movements, in Sutherland, S., Houlden, L., Clarke, P. and Hardy, F. (eds) *The World's Religions*, G.K. Hall & Co., Boston, Mass., 907–11

Claval, P. (1994) From Michelet to Braudel: personality, identity and organization, in Hooson, D. (ed.) *Geography and National Identity*, Blackwell, Oxford, 39–57

Close, P. (1995) *Citizenship, Europe and Change*, Macmillan, Basingstoke

Clout, H. (1996) Restoring the ruins: the social context of reconstruction in the countrysides of northern France in the aftermath of the Great War, *Landscape Research*, **21**(3), 213–30

Coddington, P. (1993) The impact of videoconferencing on airline business traffic, *Journal of Travel Research*, **32**(2), 64–6

Cole, J. and Cole, F. (1993) *The Geography of the European Community*, Routledge, London

Coleman, D. (1996) New patterns and trends in European fertility: international and sub-national comparisons, in Coleman, D. (ed.) *Europe's Population in the 1990s*, Oxford University Press, Oxford, 1–61

Coles, P. (1968) *The Ottoman Impact on Europe*, Thames & Hudson, London

Collaborative Research Plan for Intervening in Emerging Infectious Diseases Epidemic in St. Petersburg District (1996) unpublished report by the National Public Health Institute, Helsinki, Finland

Colley, L. (1995) Britishness and Otherness: an argument, in O'Dea, M. and Whelan, K. (eds) *Nations and Nationalisms: France, Britain, Ireland and the Eighteenth-Century Context*, Voltaire Foundation, London, 39–54

Collins, D. (ed.) (1975) *The Origins of Europe*, Allen & Unwin, London

Collins, R. (1990a) *The Basques*, 2nd edition, Blackwell, Oxford

Collins, R. (1990b) The ethnogenesis of the Basques, in Wolfram, H. and Pohl, W. (eds) *Typen der Ethnogenese unter besonder Berücksichtigung der Bayern, Volume 1*, Institut für österreichische Geschichtsforschung, Vienna, 35–44

Collins, R. (forthcoming) Les Basques dans l'histoire. Récurrences et fractures, in Laborde, D. (ed.) *Les Basques, un peuple hors-la-loi?*, L'Harmattan, Paris

Commission Européenne (1993a) *Le Marché Unique Européen, Documentation Européenne*, Commission Européenne, Brussels

Commission Européenne (1993b) *Livre Blanc: Croissance, Compétivité, Emploi, Les Défis et les Pistes pour Entrer dans le XXIe Siècle*, Commission Européenne, Brussels

Commission Européenne (1994) *Compétitivité et Cohésion: Tendances dans les Régions, Cinquième Rapport*, Commission Européenne, Brussels

Commission of the European Communities (1985) COM (85), 310 final, 14 June 1985

Commission of the European Communities (1986) *Europe's Common Agricultural Policy*, Commission of the European Communities, Brussels

Comrie, B. (1990a) Slavonic languages, in Comrie, B. (ed.) *The Major Languages of Eastern Europe*, Routledge, London and New York, 56–62

Comrie, B. (ed.) (1990b) *The Major Languages of Eastern Europe*, Routledge, London and New York

Comrie, B. and Corbett, G. (eds) (1993) *The Slavonic Languages*, Routledge, London and New York

Connell, R.W. (1995) *Masculinities*, Polity Press, Cambridge

Connor, J. (ed.) (1968) *Lenin on Politics and Revolution*, Pegasus, New York

Connor, W. (1978) A nation is a nation, is an ethnic group, is a ..., *Ethnic and Racial Studies*, **1**, 379–88

Conradi, P. (1992a) Crisis in Russian health service, *British Medical Journal*, **304**, 937

Conradi, P. (1992b) Russia's doctors plan strike, *British Medical Journal*, **304**, 1201

Conti, S. and Enrietti, A. (1995) The Italian automobile industry and the case of Fiat: one country, one company, one market, in Hudson, R. and Schamp, E.W. (eds) *Towards a New Map of Automobile Manufacturing in Europe? New Production Concepts and Spatial Restructuring*, Springer, Berlin, 117–46

Cook, R. (1996) Interview, *New Statesman*, 6 September, 16–18

Cooke, P. (1995) Keeping to the high road: learning, reflexivity and associative governance in regional economic development, in Cooke, P. (ed.) *The Rise of the Rustbelt*, University of London Press, London, 231–46

Coon, C.S. (1939) *The Races of Europe*, Macmillan New York

Coppock, V., Haydon D. and Richter, I. (1995) *The Illusions of 'Post-feminism': New Women, Old Myths*, Taylor and Francis, London

Coriat, B. (1991) Technical flexibility and mass production : flexible specialisation and dynamic flexibility, in Benko, G. and Dunford, M. (eds) *Industrial Change and Regional Development: the Transformation of New Industrial Spaces*, Belhaven, London, 134–58

Cornell, T. and Matthews, J. (1982) *Atlas of the Roman World*, Phaidon, Oxford

Corrin, C. (1992) *Superwomen and the Double Burden*, Scarlet Press, London

Cosgrove, D. (1993) *The Palladian Landscape: Geographical Change and its Representations in Sixteenth-Century Italy*, Leicester University Press, Leicester

Cosgrove, D. (1996) Ideas and culture: a response to Don Mitchell, *Transactions of the Institute of British Geographers*, **21**(3), 574–75

Cosgrove, D. (1997) *Social Formation and Symbolic Landscape*, 2nd edition, University of Wisconsin Press, Madison

Council of Europe (1995) *The Council of Europe: Achievements and Activities*, Directorate of Information, Council of Europe, Strasbourg

Council of Europe (1996) *Recent Demographic Developments in Europe 1996*, Council of Europe Publishing, Strasbourg

Courtney, C. and Thompson, P. (1996) *City Lives. The Changing Voices of British Finance*, Methuen, London

Cousens, S. (1967) Changes in Bulgarian agriculture, *Geography*, **52**, 11–22

Creese, A. (1994) Global trends in healthcare reform, *World Health Forum*, **15**, 317

Cressy, D. (1992) The fifth of November remembered, in Porter, R. (ed.) *Myths of the English*, Polity, Cambridge, 68–90

Crewe, L. and Davenport, E. (1992) The puppet show: changing buyer–supplier relations within clothing retailing, *Transactions of the Institute of British Geographers*, **NS 17**, 183–97

Crompton, R. and Sanderson, K. (1990) *Gendered Jobs and Social Change*, Unwin Hyman, London

Crook, S., Pakulski, J. and Waters, M. (1992) *Postmodernization: Change in Advanced Society*, Sage, London

Cruijsen, H. (1996) Preface, in Rees, P., Stillwell, J., Convey, A. and Kupiszewski, M. (eds) *Population Migration in the European Union*, Wiley, Chichester, xix

Csaki, C. (1990) Agricultural changes in eastern Europe at the beginning of the 1990s, *American Journal of Agricultural Economics*, **12**, 1233–42

Curtis, S. and Taket, A. (1996) *Health and Societies: Changing Perspectives*, Edward Arnold, London

Curtis, S., Petukhova, N. and Taket, A. (1995) Health care reforms in Russia: the example of St Petersburg, *Social Science and Medicine*, **40**, 755–65

Cutler, T., Haslam, C., Williams, J. and Williams, K. (1989) *1992: The Struggle for Europe: A Critical Evaluation of the European Community*, Berg, New York

Czigány, L. (1984) *The Oxford History of Hungarian Literature*, Clarendon Press, Oxford

Dahl, R.A. (ed.) (1966) *Political Oppositions in Western Democracies*, Yale University Press, New Haven, Conn., and London

Dahlström, M. (1996) Young women in a male periphery – experiences from the Scandinavian north, paper presented at European Urban and Regional Studies Conference, Exeter, UK, April 1996

Dale, R. (1992) *International Banking Deregulation: the Great Banking Experiment*, Blackwell, Oxford

Dalton, R.J. (1988) *Citizen Politics in Western Democracies*, Chatham House, Chatham, NJ

Dalton, R.J. (1991) The dynamics of party system change, in Reif, K. and Ingelhart, R. (eds) *Eurobarometer: the Dynamics of European Public Opinion. Essays in Honour of Jacques-Réné Rabier*, Macmillan, London, 215–31

Daniels, H., Lucas, N., Toterdell, M. and Fomina, O. (1995) Humanization in Russian education – a transition between state determinism and individualism, *Educational Studies*, **21**(1), 29–39

Dannhaeuser, N. (1994) Concentration of trade and its urban impact under capitalism and socialism: former West Germany (Hassfurt) and East Germany (Hildburghausen) compared, *Urban Studies*, **31**(1), 79–98

Dansero, E. (1996) *Eco-sistemi Locali. Valori dell'Economia e Ragioni dell'Ecologia in un Distretto Industiale Tessile*, Franco Angeli, Milan

Davidson, H.R.E. (1982) *Scandinavian Mythology*, Hamlyn, London

Davies, N. (1982) *God's Playground: A History of Poland*, Columbia University Press, New York

Dawson, A.H. (1991), Poland, in Hall, D.R. (ed.) *Tourism and Economic Development in Eastern Europe and the Soviet Union*, Belhaven Press, London, 190–202

de Mouzon, A. *et al.* (1980) HLA-A, B typing in Basque and other Pyrenean populations, *Tissue Antigens*, **15**, 11–18

Delamont, S. (1995) *Appetites and Identities: an Introduction to the Social Anthropology of Western Europe*, Routledge, London

Den Boer, P. (1993) Europe to 1914: the making of an idea, in Wilson, K. and van der Dussen, J. (eds) *The History of the Idea of Europe*, Routledge, London, 13–82

Denecke, D. (1992) Ideology in the planned order upon the land: the example of Germany, in Baker, A. and Biger, G. (eds) *Ideology and Landscape in Historical Perspective*, Cambridge University Press, Cambridge, 303–29

Department of Health (1992) *The Health of the Nation*, HMSO, London, Cm1986

Department of Social Security (1994) *Households Below Average Income: a Statistical Analysis 1979–1991/2*, HMSO, London

Department of the Environment (1995) *Projections of Households in England to 2016*, HMSO, London

Deutsche Bundesbank (1992) *Impact de la Réunification Allemande sur les Echanges de l'Allemagne avec ses Partenaires Européens, Rapports Mensuels*, Deutsche Bundesbank, Frankfurt

Diaz Olivera, L. Le Nir, M., Plat, D. and Raux, Ch. (1995) *Les Effets-frontières: Evidence Empiriques, Impasses Théoriques*, Laboratoire d'Economie des Transports, Etudes et Recherches, Lyon

Dicken, P. and Oberg, S. (1996) The global context: Europe in a world of dynamic economic and population change, *European Urban and Regional Studies*, **3**(2), 101–20

Dicken, P., Forsgren, M. and Malmberg, A. (1994) The local embeddedness of transnational corporations, in Amin, A. and Thrift, N. (eds) *Globalization, Institutions, and Regional Development in Europe*, Oxford University Press, Oxford, 23–45

Dicken, P., Hudson, R. and Schamp, E. (1995) New challenges to the automobile production systems in Europe, in Hudson, R. and Schamp, E. (eds) *Towards a New Map of Automobile Manufacturing in Europe? New Production Concepts and Spatial Restructuring*, Springer, London, 1–20

Diez de Salazar, L.M. (1983) *Ferrerías en Guipúzcoa (siglos XIV–XVI)*, Haranburu, San Sebastián

Dingsdale, A. and Kovacs, Z. (1996) A return to socialism. The Hungarian General Election of 1994, *Geography*, **81**(3), 267–8

Dittus, P. (1994) *Corporate Governance in Central Europe: The Role of the Banks*, BIS Economic Papers 42, Bank for International Settlements, Basle

Dixon, R. (1991) *Banking in Europe: the Single Market*, Routledge, London

Dodd, C.H. (1983) *The Crisis of Turkish Democracy*, Eothen Press, London

Domosh, M. (1996) The feminized retail landscape: gender, ideology and consumer culture in nineteenth century New York City, in Wrigley, N. and Lowe, M. (eds) *Retailing, Consumption and Capital: Towards the New Retail Geography*, Longman, Harlow, 257–70

Dornbusch, R. (1991) Problems of European monetary integration, in Giovannini, A. and Mayer, C. (eds) *European Financial Integration*, Cambridge University Press, Cambridge, 305–27

Doughty, R.W. (1981) Environmental theology: trends and prospects in Christian thought, *Progress in Human Geography*, **5**(2), 234–48

Douglas, M. and Isherwood, B. (1979) *The World of Goods*, Allen Lane, London

Douglass, W.A. and Bilbao, J. (1975) *Amerikanuak: Basques in the New World*, University of Nevada Press, Reno

Dow, S. (1994) European monetary integration and the distribution of credit availability, in Corbridge, S., Martin, R. and Thrift, N. (eds) *Money, Power and Space*, Blackwell, Oxford, 149–64

Doyle, W. (1989) *The Origins of the French Revolution*, Oxford University Press, Oxford

Duke, S. (1996) The second death (or the second coming?) of the WEU, *Journal of Common Market Studies*, **34**, 167–90

Dukes, P. (1979) *October and the World: Perspectives on the Russian Revolution*, Macmillan, London

Duncan, C., Jones, K. and Moon, G. (1993) Do places matter? A multi-level analysis of regional variations in health-related behaviour in Britain, *Social Science and Medicine*, **37**(7), 725–33

Duncan, S. (1994a) Theorising differences in patriarchy, *Environment and Planning A*, **26**, 1177–94

Duncan, S. (1994b) Women's and men's lives and work in Sweden, *Gender, Place and Culture*, **1**(2), 261–68

Duncan, S. (1996) The diverse worlds of European patriarchy, in Garcia-Ramon, M.D. and Monk, J. (eds) *Women of the European Union*, Routledge, London, 74–110

Dunford, M. (1994) Winners and losers: the new map of inequality in the European Union, *European Urban and Regional Studies*, **1**(2), 95–114

Dunford, M. and Hudson, R. (1996) *Successful European Regions: Northern Ireland Learning from Others*, Northern Ireland Economic Council, Belfast

Dunford, M. and Perrons, D. (1994) Regional inequality, regimes of accumulation and economic development in contemporary Europe, *Transactions of the Institute of British Geographers*, **19**(2), 163–82

Durkheim, E. (1915) *The Elementary Forms of the Religious Life*, George Allen & Unwin, London

Duverger, M. (1964) *Political Parties*, 3rd edition, Methuen, London

Dyson, K. (1994) *Elusive Union: the Process of Economic and Monetary Union in Europe*, Longman, Harlow

Eadington, W.R. and Smith, V.L. (1992) The emergence of alternative forms of tourism, in Smith, V.L. and Eadington, W.R. (eds) *Tourism Alternatives,* University of Pennsylvania Press, Philadelphia, 1–12

Edmond, H. and Crabtree, R. (1994) Regional variations in Scottish pluriactivity: the socio-economic context for different types of non-farming activity, *Scottish Geographical Magazine*, **110**, 76–84

Edwards, A. (1981) *Leisure Spending in the European Community – Forecasts to 1990*, Economist Intelligence Unit, London

Efstratoglou-Todoulou, S. (1990) Pluriactivity in different socio-economic contexts: a test of the push–pull hypothesis in Greek farming, *Journal of Rural Studies*, **6**, 407–13

Einhorn, B. (1993) *Cinderella goes to Market: Citizenship, Gender and Women's Movements in East Central Europe*, Verso, London

Eisenstadt, S.N. (1992) The breakdown of Communist regimes, *Daedalus* **121**(2), 21–42

Eisenstein, E.L. (1979) *The Printing Press as an Agent of Change: Communications and Cultural Transformations in Early Modern Europe*, Cambridge University Press, Cambridge

EIU (1993) Austria, *International Tourism Report*, 2, Economist Intelligence Unit, London

Ekstedt, E., Henning, R., Andersson, R., Elvander, N., Forsgren, M., Malmberg, A. and Norgren, L. (1995) *Kulturell friktion.Konfliktkälla och förnyelsekraft i en integrerad ekonomi*, SNS Förlag, Stockholm

Elcock, W.D. (1975) *The Romance Languages*, Faber & Faber, London

Eliade, M. (1964) *Shamanism: Archaic Techniques of Ecstasy*, Princeton University Press, Princeton, NJ

Elola, J. (1996) Health care system reforms in western European countries: the relevance of health care organization, *International Journal of Health Services*, **26**, 239–51

Elton, G. (1971) *The Revolutionary Idea in France 1789–1871*, AMS Press, New York, reprint of 2nd 1931 edition

Embacher, H. (1994) Marketing for agri-tourism in Austria: strategy and realisation in a highly developed tourist destination, *Journal of Sustainable Tourism*, **2** (1&2), 61–76

Emerson, M., Gros, D., Italianer, A., Pisani-Ferry, J. and Reichenbach, H. (1992) *One Market, One Money: an Evaluation of the Potential Benefits and Costs of Forming an Economic and Monetary Union*, Oxford University Press, Oxford

Enders-Dragasser, U. (1991) Childcare: love, work and exploitation, *Womens Studies International Forum* (special edition, edited by Petrie, P., Meijvogel, R. and Enders-Draggasser, U.), **14**(6)

Engel, J. (ed.) (1970) *Grosserhistorischer Welt Atlas*, Bayerische Schulbuch Verlag, Munich

Engels, F (1977) *Socialism: Utopian and Scientific, with the Essay on "the Mark"*, Greenwood Press, Westport, Conn.

Ensor, T. (1993) Health system reform in former socialist countries of Europe, *International Journal of Health Planning Management*, **8**, 169

Eralp, A. *et al.* (eds) (1994) *The Political and Socioeconomic Transformation of Turkey*, Praeger, London

Eriksson, A.W. (1973) Genetic polymorphism in Finno-Ugrian populations, *Israel Journal of Medical Sciences*, **9**, 9–10, 1156–70

Ermisch, J. (1990) *Fewer Babies, Longer Lives*, Joseph Rowntree Foundation, York

Euromonitor (1996) *European Marketing and Data Statistics*, Euromonitor, London, 31st edition

European Commission (1994) *23rd Report on Competition Policy 1993*, European Commission, Brussels

European Commission (1996) *Economic and Monetary Union*, Office for Official Publications of the European Communities, Luxembourg

Eurostat (1993) *Retailing in the European Single Market*, Eurostat, Brussels

Eurostat (1994) *REGIO: Regional Data Bank*, Eurostat, Luxembourg

Eurostat (1995) *Vue Statistique sur l'Europe*, Eurostat, Brussels

Eurostat (1996) *Demographic Statistics*, Eurostat, Brussels

Evans, E. (1958) The Atlantic Ends of Europe, *Advancement of Science*, **15**, 54–64

Evans, T. (1988) Money makes the world go around, in Harris, L., Coakley, J., Croasdale, M. and Evans, T. (eds) *New Perspectives on the Financial System*, Croom Helm, London, 41–68

Evin, A. and Denton, G. (eds) (1990) *Turkey and the European Community*, Leske and Budrich, Opladen

Ewen, S. and Ewen, E. (1982) *Channels of Desire*, McGraw-Hill, New York

Fanon, F. (1965) *A Dying Colonialism*, Grove, New York

FAO-UNESCO (1974) *FAO/UNESCO Soil Map of the World, Legend*, UNESCO, Paris

Farrant, W. (1991) Addressing the contradictions: health promotion and community health action in the UK, *International Journal of Health Services*, **21**, 423–39

Faus-Pujol, M.C. (1995) Changes in the fertility rate and age structure of the population of Europe, in Hall, R. and White, P. (eds) *Europe's Population: Towards the Next Century*, UCL Press, London, 17–33

Featherstone, M. (1991) *Consumer Culture and Postmodernism*, Sage, London

Federal Department of Internal Affairs (1989) *Quadrilingualism in Switzerland*, Federal Chancellery, Berne

Ferguson, J. (1980) *Greek and Roman Religion: a Source Book*, Noyes Press, Park Ridge, NJ

Ferrão, J. (1987) *Indústria e Valorização do Capital. Uma Análise Geográfica*, CEG, Lisbon

Ferrão, J. and Vale, M. (1995) Multi-purpose vehicles, a new opportunity for the periphery? Lessons from the Ford/VW project (Portugal), in Hudson, R. and Schamp, E. (eds) *Towards a New Map of Automobile Manufacturing in Europe? New Production Concepts and Spatial Restructuring*, Springer, London, 195–217

Ferrill, A. (1986) *The Fall of the Roman Empire: the Military Explanation*, Thames & Hudson, London

Fielding, A.J. (1982) Counterurbanization in western Europe, *Progress in Planning*, **17**, 1–52

Figueiredo, A. de (1975) *Portugal: Fifty Years of Dictatorship*, Penguin, Harmondsworth

Findlay, A.M. (1995) The future of skill exchanges within the European Union, in Hall, R. and White, P. (eds) *Europe's Population: Towards the Next Century*, UCL Press, London, 130–41

Findlay, A.M. (1996) Extra-Union migration: the South–North perspective, in Rees, P., Stillwell, J., Convey, A. and Kupiszewski, M. (eds) *Population Migration in the European Union*, Wiley, Chichester, 39–50

Fischer, J. (1957) *Oriens-Occidens-Europa: Begriff und Gedanke 'Europa' in der späten Antike und im frühen Mittelalter*, Steiner, Weisbaden

Fladmark, J.M. (ed.) (1994) *Cultural Tourism*, Donhead, London

Focsa, G. (1970) *Muzeul Satului din Bucuresti*, Meridiane, Bucharest

Fodor, I. (1982) *In Search of a New Homeland: The Prehistory of the Hungarian People and the Conquest*, translated by Helen Tarnoy, Corvina, Budapest

Fogarty, M.P. (1957) *Christian Democracy in Western Europe 1820–1953*, Routledge & Kegan Paul, London

Fontana, J. (1995) *The Distorted Past: a Reinterpretation of Europe*, Blackwell, Oxford

Forum Europe (1992) *Turkey and the European Community*, Proceedings of the Forum Europe Conference (October 30–31, Brussels, 1991)

Forum Europe (1994) *Partners for Growth: New Trends in EC–Turkish Cooperation*, Proceedings of the Forum Europe Conference (May 13–14, Brussels, 1993)

Foucault, M. (1967) *Madness and Civilization: a History of Insanity in the Age of Reason*, Tavistock, London

Foucault, M. (1972) *The Archaeology of Knowledge*, Tavistock, London

Fox, J. (ed.) (1989) *Health Inequalities in Europe*, Gower, Aldershot

Frankland, E.G. (1995) Green revolutions? The role of Green parties in Eastern Europe's transition, 1989–1994, *East European Quarterly*, **29**(3), 315–45

Fraters, D. (1994) Generalized soil map of Europe: aggregation of the FAO-Unesco soil units based on the characteristics determining the vulnerability to soil degradation processes, RIVM (Report 71240300), Bilthoven, Netherlands

Frazer, J.G. (1981) *The Golden Bough: the Roots of Religion and Folklore*, Avenel Books, New York

Frend, W. (1988) Christianity in the first five centuries, in Sutherland, S., Houlden, L., Clarke, P. and Hardy, F. (eds) *The World's Religions*, G.K. Hall & Co., Boston, Mass., 142–66

Friedlander, A.H. (1988) Judaism, in Sutherland, S., Houlden, L., Clarke, P. and Hardy, F. (eds) *The World's Religions*, G.K. Hall & Co., Boston, Mass., 111–41

Führ, C. (1995) On the education system in the five new *Länder* of the Federal Republic of Germany, in Phillips, D. (ed.) *Education in Germany: Tradition and Reform in Historical Context*, Routledge, London, 259–83

Fujs, V. and Krašovec, M. (1996) *Wine Journeys in Slovenia*, Vas Travel Agency and Republic of Slovenia Ministry of Agriculture, Forestry and Food, Ljubljana

Fukuyama, F. (1992) *The End of History and the Last Man*, Free Press, New York

Fuller, A.M. (1990) From part-time farming to pluriactivity: a decade of change in rural Europe, *Journal of Rural Studies*, **6**, 361–73

Funk, N. and Mueller, M. (eds) (1993) *Gender Politics and Post-Communism: Reflections from Eastern Europe and the Former Soviet Union*, Routledge, London

Furter, P. (1995) *Mondes Rêvés*, Delachaux et Niestlé, Neuchâtel

Fusi, J.P. (1979) *El problema vasco en la II Republica*, Ediciones Turner, Madrid

Fusi, J.P. (1984) The Basque question 1931–7, in Preston, P. (ed.) *Revolution and War in Spain 1931–1939*, Routledge, London, 182–201

Gamkrelidze, T.V. and Ivanov, V.V. (1990) The early history of the Indo-European languages, *Scientific American*, **263**(3), 110–16

Garcia-Ramon, M.D. and Monk, J. (eds) (1996) *Women of the European Union*, Routledge, London

Gardener, P.M. and Molyneux, P. (1990) *Changes in Western European Banking*, Unwin Hyman, London

Garofoli, G. (1986) Le développement périphérique en Italie, *Economie et Humanisme*, **289**, 30–6

Garrahan, P. and Stewart, P. (1992) *The Nissan Enigma: Flexibility at Work in a Local Economy*, Mansell, London

Gaspar, J. (1995) Lisbon Metropolitan Area: structure, function and urban policies, in Fonseca, L. (ed.) *Lisboa: Abordagens Geográficas*, CEG, EPRU 42, Lisbon, 81–104

Gaspar, J., Fonseca, L. and Vale, M. (1996) Permanence and innovation in a traditionally dynamic region: the Setúbal Peninsula, paper presented to the *EUNIT Dortmund Seminar*, Dortmund

Gaspard, F. and Khosrokhavar, F. (1995) *Le Foulard et la Republique*, La Découverte, Paris

Gay, J.D. (1971) *The Geography of Religion in England*, Duckworth, London

Geertz, C (1966) Religion as a cultural system, in Banton, M. (ed.) *Anthropological Approaches to the Study of Religion*, Tavistock, London, 1–46

Gellner, E. (1983) *Nations and Nationalism*, Blackwell, Oxford

Gerth, H.H. and Wright-Mills, C. (eds) (1948) *From Max Weber: Essays in Sociology*, Routledge & Kegan, London

Getimis, P. and Kafkala, G. (1992) Local development and forms of regulation: fragmentation and hierarchy of spatial policies in Greece, *Geoforum*, **23**, 73–83

Getz, D. (1981) Tourism and rural settlement policy, *Scottish Geographical Magazine,* **97**, 158–68

Gibb, R. (ed.) (1994) *The Channel Tunnel: a Geographical Perspective*, John Wiley, Chichester and New York

Gibson, M. (1996) From Russia with years of expertise, *The European*, 17 October

Giddens, A. (1985) *A Contemporary Critique of Historical Materialism: the Nation-state and Violence*, Polity Press, Cambridge

Giddens, A. (1996) Affluence, poverty and the idea of a post scarcity society, *Development and Change*, **27**, 365–77

Gilbert, D.C. (1994) The European Community and leisure lifestyles, in Cooper, C.P. and Lockwood, A. (eds) *Progress in Tourism, Recreation and Hospitality Management, Volume 5*, John Wiley, Chichester, 116–31

Gill, L. and Rathwell, T. (1989) The effect of health services on mortality: amenable and non-amenable causes in Spain, *International Journal of Epidemiology*, **18**, 652–7

Gilley, S. (1988) Christianity in Europe: Reformation to today, in Sutherland, S., Houlden, L., Clarke, P. and Hardy, F. (eds) *The World's Religions*, G.K. Hall & Co., Boston, Mass., 216–42

Glacken, C.J. (1967) *Traces on the Rhodian Shore: Nature and Culture in Western Thought from Ancient Times to the End of the Eighteenth Century*, University of California Press, Berkeley and Los Angeles

Glasner, A. (1992) Gender and Europe: cultural and structural impediments to change, in Bailey, J. (ed.) *Social Europe*, Longman, Harlow, 70–105

Glenny, M. (1993) *The Rebirth of History: Eastern Europe in the Age of Democracy*, Penguin, Harmondsworth

Glyn, A. and Miliband, D. (1994) *Paying for Inequality: the Economic Cost of Social Injustice*, IPPR/Rivers Oram Press, London

Gómez-Ibañez, D.A. (1975) *The Western Pyrenees*, Oxford University Press, Oxford

Goodwin, M., Duncan, S. and Halford, S. (1993) Regulation theory, the local state, and the transition of urban politics, *Environment and Planning D: Society and Space*, **11**(1), 67–88

Gowan, P. (1995) Neo-liberal theory and practice for eastern Europe, *New Left Review*, **213**, 3–60

Gradmann, R. (1931) *Süd Deutschland*, J. Engelhorns Nachf., Stuttgart

Graham, L.S. and Makler, H.M. (1979) *Contemporary Portugal: the Revolution and Its Antecedents*, University of Texas Press, Austin

Grahl, J. and Teague, P. (1990) *1992 The Big Market: the Future of the European Community*, Lawrence and Wishart, London

Granada Television (1987) *The Basques of Santasi* (documentary film), Granada Television, Manchester

Grant, M. (1976) *The Fall of the Roman Empire*, Annenburg, Radnor

Gratton, C. and Taylor, P.D. (1995) Impacts of festival events: a case-study of Edinburgh, in Ashworth, G.J. and Dietvorst, A.G.J. (eds) *Tourism and Spatial Transformations*, CAB International, Wallingford, 225–38

Gratton, C. and van der Straaten, J. (1994) The environmental impact of tourism in Europe, in Cooper, C.P. and Lockwood, A. (eds) *Progress in Tourism, Recreation and Hospitality Management, Volume 5*, Chichester, John Wiley, 147–61

Green, A. and Owen, D. (1995) The labour market aspects of population change in the 1990s, in Hall, R. and White, P. (eds) *Europe's Population: Towards the Next Century*, UCL Press, London, 51–68

Green, J.N. (1994) France: Language situation, in Asher, R.E. (ed.) *The Encyclopedia of Language and Linguistics, Volume 3*, Pergamon Press, Oxford, 1295–6

Green, N. (1990) *The Spectacle of Nature: Landscape and Bourgeois Culture in Nineteenth-Century France*, Manchester University Press, Manchester

Greene, D. (1964) The Celtic languages, in Raftery, J. (ed.) *The Celts*, Mercier Press, Cork, 9–21

Grundy, E. (1996) Population ageing in Europe, in Coleman, D. (ed.) *Europe's Population in the 1990s*, Oxford University Press, Oxford, 267–96

Gual, J. and Neven, D. (1993) Banking, in *Social Europe – Market Services and European Integration*, European Economy, Report and Studies No. 3

Guardian, 18 September 1996

Güell, P.I. (1987) *La evolución estratégica de ETA (1963–1987)*, Kriselu, San Sebastián

Guirand, F. (1968) Greek mythology, in *New Larousse Encyclopedia of Mythology*, 2nd edition, Paul Hamlyn, London, 85–198

Guirand, F. and Pierre, A.V. (1968) Roman mythology, in *New Larousse Encyclopedia of Mythology*, 2nd edition, Paul Hamlyn, London, 199–221

Gumpel, W. (ed.) (1992) *Turkey and the European Community: An Assessment*, Verlag, Munich

Habermas, J. (1976) *Legitimation Crisis*, Heinemann, London

Habermas, J. (1990) What does Socialism mean today? The rectifying revolution and the need for new thinking on the Left, *New Left Review*, **183**, 3–21

Habermas, J. (1994a) Europe's second chance, in Pensky, M. (ed.) *The Past as Future: Jürgen Habermas Interviewed by Michael Haller*, Polity Press, Cambridge, 73–98

Habermas, J. (1994b) The past as future, in Pensky, M. (ed.) *The Past as Future: Jürgen Habermas Interviewed by Michael Haller*, Polity Press, Cambridge, 55–72

Hajdú, P. (ed.) (1976) *Ancient Cultures of the Uralian Peoples*, Corvina, Budapest

Halasz, G. (1993) The policy of school autonomy and the reform of educational administration: Hungarian changes in an east European perspective, *International Review of Education*, **39**(6), 489–97

Hall, D. and Brown, F. (1996) Towards a welfare focus for tourism research, *Progress in Tourism and Hospitality Research,* **2**(1), 41–57

Hall, D. and Danta, D. (eds) (1996) *Reconstructing the Balkans*, John Wiley, Chichester and New York

Hall, D. and Kinnaird, V. (1994) Ecotourism in Eastern Europe, in Cater, E. and Lowman G. (eds) *Ecotourism: a Sustainable Option?*, John Wiley, Chichester and New York, 111–36

Hall, D.R. (ed.) (1991) *Tourism and Economic Development in Eastern Europe and the Soviet Union*, Belhaven Press and Halsted Press, London and New York

Hall, D.R. (1995) Tourism change in Central and Eastern Europe, in Montanari, A. and Williams, A.M. (eds) *European Tourism: Regions, Spaces and Restructuring*, John Wiley, Chichester, 221–44

Hall, D. R. (1996) Recovering the Danube Delta, *Environmental Scientist,* **5**(1), 10–11

Hall, R. (1995) Households, families and fertility, in Hall, R. and White, P. (eds) *Europe's Population: Towards the Next Century*, UCL Press, London, 34–50

Hamill, J. (1993) Cross-border mergers, acquisitions and strategic alliances, in Bailey, P., Parisotto, A. and Renshaw, G. (eds) *Multinationals and Employment: the Global Economy of the 1990s*, ILO, Geneva, 95–123

Hanley, D. (1994) *Christian Democracy in Europe*, Pinter, London

Hannigan, K. (1994) National policy, European structural funds and sustainable tourism: the case of Ireland, *Journal of Sustainable Tourism*, **2**, 179–92

Hansen, J.C. (1996) Les pays nordiques et l'Union européenne: intégration ou isolement?, *Cahiers de Géographie du Québec*, **40**, 255–65

Hargreaves, A. (1995) *Immigration, 'Race' and Ethnicity in Contemporary France*, Routledge, London

Harper-Bill, C. (1988) Christianity in the West to the Reformation, in Sutherland, S., Houlden, L., Clarke, P. and Hardy, F. (eds) *The World's Religions*, G.K. Hall & Co., Boston, Mass., 193–215

Harris, N. (1995) *The New Untouchables: Immigration and the New World Worker*, I.B. Tauris, London

Harrison, J. (1983) Heavy industry, the state and economic development in the Basque region 1876–1936, *Economic History Review*, **36**, 535–51

Hartmann, M.V. (1977) *Raumwirtschaftliche Implikationen der Organisation der Kreditwirtschaft*, Untersuchungen über das Spar-, Giro- und Kreditwesen 92, Duncker & Humblot, Berlin

Harvey, D. (1989) *The Condition of Postmodernity*, Basil Blackwell, Oxford

Hayuth, Y. (1987) *Intermodality: Concept and Practice. Structural Changes in the Ocean Freight Transport Industry*, Lloyd's of London Press, London

Hedges, C. (1996) In the Balkans, three languages now fight it out, *New York Times*, 15 May, A4

Heelas, P. (1988) Western Europe: Self-religions, in Sutherland, S., Houlden, L., Clarke, P. and Hardy, F. (eds) *The World's Religions*, G.K. Hall & Co., Boston, Mass., 925–31

Heiberg, M. (1989) *The Making of the Basque Nation*, Cambridge University Press, Cambridge

Held, D. (1991) Democracy, the nation-state and the global system, *Economy and Society*, **20**, 138–72

Held, D. (1995) *Democracy and Global Order*, Polity, Cambridge

Hemingway, E. (1955) *For Whom the Bell Tolls*, Penguin, Harmondsworth

Henriksen, C. (1990) The Scandinavian languages and the European Community, *Scandinavian Studies*, **64**(4), 685–98

Heper, M. *et al.* (eds) (1993) *Turkey and the West. Changing Political and Cultural Identities*, I.B. Tauris, London

Herder, J. G. (1791) *Ideen zur Philosophie der Geschichte der Menschheit, Volume IV*, Johann Friedrich Hartknoch, Riga-Leipzig, 20

Herodotus (1954) *The Histories*, translated by de Sélincourt, A., Penguin, Harmondsworth

Herzman, C. (1994) *Environment and Health in Central and Eastern Europe*, World Bank, Washington, DC

Hewison, R. (1987) *The Heritage Industry: Britain in a Climate of Decline*, Methuen, London

Hill, C. (1961) *The Century of Revolution, 1603–1714*, Nelson, Edinburgh

Hill, C. (1972) *The World Turned Upside Down: Radical Ideas during the English Revolution*, Temple-Smith, London

Hill, C. (1980) A bourgeois revolution?, in Pocock, J.G.A. (ed.) *Three British Revolutions: 1641; 1688; 1776*, Princeton University Press, Princeton, 109–39

Hill, C. (1986) The word 'revolution' in seventeenth-century England, in Ollard, R. and Tudor-Craig, P. (eds) *For Veronica Wedgwood These: Studies in Seventeenth-Century History*, Collins, London, 134–51

Hillerbrand, H.J. (1987) Reformation, in Eliade, M. (ed.) *The Encyclopedia of Religion, Volume 12*, Macmillan, New York, 244–54

Hirst, P. and Thompson, G. (1996a) Globalisation: ten frequently asked questions and some surprising answers, *Soundings*, **4**, 47–66

Hirst, P. and Thompson, G. (1996b) *Globalization in Question*, Polity, Cambridge

HIV/AIDS Surveillance in Europe (1996) European Centre for the Epidemiological Monitoring of AIDS, Quarterly Report No. 49

Hobsbawm, E. (1983) Introduction: inventing traditions, in Hobsbawm, E. and Ranger, T. (eds) *The Invention of Tradition*, Cambridge University Press, Cambridge, 1–14

Hobsbawm, E. and Ranger, T. (eds) (1983) *The Invention of Tradition*, Cambridge University Press, Cambridge

Hodder, I. (1990) *The Domestication of Europe*, Blackwell, Oxford

Hodges, M. (1981) Liberty, equality, divergency: the legacy of the Treaty of Rome?, in Hodges, M. and Wallace, W. (eds) *Economic Divergency in the European Community*, George Allen & Unwin, London, 1–15

Hoggart, K. and Buller, H. (1995) British home owners and housing change in rural France, *Housing Studies,* **10**, 179–98

Hoggart, K., Buller, H. and Black, R. (1995) *Rural Europe – Identity and Change*, Edward Arnold, London

Höll, A. and Von Meyer, H. (1996) Germany, in Whitby, M. (ed.) *The European Environment and CAP Reform: Policies and Prospects for Conservation*, CAB International, Wallingford, 70–85

Honkala, T., Leporanta-Morley, P., Liukka, L. and Rougle, E. (1988) Finnish children in Sweden go on strike for better education, in Skutnabb-Kangas, T. and Cummins, J. (eds) *Minority Education*, Multilingual Matters Ltd, Clevedon, 239–50

Hooson, D. (1994a) Introduction, in Hooson, D. (ed.) *Geography and National Identity*, Blackwell, Oxford, 1–11

Hooson, D. (ed.) (1994b) *Geography and National Identity*, Blackwell, Oxford

Howard, M.C. and King, J.E. (1985) *The Political Economy of Marx*, 2nd edition, Longman, Harlow

Hoyle, B.S. (ed.) (1996) *Cityports, Coastal Zones and Regional Change*, John Wiley, Chichester

HRFT (1994) *Turkey. Human Rights Report*, The Human Rights Foundation of Turkey (HRFT), Ankara (annual publication)

Hualde, J.I., Lakarra, J.A. and Trask, R.L. (eds) (1996) *Towards a History of the Basque Language*, J. Benjamins Publishing Co., Amsterdam

Hudson, R. (1983) Capital accumulation and chemicals production in Western Europe in the postwar period, *Environment and Planning A*, **15**, 105–22

Hudson, R. (1988) Uneven development in capitalist societies: changing spatial divisions of labour, forms of spatial organization of production and service provision, and their impact upon localities, *Transactions of the Institute of British Geographers*, **NS 13**, 484–96

Hudson, R. (1989) Labour market changes and new forms of work industrial regions: maybe flexibility for some but not flexible accumulation, *Society and Space*, **7**, 5–30

Hudson, R. (1994a) New production concepts, new production geographies? Reflections on changes in the automobile industry, *Transactions of the Institute of British Geographers*, **NS 19**, 331–45

Hudson, R. (1994b) East meets west: the regional implications within the European Union of political and economic changes in Eastern Europe, *European Urban and Regional Studies*, **1**(1), 79–83

Hudson, R. (1995a) The role of foreign investment, in Darnell, A., Evans, L., Johnson, P. and Thomas, B. (eds) *The Northern Region Economy: Progress and Prospects*, Mansell, London, 79–95

Hudson, R. (1995b) Towards sustainable industrial production: but in what sense sustainable?, in Taylor, M. (ed.) *Environmental Change: Industry, Power and Place*, Avebury, Winchester, 37–56

Hudson, R. (1997) Regional futures: industrial restructuring, new high volume production concepts and spatial development strategies in the new Europe, *Regional Studies* **31**(5), 467–78

Hudson, R. and Sadler, D. (1989) *The International Steel Industry*, Routledge, London

Hudson, R. and Schamp, E.W. (eds) (1995) *Towards a New Map of Automobile Manufacturing in Europe? New Production Concepts and Spatial Restructuring*, Springer, Berlin

Huma (1996) *Estonia for Tourists*, Huma, Tallinn

Humblett, P., Lagasse, R., Moens, G., Van de Voorde, H. and Wollast, E. (1986) *Atlas de la Mortalité Evitable en Belgique*, École de Santé Publique, Leuven

Hummelbrunner, R. and Miglbauer, E. (1994) Tourism promotion and potential in peripheral areas: the Austrian case, *Journal of Sustainable Tourism*, **2**(1&2), 41–50

Hundeckova, H. and Lostak, M. (1992) Attitudes to privatisation in Czechoslovakia: results of a 1990 sociological survey, *Sociologia Ruralis*, **32**, 287–304

Huntington, E. (1924) *Civilisation and Climate*, Yale University Press, New York

Huntington, E. (1925) *The Character of Races*, Charles Scribner's Sons, New York

Hutchinson, J. and Smith, A.D. (eds) (1994) *Nationalism*, Oxford University Press, Oxford

Ilbery, B.W. (1990a) The challenge of land redundancy, in Pinder, D. (ed.) *Western Europe: Challenge and Change*, Belhaven Press, London, 211–25

Ilbery, B.W. (1990b) Adoption of the arable set-aside scheme in England, *Geography*, **75**, 69–73

Ilbery, B.W. (1992) Agricultural policy and land diversion in the European Community, in Gilg, A. (ed.) *Progress in Rural Policy and Planning*, **2**, 153–66

Ilbery, B.W. and Bowler, I.R. (1993) The Farm Diversification Grant Scheme: adoption and non-adoption in England and Wales, *Environment and Planning C*, **11**, 161–70

Ilbery, B.W. and Bowler, I.R. (1998) From agricultural productivism to post-productivism, in Ilbery, B.W. (ed.) *The Geography of Rural Change*, Longman, London

İnalcık, H. (1973) *The Ottoman Empire: the Classical Age 1300–1600*, Weidenfeld & Nicolson, London

Ireland, M. (1993) Gender and class relations in tourism employment, *Annals of Tourism Research*, **20**(4), 666–84

Isachsen, F. (1968) Norden, in Sømme, A. (ed.) *A Geography of Norden*, Cappelens Forlag, Oslo, 13–19

Isaksen, A. (1994) New industrial spaces and industrial districts in Norway: productive concepts in explaining regional development?, *European Urban and Regional Studies*, **1**, 31–48

Israel, J.I. (1991) General introduction, in Israel, J.I. (ed.) *The Anglo-Dutch Moment: Essays on the Glorious Revolution and its World Impact*, Cambridge University Press, Cambridge, 1–43

Izquierdo, A. (1991) La inmigración ilegal en España, *Revista Economía y Sociología del Trabajo*, **11**(March), 18–38

Jackson, P. (1989) *Maps of Meaning: an Introduction to Cultural Geography*, Allen & Unwin, London

Jackson, P. (1991) The cultural politics of masculinity: towards a social geography, *Transactions of the Institute of British Geographers*, **16**, 199–213

Jackson, P. (1993) Identity and the cultural politics of difference, paper presented at the British–Georgian Geographical Seminar at Tbilisi State University, 12–19 September 1993

Jackson, P. (1996) The idea of culture: a response to Don Mitchell, *Transactions of the Institute of British Geographers*, **21**(3), 572–3

Jacob, J.E. (1994) *Hills of Conflict: Basque Nationalism in France*, University of Nevada Press, Reno

Jakubovich, E. and Pais, D. (eds) (1929) *Ó-magyar olvasókönyv*, Danubia, Pécs

Jansen, A. and Hetson, H. (1991) Agricultural development and spatial organisation in Europe, *Journal of Rural Studies*, **7**, 143–51

Jansen-Verbeke, M. and van de Wiel, E. (1995) Tourism planning in urban revitalization projects: lessons from the Amsterdam waterfront development, in Ashworth, G.J. and Dietvorst, A.G.J. (eds) *Tourism and Spatial Transformations*, CAB International, Wallingford, 129–45

Jehlicka, P. and Kostelecky, T. (1995) Czechoslovakia: Greens in a post-Communist society, in Richardson, D. and Rootes, C. (eds) *The Green Challenge: the Development of Green Parties in Europe*, Routledge, London, 208–31

Jenner, P. and Smith, C. (1993) *Tourism in the Mediterranean*, Economist Intelligence Unit, London

Johansson, M. (1995) Nej till EU – Ja till staten, *Nordrevy*, **1995**(2), 7

Johnson, N. (1994) Sculpting heroic histories: celebrating the centenary of the 1798 rebellion in Ireland, *Transactions of the Institute of British Geographers*, **NS 19**, 78–93

Johnston, R.J. (1987a) I. The geography of the working class and the geography of the Labour vote in England, 1983. A prefatory note to a research agenda, *Political Geography Quarterly*, **6**(1), 7–16

Johnston, R.J. (1987b) VII. What price place? *Political Geography Quarterly*, **6**(1), 51–2

Johnston, R.J. and Pattie, C.J. (1987) Family background, ascribed characteristics, political attitudes and regional variations in voting within England, 1983: a further contribution, *Political Geography Quarterly*, **6**(4), 347–49

Johnston, R.J., Pattie, C.J. and Allsopp, J.G. (1988) *A Nation Dividing?* Longman, London

Jones, A. (1991) The impact of the EC's set-aside programme: the response of farm businesses in Rendsburg-Eckernforde, Germany, *Land Use Policy*, **8**, 108–24

Jones, A. (1994) *The New Germany: a Human Geography*, Wiley, Chichester

Jones, A. and Budd, S. (1994) *The European Community: a Guide through the Maze*, 5th edition, Kogan Page, London

Jones, A., Fasterding, F. and Plankl, R. (1993) Farm household adjustments to the European Community's set-aside policy: evidence from Rheinland-Pfalz (Germany), *Journal of Rural Studies*, **9**, 65–80

Jones, K. and Duncan, C. (1995) Individuals and their ecologies: analysing the geography of chronic illness within a multi-level modelling framework, *Health and Place*, **1**, 27–40

Jones, K. and Moon, G. (1992) Medical geography: global perspectives, *Progress in Human Geography*, **16**, 563–72

Jones, N. (1990) The single market for financial services, in Gillespie, I. (ed.) *Banking 1992*, Eurostudy Special Report, Eurostudy, London, 16–34

Jordan, T.G. (1996) *The European Culture Area*, 3rd edition, HarperCollins, New York

Joshi, H. (1996) Projections of European population decline: serious demography or false alarm?, in Coleman, D. (ed.) *Europe's Population in the 1990s*, Oxford University Press, Oxford, 222–66

Jutikkala, E. (1973) *An Atlas of Settlement in Finland in the Late 1560s*, Helsinki

Jutikkala, E. and Pirinen, K. (1979) *A History of Finland*, Heinemann, London

Kaiser, R.J. (1995) Czechoslovakia: the disintegration of a binational state, in Smith, G. (ed.) *Federalism: the Multiethnic Challenge*, Longman, London

Kamen, H. (1985) *Inquisition and Society in Spain in the Sixteenth and Seventeenth Centuries*, Weidenfeld & Nicolson, London

Katkov, G. (1967) *Russia 1917: the February Revolution*, Longman, London

Katz, R. and Mair, P. (1992) Changing models of party organization: the emergence of the cartel party, paper presented at the European Consortium for Political Research, Joint Session, Limerick, Ireland. Cited in Hayward, J. and Page E.C. (eds) *Governing the New Europe*, Polity Press, Cambridge, 193–4

Kazancigil, A. and Özbudun, E. (eds) (1981) *Atatürk: Founder of a Modern State*, C. Hurst, London

Kearney, A.T. (1995) *Cooperation Improves the Competitive Advantage of Ports and Has been Implemented by Port Related Businesses*, Ministry of Transport, Public Works and Water Management, The Hague

Kearney, H. (1989) *The British Isles: a History of Four Nations*, Cambridge University Press, Cambridge

Keating, M. (1995) Europeanism and regionalism, in Jones, B. and Keating, M. (eds) *The European Union and the Regions*, Clarendon Press, Oxford

Kedourie, E. (1960) *Nationalism*, Hutchinson, London

Keen, M. (1984) *Chivalry*, Yale University Press, New Haven, Conn.

Kennedy, P. (1993) *Preparing for the Twenty-first Century*, HarperCollins, London

Kennedy, S., Whiteford, P. and Bradshaw, J. (1996) The economic circumstances of children in ten countries, in Brannen, J. and O'Brien, M. (eds) *Children in Families*, Falmer Press, London, 145–70

Kibbee, D.A. (1994) Renaissance linguistics: French tradition, in Asher, R.E. (ed.) *The Encyclopedia of Language and Linguistics, Volume 7*, Pergamon Press, Oxford, 3536–40

Kickbusch, I. (1993) Health promotion and disease prevention: the implications for health promotion, in Normand, C. and Vaughan, P. (eds) *Europe without Frontiers: the Implications for Health*, Wiley, Chichester, 47–53

Kiernan, K. (1996) Partnership behaviour in Europe: recent trends and issues, in Coleman, D. (ed.) *Europe's Population in the 1990s*, Oxford University Press, Oxford, 62–91

King, A. (1997) The Euro debate: four nations united in uncertainty, *The Daily Telegraph*, 10 January 1997, 13

King, R. (1991) Italy: multifaceted tourism, in Williams, A.M. and Shaw, G. (eds) *Tourism and Economic Development: Western European Experiences*, 2nd edition, Belhaven, London, 61–83

King, R.L. and Rybaczuk, C. (1993) Southern Europe and the international division of labour: from emigration to immigration, in King, R.L. (ed.) *The New Geography of European Migrations*, Belhaven, London, 175–206

King, S. (1996) Montenegro and Slovenia seek new face in tourism, *The European*, 17 October

King, W.L. (1987) Religion, in Eliade, M. (ed.) *The Encyclopedia of Religion, Volume 12*, Macmillan, New York, 282–93

Kinnaird, V. H. and Hall, D. R. (eds) (1994) *Tourism: a Gender Analysis*, John Wiley, Chichester and New York

Kinross, P. (1964) *Atatürk: the Rebirth of a Nation*, Weidenfeld & Nicolson, London

Kirchheimer, O. (1966) The transformation of West European party systems, in LaPalombara, J. and Weiner, M. (eds) *Political Parties and Development*, Princeton University Press, Princeton, NJ, 177–200

Kirchner, E.J. (1992) *Decision Making in the European Community: the Council Presidency and European Integration,* Manchester University Press, Manchester

Kitaev, I.V. (1994) Russian education in transition – transformation of labor-market, attitudes of youth and changes in management of higher and lifelong learning, *Oxford Review of Education*, **20**(1), 111–30

Klagge, B. (1995) Strukturwandel im Bankenwesen und regionalwirtschaftliche Implikationen. Konzeptionelle Ansätze und empirische Befunde, *Erdkunde*, **49**, 285–304

Klingemann, H-D. and Fuchs, D. (eds) (1995) *Citizens and the State*, Oxford University Press, Oxford

Knickel, K. (1990) Agricultural structural change: impact on the rural environment, *Journal of Rural Studies*, **6**(4), 383–93

Kofman, E. and Sales, R. (1996) The geography of gender and welfare in Europe, in Garcia-Ramon, M.D. and Monk, J. (eds) *Women of the European Union*, Routledge, London, 31–60

Kokkonen, P. and Kekomaki, M. (1993) Legal and economic issues in European public health, in Normand, C. and Vaughan, P. (eds) *Europe without Frontiers: the Implications for Health*, Wiley, Chichester, 35–42

Kopeva, D., Mishev, P. and Jackson, M. (1994) Formation of land market institutions and their impacts on agricultural activity, *Journal of Rural Studies*, **10**, 377–85

Kors, A.C. and Peters, E. (1972) *Witchcraft in Europe 1100–1700: a Documentary History*, University of Pennsylvania Press, Philadelphia

Kossman, E.H. (1991) Freedom in seventeenth-century Dutch thought and practice, in Israel, J.I. (ed.) *The Anglo-Dutch Moment: Essays on the Glorious Revolution and its World Impact*, Cambridge University Press, Cambridge, 281–98

Koulov, B. (1995) Geography of electoral preferences: the 1990 Great National Assembly elections in Bulgaria, *Political Geography*, **14**(3), 241–58

Koulov, B. (1996) Market reforms and environmental protection in the Bulgarian tourism industry, in Hall, D. R. and Danta, D. (eds) *Reconstructing the Balkans,* John Wiley, Chichester and New York, 187–96

Kousis, M. (1989) Tourism and the family in a Cretan community, *Annals of Tourism Research*, **16**(3), 318–32

Krause, A. (1991) *Inside the New Europe*, Cornelia and Michael Bessie Books, New York

Krippendorf, J. (1987) *The Holiday Makers: Understanding the Impact of Leisure and Travel,* Heinemann, Oxford

Krueger, A.O. and Aktan, O.H. (1992) *Swimming Against the Tide. Turkish Trade Reform in the 1980s*, ICS Press, San Francisco, (Publication of the International Center for Economic Growth)

Krugman, P. (1990) Policy problems of a monetary union, in de Grauwe, P. and Papademos, L. (eds) *The European Monetary System in the 1990s*, Longman, Harlow, 48–64

Kunzmann, K.R. and Wegener, M. (1991) *The Pattern of Urbanization in Western Europe 1960–1990*, Universität Dortmund, Dortmund (Berichte aus dem Institut für Raumplanung 28)

Kupiszewski, M. (1996) Extra-Union migration: the East–West perspective, in Rees, R., Stillwell, J., Convey, A. and Kupiszewski, M. (eds) *Population Migration in the European Union*, Wiley, Chichester, 13–38

Kynaston, D. (1994) *The City of London. Volume 1: A World of its Own 1815–1950*, Chatto & Windus, London

Kynaston, D. (1996). *The City of London. Volume 2: Golden Years 1890–1914*, London, Chatto & Windus, London

Labba, N.G. (1996) Tankar om en samisk framtid i EU, http://www.sametinget.se/st/euframt.html

Lafferty Business Research (1992) *Financial Revolution in Europe II: the Revolution Deepens and Widens*, Lafferty Publications, Dublin

Lafourcade, M. (1983) Le particularisme juridique, in Haritschelar, J. (ed.) *Etre Basque*, Privat, Toulouse, 163–91

Langland, W. (1966) *Piers the Plowman*, Penguin, Harmondsworth

Law, C. (1994) *Urban Tourism*, Mansell, London

Lawrence, B.E. (1990) *Defenders of God: the Fundamentalist Revolt against the Modern Age*, I.B. Tauris, London

Laws, E. (1995) *Tourist Destination Management*, Routledge, London

Le Roy Ladurie, E. (1978) *Montaillou: Cathars and Catholicism in a French Village 1294–1324*, Scolar Press, London

Lebovics, H. (1992) *True France: the Wars over Cultural Identity 1900–1945*, Cornell University Press, London

Lee, R. (1995) Look after the pounds and the people will look after themselves: social reproduction, regulation, and social exclusion in western Europe, *Environment and Planning A*, **27**, 1577–94

Leger, J. (1669) *Histoire Générale des Eglises Evangéliques des Vallées du Piémont ou Vaudoises*, Jean Carpentier, Leyden

Lenin, V.I. (1964) *Selected Works*, Lawrence and Wishart, London

Leroi-Gourhan, A. (1968) *The Art of Prehistoric Man in Western Europe*, Thames & Hudson, London

Levack, B.P. (1987) *The Witch-hunt in Early Modern Europe*, Longman, London

Lever, A. (1987) Spanish tourism migrants: the case of Lloret de Mar, *Annals of Tourism Research*, **14**(4), 449–70

Levitt, T. (1983) The globalisation of markets, *Harvard Business Review*, **61**(3), 92–102

Lewis, I.M. (1988) Shamanism, in Sutherland, S., Houlden, L., Clarke, P. and Hardy, F. (eds) *The World's Religions*, G.K. Hall & Co., Boston, Mass.

Leyshon, A. (1993) Crawling from the wreckage: speculating on the future of the European Exchange Rate Mechanism, *Environment and Planning A*, **25**(11), 1553–7

Leyshon, A. and Thrift, N. (1992) Liberalisation and consolidation: the Single European Market and the remaking of European financial capital, *Environment and Planning A*, **24**, 49–81

Leyshon, A. and Thrift, N. (1995) European financial integration: the search for an 'island of monetary stability' in the seas of global financial turbulence, in Hardy, S., Hart, M., Albrechts, L. and Katos, A. (eds) *An Enlarged Europe: Regions in Competition?* Regional Studies Association, Jessica Kingsley Publishers, London, 109–43

Leyshon, A. and Thrift, N.J. (1997) *Money/Space. Geographies of Monetary Transformation*, Routledge, London

Lieven, A. (1994) *The Baltic Revolution: Estonia, Latvia, Lithuania and the Path to Independence*, 2nd edition, Yale University Press, New Haven, Conn.

Lipietz, A. (1996) Social Europe: the post-Maastricht challenge, *Review of International Political Economy*, **3**, 369–79

Livingstone, D.N. (1992) *The Geographical Tradition: Episodes in the History of a Contested Enterprise*, Blackwell, Oxford

Llewellyn, D.T. (1992) Banking and financial services, in Swann, D. (ed.) *The Single European Market and Beyond: a Study of the Implications of the Single European Act*, Routledge, London, 106–45

Locke, J. (1967) *Two Treatises of Government: a Critical Edition with an Introduction and Apparatus Criticus by Peter Laslett*, 2nd edition, Cambridge University Press, Cambridge

Locke, J. (1975) *Essay Concerning Human Understanding*, Clarendon Press, Oxford

Lodge, J. (1991) *The Democratic Deficit and the European Parliament*, Fabian Society, London (Fabian Society Discussion Paper No. 4)

Lodge, J. (ed.) (1996) *The 1994 Elections to the European Parliament*, Pinter, London

Lofman, B. (1994) Polish society in transformation: the impact of marketization on business, consumption and education, in Schultz, C.J. III, Belk, R. and Gur, G. (eds) *Research in Consumer Behaviour, Volume 7*, JAI Press, New York, 29–55

London Business School (1995) *City Research Project. Final Report*, London Business School, London

Lopez Garcia, B. (ed.) (1996) *Atlas de la Inmigración Magrebí en España*, Universidad Autónoma de Madrid y Ministerio de Asuntos Sociales, Madrid

Lorwin, V.R. (1966) Belgium: religion, class, and language in national politics, in Dahl, R.A. (ed.) *Political Oppositions in Western Democracies*, Yale University Press, New Haven, Conn., and London, 147–87

Lowe, P., Murdoch, J., Marsden, T., Munton, R. and Flynn, A. (1993) Regulating the new rural spaces: the uneven development of land, *Journal of Rural Studies*, **9**, 205–22

Löytönen, M. and Maasilta P. (1996) Multi-drug resistant tuberculosis in Finland – a forecast, *Social Science and Medicine* (in press)

Löytönen, M. (1995) The effects of the HIV epidemic on the population of Europe, in Hall, R. and White, P. (eds) *Europe's Population: Towards the Next Century*, UCL Press, London, 83–98

Lubbock, J. (ed.) (1868) *Nilsson on the Stone Age*, Longman, Green & Co., London

Lumsden, G.I. (ed.) (1994) *Geology and the Environment in Western Europe: a Co-ordinated Statement by the Western European Geological Surveys*, Clarendon Press, Oxford

Lury, C. (1996) *Consumer Culture*, Polity Press, Cambridge

Luxembourg, R. (1961) *The Russian Revolution and Leninism or Marxism?*, University of Michigan Press, Ann Arbor

MacAulay, D. (ed.) (1992) *The Celtic Languages*, Cambridge University Press, Cambridge

Macintyre, S. and Tribe, K. (1975) *Althusser and Marxist Theory*, 2nd edition, the authors, Cambridge

Mackie, T. (1995) Parties and elections, in Hayward, J. and Page, E.C. (eds) *Governing the New Europe*, Polity Press, Cambridge, 166–95

Mailer, P. (1977) *Portugal: the Impossible Revolution*, Solidarity, London

Mair, A., Florida, R. and Kenney, M. (1988) The new geography of automobile production: Japanese transplants in North America, *Economic Geography*, **64**(4), 352–73

Manisali, E. (ed.) (1988) *Turkey's Place in Europe. Economic, Political and Cultural Dimensions*, The Middle East Business and Banking Magazine Publications, Ankara

Manning, N. (1992) Social policy in the Soviet Union and its successors, in Deacon, R. *et al.* (eds) *The New Eastern Europe: Social Policy Past, Present and Future*, Sage, London, 31–66

Markoff, J. (1996) *Waves of Democracy: Social Movements and Political Change*, Sage, London

Martin, P.L. (1991) *The Unfinished Story: Turkish Labour Migration to Western Europe with Special Reference to the Federal Republic of Germany*, International Labour Office (ILO), Geneva

Martínez-Montoya, J. (1995) Pour une anthropologie de la montagne basque. Enjeux contemporains d'une discipline, *Ethnologie Française*, **25**, 294–303

Marx, K. (1960) *The Class Struggles in France: 1848 to 1850*, Progress, Moscow

Marx, K. (1963) *The Eighteenth Brumaire of Louis Bonaparte*, International, New York

Marx, K. (1971) *Critique of Political Economy*, Lawrence and Wishart, London

Marx, K. (1976) *Capital Volume 1*, Penguin, Harmondsworth

Marx, K. and Engels, F. (1975a) *Manifesto of the Communist Party*, Progress Publishers, Moscow

Marx, K. and Engels, F. (1975b) The Holy Family, or Critique of Critical Criticism: against Bruno Bauer and Company, in *Karl Marx and Frederick Engels Collected Works, Volume 4: Marx and Engels 1844–1845*, Lawrence and Wishart, London, 5–211

Masera, R.S. and Portes, R. (1991) Foreword, in Giovannini, A. and Mayer, C. (eds) *European Financial Integration*, Centre for Economic Policy Research, Cambridge University Press, xix–xxi

Masser, I., Svidén, O. and Wegener, M. (1992) *The Geography of Europe's Futures*, Belhaven, London

Massey, D. (1994) *Space, Place and Gender*, Polity Press, Cambridge

Massey, D. (1995) Masculinity, dualisms and high technology, *Transactions of the Institute of British Geographers*, **20**(4), 487–99

Mathieson, A. and Wall, G. (1982) *Tourism: Economic, Physical and Social Impacts*, Longman, London

Matless, D. (1992) Regional surveys and local knowledges: the geographical imagination in Britain, 1918–39, *Transactions of the Institute of British Geographers*, **NS**, **17**, 1–21

Mattick, P. (1971) *Marx and Keynes: the Limits of the Mixed Economy*, Merlin, London

Mayhew, A. (1973) *Rural Settlement and Farming in Germany*, Batsford, London

Mazey, S. (1989) *European Community Social Policy*, Polytechnic of North London, London (European Dossier Series, No. 14)

McAllister, I. (1987a) II. Social context, turnout, and the vote: Australian and British comparisons, *Political Geography Quarterly*, **6**(1), 17–30

McAllister, I. (1987b) VI. Comment on Johnston, *Political Geography Quarterly*, **6**(1), 45–9

McCannon, J. (1995) To storm the Arctic: Soviet polar exploration and public visions of nature in the USSR 1932–39, *Ecumene*, **2**(1), 15–32

McCrum, R., Cran, W. and MacNeil, W. (1986) *The Story of English*, Viking Publishing Company, New York

McDowell, L. (1995) Body work: heterosexual gender performances in city workspaces, in Bell, D. and Valentine, G. (eds) *Mapping Desire*, Routledge, London, 245–63

McDowell, L. (1997) *Capital Culture*, Routledge, London

McDowell, L. and Massey, D. (1984) A woman's place?, in Massey, D. and Allen, J. (eds) *Geography Matters*, Cambridge University Press, Cambridge, 128–47

McKee, M. (1991) Health services in central and eastern Europe: past problems and future prospects, *Journal of Epidemiology and Community Health*, **45**, 260–65

McKeown, T. (1979) *The Role of Medicine: Dream, Mirage or Nemesis?*, Blackwell, Oxford

McLean, M. (1995) *Educational Traditions Compared: Content, Teaching and Learning in Industrialised Countries*, David Fulton Publishers, London

McLeod, H. (ed.) (1995) *European Religion in the Age of the Great Cities 1830–1930*, Routledge, London

McRae, K.D. (1984) *Conflict and Compromise in Multilingual Societies: Switzerland*, Wilfrid Laurier University Press, Waterloo, Ontario

McRae, K.D. (1984) *Conflict and Compromise in Multilingual Societies: The Case of Switzerland*, Wilfrid Laurier University Press, Waterloo

Mead, W.R. (1982) The discovery of Europe, *Geography*, **67**(3), 193–202

Medved, F. (1995) A path towards the cartography of Slovene national identity, *Razprave in Gradivo (Documents and Treatises)*, Institute for Ethnic Studies, Ljubljana, 177–210

Medvedev, Z.A. (1990a) Evolution of AIDS policy in the Soviet Union I. Serological screening 1986–7, *British Medical Journal*, **300**, 860–1

Medvedev, Z.A. (1990b) Evolution of AIDS policy in the Soviet Union II. The AIDS epidemic and emergency measures, *British Medical Journal*, **300**, 932–4

Merenne-Schoumaker, B. (1995) Retail planning policy in Belgium in Davies, R.L. (ed.) *Retail Planning Policies in Western Europe*, Routledge, London, 31–50

Meslé, F. (1996) Mortality in eastern and western Europe: a widening gap, in Coleman, D. (ed.) *Europe's Population in the 1990s*, Oxford University Press, Oxford, 127–43

MESS (1988, 1993) *Quadros de Pessoal, 1988 e 1993*, MESS, Departamento de Estatistica, Lisbon

Messenger, J.C. (1969) *Inis Beag: Isle of Ireland*, Holt, Rinehart and Winston, New York

Mezentseva, E. and Rimachevskaya, N. (1990) The Soviet country profile: health of the USSR population in the 70s and 80s – an approach to a comprehensive analysis, *Social Science and Medicine*, **31**, 867–77

Michie, R. (1992) *The City of London. Continuity and Change, 1850–1990*, Macmillan, London

Miles, I. (1989) Masculinity and its discontents, *Futures*, **21**(1), 47–59

Miller, J. (1983) *The Glorious Revolution*, Longman, London

Milliband, R. (1991) What comes after communist regimes?, in *Socialist Register 1991*, Merlin Press, London, 375–89

Misiti, M., Muscarà, C., Pumares, P., Rodriguez, V. and White, P. (1995) Future migration into southern Europe, in Hall, R. and White, P. (eds) *Europe's Population: Towards the Next Century*, UCL Press, London, 161–87

Mitchell, D. (1995) There's no such thing as culture: towards a reconceptualization of the idea of culture in geography, *Transactions of the Institute of British Geographers*, **20**(1), 102–16

Moda Industria (1996) *Rapporto di Settore 1995*, Moda Industria, Milan

Molyneux, P., Altunbas, Y. and Gardener, E. (1996) *Efficiency in European Banking*, John Wiley, Chichester

Mongait, A.L. (1955) *Archaeology in the USSR*, Penguin Books, London

Montanari, A. (1995) The Mediterranean region: Europe's summer leisure space, in Montanari, A. and Williams, A.M. (eds) *European Tourism: Regions, Spaces and Restructuring*, John Wiley, Chichester, 41–65

Montanari, A. and Williams, A.M. (eds) (1995) *European Tourism: Regions, Spaces and Restructuring*, John Wiley, Chichester

Monter, W. (1990) *Frontiers of Heresy: the Spanish Inquisition from the Basque Lands to Sicily*, Cambridge University Press, Cambridge

Moon, G. (1994) Health trends in Eastern Europe: a comparative analysis, in Vaishar, A. (ed.) *Health, Environment and Development*, Regiograph, Brno, 61–78

Moran, M. (1991) *The Politics of the Financial Services Revolution*, Macmillan, London

Moran, M. and Prosser, T. (eds) (1994) *Privatization and Regulatory Change in Europe*, Open University Press, Buckingham

Morgan, D.H.J. (1992) *Discovering Men*, Routledge, London

Morgan, R. (1995) *The Times Guide to the European Parliament 1994*, The Times, London

Morgan, W. (1992) Economic reform, the free market and agriculture in Poland, *Geographical Journal*, **158**, 145–56

Morland, S. (1658) *The History of the Evangelical Churches of the Valleys of Piedmont*, Henry Hill, London

Mortillet, G. de (1881) *Musée Préhistorique*, Paris

Mortimer, E. (1995) Euro-structures under one roof, *Financial Times*, 3 May 1995

Moruzzi, C. (1994) A problem with headscarves: contemporary complexities of political and social identity, *Political Theory*, **22**(4), 653–72

Moruzzi, N. (1993) Veiled agents: political action and the feminine, in *The Battle of Algiers*, in Fisher, S. and Davis, K. (eds) *Negotiating at the Margins*, Rutgers University Press, New Brunswick, 255–77

Mossé, G.L. (1975) *The Nationalization of the Masses*, Howard Fertig, New York

Murphy, A.B. (1988) *The Regional Dynamics of Language Differentiation in Belgium: a Study in Cultural–Political Geography*, University of Chicago Geography Research Paper No. 227, Chicago

Murphy, A.B. (1995) Belgium's regional divergence: along the road to federation, in Smith, G. (ed.) *Federalism: The Multiethnic Challenge*, Longman, London and New York, 73–100

Murphy, A.B. and Huderi-Ely, A. (1996) The geography of the 1994 Nordic vote on European Union membership, *Professional Geographer*, **48**, 284–97

Murphy, P. (1994) Tourism and sustainable development, in Theobold, W.F. (ed.) *Global Tourism: the Next Decade*, Butterworth Heinemann, London

Muston, A. (1851) *L'Israël des Alpes*, Ducloux, Paris

Myklebost, H. and Gläßer, E. (1996) Das norwegische Nein zur EU, *Geographische Rundschau*, **48**, 285–91

Nairn, T. (1977) *The Break-up of Britain: Crisis and Neo-Nationalism*, New Left Books, London

Nash, C. (1994) Re-mapping the body/land: new cartographies of identity by Irish women artists, in Blunt, A. and Rose, G. (eds) *Writing Women and Space*, Guilford Press, London, 227–50

Nash, D. (1996) *Anthropology of Tourism*, Pergamon, Oxford

Needham, J. and Wang Ling (1956) *Science and Civilisation in China, Volume 2, History of Scientific Thought*, Cambridge University Press, Cambridge

Needham, J. and Wang Ling (1959) *Science and Civilisation in China, Volume 3, Mathematics and the Sciences of the Heavens and the Earth*, Cambridge University Press, Cambridge

Newburn, T. and Stanko, E. (eds) (1994) *Just Boys Doing Business?*, Routledge, London

Newman, M. (1993) *The European Community: Where Does the Power Lie?*, University of North London, London (European Dossier Series No. 25)

Nilsson, S. (1838) *The Primitive Inhabitants of Scandinavia*, Longman, Green & Co., London

Noin, D. (1995) Spatial inequalities of mortality in the European Union, in Hall, R. and White, P. (eds) *Europe's Population: Towards the Next Century*, UCL Press, London, 69–82

Nolan, M.L. and Nolan, S. (1989) *Christian Pilgrimage in Modern Western Europe*, University of North Carolina Press, Chapel Hill

Nolte, E. (1991) *Geschichtsdenken im 20. Jahrhundert*, Propylaën, Berlin

Nussbaum, J.M. (1951) Histoire d'un mouvement ouvrier, *Cahiers Suisses*, **3–4**, 83–99

Nyberg, L. (1995) Scandinavia: tourism in Europe's northern periphery, in Montanari, A. and Williams, A.M. (eds) *European Tourism: Regions, Spaces and Restructuring*, John Wiley, Chichester, 87–107

O'Brien, R. (1992) *Global Financial Integration: the End of Geography*, Royal Institute of International Affairs, Pinter, London

Oberreit, J. (1996) Destruction and reconstruction: the case of Dubrovnik, in Hall, D.R. and Danta, D. (eds) *Reconstructing the Balkans*, John Wiley, Chichester and New York, 67–77

OECD (1994a) *Review of Agricultural Policies: Hungary*, OECD, Paris

OECD (1994b) *Review of Agricultural Policies: Poland*, OECD, Paris

OECD (1994c) *Trends in International Migration*, Organisation for Economic Co-operation and Development, Paris

OECD (1995a) *Etudes Economiques de l'OCDE. Turquie 1994–1995*, OECD, Paris

OECD (1995b) *Review of Agricultural Policies: Czech Republic*, OECD, Paris

Ohmae, K. (1990) *The Borderless World*, Collins, London

Ohmae, K. (1995) *The End of the Nation State*, Free Press, New York

Olwig, K. (1996a) Recovering the substantive nature of landscape, *Annals of the Association of American Geographers*, **86**(4), 630–53

Olwig, K.R. (1996b) Nature – mapping the ghostly traces of a concept, in Earle, C., Mathewson, K. and Kenzer, M.S. (eds) *Concepts in Human Geography*, Rowman and Littlefield, Lanham, Mass., 63–96

Omran, A. (1971) The epidemiological transition: a theory of the epidemiology of population change, *Millbank Memorial Quarterly*, **64**, 355–91

OPCS (1981) *Area Mortality: Decennial Supplement*, HMSO, London, DS no. 4

Orosz, E. (1990) Regional inequalities in the Hungarian health system, *Geoforum*, **21**, 245–59

Ostrowska, A. (1993) From totalitarianism to pluralism in Poland: problems of transformation on the health scene, *European Journal of Public Health*, **3**, 43–7

Ovid (1986) *Metamorphoses*, translated by Melvill, A.D., introduction by Kenney, E.J., Oxford University Press, Oxford

Oxfam (1996) *Women and Land*, Oxfam/Virago, London

Padoa-Schioppa, T. (1988) The European Monetary System: a long term view, in Givazzi, F., Micossi, S. and Miller, M. (eds) *The European Monetary System*, Cambridge University Press, Cambridge, 369–84

Page, S.J. and Sinclair, T. (1992) The Channel Tunnel and tourism markets in the 1990s, *Travel and Tourism Analyst*, **1**, 5–32

Pallaruelo, S. (1988) *Pastores del Pirineo*, Ministerio de Cultura, Madrid

Panebianco, A. (1988) *Political Parties: Organization and Power*, Cambridge University Press, Cambridge

Park, C.C. (1994) *Sacred Worlds: an Introduction to Geography and Religion*, Routledge, London

Parker, K.J. (1996) Pride in Place: a Report on a 1995 Churchill Fellowship Study (unpublished manuscript)

Pearce, D.G. (1987a) Spatial patterns of package tourism in Europe, *Annals of Tourism Research*, **14**(2), 183–201

Pearce, D.G. (1987b) Mediterranean charters: a comparative geographic prospective, *Tourism Management*, **8**, 291–305

Pearce, D.G. (1988) Tourism and regional development in the European Community, *Tourism Management*, **9**, 13–22

Pearce, D.G. (1995) *Tourism Today*, 2nd edition, Longman, Harlow

Pearce, D.G. (1997) Tourism and the autonomous communities in Spain, *Annals of Tourism Research*, **24**(1), 156–77

Perrons, D. (1994) Measuring equal opportunities in European employment, *Environment and Planning A*, **26**, 1195–220

Perrons, D. (1996) Gender as a form of social exclusion: gender inequality in the regions of Europe, paper presented at European Urban and Regional Studies Conference, Exeter, UK, 11–14 April 1996

Petrie, P. (1996) School-age childcare and the school: recent European developments, in Bernstein, B. and Brannen, J. (eds) *Children, Research and Policy*, Taylor and Francis, London, 220–41

Pettersen, P.A., Jenssen, A.T. and Listhaug, O. (1996) The 1994 EU referendum in Norway: continuity and change, *Scandinavian Political Studies*, **19**, 257–81

Pevetz, W. (1991) Agriculture and tourism in Austria, *Tourism Recreation Research*, **16**(1), 57–60

Pfau-Effinger, B. (1994) The gender contract and part-time paid work by women — Finland and Germany compared, *Environment and Planning A*, **26**, 1355–76

Phillimore, P. and Morris, D. (1991) Discrepant legacies: premature mortality in two industrial towns, *Social Science and Medicine*, **33**(2) 139–52

Phillimore, P. and Reading, R. (1992) A rural disadvantage? Urban–rural health differences in Northern England, *Journal of Public Health Medicine*, **14**, 290–9

Phillips, D. (ed.) (1995) *Education in Germany: Tradition and Reform in Historical Context*, Routledge, London

Pickles, J. and Smith, A. (eds) (1997) *Theorising Transition*, Routledge, London

Pike, A. and Vale, M. (1996) 'Greenfields' and 'Brownfields': automotive industrial development in the UK and in Portugal, *Finisterra*, **62**, 97–119

Pilbeam, P. (1990) *The Middle Classes in Europe 1789–1914: France, Germany, Italy and Russia*, Macmillan, Basingstoke

Pilbeam, P. (1991) *The 1830 Revolution in France*, Macmillan, London

Pinch, S. (1984) Inequality in pre-school provision: a geographical perspective, in Kirby, A. (ed.) *Public Service Provision and Urban Development*, Croom Helm, London, 231–82

Pinder, J. (1991) *European Community: the Building of a Union*, Oxford University Press, London

Pine, B.J. (1993) *Mass Customization: the New Frontier in Business Competition*, Harvard University Press, Harvard, Mass.

Plato (1974) *The Republic*, 2nd edition translated with an introduction by Desmond Lee, Penguin, Harmondsworth

Pogonowski, I.C. (1988) *Poland: a Historical Atlas*, revised edition, Dorset Press, New York

Poiani, M. (1994) *Alti Consumatori. Il Marketing dei Beni ad Alto Valore Simbolico*, Lupetti, Milan

Poikolainen, K. and Eskola, J. (1986) The effect of health services on mortality: decline in death rates from amenable and non-amenable causes in Finland, 1969–81, *Lancet*, **1**, 8474

Pompl, W. and Lavery, P. (eds) (1993) *Tourism in Europe: Structures and Developments*, CAB International, Wallingford

Poon, A. (1993) *Tourism, Technology and Competitive Strategies*, CAB International, Wallingford

Popesku, J. and Milojević, L. (1996) *Serbia: Landscape Painted from the Heart*, National Tourism Organization of Serbia, Belgrade

Population Reference Bureau (1996) *World Population Data Sheet 1996*, Population Reference Bureau, Washington, DC

Pounds, N. (1973) *An Historical Geography of Europe, 450 BC – AD 1330*, Cambridge University Press, Cambridge

Powell, T.G.E. (1958) *The Celts*, Thames & Hudson, London

Preston, P. (1994) *The Coming of the Spanish Civil War: Reform, Reaction and Revolution in the Second Republic*, Routledge, London

Preston, P. (ed.) (1984) *Revolution and War in Spain 1931–1939*, Methuen, London

Pridham, G. and Lewis, P.G. (1996) *Stabilising Fragile Democracies: Comparing New Party Systems in Southern and Eastern Europe*, Routledge, London

Priestly, G.K. (1995a) Evolution of tourism on the Spanish coast, in Ashworth, G.J. and Dietvorst, A.G.J. (eds) *Tourism and Spatial Transformations*, CAB International, Wallingford, 37–54

Priestly, G.K. (1995b) Sports tourism: the case of golf, in Ashworth, G.J. and Dietvorst, A.G.J. (eds) *Tourism and Spatial Transformations*, CAB International, Wallingford, 205–23

Priimägi, L. (1991) Personal communication. Institute of Preventive Medicine, Tallinn, Estonia

Puijk, R. (1996) Dealing with fish and tourists: a case study from Northern Norway, in Boissevain, J. (ed.) *Coping with Tourists,* Berghahn, Providence, RI, and Oxford, 204–26

Quack, S. and Maier, F. (1994) From state socialism to market economy – women's employment in East Germany, *Environment and Planning A*, **26**, 1257–76

Radcliffe, S. (1993) Women's place/el lugar de mujeres: Latin America and the politics of gender identity, in Keith, M. and Pile, S. (eds) *Place and the Politics of Identity*, Routledge, London, 102–16

Radcliffe, S. (1996) Gendered nations: nostalgia, development and territory in Ecuador, *Gender, Place and Culture*, **3**(1), 5–21

Radical Statistics Health Group (1991) Missing: a strategy for *Health of the Nation*, *British Medical Journal*, **303**, 299–302

Ramet, P. (ed.) (1984) *Religion and Nationalism in Soviet and East European Politics*, Duke University Press, Durham, NC

Ramirez, A. and Gregorio, C, (1994) Un pays providentiel, in Basfao, K. and Taarji, H. (eds) *Annuaire de l'Emigration Maroc*, Fondation Hassan II, Rabat, Morocco, 602–5

Ramsay, H. (1990) *1992: the Year of the Multinational?*, Warwick Papers in Industrial Relations, University of Warwick

Rathwell, T. (1991) Too many bits and pieces, *Health Service Journal*, 14 March 1991, 22–3

Rees, P. (1996) Projecting the national and regional populations of the European Union using migration information, in Rees, P., Stillwell, J., Convey, A. and Kupiszewski, M. (eds) *Population Migration in the European Union*, Wiley, Chichester, 331–64

Rees, P., Stillwell, J., Convey, A. and Kupiszewski, M. (eds) (1996) *Population Migration in the European Union*, Wiley, Chichester

Renfrew, C. (1987) *Archaeology and Language: the Puzzle of Indo-European Origins*, Cape, London

Renfrew, C. (1988) The origins of Indo-European languages, *Scientific American*, **261**(4), 106–14

Repassy, H. and Symes, D. (1993) Perspectives on agrarian reform in east-central Europe, *Sociologia Ruralis*, **33**, 81–91

Richards, G. (1996) Production and consumption of European cultural tourism, *Annals of Tourism Research,* **23**(2), 261–83

Rickard, P. (1989) *A History of the French Language*, 2nd edition, Unwin Hyman, London

Riley-Smith, J. (1990) *The Crusades: a Short History*, Athlone Press, London

Riley-Smith, J. (ed.) (1995) *The Oxford Illustrated History of the Crusades*, Oxford University Press, Oxford

Rinschede, G. (1986) The pilgrimage town of Lourdes, *Journal of Cultural Geography*, **7**, 21–33

Ripley, W.Z. (1900) *Races of Europe*, Kegan Paul, London

Ritter, G. and Haidu, J.G. (1989) The East–West German boundary, *Geographical Review*, **79**, 326–44

Robinson, G. (1993) Tourism and tourism policy in the European Community: an overview, *International Journal of Hospitality Management,* **12**(1), 7–20

Robinson, G. and Ilbery, B.W. (1993) Reforming the CAP: beyond MacSharry, in Gilg, A. (ed.) *Progress in Rural Policy and Planning*, **3**, 197–207

Robinson, V. (1995) The changing nature and European perceptions of Europe's refugee problem, *Geoforum*, **26**, 411–27

Robinson, V. (1996) Redefining the front line: the geography of asylum seeking in the new Europe, in Rees, P., Stillwell, J., Convey, A. and Kupiszewski, M. (eds) *Population Migration in the European Union*, Wiley, Chichester, 67–88

Robinson, W. (1996) Globalisation: nine theses on our epoch, *Race and Class*, **38**(2), 13–31

Rohrschneider, R. (1993) Impact of social movements on European party systems, *Annals of the American Academy of Political and Social Science*, **528**, 157–70

Rollins, W.H. (1995) Whose landscape? Technology, Fascism and environmentalism on the National Socialist *Autobahn*, *Annals of the Association of American Geographers*, **85**(3), 494–520

Romanus, V., Tala, E., Blöndal, T., Heldal, E. and Poulsen, S. (1995) Contending with tuberculosis in the Nordic countries, *Nordisk Medicin*, **10**(45), 1995

Rose, D.B. (1992) *Dingo Makes Us Human: Life and Land in an Aboriginal Australian Culture*, Cambridge University Press, Cambridge

Rosetti, A. (1973) *Brève Histoire de la Langue Roumaine des Origines à Nos Jours*, Mouton, The Hague

Ross, G. (1991) Confronting the new Europe, *New Left Review*, **191**, 49–68

Ross, J. (1993) The changing relationships between the public, private and voluntary sectors in pre-school provision, England 1982–1987, unpublished PhD thesis, University of West of England, Bristol

Rougier, H. and Sanguin, A-L. (1991) *The Rumanschs or the Fourth Switzerland*, Peter Lang Publishers, Berne

Rueschemeyer, D., Stephens, E.H. and Stephens, J.D. (1992) *Capitalist Development and Democracy*, Polity Press, Cambridge

Rugg, D. (1985) *Eastern Europe*, Longman, London

Ruukel, A. (ed.) (1996) *Estonia – the Natural Way*, Kodukant Ecotourism Association of Estonia, Pärnu

Ryan, M. (1966) *Clothing: a Study of Human Behavior*, Holt, Rinehart and Winston, New York

Ryan, M. (1978) *The Organization of Soviet Medical Care*, Blackwell, Oxford

Sabaté-Martinez, A. (1996) Women's integration into the labour market and rural industrialization in Spain: gender relations and the global economy, in Garcia-Ramon, M.D. and Monk, J. (eds) *Women of the European Union*, Routledge, London, 263–81

Sachs, J. (1990) What is to be done?, *Economist*, 13 January 1990

Said, E. (1978) *Orientalism*, Routledge & Kegan Paul, London

Salmela, M. (1994) *Suomen Perinne-Atlas, II*, Suomalaisen Kirjallisuuden Sevura, Helsinki (captions in English)

Salt, J. (1996) Migration pressures in western Europe, in Coleman, D. (ed.) *Europe's Population in the 1990s*, Oxford University Press, Oxford, 92–126

San Sebastián, K. (1984) *Historia del Partido Nacionalista Vasco*, Editorial Txertoa, San Sebastián

Sandner, S. (1994) In search of identity: German nationalism and geography 1871–1910, in Hooson, D. (ed.) *Geography and National Identity*, Blackwell, Oxford, 71–91

Sanguin, A-L. (1983) *Switzerland, An Attempt at Political Geography*, Ophrys Publishers, Paris

Savage, M. (1987) Understanding political alignments in contemporary Britain: do localities matter?, *Political Geography Quarterly*, **6**(1), 53–76

Sayer, A. and Walker, R. (1992) *The New Social Economy*, Blackwell, Oxford

Schama, S. (1989) *Citizens: a Chronicle of the French Revolution*, Penguin, Harmondsworth

Schama, S. (1995) *Landscape and Memory*, Alfred Knopf, New York

Schamp, E.W. (1995) The German automobile industry going European, in Hudson, R. and Schamp, E.W. (eds) *Towards a New Map of Automobile Manufacturing in Europe? New Production Concepts and Spatial Restructuring*, Springer, Berlin, 93–116

Scheiber, G. (1993) Health care financing reform in Russia and Ukraine, *Health Affairs*, **12**, supp. 93, 294–99

Schlüter, O. (1952) Die Siedlungsräume Mitteleuropas in Frühgeschlichiche Zeit, *Forschung zum Deutschen Landeskunde*, Remagen, Hamburg

Schmitt, H. and Holmberg, S. (1995) Political parties in decline?, in Klingemann, H.-D. and Fuchs, D. (eds) *Citizens and the State*, Oxford University Press, Oxford, 95–133

Schofield, R., Reher, D. and Bideau, A. (eds) (1991) *The Decline of Mortality in Europe*, Clarendon Press, Oxford

Schweizer, P. (1988) *Shepherds, Workers, Intellectuals: Culture and Centre–Periphery Relationships in a Sardinian Village,* University of Stockholm, Stockholm

Scott, A. (1988) *New Industrial Spaces*, Pion, London

Segal, L. (1990) *Slow Motion. Changing Masculinities, Changing Men*, Virago Press, London

Semple, E.C. (1911) *Influences of Geographic Environment on the Basis of Ratzel's System of Anthropo-Geography*, Henry Holt, New York

Sen, F. (ed.) (1993) *Turkey and the European Community*, Leske and Budrich, Opladen

Sereno, P. (1988) Flussi migratori e colonie interne negli stati sabaudi: la colonizzazione delle Valli Valdesi, 1686–1689, *Migrazioni Attraverso le Alpi Occidentali. Relazioni tra Piemonte, Provenza e Delfinato dal Medioevo ai Nostri Giorni*, Regione Piemonte, Turin, 425–70

Sereno, P. (1990) Popolazione, territorio, risorse: sul contesto geografico delle Valli Valdesi dopo la Glorieuse Rentrée, in De Lange A. (ed.) *Dall'Europa alle Valli Valdesi,* Claudiana, Turin, 293–314

Seton-Watson, H. (1977) *Nations and States*, Methuen, London

Sharp, J. (1996) Gendering nationhood: a feminist engagement with national identity, in Duncan, N. (ed.) *Bodyspace: Destabilising Geographies of Gender and Sexuality*, Routledge, London, 97–108

Shaw, G. and Williams, A. (1993) *Critical Issues in Tourism: a Geographical Perspective*, Blackwell, Oxford

Shelley, M. and Winck, M. (eds) (1995) *Aspects of European Cultural Diversity*, Routledge, London; Open University Press, Milton Keynes

Shouls, S., Congdon, P. and Curtis, S. (1996) Modelling inequality in reported long term illness in the UK: combining individual and area characteristics, *Journal of Epidemiology and Community Health*, **50**, 366–76

Shucksmith, M. (1993) Farm household behaviour and the transition to post-productivism, *Journal of Agricultural Economics*, **44**, 466–78

Sjøholt, P. (1990) Marginality, crisis and the response to crisis: some development issues in the Scandinavian Northlands, Department of Geography, University of Bergen, Bergen (Geografi i Bergen, Nr 135)

Smart, N. (1989) *The World's Religions: Old Traditions and Modern Transformations*, Cambridge University Press, Cambridge

Smith, A. (1996) Industrial restructuring and uneven regional development in Slovakia: a regulationist approach to 'the transition' in central and eastern Europe, Unpublished D.Phil. thesis, University of Sussex

Smith, A. and Jacobson, B. (1988) *The Nation's Health*, King's Fund, London

Smith, A.D. (1986) *The Ethnic Origins of Nations*, Blackwell, Oxford

Smith, A.D. (1992) National identity and the idea of European identity, *International Affairs*, **68**(1), 55–76

Smith, M. (1996) The European Union and a changing Europe: establishing the boundaries of order, *Journal of Common Market Studies,* **34**, 5–28

Soboul, A. (1988) *Understanding the French Revolution*, Merlin Press, London

Social Science and Medicine (1990) Special issue: Health inequities in Europe, *Social Science and Medicine*, **36**(10)

Sogner, I. and Archer, C. (1995) Norway and Europe: 1972 and now, *Journal of Common Market Studies*, **33**, 389–410

Soininen, A. (1961) Pohjois-Savon asuttaminen kaski- ja uuden ajan vaihtessa, *Historiallinen tutkimuksia*, 18, Helsinki (English summary)

Sommers, J. (1996) 'A national disgrace': masculinity and the politics of place in Vancouver, paper presented at the Annual Conference of the Association of American Geographers, Charlotte, NC, April 1996

Sopher, D.E. (1967) *Geography of Religions*, Prentice-Hall, Englewood Cliffs, NJ

Speck, W.A. (1988) *Reluctant Revolutionaries: Englishmen and the Revolution of 1688*, Oxford University Press, Oxford

Sperber, J. (1994) *The European Revolutions, 1848–1851*, Cambridge University Press, Cambridge

Sporton, D. (1993) Fertility: the lowest level in the world, in Noin, D. and Woods, R. (eds) *The Changing Population of Europe*, Blackwell, Oxford, 49–61

Stadel, C., Sluppetzky, H. and Kremser, H. (1996) Nature conservation, traditional living space, or tourist attraction? The Hohe Tauern National Park, Austria, *Mountain Research and Development,* **16**(1), 1–16

Stanners, D. and Bourdeau, P. (eds) (1995) *Europe's Environment: the Dobříš Assessment*, European Environment Agency, Copenhagen

Starr, C.G. (1982) *The Roman Empire, 27 B.C. – A.D. 476: a Study in Survival*, Oxford University Press, Oxford

Statistical Office of the European Communities (1996) *1995 Regions Statistical Yearbook*, Office des Publications Officielles des Communautés Européennes, Luxembourg

Stein, H. (1993) Can supra-national public health strategies improve the health status of national populations?, *European Journal of Public Health*, **3**, 3–7

Steinecke, A. (1993) The historical development of tourism in Europe, in Pompl, W. and Lavery, P. (eds) *Tourism in Europe: Structures and Developments,* CAB International, Wallingford, 3–12

Steiner, J. (1983) Conclusion: Reflections on the consociational theme, in Penniman, H.R. (ed.) *Switzerland at the Polls: the National Elections of 1979*, American Enterprise Institute for Public Policy, Washington, DC, 161–77

Stephens, M. (1978) *Linguistic Minorities in Western Europe*, Gomer Press, Llandysul, Wales

Stevenson, E.L. (ed.) (1991) *Claudius Ptolemaeus, The Geography*, Dover Publications, New York

Stone, G. (1990) Polish, in Comrie, B. (ed.) *The Major Languages of Eastern Europe*, Routledge, London and New York, 82–100

Storper, M. (1995) The resurgence of regional economies, ten years after: the region as a nexus of untraded dependencies, *European Urban and Regional Studies*, **2**(3), 191–223

Storper, M. and Scott, A. (1989) The geographical foundations and social regulation of flexible production complexes, in Dear, M. and Wolsh, J. (eds) *The Power of Geography*, Unwin Hyman, London

Strabo (1949–54) *The Geography of Strabo*, Heinemann, London

Strange, S. (1986) *Casino Capitalism*, Blackwell, Oxford

Strange, S. (1988) *States and Markets*, Frances Pinter, London

Stratigaki, M. and Vaiou, D. (1994) Women's work and informal activities in Southern Europe, *Environment and Planning A*, **26**, 1221–34

Sugar, P.F. (1977) *Southeastern Europe under Ottoman Rule, 1354–1804*, University of Washington Press, Seattle

Sullivan, J. (1988) *ETA and Basque Nationalism: the Fight for Euskadi 1890–1986*, Routledge, London

Sutton, P. (1988) Dreamings, in Sutton, P. (ed.) *Dreamings: the Art of Aboriginal Australia*, The Asia Society Galleries and George Braziller, New York, 13–32

Swain, A. (1996) A geography of transformation: the automotive industry in eastern Germany and Hungary, 1989–94, Unpublished Ph.D. thesis, University of Durham

Swain, M.B. (1995) Gender in tourism, *Annals of Tourism Research*, **22**(2), 247–66

Sykes, B. *et al.* (1996) Paleolithic and Neolithic lineages in the European mitochondrial gene pool, *American Journal of Human Genetics*, **59**, 185–203

Symes, D. (1993) Agrarian reform and the restructuring of rural society in Hungary, *Journal of Rural Studies*, **9**, 291–98

Taylor, I. and Jamieson, R. (1996) 'Proper Little Mesters' – nostalgia and protest masculinity in de-industrialised Sheffield, in Westwood, S. and Williams, J. (eds) *Imagining Cities*, Sage, London, 152–78

Taylor, I., Evans, K. and Fraser, P. (1996) *A Tale of Two Cities*, Routledge, London

Taylor-Gooby, P. (1996) The future of health care in six European countries: the views of policy elites, *International Journal of Heath Services*, **26**, 203–19

Telyukov, A. (1991) A concept of health financing reform in the Soviet Union, *International Journal of Health Services*, **21**, 493–504

Therborn, G. (1991) Staten och människornas välfärd, in *Utsikt mot Europa*, Utbildningsradion och Bokförlaget Bra Böcker, Stockholm

Thrift, N. (1996) *Spatial Formations*, Sage, London

Thrift, N. and Leyshon, A. (1994) A phantom state? The de-traditionalization of money, the international financial system and international financial centres, *Political Geography*, **13**(4), 299–327

Thrift, N.J. (1994) On the social and cultural determinants of international financial centres: the case of the City of London, in Corbridge, S., Martin, R.L. and Thrift, N.J. (eds) *Money, Power and Space*, Oxford, Blackwell, 327–55

Tickle, A. and Welsh, I. (eds) (1998) *Environment and Society in Transition: Central and Eastern Europe*, Addison Wesley Longman, Harlow (forthcoming)

Tilly, C. (1993) *European Revolutions 1492–1992*, Blackwell, Oxford

Tolson, A. (1977) *The Limits of Masculinity*, Tavistock Publications, London

Tomiak, J.J. (ed.) (1986a) *Western Perspectives on Soviet Education in the 1980s*, Macmillan, Basingstoke

Tomiak, J.J. (1986b) Introduction: the dilemmas of Soviet education in the 1980s, in Tomiak, J.J. (ed.) *Western Perspectives on Soviet Education in the 1980s*, Macmillan, Basingstoke, 1–18

Tonnellier, F. (1990) *Géographie des soins, géographie économique: étude de divers contours géographiques en France*, CREDES, Paris

Tordjman, A. (1995) European retailing: convergences, differences and perspectives, in McGoldrick, P.J. and Davies, G.J. (eds) *International Retailing: Trends and Strategies*, Pitman, London, 17–50

Tourn, G. (ed.) (1994) *Viaggiatori Britannici alle Valli Valdesi (1753–1899)*, Claudiana, Turin

Townsend, P., Davidson, N. and Whitehead, M. (1992) *Inequalities in Health*, Penguin, Harmondsworth

Townson, D. (1990) *France in Revolution*, Hodder & Stoughton, London

Treadgold, A. (1988) Retailing without frontiers, *Retail and Distribution Management*, **16**(6), 8–12

Trollope, A. (1983) *Phineas Redux*, Oxford University Press, Oxford

Trotsky, L. (1973) *The Spanish Revolution (1931–39)*, Pathfinder Press, New York

Troughton, M.J. (1986) Farming systems in the modern world, in Pacione, M. (ed.) *Progress in Agricultural Geography*, Croom Helm, London, 93–123

Tuck, J.A. *et al.* (1985) Sixteenth century Basque whaling in America, *National Geographic*, **168** (July), 40–71

Tucker, R.C. (1970) *The Marxian Revolutionary Idea*, George Allen & Unwin, London

Turner, B.S. (1991) *Religion and Social Theory*, 2nd edition, Sage, London

Turner, R. (1971) Sponsored and contest mobility in the school system, in Hopper, E. (ed.) *Readings in the Theory of Educational Systems*, Hutchinson, London, 71–90

Turnock, D. (1988) *The Making of Eastern Europe*, Routledge, London

Turok, B. (ed.) (1980) *Revolutionary Thought in the 20th Century*, Zed Books, London

Ucko, P.J. and Rosenfeld, A. (1967) *Palaeolithic Cave Art*, Weidenfeld & Nicolson, London

UN (1991) *World Population Prospects 1990*, United Nations, New York

UN (1992) *Changing Population Age Structures 1990–2015: Demographic and Economic Consequences and Implications*, United Nations Economic Commission for Europe, Geneva

UNDP (1996) *Human Development Report 1996*, Oxford University Press, New York and Oxford

United States Government (1993) *Economic Report of the President 1992*, US Government Printing Office, Washington, DC

Unwin, P.T.H. (1987) *Portugal*, Clio Press, Oxford

Unwin, T. (1992) *The Place of Geography*, Longman, Harlow

Unwin, T. (1994) Agrarian change and integrated rural development in a continuum: the case of Estonia, *European Urban and Regional Studies*, **1**, 180–85

Urry, J. (1990) *The Tourist Gaze: Leisure and Travel in Contemporary Societies,* Sage, London

Vaiou, D. (1995) Women of the South after, like before, Maastricht?, in Hadjimichalis, C. and Sadler, D. (eds) *Europe at the Margins: New Mosaics of Inequality*, Wiley, Chichester, 35–49

Vaiou, D. (1996) Women's work and everyday life in southern Europe in the context of European integration, in Garcia-Ramon, M.D. and Monk, J. (eds) *Women of the European Union*, Routledge, London, 61–73

van de Kaa, D.J. (1987) Europe's second demographic transition, *Population Bulletin*, **42**(1), 1–57

Van de Velde, R.J., Faber, W., van Katwijk, V., Scholten, H.J., Thewessen, T., Verspuy, M. and Zevenbergen, M. (1994) *The Preparation of a European Land Use Database*, RIVM (Report 712401001), Bilthoven, Netherlands

van den Berg, L., van der Borg, J. and van der Meer, J. (1995) *Urban Tourism*, Avebury, Aldershot

Van der Auwera, J. and Konig, E. (eds) (1994) *The Germanic Languages*, Routledge, London and New York

van Tulder, R. and Ruigrok, W. (1993) Regionalisation, globalisation or glocalisation: the case of the world car industry, in Humbert, M. (ed.) *The Impact of Globalisation on Europe's Firms and Industries*, Pinter, London, 22–33

Vaughan, R.E. (1996) Procurement and capital projects, paper presented to the conference on Supply Chain Management – The Challenges for the 21st Century, Durham University Business School, 9–10 May

Veltz, P. (1991) New models of production organisation and trends in spatial development, in Benko, G. and Dunford, M. (eds) *Industrial Change and Regional Development: the Transformation of New Industrial Spaces*, Belhaven, London, 193–204

Verwers, A. (1992) Towards a new EC health policy?, in European Public Health Association, *Uniting Health in Europe*, proceedings of founding meeting, European Public Health Association, Amsterdam, 11–14

Vine, S. (1994) 'That mild beam' Enlightenment and enslavement in William Blake's *Visions of the Daughters of Albion*, in Plasa, C. and Ring, B. (eds) *The Discourse of Slavery*, Routledge, London, 40–63

Virganskaya, I. and Dimitriev, V. (1992) Some problems of medicodemographic development in the former USSR, *World Health Statistics Quarterly*, **45**(1), 4–14

Vuilleumier, M. (1988) *Horlogers de l'Anarchisme*, Payot, Lausanne

Wæver, O. (1993) Europe since 1945: crisis to renewal, in Wilson, K. and van der Dussen, J. (eds) *The History of the Idea of Europe*, Routledge, London, 151–214

Walby, S. (1990) *Theorising Patriarchy*, Basil Blackwell, Oxford

Walby, S. (1994) Methodological and theoretical issues in the comparative analysis of gender relations in Western Europe, *Environment and Planning A*, **26**, 1139–54

Walker, A. and Walker, C. (1997) *Britain Divided: the Growth of Social Exclusion in the 1980s and 1990s*, Child Poverty Action Group, London

Wallerstein, I. (1974) *The Modern World-System: Capitalist Agriculture and the Origins of the European World Economy in the Sixteenth Century*, Academic Press, London

Wallerstein, I. (1979) *The Capitalist World-Economy*, Cambridge University Press, Cambridge

Wallerstein, I. (1980) *The Modern World-System II: Mercantilism and the Consolidation of the European World Economy, 1600–1750*, Academic Press, London

Warner, M. (1985) *Monuments and Maidens: the Allegory of the Female Form*, Picador, London

Warnes, A.M. (1993) Demographic ageing: trends and policy responses, in Noin, D. and Woods, R. (eds) *The Changing Population of Europe*, Blackwell, Oxford, 82–99

Warnke, M. (1995) *Political Landscape*, Reaktion, London

Warriner, D. (1969) *Land Reform in Principle and Practice*, Oxford University Press, Oxford

Weber, E. (1976) *Peasants into Frenchmen: the Modernization of Rural France, 1870–1914*, Stanford University Press, Calif.

Weber, M. (1965) *The Sociology of Religion*, 4th edition, Methuen, London

Weber, M. (1992) *The Protestant Ethic and the Spirit of Capitalism*, 2nd edition, Routledge, London

Weil, S. (1949) *L'Enracinement. Prélude à une Déclaration des Devoirs Envers l'Être Humain*, Editions Gallimard, Paris

Weinstein, B. (ed.) (1990) *Language Policy and Political Development*, Ablex Publishing, Norwood, NJ

Wells, H.G. (1920) *Outline of History*, Cassell, London

Werner International Inc. (1991) *Situation and Perspective of Technical Textiles in the European Community*, Commission of the European Communities, Brussels

Werwicki, A. and Poweska, H (1993) Rejony przejsc granicznych jako obszary koncentracji handlu i uslug – granica zachodnia, in *Problematyka Zachodniego Obszaru Pogranicza, Biul. No. 1, Podstawy Rozwoju Zachodnich i Wschodnich Obszarów Przygranicznych Polski* (Foundations for the Development of Western and Eastern Border Areas of Poland), Institute of Geography and Spatial Organisation, Polish Academy of Sciences, Warsaw, 61–85

Westlake, M. (1994) *A Modern Guide to the European Parliament*, Pinter, London

WGSG (Women and Geography Study Group) (ed.) (1997) *Feminist Geographies: Explorations in Diversity and Difference*, Longman, Harlow

Wheelock, J. (1991) *Husbands at Home*, Routledge, London

Whitby, M. (ed.) (1996) *The European Environment and CAP Reform: Policies and Prospects for Conservation*, CAB International, Wallingford

White, P. (1993a) The social geography of immigrants in European cities: the geography of arrival, in King, R. (ed.) *The New Geography of European Migrations*, Belhaven, London, 47–66

White, P. (1993b) Ethnic minority communities in Europe, in Noin, D. and Woods, R. (eds) *The Changing Population of Europe*, Blackwell, Oxford, 206–25

White, P. and Sporton, D. (1995) East–west movement: old barriers, new barriers?, in Hall, R. and White, P. (eds) *Europe's Population: Towards the Next Century*, UCL Press, London, 142–60

Whitehead, M. (1994) Equity and ethics in health, in *WHO Regional Committee for Europe 44th Session 1994: Technical Discussions*, WHO, Copenhagen, 51–69

WHO (World Health Organization) (1993) *Health for All Targets: the Health Policy for Europe*, WHO, Copenhagen

WHO (World Health Organization) (1996) *European Health Policy Conference: Opportunities for the Future*, WHO, Copenhagen

Wickens, P. (1986) *The Road to Nissan*, Macmillan, London

Wickham, C. (1985) Pastoralism and underdevelopment in the early Middle Ages, *Settimane di Studio sull'Alto Medioevo*, **31**, 401–51

Williams, A.M. (1994) *The European Community: the Contradictions of Integration*, 2nd edition, Blackwell, Oxford

Williams, A.M. (1996) The Balkans: a European challenge, in Hall, D. and Danta, D. (eds) *Reconstructing the Balkans: A Geography of the New Southeast Europe,* Wiley, Chichester, 211–26

Williams, A.M. and Shaw, G. (eds) (1991) *Tourism and Economic Development: Western European Experiences*, 2nd edition, Belhaven, London

Williams, A.M. and Shaw, G. (1994) Tourism: opportunities, challenges and contradictions in the EC, in Blacksell, M. and Williams, A.M. (eds) *The European Challenge*, Oxford University Press, Oxford, 301–20

Williams, A.M. and Shaw, G. (1996) *Tourism, Leisure, Nature Protection and Agri-Tourism: Principles, Partnerships and Practice*, Tourism Research Group, University of Exeter

Williams, B. (1987) *The Russian Revolution 1917–1921*, Basil Blackwell, Oxford

Williams, G.A. (1989) *Artisans and Sans-Culotte: Popular Movements in France and Britain during the French Revolution*, 2nd edition, Libris, London

Williams, R. (1982) *The Sociology of Culture*, Schocken, New York

Wilson, G.A. (1994) German agri-environmental schemes I – a preliminary review, *Journal of Rural Studies*, **10**(1), 27–45

Wilson, G.A. (1995) German agri-environmental schemes II – the MEKA programme in Baden-Württemberg, *Journal of Rural Studies*, **11**(2), 149–59

Wilson, O.J. (1996) Emerging patterns of restructured farm businesses in Eastern Germany, *GeoJournal*, **38**(2), 157–60

Wise, M. and Chalkley, B. (1990) Unemployment: regional policy defeated?, in Pinder, D. (ed.) *Western Europe: Challenge and Change*, Belhaven, London, 179–94

Witak, A. and Lewandowska, U. (eds) (1996) *Poland: the Natural Choice*, Sport i Turystyka, Warsaw

Wittgenstein, L. (1961) *Tractatus Logico-philosophicus*, Routledge & Kegan Paul, London

Wittgenstein, L. (1967) *Philosophical Investigations*, Basil Blackwell, Oxford

Womack, J.P., Jones, D.T. and Roos, D. (1990) *The Machine that Changed the World*, Macmillan, New York

Wood, A. (1986) *The Russian Revolution*, 2nd edition, Longman, London

Wood, A.W. (1987) The Enlightenment, in Eliade, M. (ed.) *The Encyclopedia of Religion, Volume 5*, Macmillan, New York, 109–13

Woolf, S.J. (ed.) (1968) *European Fascism*, Weidenfeld & Nicolson, London

World Almanac (1993) *The World Almanac and Book of Facts 1993*, World Almanac, New York

World Bank (1993) *World Development Report: Investing in Health*, Oxford University Press, Oxford

WTO (World Tourism Organization) (1994) *Seminar on Tourism Statistics in the Countries of Central and Eastern Europe*, WTO, Madrid

WTO (World Tourism Organization) (1996a) *Compendium of Tourism Statistics 1990–1994*, 16th edition, WTO, Madrid

WTO (World Tourism Organization) (1996b) Rural tourism to the rescue of Europe's countryside, *WTO News,* **3**, 6–7

WTO (World Tourism Organization) (1996c) *Yearbook of Tourism Statistics,* 48th edition, WTO, Madrid

Wybrew, H. (1988) Eastern Christianity since 451, in Sutherland, S., Houlden, L., Clarke, P. and Hardy, F. (eds) *The World's Religions*, G.K. Hall & Co., Boston, Mass., 167–92

Wynne, D. (ed.) (1992) *The Culture Industry: the Arts in Urban Regeneration*, Avebury, Aldershot

Wynne, P. (1994) Agri-environmental schemes: recent events and forthcoming attractions, *Ecos*, **15**, 48–52

Young, G. (1973) *Tourism: Blessing or Blight?*, Penguin, Harmondsworth

Yuval-Davis, N. (1992) Fundamentalism, multiculturalism and women in Britain, in Donald, J. and Rattansi, A. (eds) *'Race', Culture and Difference*, Open University Press, Milton Keynes, 278–92

Zanetto, G., Vallerani, F. and Soriani, S. (1996) *Nature, Environment, Landscape: European Attitudes and Discourses in the Modern Period. The Italian Case 1920–1970*, Padua

Zentrum für Türkeistudien (ed.) (1989) *Die Türkei und die Europäische Gemeinschaft. Eine eventuelle EG-Vollmitgliedschaft der Türkei und ihre Alternativen*, Zentrum für Türkeistudien, Bonn

Zetterholm, S. (1994) Introduction: cultural diversity and common policies, in Zetterholm, S. (ed.) *National Cultures and European Integration*, Berg, Oxford, 1–13

Zimmermann, F. (1991) Austria: contrasting tourist seasons and contrasting regions, in Williams, A.M. and Shaw, G. (eds) (1991) *Tourism and Economic Development,* Belhaven, London, 153–72

Zukin, S. (1991) *Landscapes of Power. From Detroit to Disneyworld*, University of California Press, Los Angeles

Zvelabil, M. (1978) Substance and settlement in the north-east Baltic, in Mellars, P. (ed.) *The Early Post-glacial Settlement of Northern Europe*, Duckworth, London

Index